Radiotherapy in Practice

Radiotherapy in Practice: Brachytherapy, Second Edition
Edited by Peter Hoskin and Catherine Coyle

Radiotherapy in Practice: Radioisotope Therapy
Edited by Peter Hoskin

Radiotherapy in Practice: Imaging
Edited by Peter Hoskin and Vicky Goh

Radiotherapy in Practice: External Beam Therapy

SECOND EDITION

Edited by

Peter Hoskin

Consultant Clinical Oncologist,
Mount Vernon Cancer Centre
Professor in Clinical Oncology,
University College London and
Honorary Consultant in Clinical Oncology,
University College London Hospitals NHS Trust

OXFORD
UNIVERSITY PRESS

OXFORD

UNIVERSITY PRESS

Great Clarendon Street, Oxford, OX2 6DP,
United Kingdom

Oxford University Press is a department of the University of Oxford.
It furthers the University's objective of excellence in research, scholarship,
and education by publishing worldwide. Oxford is a registered trade mark of
Oxford University Press in the UK and in certain other countries

© Oxford University Press, 2012

The moral rights of the authors have been asserted

First Edition published in 2006
Second Edition published in 2012

Impression: 1

British Library Cataloguing in Publication Data

Data available

Library of Congress Cataloging in Publication Data

Library of Congress Control Number: 2012934409

ISBN 978–0–19–969656–7

Printed in China by
C & C Offset Printing Co. Ltd.

Oxford University Press makes no representation, express or implied, that the
drug dosages in this book are correct. Readers must therefore always check
the product information and clinical procedures with the most up-to-date
published product information and data sheets provided by the manufacturers
and the most recent codes of conduct and safety regulations. The authors and
the publishers do not accept responsibility or legal liability for any errors in the
text or for the misuse or misapplication of material in this work. Except where
otherwise stated, drug dosages and recommendations are for the non-pregnant
adult who is not breast-feeding

Links to third party websites are provided by Oxford in good faith and
for information only. Oxford disclaims any responsibility for the materials
contained in any third party website referenced in this work.

Contents

List of contributors

Edwin GA Aird
Mount Vernon Cancer Centre, Northwood, Middlesex, UK

Neil G Burnet
Neuro-Oncology Unit, University of Cambridge Department of Oncology, Addenbrooke's Hospital, Cambridge, UK

Kate E Burton
Neuro-Oncology Unit, Oncology Centre, Addenbrooke's Hospital, Cambridge, UK

Sorcha Campbell
Consultant Clinical Oncologist, Edinburgh Cancer Centre, Edinburgh, UK

Anna Cassoni
Radiotherapy Department, University College London, London, UK

Charlotte Coles
Addenbrooke's Oncology Centre, Cambridge, UK

C Corner
Mount Vernon Cancer Centre, Northwood, Middlesex, UK

Antonia Creak
Royal Marsden Hospital and Institute of Cancer Research, Sutton, UK

David P Dearnaley
Royal Marsden Hospital and Institute of Cancer Research, Sutton, UK

Patricia Diez
Mount Vernon Cancer Centre, Northwood, Middlesex, UK

Ellen Donovan
Royal Marsden Foundation Trust and Institute of Cancer Research, Sutton, UK

Sara C Erridge
Consultant Clinical Oncologist and Honorary Senior Lecturer in Radiation Oncology, Edinburgh Cancer Centre, University of Edinburgh, Edinburgh, UK

Stephen Falk
Bristol Haematology and Oncology Centre, Bristol, UK

Rob Glynne-Jones
Mount Vernon Cancer Centre, Northwood, Middlesex, UK

Mary K Gospodarowicz
Department of Radiation Oncology, University of Toronto and Princess Margaret Hospital, Toronto, Ontario, Canada

Shaista Hafeez
Royal Marsden Hospital and Institute of Cancer Research, Sutton, UK

Mark Harrison
Mount Vernon Cancer Centre, Northwood, Middlesex, UK

Fiona Harris
Neuro-Oncology Unit, Oncology Centre, Addenbrooke's Hospital, Cambridge, UK

David C Hodgson
Department of Radiation Oncology, University of Toronto and Princess Margaret Hospital, Toronto, Ontario, Canada

Peter Hoskin
Mount Vernon Cancer Centre and
University College London, UK

Nicholas James
School of Cancer Sciences, University
of Birmingham, Birmingham, UK

Sarah J Jefferies
Neuro-Oncology Unit, Oncology
Centre, Addenbrooke's Hospital,
Cambridge, UK

Raj Jena
Neuro-Oncology Unit, Oncology
Centre, Addenbrooke's Hospital,
Cambridge, UK

Christopher Nutting
Consultant and Reader in Clinical
Oncology, Head and Neck Unit, Royal
Marsden Hospital, London, UK

Melanie Powell
Department of Radiotherapy,
St Bartholomew's Hospital, London, UK

Carl Rowbottom
The Christie NHS Foundation Trust,
Manchester, UK

Michele Saunders
Emeritus Professor of Clinical Oncology,
University College London, London, UK

David Sebag-Montefiore
St James's Institute of Oncology and
University of Leeds, UK

Elizabeth Southgate
Queen Elizabeth Hospital,
Birmingham, UK

Alexandra Taylor
Department of Radiotherapy, Royal
Marsden Hospital, London, UK

Roger E Taylor
Professor of Clinical Oncology and
Honorary Consultant Clinical
Oncologist, South West Wales Cancer
Centre, Swansea, UK

Elizabeth Toy
Consultant Clinical Oncologist,
Exeter Oncology Centre, Royal
Devon and Exeter Foundation Trust,
Exeter, UK

Richard W Tsang
Department of Radiation Oncology,
University of Toronto and Princess
Margaret Hospital, Toronto, Ontario,
Canada

Karen Venables
Mount Vernon Cancer Centre,
Northwood, Middlesex, UK

Charlotte Westbury
University College Hospital,
London, UK

John Yarnold
Royal Marsden Foundation Trust
and Institute of Cancer Research,
Sutton, UK

Anjali Zarkar
Queen Elizabeth Hospital,
Birmingham, UK

List of abbreviations

2D	two-dimensional
3D	three-dimensional
3DCRT	three-dimensional conformal radiotherapy
ACTH	adrenocorticotrophic hormone
AJCC	American Joint Committee on Cancer
ALARP	as low as reasonably practicable
ALL	acute lymphoblastic leukaemia
AML	acute myeloid leukaemia
ANLL	acute non-lymphoblastic leukaemia
ASCT	autologous stem-cell transplantation
BASO	British Association of Surgical Oncology
BCC	basal cell carcinoma
BED	biologically effective dose
BEV	beam's eye view
BMT	bone marrow transplantation
CBCT	cone-beam computed tomography
CCLG	Children's Cancer and Leukaemia Group
CHART	continuous hyperfractionated accelerated radiotherapy
CHOP	cyclophosphamide, doxorubicin, vincristine, and prednisone
CI	confidence interval
cm	centimetre
COPD	chronic obstructive pulmonary disease
CRM	circumferential resection margin
CRT	chemoradiotherapy
CSA	craniospinal axis
CSF	cerebrospinal fluid
CSRT	craniospinal radiotherapy
CT	computed tomography
CTV	clinical target volume
DCIS	ductal carcinoma in situ
DFS	disease-free survival
DHAP	dexamethasone, cytarabine, cisplatinum
DLCO	diffusing capacity of the lung for carbon monoxide
DRL	diagnostic reference level
DRR	digitally reconstructed radiograph
DVH	dose–volume histogram
EBCTCG	Early Breast Cancer Trialists' Collaborative Group

EBUS	endobronchial/endoluminal ultrasound
EFRT	extended-field radiotherapy
EIC	extensive intraductal component
EORTC	European Organisation for Research and Treatment of Cancer
EPI	electronic portal imaging
EPID	electronic portal imaging device
ESR	erythrocyte sedimentation rate
ETAR	equivalent tissue–air ratio
EUA	examination under anaesthetic
EUD	equivalent uniform dose
EUS	endoluminal ultrasound
FDG	fludeoxyglucose
FEV1	forced expiratory volume in 1 second
FSD	focus-to-skin distance
FSRT	fractionated stereotactic radiotherapy
FVC	forced vital capacity
GBM	glioblastoma multiforme
G-CSF	granulocyte-colony stimulating factor
GTV	gross tumour volume
Gy	gray
HDR-ILBT	high dose rate intraluminal brachytherapy
HFRT	hyperfractionated radiotherapy
HIV	human immunodeficiency virus
HLA	human leukocyte antigen
HPV	human papilloma virus
HR	hazard ratio
HVL	half-value layer
IAM	internal auditory meatus
IBD	inflammatory bowel disease
ICRP	International Commission on Radiological Protection
ICRU	International Commission on Radiation Units and Measurements
IFRT	involved-field radiation therapy
IGRT	image-guided radiotherapy
IM	internal margin
IMAT	intensity-modulated arc radiotherapy
IMC	internal mammary chain
IMRT	intensity-modulated radiotherapy
IORT	intraoperative radiotherapy
IPI	International Prognostic Index
IR	irradiated volume
ITV	internal target volume
KCO	corrected transfer factor
kV	kilovoltage

LDH	lactate dehydrogenase
LGG	low-grade glioma
LSCLC	limited small cell lung cancer
MALT	mucosa associated lymphoid tissue
MCC	Merkel cell tumour
MDT	multidisciplinary team
MIBG	meta-iodobenzyl guanidine
MIP	maximum intensity projection
MLC	multileaf collimator
mm	millimetre
MMC	mitomycin C
MR	magnetic resonance
MRF	mesorectal fascia
MRI	magnetic resonance imaging
MU	monitor unit
MV	megavoltage
NB	neuroblastoma
NF	neurofibromatosis
NTCP	normal tissue complication probability
OAR	organ at risk
OEV	operator's eye view
OS	overall survival
OTT	overall treatment time
Pb	lead
PCI	prophylactic cranial irradiation
PCNSL	primary central nervous system lymphoma
pCR	pathological complete response
PET	positron emission tomography
PNET	primitive neuroectodermal tumour
PORT	postoperative radiotherapy
PPNET	peripheral primitive neuroectodermal tumour
PRV	planning-risk volume
PRV	planning organ at risk volume
PS	performance status
PSA	prostate-specific antigen
PTV	planning target volume
QA	quality assurance
QC	quality control
QUANTEC	Quantitative Analyses of Normal Tissue Effects in the Clinic
RMS	rhabdomyosarcoma
RTCT	radiotherapy and chemotherapy
RTOG	Radiation Therapy Oncology Group
RVR	remaining risk volume

SBRT	stereotactic body radiotherapy
SCC	squamous cell carcinoma
SCF	supraclavicular fossa
SCPRT	short-course preoperative radiotherapy
SIB	simultaneous integrated boost
SIOP	International Society of Paediatric Oncology
SLNB	sentinel lymph node biopsy
SM	set-up margin
SMART	simultaneous modulated accelerated radiotherapy
SRS	stereotactic radiosurgery
SRT	stereotactic radiotherapy
SSD	source-to-surface distance
Sv	sievert
TBI	total body irradiation
TCP	tumour control probability
TLD	thermoluminescent dosimetry
TME	total mesorectal excision
TMZ	temozolomide
TNM	tumour, node, metastasis
TP	treatment planning
TPR	tissue phantom ratio
TPS	treatment planning system
TRH	thyrotropin-releasing hormone
TURP	transurethral resection of the prostate
TV	treated volume
UICC	International Union Contra Cancer
UK	United Kingdom
USA	United States of America
VMAT	volumetric-modulated arc radiotherapy
VNPI	Van Nuys Prognostic Index
WHO	World Health Organization

1

Introduction

Peter Hoskin

Radiotherapy remains the most important non-surgical treatment in the management of cancer. Over 50% of patients will receive treatment at some time during the management of their malignant disease. In recent years rapid advances in the technology available to radiotherapy have been made and there is a challenge to the practising clinician to remain abreast of these and harness them to their best use in the management of patients. For most patients receiving radiotherapy this will mean treatment delivered with external X-ray or electron beams. The processes required for the safe delivery of modern radiotherapy comprise a lengthy pathway from treatment decision to treatment delivery and verification. For the more complex treatments this will involve sophisticated immobilization devices, high-precision computed tomography (CT), magnetic resonance imaging (MRI) and positron emission tomography (PET) image-guided volume localization, complex and increasingly accurate physics planning systems with state-of-the-art algorithms to account for tissue inhomogeneities and beam variables, and, finally, the widespread use of high-energy linear accelerators with multileaf collimators, the capacity for conformal and intensity-modulated radiation therapy, and the ability to provide on-line image guidance of treatment delivery. New approaches now include the use of tomotherapy and stereotactic radiotherapy for more precise high-dose delivery. Despite this, however, the basic principles of radiotherapy remain unchanged. Radiotherapy is a locoregional treatment suitable for radical treatment of tumours in their early stages with high success rates where there has been no metastatic spread. The basic steps of treatment delivery remain: defining the patient position with a means of reproducing that position day to day with appropriate immobilization, followed by accurate localization and definition of the volume to be covered by the high-dose envelope, and then collaboration with medical physicists to identify the optimal means of doing this using available beams with appropriate modifications. The process of daily implementation of the treatment plan is often neglected but of vital importance in ensuring accurate and effective radiotherapy together with verification that treatment delivery is reproducing the expected beam as defined in the planning process.

1.1 External beam sources

Linear accelerators are the common source of high-energy X-ray beams producing megavoltage photons of between 4 and 20 million volt energy able to penetrate to the most deep-seated tumours in the largest of patients. Clinically, 4–8 MV beams are the most useful, providing a balance between penetration and adequate surface dose. The fundamental property of megavoltage beams to have skin sparing is both beneficial in terms of reducing skin reaction but also potentially hazardous in reducing dose to surface or superficial tumour. Modern linear accelerators are highly sophisticated machines working within a high precision of 2–3%. Recent years have seen widespread implementation of the multileaf collimator to provide complex beam shaping and in intensity-modulated radiotherapy (IMRT) applications varying beam transmission. Additional beam modification using motorized wedges has largely replaced the manual wedged-shape filters which were placed in the beam by the machine operators. The smooth and reliable running of such machines requires careful maintenance and quality assurance, the importance of which cannot be emphasized too much, often carried out unseen out of hours by the dedicated band of bioengineers and physicists who attend to the processes required to enable safe delivery of radiotherapy to patients.

Cobalt machines were widely used in the past and in some centres still have their place. They require less maintenance comprising a cobalt source which releases gamma rays with energy equivalent to 2.5 MV photons and a relatively simple mechanism shielding and exposing the source to provide the beam. The penetration of the beam, however, is relatively poor and because it arises from a source of finite size the penumbra of the beam is quite large. Such considerations have led many centres to decommission cobalt units to be replaced by modern linear accelerators or, where retained, for palliative treatments and out-of-hours work where their simplicity does have advantages. However, their robust construction and function means that they still have an important place, particularly in less developed health care systems.

Particle therapy is not addressed in any detail in this book other than electron treatment which is widely used for superficial tumours being produced from the standard linear accelerator. There are, however, many other particles which can be used in therapy. Neutrons have been evaluated over many years and their clinical utility remains limited and they cannot be regarded as part of routine clinical practice. Protons, in contrast, have excited increasing interest in recent years. Their main advantage is that their energy deposition follows the Bragg peak with a high-intensity, highly localized deposition of energy at a fixed depth. This has advantages in the treatment of certain sites, for example, retinal tumours and tumours of the brainstem where highly localized energy deposition avoiding surrounding structures is required. They have also been used in other sites, for example, prostatic carcinoma, as a means of enabling dose escalation within normal tissue tolerance. There are an increasing number of facilities available across the world although as yet none in the United Kingdom, but it is likely that in future years proton therapy will become more prominent in the management of selected sites with radiotherapy.

1.2 Radiotherapy planning

Planning is a critical step in the delivery of clinical radiotherapy. For any treatment to be effective it must be delivered accurately to the region of interest. The identification of the GTV (gross tumour volume), CTV (clinical target volume), and PTV (planning target volume) representing sequential volume expansions from the macroscopic identifiable tumour to an expansion including areas where there is risk of tumour spread even if not identified, to a larger volume which takes into account patient movement and other variations in day-to-day set-up of a radiation beam during a fractionated course of treatment is now embedded in the practice of modern radiotherapy.

Alongside this, major developments in imaging technology have allowed us to identify tumours with far more accuracy and certainty than before. It is now routine practice to identify internal tumours with CT planning. Increasingly, where appropriate, MRI and PET images are also imported into the planning system and image registration used to provide greater certainty and clarification of the anatomy. Functional imaging techniques with MRI can enhance this further to provide ever more sophisticated information on the tumour and its surrounding areas, alongside the equally critical identification of the organs at risk where dose should be minimized.

Effective treatment planning ultimately depends upon complex computer algorithms to simulate the effect of a beam passing through the designated area and the amount of radiation energy deposited at any one site. The mathematical accuracy of such algorithms has increased considerably in recent years and combined with the use of CT imaging to provide accurate inhomogeneity data across different tissues, far greater accuracy in dose distribution is now achievable. This is of particular importance where large areas of lung or other air-filled cavities such as the paranasal sinuses are present in the treatment volume.

1.3 **Treatment implementation and verification**

The delivery of high-dose fractionated radiotherapy requires implementation on the treatment machine of a complex multistep process which may be repeated on a daily basis for 30 or 40 fractions. This presents a significant challenge in devising quality assurance processes to ensure that safe and reproducible radiotherapy is the standard. The increasing use of computerized data sets with electronic transfer of treatment parameters from the planning software to the linear accelerator minimize the risk of human error in transferring data. The use of mechanized wedges and multileaf collimators takes out the risk of the machine operator placing the wrong wedge in position or when using lead shielding on a lead tray having this wrongly positioned. The importance of immobilization to reproduce patient set-up is now widely recognized and facilitated by the use of immobilization devices ranging from simple head shells and vacuum bags to stereotactic frames.

The megavoltage beams used for radiotherapy produce images having poor definition between bone and soft tissue and are therefore more difficult to interpret than kilovoltage (kV) energy beams. The advent of modern linear accelerators with silicone diode array detectors, electronic portal imaging devices (EPIDs), and on-board kV imaging equipment has greatly enhanced the ability to accurately verify beam position. Many patients are now treated with implanted fiducial markers in the treatment region to improve identification with radiotherapy imaging. More frequent validation with EPIDS can be achieved and presentation of these on the computer screen alongside the planning images with software tools to facilitate comparison greatly improves the accuracy of treatment and enables systematic errors to be readily identified and corrected.

1.4 Radiation dose prescriptions

Radiotherapy prescriptions have long carried an air of mystery and confusion. This is hardly surprising when three or four different total doses given in a similar variation of fractions of treatment over differing times are recommended for exactly the same tumour. Much of this relates to history and legend rather than systematic evaluation. The problem is now compounded by new concepts in which acceleration of fraction is recommended by some, hyperfractionation by others, the use of a concomitant boost by yet others, whilst in most centres daily fractionation Monday to Friday remains the standard. For radical treatment the following schedules may be encountered:

- *Conventional fractionation* usually refers to daily treatment on a Monday to Friday basis.
- *Accelerated fractionation* means that the overall total dose is given in a shorter time than would be achieved with conventional fractionation. This results in greater toxicity and therefore only limited acceleration is possible without altering fraction size. An example of this is the DAHANCA regimen in which six fractions are given over 5 days so that a conventional 6-week treatment schedule is delivered in 5 weeks. This modest acceleration has been shown to improve the results in head and neck cancer.
- *Hyperfractionation* refers to the practice of reducing the fraction size of a conventional regimen, often delivering treatment twice or even three times a day in the smaller fraction sizes to enable a higher dose overall to be delivered. This is possible because the toxicity, in particular the late toxicity, is reduced when the fraction size is reduced for a given total dose. This approach has been investigated in many sites including head and neck cancer and non-small cell lung cancer.
- *CHART* (continuous hyperfractionated accelerated radiotherapy) is a schedule which encompasses both acceleration and hyperfractionation delivering the total dose in a shorter overall time (acceleration) and in smaller individual fractions (hyperfractionation). The original CHART schedule delivered 54 Gy in 36 fractions of 1.5 Gy over 12 days.
- *Hypofractionation* refers to giving a treatment in a shorter time than conventional treatment using bigger doses per day and in order to do so safely reducing the total dose. In the radical setting examples are the delivery of 55 Gy in 20 daily fractions or 50 Gy in 16 daily fractions which are considered equivalent to a radical conventional dose of 65 Gy in 6½ weeks. There is increased interest in such schedules following the observation that some tumours such as prostate cancer may have radiation response characteristics with low alpha beta ratios. As a result, large doses per fraction are biologically more effective. Concerns relating to normal tissue

effects even with highly accurate IMRT remain, however, hence formal evaluation in randomized trials is awaited.

♦ *Palliative radiotherapy* is one area where hypofractionation is indicated. In symptom control the aim is not to deliver a high dose to eradicate tumour but a sufficient dose to enable symptom control. It has been widely shown that single doses of 8 Gy or thereabouts are sufficient to improve bone pain and single doses of 10 Gy will improve symptoms from non-small cell lung cancer. Other common schedules in use for palliation are 21 Gy in three fractions, 20 Gy in five fractions and 30 Gy in 10 fractions.

The second edition of this book continues its aim to provide a practical guide to the use of external beam radiotherapy incorporating the substantial technological advances that have been made in recent years. It will provide a firm background in the physics of external beam radiotherapy and then deals with each anatomical site in turn with details of the indications and techniques used for radiotherapy delivery.

2

Basic physics

Karen Venables

2.1 Introduction

The distribution of radiation within the patient will be affected by many factors. These include the energy and modality of the beam, the density of the tissue, the use of beam modifiers such as wedges and compensators, and the distance of the patient from the machine. The apparent distribution will also be affected by the accuracy of the algorithm used in the planning system.

2.2 Interaction processes

The deposition of dose within the patient is dependent on the interaction process or processes involved. Dose is the energy deposited in the material as a result of interactions of photons and electrons with the material. When photons undergo an interaction, energy is transferred to electrons, which will then deposit their energy in the medium. At low energies photons interact predominantly by the photoelectric effect in which a tightly bound electron is ejected from the atom. The dominant interaction process in tissue for photons produced from linear accelerators (1–20 MeV) is the Compton process whereby the photon interacts with a loosely bound electron, resulting in a free electron and a scattered photon of reduced energy. Pair production also occurs above 1.02 MeV whereby a photon interacts within the nucleus of the atom producing an electron and a positron; the positron will travel a short distance and then annihilate producing two further photons of energy 0.511 MeV. In contrast to photons in which the probability of an interaction occurring is governed by a chance process, electrons deposit energy continuously along the length of their path by collision with atomic electrons. They also lose energy through Bremsstrahlung; when electrons pass close to a nucleus (which has a positive charge) it attracts the negatively charged electron, changing its direction, and an X-ray photon is emitted. The range of an electron in a medium is dependent upon its initial energy and the density of material through which it is travelling. Interactions produced from photon beams are illustrated in Fig. 2.1. Monte Carlo is a powerful computing tool for determining the result of irradiating a material with a beam of photons or electrons. It uses statistical methods to determine the outcome of interactions and can be used to follow the history of individual particles in a beam. It is used in some treatment planning system (TPS) algorithms to generate dose at a point. Diagrams illustrating the deposition of dose are shown in Figs 2.2–2.7. The deposition of dose within a medium can be described by a dose deposition kernel.

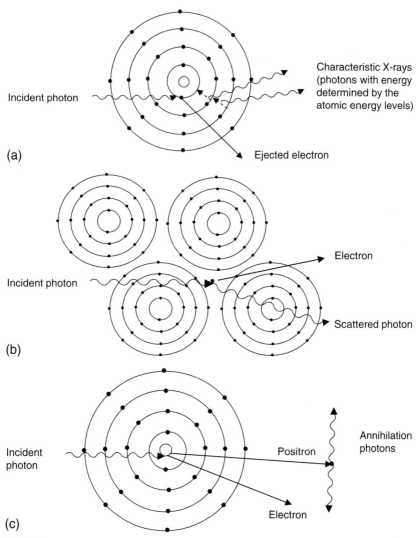

Fig. 2.1 Photon interaction processes. (a) Photoelectric effect: the incident photon interacts with a bound electron which is ejected from the atom. Other electrons of higher energy take its place and their excess energy is emitted as characteristic X-rays. (b) Compton effect: the incident photon interacts with a loosely bound electron producing a scattered photon and scattered electron. (c) Pair production: the incident photon interacts with the nucleus of the atom.

2.3 **Dose deposition within the patient**

The penetration of X-rays or electrons will be dependent on the effective accelerating potential to which the electrons in the waveguide have been subjected, although the design of the treatment machine head will also affect this and photons or electrons with a nominal beam energy from one machine may not have the same properties as

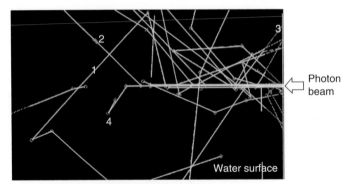

Fig. 2.2 Monte Carlo calculation of 25, 100-kV photons incident on a 20-cm water slab. The yellow lines (e.g. 1) show the photon and the blue circles (e.g. 2) the electrons. At this energy, the electron range is below the resolution of the diagram and their energy is absorbed at the points of interactions. All of the photons interact before leaving the slab and a number have been scattered back towards the surface (e.g. 3). Large angle scattering is common (e.g. 4).

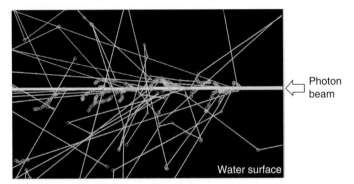

Fig. 2.3 Monte Carlo calculation of 25, 6-MV photons incident on a 20-cm water slab. The yellow lines show the photons, the blue lines and circles the electrons, and the red lines and crosses the positrons. Most of the electrons travel a short distance before losing their energy and being reabsorbed but as shown by the number of blue circles, they interact many times each time losing a little of their energy, in contrast to the photons most of which travel a much larger distance between interactions. Some of the photons do not interact before leaving the slab and the number scattered back towards the surface is reduced compared to the 100kV beam. Pair production is possible at this energy as illustrated by the production of a positron (1) which travels a short distance before annihilating. Compton interactions are governed by chance and dependent upon energy; for the photons shown in this diagram no photon interactions occurred in the first 2.5 cm, and this explains the build-up in dose that occurs for high-energy accelerators. Before dose can be deposited, the photons must have interacted, electrons produced in photon interactions gradually deposit dose along their path length.

those with the same nominal energy from another machine. The photons incident on the patient will have a spectrum of energies with the maximum possible energy being that of the accelerating potential. The mean energy will be much lower, usually one-quarter to one-third of the maximum.

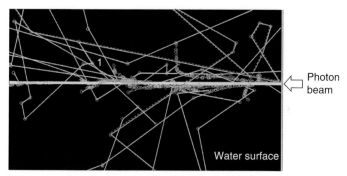

Fig. 2.4 Monte Carlo calculation of 25, 20-MV photons incident on a 20-cm water slab. The yellow lines show the photons, the blue lines and circles the electrons, and the red lines and crosses the positrons. Some of the photons do not interact before leaving the slab and the scattered photons and electrons are predominately in the same direction as that of the incident photon. There is an increase in pair production and the positrons travel further before being annihilated. Many of the photons outside of the beam are the result of this process (e.g. the track traced in green is the result of a single annihilation process).

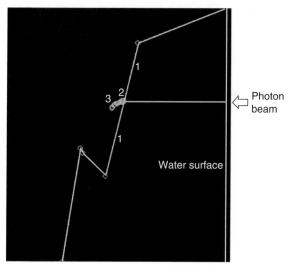

Fig. 2.5 Monte Carlo simulation of a positron from 6-MV beam, showing the two 511-keV photons almost in opposing directions (e.g. 1), the positron (e.g. 2), and electron (e.g. 3).

The deposition of dose within a material is often described in terms of either percentage depth dose or tissue phantom ratios. Percentage depth dose values relate the dose at a given depth to that at the depth of maximum dose for the same distance to the surface. They are dependent on the treatment machine energy, distance from the source, and irradiated area as well as the material in which the dose is deposited. Tissue phantom ratios (TPRs) relate the dose at a reference depth in a phantom to the

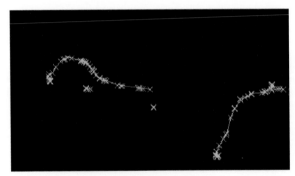

Fig. 2.6 Monte Carlo simulation showing electrons produced from interactions of 6-MV photons. Note the tortuous paths and the increased number of interactions (each shown by a blue cross) as the electron reaches the end of its range and has only a low energy.

Photon beam

Fig. 2.7 Monte Carlo simulation of a narrow beam of 100, 6-MV photons on a water phantom, illustrating that most of the electrons deposit their energy close to the position of the track of the incident photon beam. Generation of pencil beams for planning systems using Monte Carlo calculated pencil beams is performed in this way and the resultant dose distribution characterized. In the case of planning system pencil beams, the incident photon beam is not a single energy but a spectrum of energies to represent those found clinically in linear accelerator beams.

dose at a point the same distance from the source but with a different depth of material above the point. Tissue maximum ratios are a special case of TPR where the reference depth is taken to be the depth at which maximum dose is deposited. TPRs are dependent on field size, machine energy, and material in which the dose is deposited but have only a very small dependence on distance from the source of radiation. TPRs are often used for quick calculation of isocentric treatments, whereas percentage depth doses are preferred in centres that treat patients at a fixed focus to surface distance. This is illustrated in Fig. 2.8.

2.4 **Sources of high-energy X-rays**

Historically patients were treated with orthovoltage and superficial X-ray units (up to 300 kV). These deliver high dose to the surface whilst still contributing dose at depth. They are still used to treat some superficial lesions, particularly in the head and

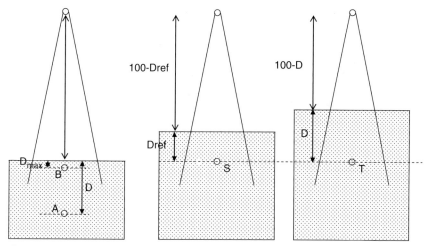

Fig. 2.8 Diagram to illustrate the difference between percentage depth dose and TPR. The percentage depth dose at point A would be found by dividing the dose at A by the dose at B and multiplying by 100 to convert to a percentage. In contrast the TPR for a depth of D would be found by dividing the dose measured at T and dividing by the dose measured at S. Note that in both these cases the same field size (jaw settings) has been used throughout. The variation of the machine output with field size must also be incorporated.

neck region. Cobalt 60 machines were developed in the 1950s and deliver a higher dose at depth due to the energy of the photons (1.17 MeV and 1.33 MeV). They are usually reliable machines and still have a place in many radiotherapy departments for simple treatments. The photons are produced from the radioactive decay of the source. The strength of the source (and therefore the intensity of the radiation) decreases with time. The source must be changed approximately every 5 years to prevent treatment times becoming too long. Depth dose curves for these machines are shown in Fig. 2.9. Modern high-energy linear accelerators offer a choice of photon and electron energies. The production of high-energy photons can be described briefly as follows. Electrons are emitted from the heated gun filament, and their energy is gradually increased as they move through the waveguide, transported by high-power radio waves. The beam of electrons is focused and steered through an angle of between 90–270° (depending on manufacturer's design) (if necessary) to hit a high atomic number target. The resultant X-ray beam is collimated and the intensity of the radiation modulated using a metal cone, known as the flattening filter, which is thickest in the centre to produce a beam with a more nearly uniform intensity within the treatment machine head. The beam is collimated using two pairs of diaphragms or one pair of diaphragms and a set of multileaf collimator (MLC) leaves. The components of the conventional linear accelerator are illustrated in Fig. 2.10.

Speciality linacs vary this basic design; both tomotherapy and cyberknife use compact linacs with shorter waveguides which can still operate at about 6 MV. In tomotherapy units, the linac is mounted on a CT type gantry system and the collimation is provided

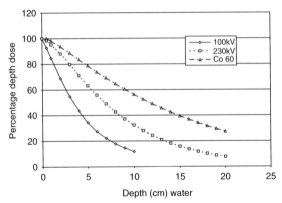

Fig. 2.9 Depth dose curves for superficial and orthovoltage units. The 100-kV curve is for a 30-cm focus-to-skin distance (FSD) unit, with a half-value layer (HVL) of 3 mm and a 10-cm diameter field, the 230-kV curve is for a 50-cm FSD unit with an HVL of 2 mmCu and a closed end applicator, field size 10×10. The cobalt-60 data is for an FSD of 80 cm and a field size of 10×10 cm. Data taken from *British Journal of Radiology* supplement 25.

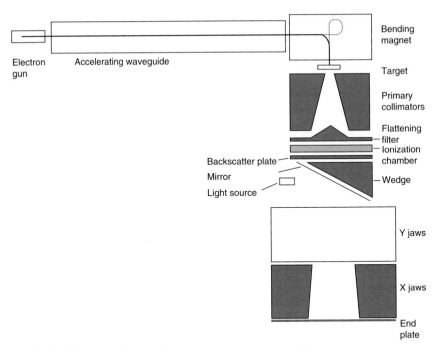

Fig. 2.10 Block diagram showing the components of a linear accelerator.

by a binary MLC (leaves are either open or closed at any point in time). In cyberknife the linac is fitted to a robotic arm allowing many degrees of freedom in the direction in which the radiation can enter the patient. Collimation is provided by a selection of fixed collimators from 0.5–6 cm radius or a variable aperture collimator. Both of these linacs operate without a flattening filter and at higher dose rates than a standard accelerator.

Electron beams used for treatment can be produced either by rapidly scanning the narrow beam of electrons across the desired area or more commonly the beam is broadened by the use of a scattering foil in place of the X-ray target. In normal use, a series of openings in an electron 'applicator' are used to collimate the beam down to or close to the patient's skin.

Typical depth dose curves for photons and electrons are shown in Fig. 2.11 and Fig. 2.12. Dose is not deposited directly by the photons but rather by electrons set in motion through interaction processes, therefore for megavoltage photons, the maximum dose does not occur at the surface but at a depth of 1–4 cm (D_{max}). The number of photons in the beam will begin to decrease immediately the beam enters the patient; however, even those photons that interact in the first millimetre of tissue will set in motion electrons which will travel a short distance before they have deposited all of their energy. A 'build-up effect' occurs with increasing dose deposited with depth until a condition is met whereby the energy transferred to electrons generated from interactions is matched by the energy deposited by electrons already set in motion. The dose at the surface is typically between 10–30% of the dose at D_{max}, dependent on beam energy, field size, linac design, and the presence of scattering materials such as wedges in the beam. The depth at which the maximum dose occurs is dependent primarily on the beam energy. After D_{max}, a gradual decrease in the dose deposited occurs as the number of photons in the beam is reduced. Two effects contribute to this: the reduction in intensity due to the larger area that the photons cover as the distance is increased and the decrease due to attenuation. For a very narrow beam of monoenergetic photons, at a large distance from the source, the decrease due to attenuation would be exponential. Deviations from exponential decrease occur for two reasons: the beams from linear accelerators are not monoenergetic but comprise a spectrum of radiation and for the majority of cases in radiotherapy a broad beam is used and therefore scatter from the medium will also affect the beam intensity. The irradiated area will affect the number of scatter photons generated. As the field size is increased from zero there is initially a rapid increase in dose to a point at the centre of the beam. This rate of increase slows as larger field sizes are reached.

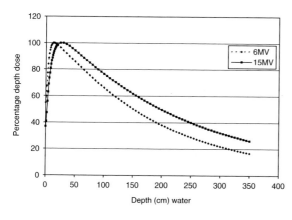

Fig. 2.11 Percentage depth dose curves for 6-MV and 15-MV photon beams at 100 cm FSD, 10 × 10 cm.

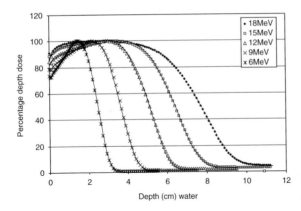

Fig. 2.12 Electron depth dose curves.

In contrast to photons, electron beams begin to deposit energy immediately on entering the patient. There is a small build-up as the electrons will travel a short distance before finally being absorbed into the medium. The range of electrons within tissue will determine the distance into the material which they can penetrate and once this distance has been reached there is a rapid decrease in the depth dose curve, beyond which the only significant dose deposited is that from contaminant photons within the beam. Correcting the intensity of a beam of electrons using the inverse square law is complex, the source of electrons will not be the radiation target but an effective scatter source within the accelerator head. It is usually advisable to measure outputs at non-standard distances.

The radial profile of the beam is dependent primarily on the shape of the flattening filter. When the machine is purchased, a depth for which the beam intensity will be uniform is stated and the manufacturer will make any required adjustments to the flattening filter. This depth is typically 5 cm or 10 cm. For large field sizes, at shallower depths, the profile will have 'horns' or areas of increased intensity whereas at greater depths the intensity at the edges of the beam are decreased. Two factors contribute to this; a non-equilibrium of scatter from the edge of the beam and a small change in mean energy as the distance from the centre of the field is increased. The energy change is caused by absorption of low energy photons at the flattening filter. In the centre where the filter is thickest more low energy photons will be absorbed in comparison with the edges of the field. This absorption of low energy photons is often referred to as 'beam hardening'. This is adequately accounted for by most planning systems and can be seen on the isodose distribution shown in Fig. 2.13.

2.5 Radiation distributions within the patient

Some patients may be treated using either a single field or a parallel-opposed pair. The appropriateness of each of these is related to the patient's size and the location of the tumour. Distributions for a range of patient sizes for 6 MV are shown in Fig. 2.14. For separations of 12 cm, two opposing beams produce a uniform distribution of radiation through the patient. As the patient separation is increased, areas of increased dose relative to the dose at the centre of the volume are seen towards the surface.

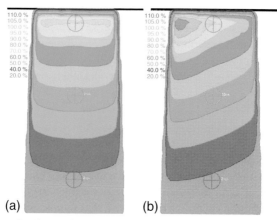

Fig. 2.13 Isodose distribution for a 10 × 10 field incident on a water phantom at 100 cm FSD. Note the change in shape of the isodose lines as the depth is increased. At 3 cm deep, the 95% isodose is deeper at the outside of the beam, in contrast to this at approximately 15cm deep, the 50% isodose is deepest at the centre of the beam. (b) Shows a 25° wedge field also incident on a flat water phantom.

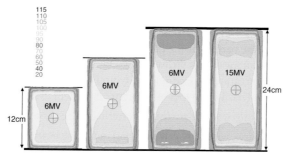

Fig. 2.14 Radiation distribution for a parallel opposed 6-MV beam for patient separations of 12 cm, 18 cm, and 24 cm and for a parallel opposed 15-MV beam for a patient separation of 24 cm. For the higher-energy beam the depth of the high-dose areas are more interior.

For a patient separation of 24 cm, these areas reach 114% of the dose at the centre and the use of higher-energy beams should be considered to give a more even dose distribution.

2.6 Machine-dependent factors affecting the dose deposition

2.6.1 Field size effects

As the radiation field size is increased, the amount of radiation to a point on the central axis per monitor unit increases. The major cause of this is an increase in the contribution of scattered radiation both from within the treatment machine head and within the patient. This will affect the depth dose curves; the depth of D_{max} will decrease and the percentage depth dose at depth will increase compared to a smaller field size. Measurements of the relative output at different field sizes and of appropriate isodose distributions are performed during commissioning. When irregular field shapes such as those produced by MLCs are used, the contribution of scatter is more difficult to assess and individual calculations for each patient may be necessary.

2.6.2 Effect of distance

As the distance of the patient from the accelerator is increased, the intensity of the radiation at the surface will decrease and the penetration of the radiation will increase due to the inverse square law, the irradiated area will increase due to the beam divergence. The increased area is utilized in treatments such as total body irradiation where the patient is placed at an increased distance (typically 4 m) from the machine. A slightly increased mean energy may be noticed, for example, when changing from a cobalt machine with an isocentre at 80 cm to one with an isocentre at 100 cm. The effect for a 6-MV beam is shown in Fig. 2.15.

2.7 Modifications to the radiation beam

2.7.1 Wedges

The intensity of the radiation can be modified by the presence of a wedge. These are used for multifield plans to compensate for the weighting of other fields, for example, in the treatment of the parotid or the prostate. They can also be used to compensate for missing tissue as in the case of breast radiotherapy, or low-density tissue in part of the field as for lung treatments. Examples of typical plans for these sites are given in the appropriate chapters. Three types of wedge are in current use:

- *Manually fitted wedges*, which are usually external to the treatment head.
- *Steep internal wedges*, which are driven in and out of the beam by a motor.
- *Dynamic wedges*, where the jaw moves across the field partway through the treatment reducing the beam intensity to give an appropriate profile, as illustrated in Fig. 2.16.

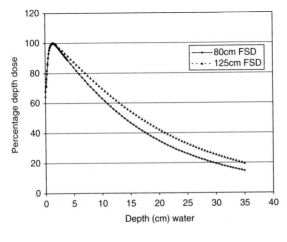

Fig. 2.15 Photon depth dose curves for 6-MV photons at different focus-to-surface distances.

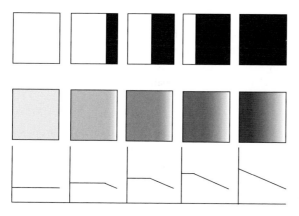

Fig. 2.16 Diagram to illustrate the dynamic wedge. The top row shows the jaw positions, the second row the beam intensity, and the third a plot of beam intensity across the field. The deeper shades represent more radiation.

Internal and dynamic wedges have the advantage that they do not have to be manually lifted by the treatment unit staff, thereby reducing staff injuries, and these are now present in the majority of machines.

The definition of wedge angle has changed over time as the design of accelerators has changed. One definition of wedge angle is the angle between the central axis of the beam and the normal to the 50% isodose. Motorized internal wedges are typically approximately 60° and are combined with an open field of the same size to produce different effective wedge angles. Physical wedges (either internal or external) will change the penetration of the beam on the central axis as well as modifying the radiation profile, whereas dynamic wedges do not change the penetration of the beam on the central axis. External wedges can also increase the patient surface dose slightly due to the production of scattered radiation.

2.7.2 Multileaf collimators

These are used to shape the radiation beam to protect organs at risk or to modify the intensity of the beam by using a segment field as simple IMRT. The width of the MLC leaves varies between machines; high-resolution MLCs used in stereotactic work have a width of 3 mm projected at isocentre, other designs have leaves which are 10 mm projected at isocentre.

2.7.3 Compensators and intensity-modulated radiotherapy

These are discussed further in Chapter 3.

2.8 Geometrical characteristics of radiation beams

2.8.1 Beam divergence

As the radiation beam leaves the treatment head it will diverge. This is illustrated in Fig. 2.17. The divergence of the beam must be considered when treating adjacent areas on the patient if overlapping areas of high dose are to be avoided.

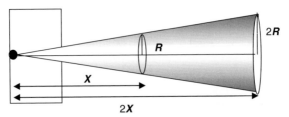

Fig. 2.17 The inverse square law. A beam of X-rays is emitted from a point and passes through a circular collimator. At a distance X from the source the radius of the beam is R, the area of the beam cross-section is $A = \pi R^2$, and the number of photons per square centimetre is I. At a distance of 2X from the source, the beam radius is 2R (by similar triangles). Thus the irradiated area $= \pi(2R)^2 = 4\pi R^2 = 4A$ and the intensity of photons must be I/4. In general, at a distance x, the intensity is $(X/x)^2 \times I$.

2.8.2 Penumbra

The intensity does not drop immediately to zero at the edge of the radiation beam.

The width of the penumbra (defined as distance between the 80% and 20% isodoses at the edge of the field) is affected by both geometric and dosimetric factors: geometric factors include the size of the focal spot and the geometry of the treatment head (in particular the collimators and their distance from the X-ray source); dosimetric factors include the width of the dose deposition kernel. The width of the kernel varies with the energy of the beam and the density of the material in which the dose is deposited. In low-density materials, the electrons will travel further and so the size of the kernel will be increased and the penumbra will be larger.

2.8.3 Asymmetry

The majority of modern accelerators have the ability for the jaws to be moved independently of each other, producing fields that are asymmetric about the beam centre. This can be useful for a non-symmetrical volume and also for producing non-divergent beam edges for use in beam matching. The width of the penumbra produced by an asymmetric jaw placed at the centre of the field may be less than that produced by the corresponding jaw placed at a distance from the central axis.

2.9 Beam matching and use of asymmetric fields

For some patients it is necessary to treat with two adjacent radiation beams either because of the size of the field or because of the geometry of the radiation volume, for example, in the treatment of lateral and anterior neck fields. In these cases if no modification is made, because of divergence the beams can only match at a single point, areas of over- and underdose will be present above and below this. Optimization will ensure that there is minimum overlap between the beams or areas of potential underdosing (Fig. 2.18). It is preferable to match the beams in an area in which there is no residual

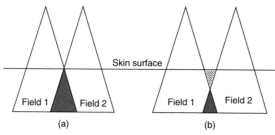

Fig. 2.18 Adjacent divergent radiation fields matched at the skin surface (a), matched at depth (b). The dark area indicates field overlap and regions of potential overdose and the hashed area region potential underdose. (Angles exaggerated for purpose of illustration.)

disease. For patients undergoing fractionated treatment moving the position of the match on alternate days may be considered. Three cases are considered as follows.

2.9.1 **Photon–photon match** (Fig. 2.19)

Use of a single isocentre is often considered as the gold standard, allowing the smoothest match between fields. The length of the either of the fields is limited to 20 cm in most modern accelerators. The quality control of the asymmetric jaws must be excellent otherwise a systematic overlap or underdosing could occur, and the sharp penumbra obtained at the centre of the field means that if there is an overlap, the dose would increase rapidly.

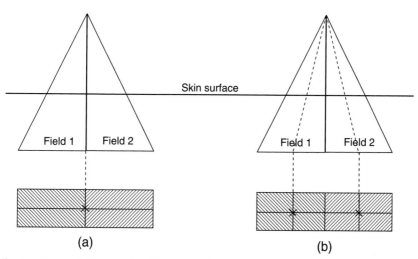

Fig. 2.19 Methods for removing divergence. The centres of the fields are shown by the dotted line and the cross. In (a), a single isocentre is used and each field uses an asymmetric jaw to remove the divergence. In (b), a gantry rotation is used to remove the divergence.

An alternative technique is to use one asymmetric jaw with collimator and couch rotation to remove the divergence from the other beam or beams. The required couch, collimator, and floor rotation can be calculated using established formula and tables of these values are often produced. This method allows for longer fields and also broadens the penumbra of one beam meaning that a slight systematic error in jaw position is less likely to lead to a hotspot. However it is much less elegant than the single isocentre technique and the different widths of the penumbra will lead to a slight overdose in one field and a slight underdose in the other.

It is also possible to match photon beams without the use of any asymmetric jaws. Casebow[1] equations can be used to compute the required angles.

2.9.2 Photon–electron match

When matching an electron beam to a photon beam the divergence from the electron beam is usually accounted for by couch collimator and floor rotation. The bulging of the low-dose isodoses means that it is not possible to achieve an exact match and there is usually a small area of slightly increased dose in the photon field.

2.9.3 Electron–electron match

The matching of electron fields is complex particularly on curved surfaces due to the bulging of the low-dose isodoses. On a flat surface an approximate match can be obtained by using appropriate gantry and collimator twist to remove divergence; however, there will be areas of non-uniform dose in the junction area.

2.10 Patient-dependent factors affecting the dose deposition

2.10.1 Inhomogeneity

The deposition of dose is dependent on the electron density of the material involved. Within tissues electron density is directly proportional to the physical density. Information about the relative electron density of a material can be obtained from a CT scan which can be calibrated using tissue-like materials to relate the Hounsfield units of the scanner to the density of material. Some planning systems, particularly those using convolution algorithms, require calibration in terms of physical density. The majority of planning systems are able to correct for density on a pixel-by-pixel basis although correction for large areas of tissue using a bulk density correction (where a single typical density is chosen for a whole region of the body, for example, setting the relative density of the lung to 0.3) is also applied. Figure 2.20 illustrates the effects of different densities of tissue on the radiation beam. In low-density materials the radiation will travel further before depositing dose, whereas in high-density materials it will be attenuated more rapidly. The increased lateral spread of the radiation into low-density materials such as lung is also illustrated on this figure. Note that there is also a decrease in dose in the central water column at the junction with the

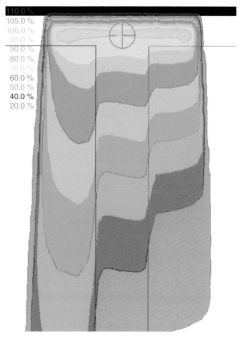

Fig. 2.20 A 12×12-cm 6-MV photon field incident on a phantom consisting of a 2-cm slab of water followed by inhomogeneities. The region on the left has a density of 0.3 g/cm^3, whilst that on the right has a density of 1.6 g/cm^3. The central column is water. Note the broader penumbra in the low density region, the high value isodoses curve towards the centre of the beam whilst the low value isodoses spread away from the beam edge.

low-density material. Many of the simple inhomogeneity correction algorithms are unable to account for changes in lateral scatter due to the presence of inhomogeneities. The dose in the region of the interface is complex and an in-depth study of this is beyond the scope of this chapter.

2.11 **Treatment planning systems**

A wide range of algorithms are used in treatment planning systems. The majority will accurately calculate doses (to within 1–2%) for square and rectangular fields incident perpendicular to a flat homogeneous phantom at the reference distance. Limitations become apparent when irregular fields, inhomogeneities, surface curvature, wedges, and different focus to surface distances are incorporated. Tables 2.1 and 2.2 summarize the calculation methods and inhomogeneity corrections used in most commercial treatment planning systems.

Table 2.1 Summary of main differences between planning systems

Algorithm type	Comments
Stored beam	Based on the work of Bentley-Milan[2] further developed by Redpath[3]. Data are stored on an array of fan lines; intermediate data are interpolated; very accurate for square fields on homogeneous media, however, less accurate for field shapes and sizes where measured data are not available such as conformally shaped fields or IMRT. A variety of algorithms are used to correct for inhomogenities which vary in their accuracy
Beam model systems	Dose divided into primary and scatter component, which are considered separately. More accurate than stored beam models for irregular fields, however, there are often limitations for fields in which the scatter is uneven (steep wedges and inhomogeneities)
Convolution algorithms	In these algorithms the kernel (dose deposition from an individual photon) is 'convolved' with the TERMA (energy removed from the beam)
Convolution algorithms: pencil beams	When convolution algorithms were first developed[4] the computing power available made it unrealistic to convolve dose deposition kernels with TERMA at every point in the calculation matrix. In order to reduce the computing requirements, kernels were defined to represent the entire dose distribution due to an incident small ('pencil') beam. The intensity of these pencil beams across the field could be varied to allow for changes in intensity caused by wedges or compensators, and they could be scaled to allow for inhomogeneities; however, accounting for scatter from inhomogeneities lateral to the pencil beam was not possible. In some implementations the pencil beam was calculated using appropriate Monte Carlo calculated kernels, in others it was derived from the measured data. Modern versions of these algorithms such as the anisotropic analytical algorithm (AAA) split the pencil beam kernels into primary and scatter and allow the distortion of the scatter kernel to better correct for the distortion of the dose distribution close to inhomogeneities
Convolution algorithms: collapsed cone	Individual kernels are retained and convolved with TERMA. To reduce the computational power needed, radiation transport to or from a point, viewed in polar coordinates, is 'collapsed' onto the axes of a set of cones centred on that point to generate the polar equivalent of large pixels. The retention of individual kernels makes it possible to account for the lateral disequilibrium in regions of low or high-density material; when computing kernels account must be taken of the polyenergetic nature of the beam
Multigrid superposition	'Multigrid' or 'adaptive' techniques reduce computation time, and are usually used with superposition algorithms (though they can be used with any calculation algorithm). In this approach the spatial frequency of dose calculation is varied depending on the dose gradient at each point within the volume being calculated. Interpolation, usually linear, is used to estimate the dose between calculated points. This allows the dose calculation to be done at fewer points (for a given accuracy) than is the case if a uniform rectilinear grid of calculation points is used
Monte Carlo	True Monte Carlo calculations follow the histories of millions of particles using sound physical principles combined with statistical methods to determine the outcome at each point. They are currently regarded as the gold standard for calculations; however, they have not been widely clinically implemented due to the intensive computer requirements to generate a plan

Table 2.1 Summary of main differences between planning systems *continued*

Algorithm type	Comments
	in a reasonable amount of time. Methods for reducing the time required include macro Monte Carlo and voxel Monte Carlo (where the energy and fluence distribution for a smaller number of particles is used to characterize a distribution. This is then used as input for the next phase of the model)
Boltzman linear transport equation (BLTE)	In some situations it is possible to formulate differential equations for radiation transport through the volume to be considered. These may be solved numerically. In theory this allows a calculation that rivals Monte Carlo for accuracy, since the basic physics of the transport equations is the same as that used to compute histories in Monte Carlo techniques. In practice the accuracy of both depends on the implementation used, the effects of pixel sizes, and the treatment of multiple scatter. BLTE algorithms however may, in some implementations, run many times faster than Monte Carlo approaches. This approach has yet to be widely implemented clinically[5]

Table 2.2 Inhomogeneity corrections used in treatment planning systems and their accuracy

Correction method	Brief description	Limitation	Estimate of accuracy	Ref.
Equivalent path length	Radiological path lengths calculated using relative electron density	Corrects in direction of ray trace only. No account taken of size or position of inhomogeneity	Errors up to 10.4%	6, 7
Batho and modified Batho	Uses tissue phantom ratios	Poor agreement is obtained within the inhomogeneity, especially for large field sizes, because of lateral disequilibrium, but agreement beyond the inhomogeneity is good.	−0.17 for Co-60 beyond cork, 3% within	8–10
Equivalent tissue–air ratio (ETAR)	All CT slices are combined to give one effective slice for the computation of scatter. This reduced the calculation time compared with systems that retain each individual slice but reduces the accuracy	Overestimates dose within low-density inhomogeneity, particularly for small field sizes because of lateral disequilibrium	0.18% for Co-60 beyond cork Errors up to 7.2% within inhomogeneity	6, 10, 11
Convolution techniques	Dose deposition kernels are scaled according to the density of the media	Allowing spatially variant kernels increases calculation time as transform techniques cannot be used. Penumbra width in low-density regions will be increased		12
Monte Carlo	No additional correction needed			

References

1. Casebow MP. Matching of adjacent radiation beams for isocentric radiotherapy. *British Journal of Radiology* 1984; **57**: 735–40.
2. Bently RE, Milan J. An interactive digital computer system for radiotherapy treatment planning. *British Journal of Radiology* 1971; **44**: 826–33.
3. Redpath AT, Vickery BL, Duncan WA. A comprehensive radiotherapy planning system implemented in Fortran on a small interactive computer. *British Journal of Radiology* 1977; **50**: 51–7.
4. Storchi PR, van Battum LJ, Woudstra E. Calculation of a pencil beam kernel from measured photon beam data. *Physics in Medicine and Biology* 1999; **44**(12): 2917–28.
5. Failla GA, Wareing T, Archambault Y, Thompson S. *Acuros XB advanced dose calculation algorithm for the Eclipse treatment planning system*. Palo Alto, CA: Varian medical systems.
6. Butson MJ, Elferink R, Cheung T, Yu PKN, Stokes M, Quach KY, *et al.* Verification of lung dose in an anthropomorphic phantom calculated by the collapsed cone convolution method. *Physics in Medicine and Biology* 2000; **45**: 143–9.
7. Parker RP, Hobday PA, Cassell KJ. The direct use of CT numbers in radiotherapy dosage calculations for inhomogeneous media. *Physics in Medicine and Biology* 1979; **24**(4): 802–9.
8. Batho HF. Lung corrections in Cobalt 60 beam therapy. *Journal of the Canadian Association of Radiologists* 1964; **15**: 79–83.
9. Cassell KJ, Hobday PA, Parker RP. The implementation of a generalised Batho inhomogeneity correction for radiotherapy planning with direct use of CT numbers. *Physics in Medicine and Biology* 1981; **26**(4): 825–33.
10. Wong JW, Purdy JA. On methods of inhomogeneity corrections for photon transport. *Medical Physics* 1990; **17**(5): 807–14.
11. Sontag MR, Cunningham JR. The equivalent tissue air ratio method for making absorbed dose calculations in a heterogeneous medium. *Radiology* 1978; **129**: 787–94.
12. Mackie TR, Scrimger JW, Battista JJ. A convolution method of calculation dose for 15MV x-rays. *Medical Physics* 1985; **12**(2): 188–96.

Further reading

Cunningham JR. Tissue inhomogeneity corrections in photon-beam treatment planning. *Progress in Medical Radiation Physics* 1982; **1**: 103–31.

Green D, Williams PC. *Linear Accelerators for Radiation Therapy* (2nd edn). Bristol: Institute of Physics Publishing, 1997.

Hurkmans C, Knoos T, Nilsson P, Svahn-Tapper G, Danielsson H. Limitations of a pencil beam approach to photon dose calculations in the head and neck region. *Radiotherapy and Oncology* 1995; **37**: 74–80.

International Electrotechnical Commission (IEC). *60601–2-1 Ed 2.0 Medical electrical equipment—Part 2–1: Particular requirements for accelerators in the range 1MeV to 50MeV*. Geneva: IEC, 1998

Johns HE, Cunningham JR. *The Physics of Radiology*. Springfield, IL: CC Thomas, 1983.

Joint Working Party of the British Institute of Radiology and Hospital Physicists'Association, U.K. *Central Axis Depth Dose Data for Use in Radiotherapy. British Journal of Radiology (Supplement 25)*. London: British Institute of Radiology, 1996.

Metcalfe P, Kron T, Hoban P. *The Physics of Radiotherapy X-Rays from Linear Accelerators*. Madison, WI: Medical Physics Publishing, 1997.

Thwaites DI, Williams JR. *Radiotherapy Physics in Practice*. Oxford: Oxford University Press, 1993.

3

Treatment delivery, intensity-modulated radiotherapy, and image-guided radiotherapy

Carl Rowbottom

3.1 Introduction and background

The successful delivery of external beam radiotherapy involves a number of complex processes beginning with the decision by the clinical oncologist to use radiotherapy as part of the patient's cancer management, through the preparation and planning of the patient's treatment, to the verification of the patient position and radiation dose delivered at the time of treatment. Figure 3.1 highlights the major processes involved in external beam radiotherapy.

The aim of radiotherapy is to deliver a homogenous radiation dose to tumour inside the patient, whilst minimizing dose to all other parts of the body, in particular to organs which are especially radiosensitive or in close proximity to the tumour. In order to begin to achieve these aims, detailed information is required about the tumour position, size, and shape within the patient and the location of radiosensitive organs at risk. This is achieved from three-dimensional patient images, usually CT scans. The various structures are outlined on a series of axial slices to produce three-dimensional (3D) volumes. Once this has been done a radiotherapy plan can be designed for the individual patient to meet the treatment goals using a detailed computer model of the way radiation dose will be deposited within the patient anatomy.

To obtain accurate dose calculations from the patient representation, the relative electron or physical density of each voxel of the patient is required so that corrections can be applied from knowledge of the mass attenuation coefficients of different tissue types. This is usually achieved by the use of a look-up table within the treatment planning system that converts the CT numbers in the CT images to electron or physical density relative to water. The look-up table is generally derived from the CT scan of a phantom containing a number of inserts of known density. The look-up table can be described by two linear fits; the first linear fit for the CT number range −1000 to 50, and the second linear fit for CT numbers greater than 50. Some planning systems require that 1000 is added to CT numbers within a CT image so that all numbers become positive, and therefore reduce the data storage requirements. Figure 3.2 is a look-up table of CT number against physical density with 1000 automatically added to all CT numbers.

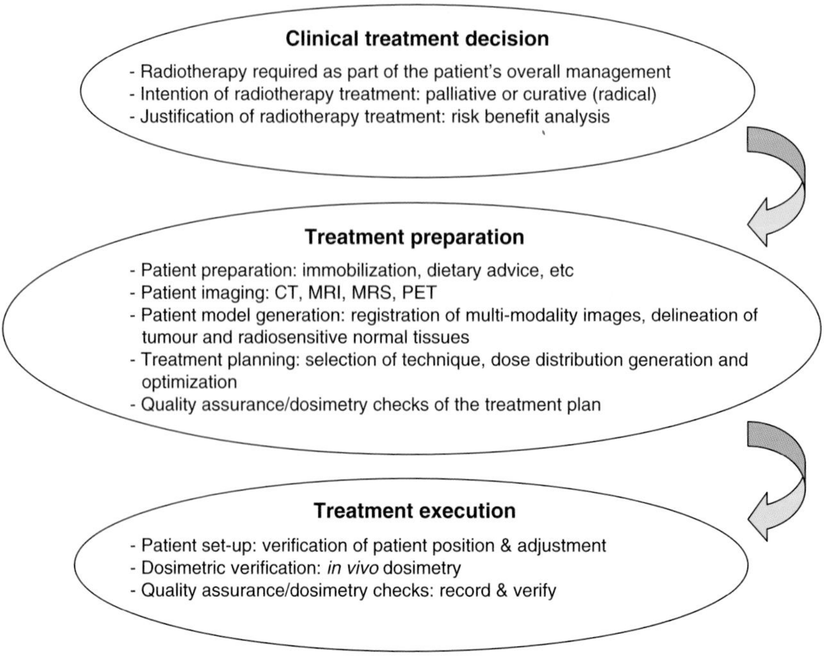

Clinical treatment decision
- Radiotherapy required as part of the patient's overall management
- Intention of radiotherapy treatment: palliative or curative (radical)
- Justification of radiotherapy treatment: risk benefit analysis
Treatment preparation
- Patient preparation: immobilization, dietary advice, etc
- Patient imaging: CT, MRI, MRS, PET
- Patient model generation: registration of multi-modality images, delineation of tumour and radiosensitive normal tissues
- Treatment planning: selection of technique, dose distribution generation and optimization
- Quality assurance/dosimetry checks of the treatment plan
Treatment execution
- Patient set-up: verification of patient position & adjustment
- Dosimetric verification: *in vivo* dosimetry
- Quality assurance/dosimetry checks: record & verify

Fig. 3.1 The radiotherapy process.

CT Number Density
0 0.000
99 0.050
1079 1.083
1226 1.156
2211 1.830
Density
2.0
1.8
1.6
1.4
1.2
1.0
0.8
0.6
0.4
0.2
0.0
0
500
1000
1500
2000
2500
CT Number

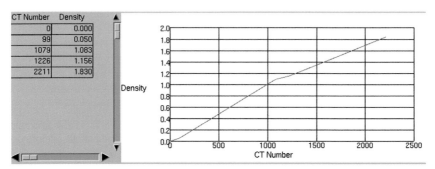

Fig. 3.2 A look-up table to convert from CT number to relative physical density used by a treatment planning system to calculate dose. All CT values have automatically been increased by 1000 so that air has a CT value of 0 and water a value of 1000.

3.1.1 ICRU 50 and 62

Radiotherapy given with the intent to cure the patient, termed radical radiotherapy, is given by small doses of radiation delivered once or twice a day for several weeks. Each radiotherapy treatment session is called a fraction of radiotherapy. When considering the development of the radiotherapy plan it is clear that the initial 3D image or images used to develop the patient anatomical model is a single snapshot of the patient concerned, whereas the resulting treatment plan has to be designed in a way that delivers an adequate radiation dose to the tumour when delivered with multiple fractions of radiotherapy over a time period of several weeks. To ensure that the treatment plan is capable of meeting this requirement, treatment margins have to be added to the initial tumour shape at the treatment planning stage. The International Commission on Radiation Units and Measurements (ICRU) have produced a number of reports[1–3] with the aim of standardizing the practice of prescribing radiotherapy and designing treatment plans that will give an adequate and reliable dose to the tumour when treating with fractionated radiotherapy.

ICRU report 50[1], published in 1993, was developed to contain recommendations on how to report a treatment in external beam radiotherapy. It introduced common terminology for reporting and prescribing the radiotherapy treatment that could be followed in all centres worldwide and intended to standardize prescribing practice. In relation to treatment planning the report gave clear guidance on the need for treatment margins to ensure that the tumour is adequately treated with a fractionated course of radiotherapy. The volumes described in the report are outlined in Table 3.1. In addition to the treatment volumes, the report outlined that treatment plans should be prescribed to a stable point of high dose and locally a low-dose gradient within the PTV; this was termed the ICRU reference point. Often the isocentre of the treatment is used as the appropriate ICRU reference point. The treatment plan should also be designed with a limit on the variation in dose within the PTV of −5% and + 7% of the ICRU reference point.

A supplement to ICRU 50 was produced in 1999[2] to address the changes in treatment planning and radiotherapy treatment that had occurred with the increased use of pre-treatment imaging, computerized treatment planning, and treatment verification. ICRU report 62 introduced the concepts of planning-risk volumes (PRVs) and the conformality index, defined organs at risk (OARs) by their tissue architecture, and refined the construction of the margin required to create the PTV. The report described OARs as having distinct tissue architectures; serial OARs, for example, the spinal cord, have a high relative seriality implying that dose above a tolerance limit, even to a small volume, impairs the function of the entire OAR; parallel OARs, for example, the lungs, have a low relative seriality where the main parameter impairing the organ's function is the proportion of the OAR receiving a dose above a specified tolerance. In reality many organs have a tissue architecture with both high and low seriality components. The report retained the basic concepts of GTV, CTV, and PTV but described new volumes for use in treatment planning. Table 3.2 outlines the additional volumes described in ICRU 62.

Figure 3.3 demonstrates the delineation of structures for a patient with a nasopharyngeal tumour and includes the delineation of the GTV, CTVs, and PTVs, as well as the spinal cord and brainstem as radiosensitive OARs in this region. The spinal cord and brainstem are both serial-like structures and so the expansion of both volumes by 5 mm to create PRVs as described in ICRU report 62 can clearly be seen.

Margin recipes have been developed to describe the required margin size to ensure the CTV is covered by a certain isodose. The margin recipes can be derived from monte carlo methods where a number of systematic and random errors are introduced and the dose to the CTV determined. Van Herk[4] presented the most commonly used

Table 3.1 The various volumes described by ICRU 50

Volume name	Description	Comments
Gross tumour volume (GTV)	The gross palpable or visible/demonstrable extent and location of the malignant growth	The GTV is usually defined from reconstructed 3D images of the patient such as CT, MR, PET. Multiple image modalities can be used to define this volume
Clinical target volume (CTV)	The tissue volume that contains a GTV and/or subclinical microscopic malignant disease, which has to be eliminated	This volume has to be treated adequately in order to achieve the aim of the therapy: cure or palliation. Although it may be described as a geometric expansion of the GTV, the CTV is constrained by anatomical boundaries such as bone
Planning target volume (PTV)	A geometrical concept, defined to select appropriate beam sizes and beam arrangements, taking into consideration the net effect of all the possible geometrical variations and inaccuracies in order to ensure that the prescribed dose is delivered to the CTV	The size of the margin from CTV to PTV can only be reduced if the likely geometric variations are reduced
Treated volume (TV)	The volume enclosed by an isodose surface, selected and specified by the radiation oncologist as being appropriate to achieve the purpose of treatment	The 90% or 95% isodose is often used to define the treated volume
Irradiated volume (IR)	The tissue volume which receives a dose that is considered significant in relation to normal tissue tolerance	The 50% isodose is often used to define the irradiated volume
Organs at risk (OARs)	Normal tissues whose radiation sensitivity may significantly influence treatment planning and/or prescribed dose	

margin recipe from a Monte Carlo study of prostate radiotherapy treatments. The margin size to ensure that the CTV receives at least 95% dose in 90% of patients is given as

$$M_{ptv} = 2.5 \sum + 1.64 \left(\sigma - \sigma_p \right) \tag{1}$$

where Σ refers to the systematic variation, σ to the random variation, and σ_p to the penumbra margin.

For the case of soft tissue and megavoltage photon beams where ≈ 3.2 mm the equation can be simplified to

$$M_{ptv} = 2.5 \sum + 0.7 \, \sigma \tag{2}$$

Table 3.2 The volumes described in ICRU 62

Volume name	Description	Comments
Internal margin (IM)	The margin that must be added to the CTV to compensate for the expected physiological movements and variations in size, shape, and position of the CTV during treatment	The motion occurs when the CTV position changes on a day-to-day level and is mainly associated with organs that are part of or adjacent to the digestive or respiratory system. Changes in the patient's condition, such as weight gain/loss, can also affect the relative position of the CTV
Set-up margin (SM)	The margin that must be added to the internal margin to compensate for the expected motion of the internal margin due to the repeated set-up of the patient during treatment	The uncertainties depend on different factors and can include variations in patient positioning, mechanical uncertainties of the equipment, dosimetric uncertainties (light-radiation field agreement), transfer set-up errors, and human-related uncertainties
Planning risk volume (PRV)	The OAR volume with a margin added to compensate for internal motion of the OAR and changes in position due to set-up errors	The PRV concept is often used for serial OARs where the maximum dose to the PRV is kept below tolerance to ensure that the OAR is not impaired if a systematic set-up variation occurs when treating the patient
Conformality index	The quotient of the treated volume and the volume of the PTV (used to compare techniques)	The conformality index can be used to compare different treatment techniques. The concept is insensitive to the shape of the treated volume compared to the PTV volume, i.e. some parts of the PTV may not receive adequate dose but the treated volume may be adequate due to the irradiation of tissue outside the PTV in other regions

The recipe highlights that the systematic variation is more than 3.5 times more important in the resulting margin required than random variations. This is because in general a systematic variation results in a shift in the dose distribution leading to an underdose to some part of the CTV whereas random variations lead to a blurring of the dose distribution with a less pronounced dosimetric effect.

Similar margin recipes have been developed for OARs. In general, if an organ is parallel and large, and the risk of complication is known to be acceptable, then the clinician may decide to ignore the geometric error. However, for dose levels that cause unacceptable complications, e.g. serial organs, then the geometric errors need to be taken into account by the application of margins to OARs to create larger PRV volumes. Dose constraints for serial OARs are then applied to the PRV volume. McKenzie et al.[5] outlined a margin recipe for small and/or serial organs at risk in low (−) or high (+) radiation dose regions as

$$M_{prv} = 1.3 \sum \pm 0.5\,\sigma \tag{3}$$

where again Σ refers to the systematic variation, σ to the random variation.

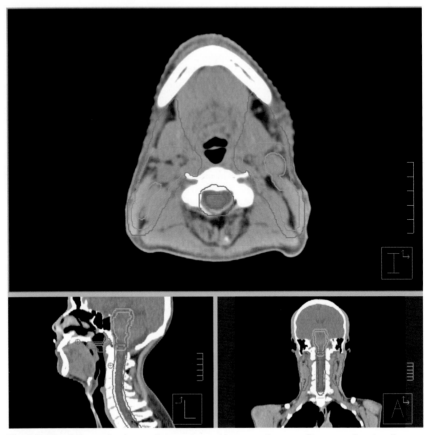

Fig. 3.3 The delineated volumes for a patient with a nasopharyngeal tumour showing the GTV (orange), CTV1(sky blue) and CTV2 (green) that will receive different radiation doses, and the spinal cord (red) and brainstem (yellow green) with their respective PRV volumes also shown.

Margins applied in practice are often based on legacy protocols or adopted from clinical trials. In reality, treatment planning margins should be based on an individual department's practices and information regarding levels of systematic and random error. Margins should only be reduced if the technique has been altered or developed to reduce either the systematic and/or random errors in the radiotherapy process. Otherwise reducing the margin to improve the dosimetric quality of the treatment plan may well result in the underdose of the CTV during the fractionated course of radiotherapy.

3.2 Forward planning process

The process of developing an individualized radiotherapy plan has traditionally been performed using forward planning. With this approach a planner starts the process by choosing the appropriate number and directions for the treatment beams to be used. In conventional radiotherapy the shape of each beam is simply square or rectangular. More commonly the beam shape is created to conform to that of the tumour from each treatment beam direction by the application of multileaf collimators (MLCs); this is referred to as conformal radiotherapy. Figure 3.4 shows a lateral beam from a conventional and conformal radiotherapy plan highlighting how the use of the MLC for each treatment beam can reduce the dose to OARs, the bladder and rectum in this case as portions of each OAR are now shielded from high levels of radiation dose by the MLCs.

The treatment planner has a number of options available in developing a clinically acceptable treatment plan such as the radiation modality used, the energy of the radiation used, the relative beam weights of each treatment beam, or the use of simple wedged beams for some or all treatment beams. The planner goes through an iterative process to alter the available treatment parameters to produce a plan that meets the ICRU limits for coverage of the PTV, and meets the dose constraints established for OARs.

Wedged beams can either be produced by the introduction of a physical wedge into the beam, or from the dynamic motion of the jaw. When the physical wedge is integrated in to the treatment head and computer controlled this is termed a motorized wedge. Figure 3.5 shows isodose plots for an open, 20°, and 55° wedged beams at 6 MV using a motorized wedge. The wedged beam creates a non-uniform intensity in one direction only. The amount of radiation leaving the radiation machine, the linac, is less at the thick end of the wedge. Similar plots to those seen in Fig. 3.5 can be produced by the dynamic motion of a jaw across the beam creating an approximately linear change in intensity. Wedged treatment beams of varying angles can be produced

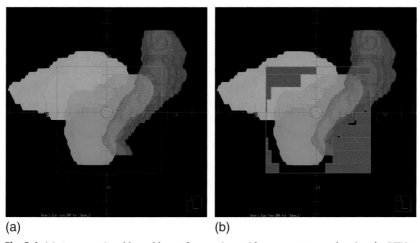

(a)   (b)

Fig. 3.4 (a) A conventional lateral beam for a patient with prostate cancer showing the PTV (sky blue), the rectum (brown) and the bladder (yellowgreen). (b) The same beam with a multi-leaf collimator (MLC) used to reduce doses to the bladder and rectum, the main organs-at-risk.

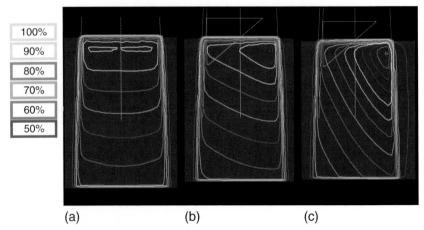

Fig. 3.5 Isodose plots for radiation beams at 6 MV for (a) an open beam, (b) a 20° wedge, and (c) a 55° degree.

either by the combination of an open and motorized wedge beam, or by altering the speed of the dynamic jaw across the treatment beam. These simple non-uniform intensity treatment beams can be used to improve the homogeneity of the dose within the PTV. In the most simplistic example the wedged beam compensates for missing patient tissue.

Figure 3.6 shows an example of wedged beams used in a simple parallel opposed pair treatment of a larynx. Less radiation is required anteriorly as the patient thickness is less in this region than posteriorly where the greater thickness of the patient will increase the attenuation of the beam. The introduction of wedged beams is able to significantly improve the homogeneity of the radiation dose delivered to the patient.

For any given treatment site a set of starting conditions for a given radiotherapy technique such as the number and approximate location of the treatment beams, the radiation modality, the radiation energy, the initial use of open and wedged beams, and the relative beam weights are often described. For a given patient the process then involves the refinement of these parameters to individualize the treatment plan for the patient. When using forward planning the skill of the planner is in the understanding of how altering the beam weight for different treatment beams, the addition of a wedged beam, or the changing of a wedge angle for a wedged beam will affect the resulting dose distribution to achieve the aims of treatment.

Once a treatment plan has been generated it needs to be evaluated to ensure that it meets the clinical requirements of the proposed radiotherapy treatment. At a basic level this is done by inspecting lines of isodose on each slice of the patient representation to determine whether the PTV receives an adequate radiation dose and determine that OARs are not receiving too much radiation. For any treatment plan there is often a significant amount of information to consider. Simple graphical tools and the reduction of the 3D dose distribution to single numbers via the use of biological models

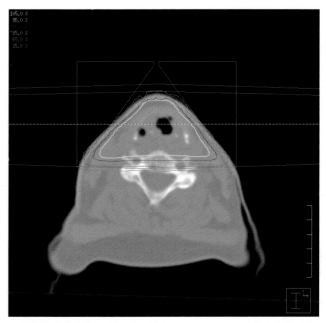

Fig. 3.6 A clinical example of use of wedged beams to compensate for missing tissue in the radiotherapy treatment of the larynx.

have been developed to aid in the process of evaluating treatment plans. For example, a plot of the cumulative dose–volume frequency for a particular volume, such as the PTV(s) or OARs can be a useful tool used at the treatment planning stage. The plot graphically summarizes the radiation dose distribution to a volume of interest of a patient and is commonly known as the dose–volume histogram (DVH)[6]. DVHs can be used to aid the evaluation of the treatment plan for a particular patient, or to compare multiple treatment plans. They can also be used to generate the tumour control probability (TCP) and normal tissue complication probability (NTCP) of OARs[7]. Figure 3.7 demonstrates a typical DVH for a patient with prostate cancer and graphically demonstrates the dose distribution to the PTV as well as the major OARs in this region, the rectum, the bladder, and the femoral heads. For any given dose on the x-axis, the ordinate on the DVH is the percentage volume of the structure receiving that dose level or more. Therefore each volume has an ordinate value of 100% for zero dose.

As well as looking at the isodoses on axial slices of the patient anatomical representation and investigating the DVHs of PTV(s) and OARs, biological models of the response of tissues to radiation can be used to evaluate treatment plans. The most commonly used biological models are the TCP, the NTCP[7], and the equivalent uniform dose (EUD)[8]. As with all models the validity of their use in radiotherapy relies heavily on the parameters used within them. The Quantitative Analyses of Normal Tissue Effects in the Clinic (QUANTEC) have published evidence-based guidelines on how 16 different organs are damaged by radiation[9].

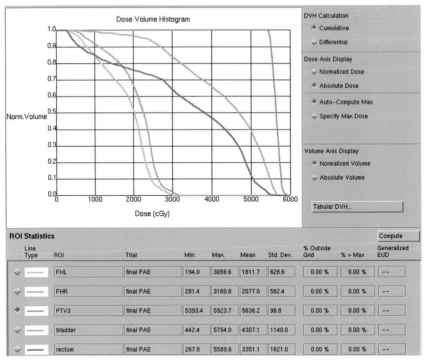

Line Type	ROI	Trial	Min.	Max.	Mean	Std. Dev.	% Outside Grid	% > Max	Generalized EUD
✓	FHL	final PAE	194.0	3056.6	1811.7	628.6	0.00 %	0.00 %	--
✓	FHR	final PAE	281.4	3160.8	2077.0	582.4	0.00 %	0.00 %	--
◆	PTV3	final PAE	5393.4	5923.7	5636.2	96.8	0.00 %	0.00 %	--
✓	bladder	final PAE	442.4	5754.0	4307.1	1140.0	0.00 %	0.00 %	--
✓	rectum	final PAE	267.8	5589.6	3351.1	1621.0	0.00 %	0.00 %	--

Fig. 3.7 A dose–volume histogram (DVH) for a patient with prostate cancer. The dose distribution to the PTV and major OARs (rectum, bladder, and femoral heads) are graphically represented by a DVH.

3.3 Inverse planning

3.3.1 Principles of inverse planning

Inverse planning is a process where instead of manually adjusting beam weights and wedges as in forward planning, the dose distribution wanted by the planner is described at the start of the process and then computer optimization is used to develop the appropriate plan. Inverse planning is used for most intensity-modulated radiotherapy (IMRT), volumetric-modulated arc radiotherapy (VMAT), and intensity-modulated arc radiotherapy (IMAT) planning for treatments such as tomotherapy. These forms of radiotherapy delivery can constrain the high radiation dose to the size and shape of the PTV. In such treatment techniques the fluence from any given beam direction does not necessarily correspond to an intuitive shape of the PTV but when combined with a number of other non-uniform treatment beams produces a high radiation dose that conforms to the shape of the PTV.

In its simplest form the inverse planning approach requires the planner to describe the dose distribution that they want to have at the end of the planning process. This could be described on a voxel-by-voxel basis, but is more often described as a series of minimum dose, maximum dose, mean dose, dose–volume limit, or biologically-based

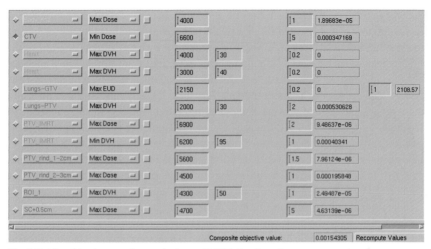

✓	Max Dose	☐	4000		1	1.89683e−05	
✦ CTV	Min Dose	☐	6600		5	0.000347169	
✓	Max DVH	☐	4000	30	0.2	0	
✓	Max DVH	☐	3000	40	0.2	0	
✓ Lungs−GTV	Max EUD	☐	2150		0.2	0	1 2108.57
✓ Lungs−PTV	Max DVH	☐	2000	30	2	0.000530628	
✓ PTV_IMRT	Max Dose	☐	6900		2	9.48637e−06	
✓ PTV_IMRT	Min DVH	☐	6200	95	1	0.00040341	
✓ PTV_rind_1−2cm	Max Dose	☐	5600		1.5	7.96124e−06	
✓ PTV_rind_2−3cm	Max Dose	☐	4500		1	0.000195848	
✓ ROI_1	Max DVH	☐	4300	50	1	2.49487e−05	
✓ SC+0.5cm	Max Dose	☐	4700		5	4.63139e−06	

Composite objective value:	0.00154305	Recompute Values

Fig. 3.8 The objectives used to optimize an IMRT plan for a lung cancer patient.

objectives for the outlined PTV and OARs. The skill in developing a treatment plan changes from the planner manually changing beam angles, weights, and wedges to improve the dose distribution, to altering the objectives to refine the dose distribution produced by the computer optimization to achieve a clinically acceptable treatment plan. Figure 3.8 demonstrates the objectives for an inverse plan for an example patient with a lung tumour.

3.3.2 An application of inverse planning: IMRT

IMRT is a radiotherapy technique where a number of non-uniform beam fluences from different beam directions are used to create a uniform radiation dose that closely conforms to the shape of the PTV. *Planning target volume*

If IMRT is taken as an example, the inverse planning process begins by the tumour and relevant OARs being outlined to create an anatomical model of the patient and the PTV created by the addition of suitable margins. The planner chooses the number of treatment beams to be used and the placement of those beams around the patient. Often equispaced beam arrangements of between five and nine beams are employed, but in some cases the application of individually designed beam directions is advantageous. The planner must also describe the final dose distribution that they want the planning system to achieve at the end of the optimization process. In simple terms this can be created by describing what the final dose–volume histogram curves look like for the PTV and OARs. The relative importance of meeting the different objectives in the optimization is also defined by the planner. An example set of objectives for a lung IMRT plan are given in Fig. 3.8.

From the set of beam directions the planning system applies starting fluences, such as open conformal fields of equal weight, and uses a cost function to see how close the resulting dose distribution is to the desired distribution described by the planner. The cost function needs to be minimized and a cost function of zero would mean that the actual and desired dose distributions were identical. The open fluence is divided in to a number of squares, called bixels, and the intensity in each bixel is allowed to vary to alter the dose distribution. Computer optimization algorithms are used to refine the fluences of each beam by changing bixel intensities to minimize the difference between the desired and actual dose distributions. This is achieved by altering the fluence and seeing if it reduces the cost function. Usually downhill optimization methods are used in this process; if the cost function of the new fluences is less than the previous iteration, the new fluences are accepted and the process is repeated, if the cost function is higher then the new fluences are rejected. By altering the bixel intensities in an iterative process the cost function is reduced and the dose distribution gets closer to the desired dose distribution given by the planner. Figure 3.9 shows the fluences for each beam of a five-field IMRT plan for a patient with prostate cancer.

In traditional inverse planning for IMRT, the optimal fluence for each beam direction is developed by the computer optimization. This 'ideal' fluence has to be converted into something that can be delivered by the linac; a two-step approach to inverse planning optimization. There are different ways in which IMRT fluences can be produced by linacs and each linac type will have different limits and constraints on their MLCs and radiation output that need to be taken in to account when converting from the 'ideal' fluence to the 'deliverable' fluence. In general, IMRT can be delivered either as a series of static subfields with the radiation switched off between subfields, or dynamic delivery where the motion of the MLCs with the radiation on provides the necessary modulation of dose. In either case the deliverable fluence is likely to be slightly different to the ideal fluence produced by the optimization and this needs to be considered when evaluating the final inverse plan.

A different way of dealing with how the radiation will be delivered in IMRT, IMAT, and VMAT is to include the delivery constraints directly in the optimization. This is usually achieved by beginning the optimization by developing the ideal fluence for a few iterations of the optimization, converting to a deliverable solution, and performing further optimization by working directly on the beam shapes and beam weights to improve the dose distribution. This approach to the inverse problem can lead to fewer segments when using a step-and-shoot approach to IMRT delivery than the two-step optimization approach.

It is clear from the process outlined here that the planner is required to define a lot of information before the inverse planning process can begin. In general, by considering a test cohort of patients for a particular treatment site it is often possible to define a set of starting conditions for the process including the number and orientation of the treatment beams and the initial set of objectives. This initial set of conditions is termed a class solution. Generally the class solution may require modification for individual patients but can significantly reduce the time taken to arrive at a clinically acceptable treatment plan.

3.3.3 Special considerations when using inverse planning

Computer optimization used in inverse planning will minimize the cost function resulting in a dose distribution that closely resembles the desired dose distribution as initially described by the planner. As the dose distribution gets closer to the desired

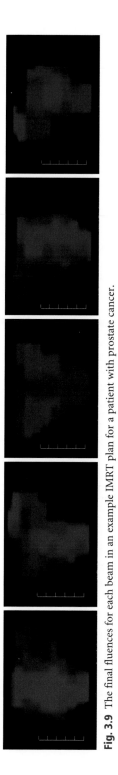

Fig. 3.9 The final fluences for each beam in an example IMRT plan for a patient with prostate cancer.

value the cost function gets smaller and the optimization works less hard to reduce the cost function further at the next iteration. This means that if the resulting plan requires 95% of the PTV volume to receive at least 95% of the prescription dose then setting this as the objective in the inverse plan will likely lead to the optimization ending with an undershoot and less than 95% of the PTV volume will receive 95% of the prescription dose. Therefore the planner will have to increase the volume limit, the dose limit, or both in the initial set of objectives, e.g. 97% of the PTV receives at least 97% of the prescription dose, to obtain a clinically acceptable result at the end of the optimization. In general, computer optimization algorithms undershoot minimum dose objectives and overshoot maximum dose objectives. An example of overshoot is that if in lung planning no more than 35% of the normal lung should receive 20 Gy or more then in setting an appropriate objective the planner will chose to try to limit the volume receiving at least 20 Gy to 25% or 30%. The distribution returned by the optimization will be slightly higher than that set in the objectives but probably less than the clinical limit of 35%. Figure 3.10 highlights the set of objectives given in Fig. 3.8 on the DVH with the final curves for some of the regions of interest. The figure clearly

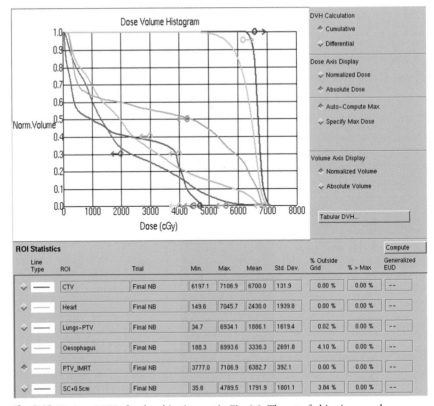

Fig. 3.10 Various DVHs for the objectives set in Fig. 3.8. The set of objectives are shown graphically on the DVH.

shows overshoot of the 'Lungs–PTV' objective and undershoot of the 'CTV' and 'PTV–IMRT' minimum dose/dose-volume objectives.

Inverse planning by computer optimization is completely dependent on the quality and quantity of information provided a priori by the planner. When minimizing the difference between the desired and actual dose distributions the optimization will not control the dose to areas that are not defined in the set of objectives. This can lead to dose distributions that look quite different to those from forward planning, and may result in adverse patient reactions if dose is inadvertently deposited in normal tissues that have not been outlined and considered in the inverse problem. It is therefore often the case when using an inverse planning approach that more OARs and normal tissues are outlined compared to forward planning. Another way to circumvent this issue is by the use of dummy volumes to control the dose distribution to look more like traditional conformal radiotherapy distributions. Dummy volumes are not anatomically based and may sometimes be automatically generated by modern treatment planning systems. Figure 3.11 demonstrates the use of dummy volumes in the treatment of a patient with a lung tumour. The rinds surrounding the PTV are used to guide the optimization to create a more conformal dose distribution. By setting a maximum dose constraint to the rind volume the optimization will attempt to push the isodose of value of the max dose constraint to the edge of the rind closest to the PTV.

Radiotherapy techniques such as IMRT, IMAT, and VMAT can produce significantly higher dose gradients than more traditional radiotherapy. In such circumstances it is possible that small systematic variations in patient position compared to the planning CT scan can lead to significantly higher doses to OARs than evaluated at

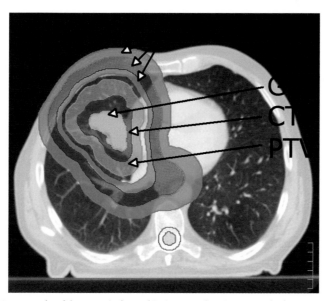

Fig. 3.11 An example of dummy rinds used in inverse planning to guide the optimization process. Such rinds can be used to limit doses to normal structures without the need for time-consuming delineation of normal structures.

the treatment planning stage. To overcome this problem PRVs as outlined in ICRU report 62 are often used for serial OARs in inverse planning to limit the dose a safe distance from the OAR. For example, if it is unlikely that a patient will have a systematic error of more than 3 mm in their treatment, then adding 3 mm to a serial organ such as the spinal cord, and limiting the dose that the PRV volume receives would make it highly unlikely that the spinal cord would receive a dose higher than that given to the PRV volume even in the presence of systematic set-up errors. Figure 3.11 highlights the use of a PRV for the spinal cord, a serial-like OAR that needs to be considered when treating thoracic tumours.

3.3.4 **ICRU 83**

ICRU report 83[3] refers specifically to the prescribing, recording, and reporting of IMRT treatments, but the principles outlined in the report are appropriate to all radiotherapy approaches employing inverse planning. The general principles of the previous ICRU reports were preserved within the report. An additional volume, the remaining risk volume (RVR), was introduced by the report. The RVR is defined by the difference between the volume enclosed by the external contour of the patient and that of the CTVs and OARs outlined. It is expected that the RVR may be a useful metric in looking at different treatment modalities and their risk of producing very late effects such as second malignancies from the radiation exposure of normal tissues during radiotherapy treatment.

Refinements of the notation of the GTV were recommended to account for changes in the delineation of the tumour and the increased use of image-guided radiotherapy. For example, the GTV could be delineated from a CT scan, or a T2-weighted MRI scan. It should be clear when reporting the radiotherapy treatment what image modality was used for delineation, and so should be given as GTV(CT) or GTV(MRI-T2) respectively. With the introduction of image-guided radiotherapy the GTV may be delineated more than once during the course of the radiotherapy treatment and this should also be clearly demonstrated in the notation. Therefore, if a treatment plan was created by taking a repeat MRI scan after 30 Gy of dose has already been delivered to the patient, then the GTV should be described as GTV(MRI-T2, 30 Gy).

The report also highlighted the need for consistent dose reporting of PTVs by reporting the median absorbed dose, and the near maximum and near minimum absorbed doses. For OARs and PRVs the report recommended the reporting of the mean absorbed dose, the near maximum dose, and V_D which if exceeded has a known probability of causing complication, e.g. $V20_{Gy}$ for lung. The report recommends the use of $D_{2\%}$ and $D_{98\%}$ for the near maximum and near minimum doses, where $D_{2\%}$ is the highest dose received by at least 2% of the volume of the PTV or OAR. A more consistent reporting may be achieved by the reporting of the highest dose or lowest dose received by at least 1 cc as this is independent of the volume of the region of interest.

In terms of prescribing the dose, ICRU report 83 recommends prescribing the required dose to the median PTV dose rather than a single point in a region of high-dose and low-dose gradient. For IMRT dose distributions the isocentre is often not a suitable prescription point as there are more local dose gradients within the PTV than for conformal dose distributions, particularly when individual treatment beams are considered rather than the dose distribution produced from the combination of all treatment beams.

3.4 **Image-guided radiotherapy**

3.4.1 **Principles of IGRT**

IGRT involves imaging the patient at the time of treatment and in the treatment position. In general, the term image-guided radiotherapy implies that information about the position, size, and shape of the tumour as well as normal tissues is gathered during the treatment course. This information can then be used to alter the radiotherapy when required. The potential of IGRT is that the additional knowledge of the patient anatomy during treatment may lead to reduced treatment margins, fewer geometric misses, and therefore improved survival and reduced side effects from radiotherapy. IGRT becomes ever more important in the radiotherapy process with increased use of conformal treatment techniques such as IMRT or VMAT that more closely conform the dose to the size and shape of the tumour. The larger conformality index of older radiotherapy techniques, such as conventional and conformal radiotherapy, made geometric miss of the tumour less likely as the CTV could move outside the designed PTV in some areas and still receive a high radiation dose. It is essential to consider the use of IGRT when using these newer radiotherapy techniques in combination with significantly reduced margins compared to conventional treatments.

3.4.2 **Implications of IGRT in treatment planning**

The traditional process of radiotherapy is generally linear. The patient is prepared for treatment by the generation of an anatomical model of the tumour and OARs, and a treatment plan developed on this static snap-shot of the patient. Patient changes and internal anatomy motion are dealt with in the planning process by the application of suitable margins to generate the PTV as described earlier. Verification of the patient set-up is performed via the comparison to the original anatomical model at the treatment unit. Traditionally the verification of patient position has been based on bony anatomy seen in planar imaging and there has been little interaction with treatment planning during the course of the patient's treatment. However with the introduction of IGRT there is much more information about the location, size, and shape of the tumour and normal tissues and this leads to a greater interaction with treatment planning to refine the original treatment plan to ensure that the patient receives the radiation dose intended whilst minimizing damage to normal tissues; this is sometimes termed adaptive planning, but may be better described as reactive planning. Figure 3.12 describes the traditional and IGRT-based radiotherapy processes.

There are a number of anatomical changes that can be observed from IGRT; they include changes in patient weight, internal organ motion, and systematic anatomical changes due to medical interventions. Where systematic changes to the patient anatomy compared to the treatment planning stage have occurred it may be necessary to modify the treatment plan to adequately treat the patient. Care needs to be taken when considering changes to the treatment plan when the anatomical changes to the patient

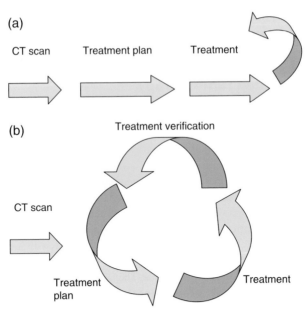

Fig. 3.12 (a) The traditional model of radiotherapy where the treatment verification process is applied at the treatment phase. (b) The IGRT model of radiotherapy where the treatment verification process feeds back in to treatment planning and may result in multiple treatment plans for an individual patient during their fractionated course of treatment.

observed by IGRT are random or transient in nature, e.g. daily changes in the amount of gas in the rectum. IGRT using cone-beam CT approaches are most applicable for direct replanning as they produce CT-like images with the patient in the treatment position. It is likely that the introduction of cone-beam-based IGRT will lead to somewhere in the region of 20% of patients requiring replanning during the course of their radiotherapy[10].

3.4.3 Concomitant dose in IGRT

When using ionizing imaging radiation approaches for IGRT it is important to carefully consider the concomitant doses given to the patient and to justify any additional exposure via a risk–benefit analysis for the patients. The methodology outlined by Harrison[11] was to consider the imaging dose to OARs and compare them to the dose received from leakage and scatter during the radiotherapy. For the case of 26 CT scans during the course of radiotherapy it was found that the imaging dose was 5–25% of the leakage and scatter dose from the radiotherapy. However Perks et al.[12] measured peripheral doses in an anthropomorphic phantom and found the IGRT dose to be as high as 50% of the leakage and scatter dose from the radiotherapy. This highlights the need to perform a risk–benefit analysis to appropriately justify the use of IGRT and the additional radiation dose given to the patient from IGRT.

Imaging protocols for IGRT should be designed to limit the radiation dose to the patient in line with the 'as low as reasonably practicable' (ALARP) principle. This results

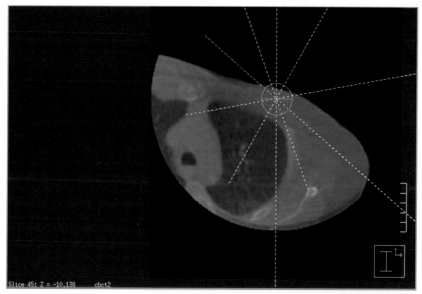

Slice 45: Z = -10.138   cbct2

Fig. 3.13 An example IGRT image using a small field of view and low-dose protocol to reduce the radiation dose received by the patient. The central axis of each treatment beam is also shown.

in imaging protocols using small fields of view and low imaging doses that produce images of sufficient quality for the purposes of verification of patient set-up. However, such imaging protocols often produce IGRT images that are difficult to use for direct dosimetric evaluation due to incomplete patient outlines, poor soft tissue differentiation, and inaccurate CT numbers. Figure 3.13 highlights an IGRT image for a sarcoma of the chest wall where a small field of view and low imaging dose protocol has been applied. The images are adequate for the purpose of patient set-up verification, but cannot be directly used for dosimetric evaluation of anatomical changes to the patient.

A solution to the conflicting requirements on the IGRT protocols for patient set-up verification and replanning is to establish two-step imaging protocols. Initially all patients receive low-dose IGRT imaging with field of view that are sufficient for the purposes of patient set-up verification. If anatomical changes are observed in such images then higher dose images with a larger field of view can be justified and applied to enable dosimetric analysis of the patient changes to be evaluated.

3.4.4 Replanning strategies using IGRT

The simplest strategy for the replanning of patients is to rescan and restart the treatment planning process if IGRT images highlight significant patient changes compared to the planning scan. This approach results in a repeat of the treatment planning process outlined in previous sections. One difficulty often encountered when replanning patients part way through their course of treatment is how to evaluate the total dose to the tumour and OARs from the separate plans with differing anatomy. The ability to accumulate dose for the same patient from multiple patient anatomical models and treatment plans is an active area of development for manufacturers of treatment planning systems and is an essential requirement for the individualization of radiotherapy treatment with adaptive/reactive radiotherapy.

Cone-beam-based IGRT provides the potential to use the IGRT information directly in the treatment planning process, particularly for simple changes in the patient anatomy such as weight loss or weight gain during treatment. However, IGRT usually involves the use of low-dose imaging with small or medium fields of view to limit the radiation doses given to the patient from the imaging procedure. Such IGRT techniques present challenges when using the images produced in the replanning process as outlined earlier. Cone-beam-based IGRT systems can be calibrated to produce appropriate CT numbers under certain conditions. However, the use of a cone-beam geometry leads to inherent inaccuracies in CT number to electron or physical density conversions due to changes in scatter conditions from patient to patient, and the use of half rotation scans to reduce patient imaging doses. These inaccuracies can be significant for IGRT images of the thorax and abdomen regions.

Replanning strategies

Patient example 1—patient weight loss

A common example of a patient change seen by IGRT is weight loss during the course of radiotherapy treatment. Figure 3.14 shows an example of a head and neck patient losing a significant amount of weight.

If the patient has lost weight, but verification of patient set-up is based on bony anatomy then the change to the patient can be approximated by altering the external patient contour on the original planning scan[13]. The assumption is that the bony anatomy is rigidly fixed in place and that changes in the proportion of muscle and fat do not greatly affect the accuracy of the dose calculation. The process involves changing the external patient outline on the planning scan via image fusion with the cone-beam IGRT images and the relative electron or physical density outside the new external patient contour being set to that of air. A heterogeneous dose calculation can be performed on the planning scan with the altered external contour. This assumes that the relative position of bony anatomy and air spaces do not change significantly

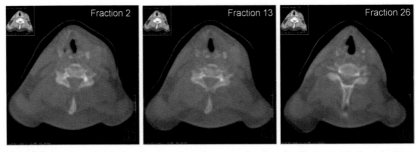

Fig. 3.14 An example of a head and neck patient experiencing significant weight loss during the course of radiotherapy treatment.

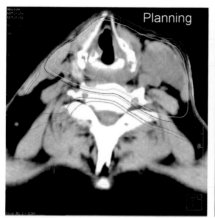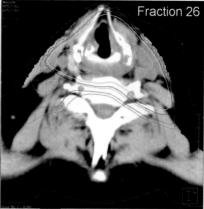

Fig. 3.15 The original dose calculation on the CT scan and the dose calculation on the CT scan but based on the external contour from the IGRT scan for fraction 26.

during weight loss. Figure 3.15 highlights this method of replanning for the patient example. The advantage of this approach is that the heterogeneity correction will be reasonably accurate as it uses the original CT scan. It is also relatively quick to recalculate and evaluate the dose distribution, particularly the appearance of significant hotspots and any changes to the dose delivered to visible OARs such as the spinal cord in this example. The disadvantage is that it is difficult to determine the dosimetric effect on soft tissues, such as the parotid glands in this example, as they may not be clearly demonstrated on the IGRT scan.

Patient example 2—patient anatomy change in a region of significant density heterogeneity

To dosimetrically evaluate the effect of the anatomical changes seen from IGRT the Hounsfield numbers in the images need to be accurate enough for direct dose calculations to be made. Figure 3.16 shows a line scan between two points in the conventional CT and IGRT images for the same patient. The Hounsfield numbers are all increased by 1000 within the treatment planning system, but it is clear from the figure that the IGRT images are not calibrated for use in treatment planning. Additionally, the use of half-rotation scans to reduce patient doses from IGRT leads to a gradient across the images and therefore different Hounsfield values for the left and lung images in the figure.

A simple strategy for the use of IGRT images for re-planning is to outline different tissue types, such as lung and soft tissue, prior to applying a bulk density override using the average density from the original treatment plan for the same tissue types. This approach ensures an accurate dosimetric evaluation, but requires manual outlining of regions of interest prior to applying the density override which can become a time-consuming step in the process. Alternatively the CT-to-density look-up table within the treatment planning system can be altered to produce a step-wise table as shown in Fig. 3.17. The table blocks ranges of Hounsfield numbers in the IGRT images

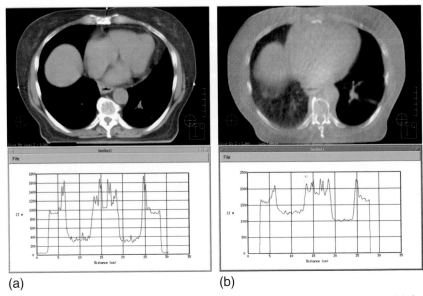

Fig. 3.16 The Hounsfield numbers between the two points marked on the CT slice for (a) the conventional CT scan and (b) the IGRT cone-beam scan.

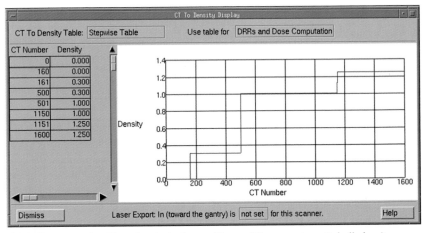

Fig. 3.17 A step-wise CT-to-density look-up table providing an automatic bulk density override for lung, soft tissue, and bone regions.

into average lung, soft tissue, and bone density regions. The strategy has the advantage of not requiring manual contouring of the images prior to bulk density overrides, and works well if the average densities of the patient tissues matches those within the look-up table. However, average lung tissue densities can vary significantly from patient to patient with average physical densities in the range 0.1–0.3 g/cm^3. Systematic errors will be introduced when calculating dose in lung regions if the average density is different to that provided by the look-up table.

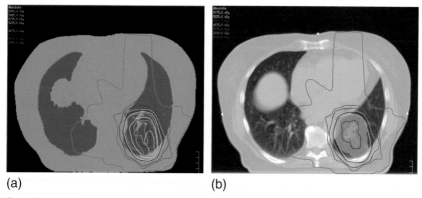

(a)   (b)

Fig. 3.18 (a) The original dose distribution, and (b) the new dose distribution due to the patient anatomical change observed with IGRT and a bulk density override approach. The new distribution shows unacceptable hotspots of 110% of the prescription dose.

Figure 3.18 shows the change in the dose distribution from changes in the patient anatomy seen from IGRT for an example patient. The dosimetric evaluation has been performed by outlining soft tissue and lung regions and overriding the density to the same value as the original planning scan. The figure shows unacceptably high hotspots of 110% of the prescription dose (5500c Gy in 20 fractions in this case) and therefore required modification.

3.5 Discussion

The use of advanced radiotherapy techniques such as IMRT, VMAT, and IMAT provide the facility to generate significantly more conformal dose distributions than those produced by simpler forward planned techniques. The introduction of IGRT provides the ability to ensure that the tumour receives the intended radiation dose on every day of treatment, and to adjust to patient changes during the course of fractionated radiotherapy, including the possibility of replanning the patient treatment when required. The combination of advanced delivery techniques and IGRT approaches has the potential to improve the outcome of patients receiving radiotherapy.

References

1. International Commission on Radiation Units and Measurements. *Prescribing, recording and reporting photon beam therapy*. Report 50. Bethesda, MD: ICRU, 1993.
2. International Commission on Radiation Units and Measurements. *Prescribing, recording and reporting photon beam therapy*. *Report 62 (Supplement to ICRU Report 50)*. Bethesda, MD: ICRU, 1999.
3. International Commission on Radiation Units and Measurements. *Prescribing, recording, and reporting intensity-modulated photon-beam therapy (IMRT)*. *Report 83*. Bethesda, MD: ICRU, 2010.
4. van Herk M, Remeijer P, Rasch C, Lebesque JV. The probability of correct target dosage: dose-population histograms for deriving treatment margins in radiotherapy. *International Journal of Radiation Oncology, Biology, Physics* 2000; **47**(4): 1121–35.

5. McKenzie A, van Herk M, Mijnheer B. Margins for geometric uncertainty around organs.at risk in radiotherapy. *Radiotherapy and Oncology* 2002; **62**(3): 299–307.

6. Drzymala RE, Mohan R, Brewster L, Shu J, Goitein M, Harms W, *et al*. Dose-volume histograms. *International Journal of Radiation Oncology, Biology, Physics* 1991; **21**(1): 71–8.

7. Kutcher GJ. Quantitative plan evaluation: TPC/NTCP models. *Frontiers of Radiation Therapy and Oncology* 1996; **29**: 67–80.

8. Niemierko A. Reporting and analyzing dose distributions: a concept of equivalent uniform dose. *Medical Physics* 1997; **24**(1): 103–10.

9. Quantitative Analyses of Normal Tissue Effects in the Clinic (QUANTEC). *International Journal of Radiation Oncology, Biology, Physics* 2010; **76**(Suppl 3): S1–S160.

10. Tanyi JA, Fuss MH. Volumetric image-guidance: Does routine usage prompt adaptive replanning? An institutional review. *Acta Oncologica* 2008; **47**(7): 1444–53.

11. Harrison RM, Wilkinson M, Shemilt A, Rawlings DJ, Moore M, Lecomber AR. Organ doses from prostate radiotherapy and associated concomitant exposures. *British Journal of Radiology* 2006; **79**: 487–96.

12. Perks JR, Lehmann J, Chen AM, Yang CC, Stern RL, Purdy JA. Comparison of peripheral dose from image-guided radiation therapy (IGRT) using kV cone beam CT to intensity-modulated radiation therapy (IMRT). *Radiotherapy and Oncology* 2008; **89**(3): 304–10.

13. van Zijtveld M, Dirkx M, Heijman B. Correction of conebeam CT values using a planning CT for derivation of the "dose of the day". *Radiotherapy and Oncology* 2007; **85**: 195–200.

4

Breast

Ellen Donovan, Charlotte Coles,
Charlotte Westbury, John Yarnold

4.1 Indications

4.1.1 Introduction

The highest level of evidence on radiotherapy effects in early breast cancer is generated by systematic overviews undertaken by the Early Breast Cancer Trialists' Collaborative Group (EBCTCG)[1]. The most recent overview reconfirms earlier reports that after surgery for early breast cancer, radiotherapy reduces absolute breast cancer mortality. The absolute benefit depends on prognosis, with radiotherapy reducing breast cancer mortality in women with node-positive disease by 8.5% in 15 years, and by an average of 3.3% at 15 years in women with node-negative disease. The reduction in node-negative subgroups analysed by age, tumour size, and grade can be as high as 7.8% and as low as 0.1%. The relative mortality reductions are similar after mastectomy and breast conservation surgery, in the presence or absence of adjuvant systemic therapies and in axillary node-negative and -positive patients[1a]. The overviews demonstrate that the prevention of four local recurrences by radiotherapy prevents one breast cancer death. As radiotherapy is a locoregional treatment, and breast cancer mortality is caused overwhelmingly by distant metastases, these data constitute powerful evidence that distant metastases can be prevented by optimizing locoregional therapies.

The systematic overviews also show that a reduction of breast cancer deaths after radiotherapy can be partially or totally offset by excess non-breast cancer mortality[1]. In node-negative patients treated by mastectomy several decades ago, the beneficial effects of radiotherapy were accompanied by an increase in non-breast cancer mortality, including second primary cancers and cardiac disease. Meta-analysis restricted to postmastectomy radiotherapy trials using modern radiotherapy techniques showed an overall survival gain with no excess non-breast cancer mortality to date[2].

4.1.2 Adjuvant radiotherapy after breast conservation surgery

Whole breast

After complete microscopic resection of primary tumour, there is no identifiable group of patients at present whose breast recurrence risk is < 10% at 10 years after surgery and endocrine therapy alone. For example, excision margins are not closely correlated with local recurrence risk provided no disease is seen at the resection margins[3–5].

Until long-term data become available from properly designed prospective observational or randomized studies, breast radiotherapy should be considered as standard of care for patients with even the most favourable features, including pT1 tumours with node-negative disease in the older age groups in which lower recurrence risks have been reported[6]. Adjuvant systemic therapies provide significant reduction in local recurrence risk, but do not yet impact on the decision to give radiotherapy[7–10]. The current evidence on the results of local excision without breast radiotherapy is summarized in Table 4.1[9–16]. Identifying patients in whom the benefits of radiotherapy are too small to justify routine use is a major research priority. Whatever the local recurrence risk, whole breast radiotherapy reduces it by a factor of 3–4. Thus, in a patient group with a local recurrence risk of 10% at 10 years after optimal surgery and adjuvant systemic therapy, breast radiotherapy prevents up to eight local recurrences and up to two breast cancer deaths.

Tumour bed boost

Breast boost policies have been strongly influenced by the European Organization for Research and Treatment of Cancer (EORTC) trial testing 16 Gy in eight fractions against no boost after complete microscopic excision, appropriate systemic therapy, and 50 Gy in 25 fractions of whole breast radiotherapy in 5569 women with early breast cancer[17]. In women randomized to the boost arm, the hazard rate for local recurrence was 0.59 (99% confidence interval 0.46–0.76) compared with the control arm, corresponding, on average, to four fewer local recurrences at 10 years for every 100 women treated (almost 20 fewer per 100 in the subgroup of women under 40 years). Premenopausal status, age < 40 years, large tumour size, absence of oestrogen

Table 4.1 Local tumour recurrence after breast conserving surgery without radiotherapy in low-risk breast cancer (a) after tumour excision and tamoxifen and (b) after tumour excision alone

(a) Tumour excision + tamoxifen for 'low risk' breast cancers trial	TNM	N	Breast relapse	Follow-up
NSABP B21[12]	pT1pN– (all ≤ 10 mm)	336	16.5%	8 years
BASO II[13]	pT1G1pN–	549	1.2% pa	35 months
CALGB[9]	cT1cN0	319 (> 70 years)	4%	5 years
Toronto[10]	pT1–2pN–	383	7.7%	5 years
Austrian[11]	pT1–2pN–	417	5.1%	5 years
(b) Trial	TNM	N	Breast relapse	Follow-up
Upsala[14]	pT1pN– (48%< 10 mm)	197	24%	10 years
Ontario[15]	pT1pN–	207 (> 50 years)	22.4%	7.6 years
Boston[16]	cT1pN– (50% < 10 mm)	87	2.8% pa	86 months

or progesterone receptors, and lack of systemic adjuvant treatment were shown to be associated with recurrence in univariate analysis[18]. In multivariate analysis, only young age, palpable primary tumour, and absence of progesterone receptor were associated with higher recurrence risk. Factors reported to predict for local recurrence in other randomized trials of breast-conserving surgery and radiotherapy include lymph node positivity, lymphovascular invasion, high grade, and extensive intraductal component (EIC) in the young[15,18–21].

The risks of moderate and severe fibrosis in the boost group were 28.1% and 4.4%, respectively, compared to 13.2% and 1.6% in women treated without boost. In principle, it is reasonable to spare patients the late adverse effects of boost therapy if risks of local recurrence after whole breast radiotherapy are < 5% at 10 years, representing a population in whom the boost prevents only one or two local recurrences per 100 women treated. On the basis of the EORTC boost trial, this applies to patients > 60 years and probably the majority of those > 50 years.

4.1.3 Adjuvant radiotherapy after mastectomy

A review of 5352 women entered in International Breast Cancer Studies Group (IBCSG) trials between 1978 and 1993 proposed an algorithm based on patient age, tumour size, node status, lymphovascular invasion, and grade to estimate 10-year risks of locoregional failure as first relapse in postmastectomy patients prescribed some form of adjuvant systemic therapy without radiotherapy (see Table 4.2)[22].

Microscopic disease at resection margins has also been reported to be a risk factor for local recurrence[23]. Whatever factors are used to estimate absolute local recurrence risk after surgery and adjuvant systemic therapies, patients with 10% risk of locoregional recurrence after optimal surgery and adjuvant systemic therapies can expect to gain 2% absolute reduction in breast cancer deaths by the prevention of eight local recurrences. This seems a reasonable threshold above which to recommend radiotherapy given that chest wall radiotherapy confers an overall survival benefit and that modern radiotherapy techniques allow reduced dose to the heart. Regional lymph node disease must also be eradicated in order to secure this survival advantage. Beyond issues of cure, the quality of life gains derived from protection against the physical morbidity and psychological consequences of chest wall recurrence are relevant.

Sources of variation in international standard practices are based on a failure to reach consensus on (i) the level of local relapse risk that justifies radiotherapy and (ii) the characteristics of the population of women treated in the current era by mastectomy and systemic therapies that are exposed to that risk. Fifteen-year follow-up data on local relapse from the EBCTCG overview show that the proportional risk reduction for local recurrence after postmastectomy radiotherapy is similar for women with node-negative and node-positive disease[1]. This equates to an absolute reduction of local recurrence risk of only 4% for node-negative disease, whereas in node-positive disease, the value is 17%. With further classification on the basis of number of nodes involved (one to three, or four or more), 15-year local recurrence reductions are 14% for women with one to three involved nodes and 20% with four or more involved nodes. For breast cancer mortality, the 15-year follow-up data from the Danish trials of postmastectomy radiotherapy suggest a significant benefit even for

Table 4.2 Ten-year risks of local-regional failure as first relapse in 5352 post-mastectomy patients receiving systemic therapy, but no radiotherapy. Axillary node positive (N +) patients received 3–12 months of classic CMF ± low-dose prednisolone and/or tamoxifen; two-thirds of axillary node negative (N–) patients received single cycle CMF chemotherapy. Overall, one-third received no adjuvant systemic therapy

Age (years)	pT	pN	LV	Grade	10-year risk
≤ 50	1	–	–		<10%
	1	–	+		15%
	> 1	–	–		15%
	> 1	–	+		15–20%
		1–3	–	1	10–15%
		1–3	+	1	10–15%
		1–3	–	2	10–15%
		1–3	+	2	15–20%
		1–3	–	3	10–15%
		1–3	+	3	25–30%
		4 +	–	1	15–20%
		4 +	+	1	25–30%
		4 +	–	2	25–30%
		4 +	+	2	35%
		4 +	–	3	25–30%
		4 +	+	3	35%
> 50	1	1–3		1	10–15%
	> 1	1–3		1	10–15%
	1	1–3		2	10–15%
	> 1	1–3		2	15–20%
	1	1–3		3	15–20%
	> 1	1–3		3	20–25%
	1	4 +		1	20–25%
	> 1	4 +		1	20–25%
	1	4 +		2	20–25%
	> 1	4 +		2	30–35%
	1	4 +		3	30–35%
	> 1	4 +		3	30–35%

Source: Data from Wallgren A, Bonetti M, Gelber RD, *et al.* Risk factors for locoregional recurrence among breast cancer patients: results from International Breast Cancer Study Group Trials I through VII. *Journal of Clinical Oncology* 2003; **21**(7): 1205–13.

intermediate risk subgroups, albeit based on trials performed before the use of modern chemotherapy regimens. Specifically, patients with one to three involved nodes benefited in terms of 10-year mortality as well as those with four or more involved nodes[24]. In fact, the survival advantage from chest wall irradiation was greatest in the intermediate risk group due to the higher rate of distant metastasis in high-risk cases[25]. The benefit for current intermediate risk cases is still uncertain, and the SUPREMO trial is evaluating the role of chest wall radiotherapy in this group of women, including patients with involvement of one to three nodes or in node-negative T2 tumours with other high-risk features[26]. Whatever level of local recurrence risk is chosen and whatever criteria are applied to estimate this risk, local guidelines should be in place to ensure treatment decisions are consistent. Regular audit of populations managed with and without radiotherapy offer a guide to the suitability of local management guidelines.

There are no international guidelines on chest wall radiotherapy in patients who have undergone neoadjuvant systemic therapy and mastectomy, so a grade B recommendation for routine chest wall radiotherapy after primary medical therapy is based on retrospective analysis of better outcomes in young women and in patients who achieve a complete pathological response when radiotherapy is given[27,28].

4.1.4 Primary breast radiotherapy

Complete response to primary medical therapy

Data relating to breast radiotherapy as an alternative to surgery in patients achieving a complete response to primary medical therapy are scanty, and do not justify its use outside a research protocol.

Inflammatory breast carcinoma

Retrospective series suggest that, as part of multimodality therapy, radiotherapy improves local control, thereby contributing to quality of life[29,30]. The data suggest that surgery should be combined with chemotherapy and radiotherapy whenever possible. Whatever treatment modalities are deployed local recurrence remains a risk after radiotherapy, reflecting residual tumour bulk in inoperable patients and dermal lymphatic infiltration beyond radiotherapy field margins.

Inoperable patient or unresectable tumour

Depending on the fitness and age of the patient, cytoreductive chemo-endocrine therapy is usually introduced first. Whole breast radiotherapy is supplemented by a boost dose to residual clinical disease. Interstitial boost therapy may enable a higher total dose to be delivered than by external beam therapy.

4.1.5 Radiotherapy to locoregional lymph nodes

Axilla

After axillary clearance (level II/III), axillary radiotherapy is not recommended if nodes are positive, even if there is extracapsular spread, due to an enhanced risk of treatment-related morbidity[31,32]. In the subgroup with microscopic tumour at the margins of axillary resection, positive excision margins do not appear to be an independent risk factor for axillary recurrence[33]. Macroscopic residual disease at the axillary apex is a different matter, and justifies cytoreductive chemo-endocrine therapy and high-dose radiotherapy in an attempt to prevent malignant brachial plexopathy. Axillary radiotherapy is indicated in patients who do not have any axillary surgery, on the basis that it reduces the morbidity of uncontrolled axillary recurrence[34,35], and contributes to cure[1]. Axillary radiotherapy versus axillary surgery is being investigated in patients with positive sentinel lymph node biopsy in the EORTC AMAROS trial.

Supraclavicular fossa

Patients with positive axillary nodes have $\geq$ 5% lifetime risk of supraclavicular fossa (SCF) node involvement[36], and SCF cancer recurrence is associated with significant morbidity[35]. In a recent large retrospective study of > 1000 patients, the number of positive nodes and tumour grade were strong independent risk factors for recurrence [37]. Radiotherapy to the SCF in patients at > 5% risk of positive SCF nodes is expected to prevent much more morbidity than it causes in terms of shoulder stiffness, arm oedema, brachial plexopathy, and vascular disease (stroke). A common practice is to give radiotherapy to the SCF in patients with four or more metastatic axillary lymph nodes, on the basis that this strikes an appropriate balance between the morbidity of local tumour recurrence and of late radiotherapy adverse effects. In primary medical therapy patients, in whom the initial stage of the axilla is unknown, SCF radiotherapy is recommended, as this subgroup is often selected on the basis of unfavourable clinical stage associated with a higher than average (30%) rate of metastatic lymph node involvement.

Internal mammary chain

Data from extended mastectomy series have shown that roughly 30% of patients with positive axillary nodes have positive internal mammary chain (IMC) nodes, and up to 10% of axillary node-negative patients have positive IMC nodes[38]. Nowadays, the rate may be less than in historical series, as patients present with smaller primary tumours and lower axillary lymph node burden. Furthermore, data from sentinel lymph node studies suggest that only a very small proportion of patients have sentinel nodes in the IMC. Few series mapping IMC sentinel lymph nodes go on to correlate lymphoscintigraphy findings with histological examination. A single multicentre study identified only 4% of 443 patients with the sentinel nodes in the IMC using 99mTc colloid[39]. The issue has been previously reviewed[40].

Clinical recurrence of internal mammary disease as a first recurrence is rare even without IMC radiotherapy. Enlarged IMC nodes may be implicated in malignant pleural effusion, but only occasionally spread directly through the anterior chest wall. Three trials of postmastectomy radiotherapy in high-risk patients receiving systemic therapy showed a survival advantage for giving radiotherapy to the IMC and SCF[41–43]. If a proportion of cancer cells gain access to the blood stream via the lymphatics, eradicating IMC disease with radiation could be an important component of curative treatment, especially when the sentinel node is located in the IMC. In practice, IMC radiotherapy is not part of standard treatment in many countries, but may be considered in patients with histologically node-positive axillary lymph nodes and/or a central/inner quadrant tumour. This question is currently addressed in EORTC trial 22922 and the Canadian MA.20 trial. Long-term toxicity data are required in the setting of modern radiotherapy delivery and adjuvant systemic therapies. The benefit from adjuvant radiation may be small and it is important to determine that this is not offset by an increased cardiovascular mortality.

4.1.6 Breast radiotherapy for duct carcinoma *in situ*

Whole breast radiotherapy reduces the risk of ipsilateral breast relapse after complete excision of pure ductal carcinoma *in situ* (DCIS) in women who do not opt for mastectomy. Four randomized trials indicate that the risks of recurrent DCIS and invasive tumours are reduced roughly by a factor of 2[44–47]. The dose prescription in all trials was 50 Gy in 25 fractions, or its near equivalent, to the whole breast without a planned boost dose to the tumour bed. A twofold reduction in local recurrence risk does not necessarily reduce absolute recurrence risks to very low levels. In women randomized to breast radiation, the rate of ipsilateral tumour recurrence was 8.9% at 15 years (NSABP B-17), 15% at a median of 10 years (EORTC 10853), 7.1% at a median of 12.7 years (UK Co-ordinating Committee on Cancer Research, UKCCCR), and 7% at 5 years (SweDCIS). About 50% of the local recurrences are invasive carcinomas. The overview confirms data from the individual trials, reporting the 10-year risk reduction of ipsilateral tumour relapse from 28.1% to 12.9%[48].

The risk of recurrence after complete microscopic resection of symptomatic DCIS is reported by Silverstein to be closely related to patient age, tumour size, grade, and margins of excision[49]. The Van Nuys prognostic Index Scoring (VNPI) system is shown in Table 4.3. The VNPI has been widely applied as a basis of treatment recommendations, but it is not independently validated as a guide to therapy.

Table 4.3 Van Nuys prognostic index scoring system for ductal carcinoma *in situ*. Reported risks of ipsilateral recurrence at 10 years are: score 4–6 = 3%; score 7–9 = 27%; score 10–12 = 66%

Score	1	2	3
Size (mm)	≤15	16–40	≥ 41
Margin width (mm)	≥10	1–9	< 1
Pathologic classification	Non-high grade without necrosis	Non-high grade with necrosis	High grade with or without necrosis
Age (years)	> 60	40–60	< 40

One to three points are awarded for each of three different predictors of local breast recurrence (size, margin width, and pathologic classification). Scores for each of the predictors are totalled to yield a Van Nuys Prognostic Index score ranging from a low of 4 to a high of 12.

Reprinted from *The American Journal of Surgery*, Volume 186, Issue 4, Melvin J. Silverstein, The University of Southern California/Van Nuys prognostic index for ductal carcinoma *in situ* of the breast, pp. 337–443. Copyright (2003), with permission from Elsevier.

Retrospective evaluations by the NSABP and EORTC do not appear to lend strong support to Van Nuys. Data from both the most recent EBCTCG overview and the Cochrane review confirm that it is correct, in principle, to recommend whole breast radiotherapy after breast conservation surgery to all patients[50]. The EBCTCG data showed absolute reduction in ipsilateral breast cancer recurrence in a low-risk group of women with negative margins and small low-grade tumours of 18%. At the other end of the prognostic scale, it is reasonable to consider a radiotherapy boost dose of 16 Gy in eight fractions after radiotherapy to the whole breast in 'high-risk' patients who refuse mastectomy, based on a high-quality retrospective study[51]. Radiotherapy to the nodal areas is not indicated for pure DCIS, nor is there any indication for radiotherapy after mastectomy.

4.1.7 Adverse effects of locoregional radiotherapy

The systematic overview of radiotherapy effects confirms an excess risk of non-breast cancer deaths in women treated with radiotherapy. The 15-year rates of non-breast cancer mortality are 15.9% for women treated with radiotherapy versus 14.6% for those treated without[1]. Cardiac mortality has been a significant component of this, more marked in the second than first decade post-treatment, and more obvious in the > 50-year age group at presentation[1,52]. The beneficial effects of radiotherapy are undermined if parts of the heart and/or major vessels are included in the treatment volume. Exposure of the left anterior descending artery in the anterior portion of the heart, included in standard tangential fields is thought to be relevant to the development of coronary artery disease[53]. Radiotherapy techniques used in the older trials delivered greater mean cardiac doses, a factor which has been shown to impact on the risk of cardiac mortality.

The overview also demonstrates an excess cancer incidence in women receiving radiotherapy, in particular contralateral breast cancer and lung cancer. Lung cancer mortality is also an important component of non-breast cancer deaths. Representative data on the frequency and severity of common adverse effects associated with current techniques and schedules are difficult to find in the literature[54,55]. Information on early side effects given to the patient might include the following: fatigue (50%); skin erythema ± irritation (100%); moist desquamation in inframammary fold or axillary fold (≥ 10% risk, in large-breasted patients or after chemotherapy). Information on late side effects might include: breast shrinkage (30% mild/moderate, ≤ 10% marked at 10 years); breast induration (30% mild/moderate, ≤ 10% marked at 10 years); breast pain and tenderness (30% mild/moderate, ≤ 10% marked at 10 years); rib fracture (1% at 10 years); symptomatic cardiac damage (left-sided radiotherapy only; < 1% at 20 years); symptomatic lung fibrosis (only after lymphatic radiotherapy; < 10% at 10 years); arm lymphoedema (only after lymphatic radiotherapy; 10% at 10 years; double this rate if combined with axillary dissection); shoulder stiffness (only after lymphatic radiotherapy; 10% at 10 years); arm paralysis (only after lymphatic radiotherapy; < 1% at 10 years); and stroke (only after lymphatic radiotherapy; < 1% at 10 years; perhaps higher at 20 years). The increased risk of contralateral breast cancer after radiotherapy of 1.8% at 15 years is likely to be technique dependent.

4.1.8 Pretreatment assessment

Multidisciplinary meetings help enormously in improving the quality of data informing clinical decisions and treatment delivery. Following these, pretreatment audit checks that treatment advice is soundly based and that the patient is capable of providing written consent. The surgical operation notes should be viewed, so that the stated surgical procedure can be correlated with the clinical, radiological, and histopathological descriptions of the tumour. One of the most important accompaniments of the primary surgery is attachment of titanium ligaclips at the anterior, posterior and four radial margins of the tumour excision cavity. This is essential for accurate localization of the tumour bed during planning of whole breast, partial breast, or tumour bed boost radiotherapy[56]. Prior neoadjuvant therapy, general fitness, and mobility are noted. The presence of comorbidities, especially cardiovascular disease and collagen vascular diseases, may be associated with an increase risk of acute and/or late toxicity and adaptation of the treatment plan and close monitoring of side effects may therefore be required[57].

Pathological review includes macroscopic clearance and the maximum diameter of the invasive component. With respect to microscopic excision margins, re-excision is needed if radial margins are incomplete (this applies to the anterior and deep margins too unless the surgeon routinely resects tissue to the skin surface and down to the pectoral fascia). Other indications for re-excision include an extensive intraduct component in a patient younger than 40 combined with narrow margins (< 1 mm)[58,59]. If the patient is older than 40, focal involvement of margins by DCIS is not an absolute indication for re-excision, although many centres advocate this[60].

4.2 Target volume definition

4.2.1 Defining the target volumes

Radiotherapy for breast cancer is usually given in the adjuvant setting, after surgical removal of the primary tumour. When considering the primary site, including the breast and tumour bed after breast conserving procedures, or the chest wall after mastectomy, the clinical target volume (CTV) is defined without a gross tumour volume (GTV). Less commonly, where the breast and tumour are intact, for instance when radiotherapy is being given as a primary radical therapy (rare) or for palliation, the GTV is present and conventional margins for CTV and planning target volume (PTV) are added. For regional lymph nodes, the same principles apply. Either way, there is a strong impetus to abandon traditional approaches that localize fields rather than target volumes

4.2.2 Whole breast clinical target volume

The whole breast CTV consists of the entire glandular and subcutaneous breast tissue, but excluding muscle, ribcage, and skin. An area of breast tissue may be compromised if this is distant from the tumour bed in order to limit the exposure of volumes of heart or lung. The skin and surgical scar are not included in the CTV, which is defined with 5 mm of skin sparing.

4.2.3 **Chest wall clinical target volume**

The chest wall CTV consists of skin, including the scar, extending posteriorly to deep fascia, but excludes muscle and ribcage. It traditionally includes the area where the breast used to be, marked as a rectangle. This is unlikely to represent the optimal CTV; an ovoid with its longitudinal axis on the scar line is likely to map relapse patterns more closely. Nevertheless, the entire ipsilateral chest wall is usually included in the CTV. The scar may be excluded if this extends across the anterior midline or posterior to the mid-axillary line, or where underlying heart or lung exposure is judged to be excessive (see section 4.6.6 Organs at risk).

4.2.4 **Breast tumour bed clinical target volume**

The breast tumour bed is an established concept that gains its relevance from being the site of highest local recurrence risk. However, the optimal volume is not well established for a number of reasons. For example, it is not clear how rapidly the risks of local recurrence fall, and late adverse effects rise, as CTV increases. Surgical factors are also relevant, including margins of excision and techniques of suturing, or not, the cavity margins. The CTV includes the wall of the surgical cavity following resection of the primary tumour. Based on pathological studies and relapse patterns from randomized trials comparing breast conservation with and without radiotherapy, a typical CTV radial margin around the surgical cavity is 1.5–2 cm. A more generous margin may be desirable in circumstances where there is a particular concern, e.g. extensive intraduct component close to the excision margins in women under 40 years of age.

4.2.5 **Axillary lymph nodes clinical target volume**

The CTV includes the interpectoral nodes and nodes that lie along the axillary vein. These are divided arbitrarily into three anatomical levels I–III. Level I lie lateral to the lateral border of pectoralis minor muscle, level II nodes lie between the medial and lateral borders of pectoralis minor, and level III are medial to the medial margin of pectoralis minor. These levels were established in the era of radical mastectomy when pectoralis minor was exposed from the front after removal of pectoralis major.

4.2.6 **Supraclavicular fossa lymph nodes clinical target volume**

The SCF nodes are continuous medially and anteriorly with level III axillary nodes and with infraclavicular nodes. SCF nodes also follow the vascular bundle. The CTV also includes the infraclavicular nodes and may be extended laterally to include the apical axillary (level III) nodes in high-risk patients in whom a level II dissection has been carried out.

4.2.7 **Internal mammary chain and medial supraclavicular lymph nodes clinical target volume**

The internal mammary nodes are in the parasternal region in the intercostal spaces. The CTV usually includes the upper IMC (commonly first to third intercostal spaces) along with the medial supraclavicular chain. This may be extended to include the lower chain (commonly fourth and fifth intercostal spaces) according to clinical judgement if the primary tumour lies in the lower inner quadrant of the breast. CT imaging of the internal mammary vessels aids localization of the IMC as use of bony landmarks can be inaccurate[61,62].

4.2.8 Planning target volumes

Set-up error and patient movement (including breast swelling and breathing) must be considered when defining the margin required for PTV for breast, tumour bed, and nodal areas for external beam radiotherapy techniques. Several studies have used electronic portal imaging devices to quantify positional error and patient movement for the whole breast[63–66]. A weighted average standard deviation for systematic variation in set-up error in the anterior–posterior direction ranged from 2.2–4.5 mm. Corresponding values in the craniocaudal distance were 3.9–6.1 mm. A comprehensive review of the set-up errors reported in the literature derived median values for population systematic and population random errors of approximately 3 mm each[67] for immobilization methods using angled boards, arm supports, and a variety of other immobilization aids. From portal imaging studies, it is difficult to determine exactly which part of the displacement is due to set-up error and what was due to patient movement. It has been suggested that the effects of breathing motion are in general about half the size of the effects of set-up error[68]. Certain centres have implemented methods to limit breathing motion such as gated radiotherapy and breath-holding techniques[69].

One institution measured the impact of patient set-up error and breathing motion to establish CTV to PTV margins[70]. This was then tested clinically for adequate coverage of treatment. The CTV-PTV margin for 'breathing only' was calculated by measuring the displacement of surgical clips during three types of CT scan: free breathing, and breath-holding at the end of normal inhalation and at the end of normal expiration using an active breathing control device. A margin of 5 mm in all directions was subsequently selected to account completely for breast motion during quiet breathing. The combined uncertainty of random patient set-up error and respiratory motion, and the distribution of systematic error across all fields and all patients, were measured. This was achieved by measuring the movement of the chest wall/ribs with portal imaging, as a surrogate for the tumour bed. A margin for set-up uncertainties of 5 mm was proposed from these data, producing a total CTV–PTV margin of 10 mm, which was tested in nine patients. Ninety-eight to 100% of the CTV was covered by the 95% isodose surface at the extremes of normal inhalation and exhalation using the 'breathing only' margin of 5 mm. The total CTV–PTV margin of 10 mm also seemed to provide coverage for most patients. The authors state that there is still uncertainty regarding the stability of the tumour cavity relative to the chest wall and that this may vary more in patients with larger breasts. Therefore, slightly larger CTV–PTV margins may be needed in this group of patients. This study did not include an analysis of the effect of strategies to reduce the set up error components of the margin. More recent studies such as those of Penninkhof et al.[71], Coles et al.[72], and Leonard et al.[73], have shown that the use of internal markers, image guidance, and detailed verification protocols enable reductions in patient population systematic errors such that PTV margins of 5–6 mm around a tumour bed CTV are achievable.

In conclusion, the CTV and PTV are routinely defined and localized as part of the planning process in many tumour types. Historically, this approach has not been utilized for breast radiotherapy, where the whole breast fields are localized by inspection and palpation, and where bony landmarks in the shoulder region are used to aid localization of anterior fields to the axilla and SCF. Partial breast radiotherapy and concomitant photon boost using more complex 3D radiotherapy techniques, however, do require the use of this concept to ensure accurate target coverage. The margins for CTV and PTV may vary between institutions depending on accuracy of localization and set-up errors respectively and guidelines on PTV margin size should be confirmed locally by assessing set-up errors, and other contributing factors. Dobbs et al. provide a framework for margin estimation in breast radiotherapy[74].

4.3 Patient position and immobilization

Good patient position and immobilization are vital for accurate radiotherapy delivery. Patient position and immobilization methods must provide good reproducibility of set-up and reasonable comfort for the patient. It is essential that the patient is not moved between the breast fields and any nodal irradiation.

4.3.1 Supine position

There are two approaches to supine patient positioning. The patient may lie either flat on the couch top or on some form of angled board. When planning and field matching were carried out on a radiotherapy simulator, the advantage of an angled board was that the sternum could be brought horizontal very easily and the need for large angulation in the treatment beam geometry was reduced. This type of set-up requires modification for use with X-ray CT scanners with standard bore size of 70 cm that limits the degree of angulation of an inclined board. CT scanners with wider bore sizes (85 cm) reduce these restrictions and now are more widely available in radiotherapy departments. An immobilization system in which the patient is flat on the couch is more easily adapted for CT scanning. It has the advantage that the breast mound moves superiorly and inframammary fold skin reactions may be reduced. One difficulty for larger-breasted women positioned flat on the couch is that this movement of the breast mound may push up skin folds in the upper regions of the breast. Skin folds here, and in the axillary region, may predispose to more acute radiation reactions.

Immobilization devices, whether in-house or commercial breast boards, tend to have the same features: a board with a number of fixed angle positions and arm supports, with some combination of a head cup, foot rest and/or bottom rest, and knee support. Examples of a typical board and patient position are shown in Fig. 4.1. Many of these devices have several degrees of movement in the arm supports. The use of the board may be customized for each patient and adapted to specific needs, e.g. a larger patient requiring set-up to one side of the couch in order to avoid beam entry through the couch. As an alternative to a breast board, the use of a vac fix bag formed around the patient's upper arms, shoulders, and neck gives good immobilization. If the vac fix bag is formed around an arm pole or plastic support, a means of daily location on the

Fig. 4.1 MT-350-N Carbon Fibre Tilting Breastboard with arm supports and indexed 'bum stop'. (Courtesy of CIVCO and Oncology Systems Ltd, Shrewsbury, UK.)

treatment set, as well as comfort for the patient, is obtained. Other systems in use include customized hemibody foam cradles[75,76] or alpha cradle casts.

The type of immobilization may influence the arm positions used which varies between both arms raised versus one arm raised. Often, the arms are abducted to the side, though they may be placed further above the head in supporting arm rests. Patients with large or pendulous breasts may require thermoplastic shells to enable reproducibility, although the partial loss of skin sparing should be remembered. Some centres treat this group of patients in bras, or bra-type supports, in an attempt to raise the breast and hence reduce skin reactions in the folds. Raising the breast also enables the lateral border of the treatment volume to move anterior, so the amount of lung in the treatment field is reduced (similarly, the volume of heart may be reduced in a left-sided volume).

4.3.2 **Prone position**

The challenge of determining a suitable position for patients with very large, pendulous breasts has been met by using a prone treatment position by some groups[77,78]. An example set-up using an in-house support system is shown in Fig. 4.2. This position allows the breast to hang freely, thereby avoiding acute skin reactions in the inframammary fold and elsewhere, although it would be inappropriate for patients requiring nodal irradiation. It is technically simple to plan the breast fields. Ipsilateral lung doses are reduced[79,80] in whole breast prone treatments, however, Varga et al. reported no improvement to heart doses, and Kirby et al. showed an improvement only for women with a whole breast CTV volume of > 1000cc . Respiratory motion is reduced [79,81] but is likely to be offset by set-up errors which are reported as larger than those from the supine position in two out of three reports on prone treatments[80–82]. These larger set-up variations imply PTV margins of 12–16 mm compared to a standard whole breast margin of 10 mm[82]. In summary, the prone position shows some evidence of benefit, particularly in the reduction of dose to non-target tissues, but improved immobilization is required before its widespread adoption for larger-breasted women.

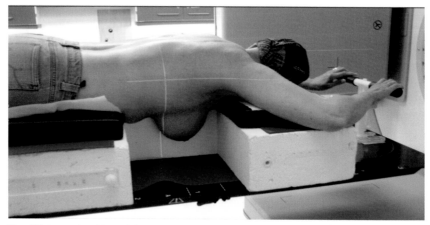

Fig. 4.2 Example of immobilization for prone positioning. (Reprinted from *Radiotherapy and Oncology*, 100(2), Anna M. Kirby *et al.* A randomised trial of Supine versus Prone breast radiotherapy (SuPrstudy): Comparing set-up errors and respiratory motion, pp. 221–6 (2011) with permission from Elsevier.)

4.4 Localization

4.4.1 Whole breast or chest wall

The development of X-ray CT scanning greatly improves localization methods for breast radiotherapy planning. Virtual simulation using CT scanning provides more detailed 3D information with better resolution of soft tissue anatomy for breast tissue, nodal areas, vessels, and OARs. In contrast, with conventional simulator planning, information is gathered in only two dimensions using limited clinical features and radiological bony landmarks. CT planning has become standard of care for breast radiotherapy, as it improves target coverage, helps to minimize exposure OARs, and allows individual localization to take account of variability in patient anatomy.

The patient should be scanned from 2 cm below the inframammary fold with a slice separation of not less than 5 mm to allow high-quality digitally reconstructed radiographs (DRRs) to be produced. The upper limit of the breast (axillary tail) is difficult to localize with any technique, and a midline bony landmark (angle of Louis) is often chosen as a surrogate. CT imaging enables better visualization of the glandular tissue compared with clinical examination, although localization can still be difficult. Marking palpable breast tissue with radio-opaque wire prior to CT scanning has been shown to be helpful[83]. When the breast falls laterally under its own weight towards the side of the patient, the medial border of the CTV may need to be moved several centimetres lateral to the anterior midline.

Conventional 2D simulation is achieved clinically using external anatomical features (breast tissue and bony landmarks) and radiological features. X-ray CT planning offers better visualization of the glandular tissue and surrounding anatomy compared with clinical examination, see Fig. 4.3. One study found that when field borders were set-up clinically, CT scanning led to adjustment of the medial or lateral borders

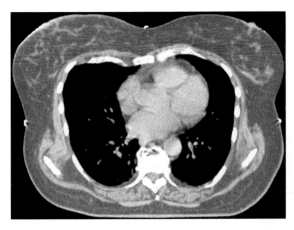

Fig. 4.3 X-ray CT scan through centre of breast with patient lying in treatment position (both arms elevated above head). Radio-opaque skin markers are visible on anterior midline and right/left mid-axillary lines. The glandular tissue and surrounding anatomy of the breast are clearly seen.

of the field in a high proportion of cases[84]. Most commonly, there appeared to be clinical underestimation of the posterior and lateral extent of glandular breast tissue. Despite the help of CT scanning, delineation of the breast target volume can present difficulties, with marked interobserver variation and lesser intraobserver variation[83]. Clinically palpating then marking the breast tissue with radio-opaque wire prior to CT scanning has shown to be helpful in reducing intraobserver variation but not interobserver variation[83]. By contrast, inspection is all that is needed to localize the postmastectomy chest wall on the skin surface, although 3D imaging using X-ray CT will accurately establish the amount of lung and heart in the treatment volume.

4.4.2 Breast boost

Localization of the postoperative tumour cavity was traditionally performed using a combination of preoperative radiological imaging, surgical annotation, palpation of a surgical defect, noting position of breast scar, and checking with the patient. Modern surgical techniques often place the scar some distance away from the site of the tumour, so it is important not to be misled. In any case, traditional techniques are crude, almost certainly inaccurate in a significant proportion of cases, and need to be improved. Many studies have reported the superiority of using surgical clips to locate the tumour bed compared with clinical methods[85–91]. All showed that the tumour cavity is underdosed using traditional planning techniques. The clinical method can also result in a substantial volume of normal tissue being irradiated unnecessarily[91].

Detailed descriptions of the planning techniques using surgical clips have been reported using both X-ray CT scanning and conventional simulator films[92,93]. A consistent policy of clip placement at the time of surgery is necessary (see Fig. 4.4). An example of this is to place a large titanium clip at the medial, lateral, superior, and inferior extent of the tumour bed, and a fifth clip at the deepest extent of the tumour bed in the direction of the surgical excision[92]. There have been reports of surgical clips becoming dislodged and tracking away from the tumour site, but this appears to be a rare occurrence[85]. Other studies have used implanted gold seeds as tumour bed fiducial markers[72,73]. Placement of tumour bed markers is essential before

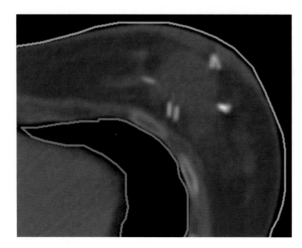

Fig. 4.4 This is an axial CT view of the left breast showing the tumour bed seroma surrounded with titanium ligaclips.

oncoplastic surgical procedures are carried out or the excision cavity is closed, as there is usually no seroma to guide localization of the tumour bed. Use of tumour bed markers are recommended as best practice by the British Association of Surgical Oncologists[94].

Breast ultrasonography has also been exploited as a method of localizing the tumour bed for radiotherapy planning where a tumour bed associated seroma is present. A study compared clinical methods with ultrasound localization and found that the full extent of the tumour cavity was underestimated in 87% of women, and the chest wall depth was incorrectly estimated in 90% using traditional methods[95]. Another study reached similar conclusions: conventional electron boost planning resulted in 55% of patients having areas of undertreatment and 20% of patients received significant overtreatment[96].

Two-dimensional ultrasound scanning is adequate for localizing a direct anterior electron boost field, as the dimensions of the cavity with a suitable margin can be marked on the patient's skin and the electron energy can be selected from measurement of the cavity depth. However, other radiotherapy techniques, such as brachytherapy interstitial implant or concomitant photon boost to the tumour bed using IMRT, require more detailed 3D information. This can be achieved with placement of radio-opaque skin markers prior to ultrasound examination and measurement of cavity depth, followed by X-ray CT scanning in the same position[97]. Alternatively, 3D methods of ultrasound can be used[98]. MRI is likely to contribute significantly to this process in future[99,100].

4.4.3 Gross tumour volume in the breast

Where the primary tumour is *in situ*, it is possible to define a GTV. However, clinical examination can be inaccurate, and additional imaging is helpful. Imaging modalities would include X-ray CT, ultrasound, or MRI, provided the patient can be scanned in the treatment position.

4.4.4 Axilla, supraclavicular fossa, and internal mammary chain

Three-dimensional imaging is now recommended for accurate localization of the nodal regions and this is usually carried out with CT scanning in the treatment position. Examples are given from the Insitut Curie breast radiotherapy planning atlas in Fig. 4.5[101,102].

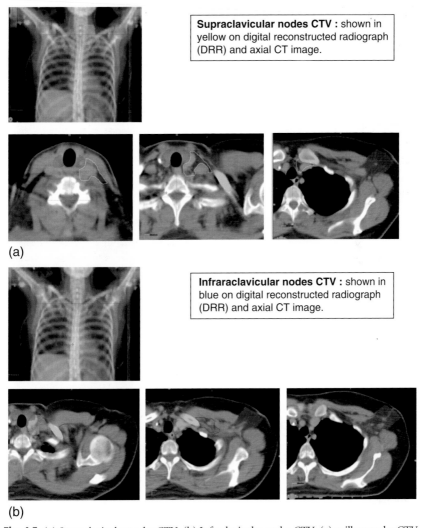

Supraclavicular nodes CTV : shown in yellow on digital reconstructed radiograph (DRR) and axial CT image.

(a)

Infraclavicular nodes CTV : shown in blue on digital reconstructed radiograph (DRR) and axial CT image.

(b)

Fig. 4.5 (a) Supraclavicular nodes CTV, (b) Infraclavicular nodes CTV, (c) axillary nodes CTV,
(Continued)

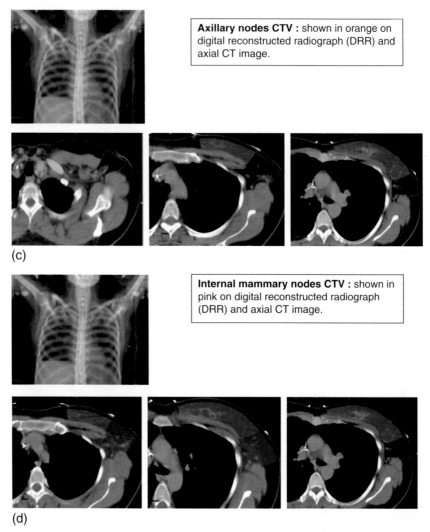

Axillary nodes CTV : shown in orange on digital reconstructed radiograph (DRR) and axial CT image.

(c)

Internal mammary nodes CTV : shown in pink on digital reconstructed radiograph (DRR) and axial CT image.

(d)

Fig. 4.5 *Continued* (d) Internal mammary nodes CTV. (Images courtesy of Professor Alain Fourquet and Dr Youlia Kirova.)

4.4.5 Linking localization, beam geometry, and dosimetry using three-dimensional techniques

Virtual simulation enables better visualization of the extent of breast glandular tissue compared with clinical examination, and therefore improves coverage of the target. It also allows assessment of 3D dose homogeneity, which can be modified with simple forward-planned IMRT techniques.

The whole breast PTV may be defined clinically and marked with wire (although superior borders are very difficult to define). Alternatively, the PTV may be explicitly defined on an X-ray CT slice set acquired throughout the volume, although the axillary tail is still difficult to identify and localize. The posterior edge of tangential beams can be applied to the central CT slice and the posterior edge of the beam viewed at all slice levels. This has the advantage of visualizing the breast and its relationship to PTV at higher and lower levels, especially the deep limit of the tumour bed PTV and to the surrounding OARs. It has been shown that this increased information allows better coverage of breast tissue compared with simulator-planned fields[84]. In particular, it is reported that medially and laterally located tumour cavities can easily be missed using simulator-planned fields (see Fig. 4.6)[87,90].

MR provides excellent definition of the breast and surrounding tissues, but its use in breast radiotherapy planning has been very limited. This is largely been due to a combination of limited MR resources and the difficulty of scanning the patient in the treatment position. Like CT imaging, MR shows how conventional breast radiotherapy planning can result in undertreatment of the target[103]. Potential problems with MR radiotherapy planning include image distortion and the difficulties of co-registration with radiotherapy planning systems.

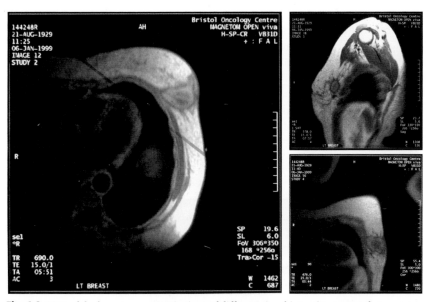

Fig. 4.6 MRI of the breast—concentric rings of different signal intensity suggest that granulation tissue is laid down within the tumour bed cavity. (Reprinted from the *International Journal of Radiation Oncology, Biology, Physics*, Volume 72, Issue 1, Elisabeth C. Whipp and Michael Halliwell, Magnetic resonance imaging appearances in the postoperative breast: the clinical target volume–tumor and its relationship to the chest wall, pp. 49–57 (2008) with permission from Elsevier.)

4.5 **Beam geometry**

4.5.1 **Whole breast or chest wall**

The basic beam geometry for the whole breast is two opposed beams at approximately 180°. The beams are usually tilted to bring the posterior edges coincident and hence to remove divergence into the lung (see Fig. 4.7). A semi-asymmetric technique, with the posterior border defined by the central axes, allows the fields to be directly opposed with no divergence into the lung. Good tumour bed localization ensures adequate coverage is achieved (see Fig. 4.8). It is recommended that fiducial markers (e.g. surgical clips) are implanted into the walls of the tumour excision cavity (tumour bed) at the time of breast conserving surgery as per British Association of Surgical Oncology (BASO) guidelines[94].

The chest wall after mastectomy can be encompassed by the same opposed beam field arrangement. Alternatively, an anterior electron field offers the advantages of simplicity with total sparing of heart and lung, provided the appropriate beam energy is used, although the clinician must be aware that stand-off will reduce the dose at lateral and medial borders of the PTV (5% less dose per centimetre of stand-off). Field matching of an electron field may not be straightforward if photon fields are given to the axilla and/or IMC. Techniques used in the Danish trials of postmastectomy radiotherapy represent a good model of what can be achieved (see Fig. 4.9)[104]. In this trial, an anterior photon field was used to encompass axilla and supra/infraclavicular fossa, with a postaxillary boost field depending on field separation. Two anterior electron fields, one to the chest wall and the other to the IMC, were matched to each other and to the anterior axilla/SCF photon field. The electron energy was selected to encompass IMC lymph nodes using ultrasound to determine the distance from the anterior skin surface to the pleura. The target depth for the intramammary chain lymph nodes was defined as the ultrasound-measured chest wall thickness plus 0.5 cm, and the target depth for the chest wall was the chest wall thickness minus 1.0 cm.

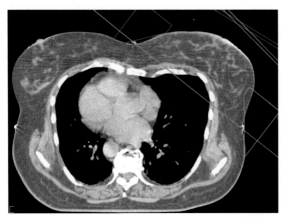

Fig. 4.7 An example of tangential field geometry. The medial field is shown in red and the lateral field in green.

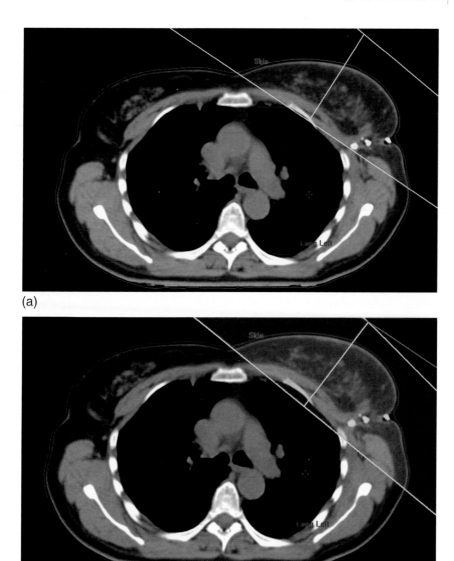

Fig. 4.8 Tumour bed localization with CT imaging (superior slice) showing tumour bed and field chosen by clinical palpation (a) and following virtual simulation to ensure coverage of the tumour bed (b).

4.5.2 Axilla and supraclavicular fossa

Irradiation of the supraclavicular and axillary nodes is delivered with an anterior or anterior oblique field. The lower border of the supraclavicular and axillary field is matched to the upper border of the tangential fields (see section 4.5.4). If the

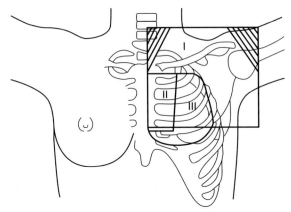

Fig. 4.9 Radiation field arrangement used in the Danish postmastectomy radiotherapy trials; I = anterior photon field; II = internal mammary nodes electron field; III = scar electron field. (Reprinted from *The Lancet*, Vol. **354**, Inger Højris *et al.* Morbidity and mortality of ischaemic heart disease in high-risk breast-cancer patients after adjuvant postmastectomy systemic treatment with or without radiotherapy: analysis of DBCG 82b and 82c randomised trials, pp. 1425–30. Copyright (1999), with permission from Elsevier.)

mid-axillary dose falls below 80%, a posterior field may be added, bearing in mind that the accuracy of beam placement appears to be limited[105]. If the axillary dissection specimen shows extensive extranodal extension and incomplete posterior excision margins, weighted anteroposterior fields to the axilla may be appropriate if an adequate dose cannot be delivered via an anterior field alone (see 'Dosimetry', section 4.6). No lung blocks should be applied to axillary fields, as they effectively shield level II/III nodes (see Fig. 4.5).

4.5.3 **Internal mammary chain**

Two different techniques can be used to treat the IMC. Modified wide tangents to cover the breast and IMC are used as part of a four-field technique. In the NCI Canada MA.20 Trial protocol testing IMC radiotherapy, the medial edge of the medial tangent is placed 3 cm from the midsternum across the midline as shown in Fig. 4.10[106]. Cardiac shielding (left-sided tumours) or lung (right-sided tumours) shielding is inserted in the inferior, posterior part of the field below the third interspace. A minimum of 2.0 cm lung thickness is recommended in the cranial half of the volume. Maximum limits of 3.5 cm and 2.0 cm, respectively, are set for the amount of lung in the upper and lower parts of the field to minimize cardiac and lung irradiation. Standard matching techniques are used for axillary and supraclavicular fields. An alternative technique in the EORTC 22922 trial protocol testing IMC radiotherapy uses direct mixed photons and electrons to the IMC and medial SCF[107]. Tangential photon fields to the breast are matched to the lateral border of the internal mammary field. A disadvantage of this technique is the difficulty in matching fields at this junction without some degree of overlap or underlap. In practice, up to 5 mm overlap is allowed on the central axis of the tangential photon fields.

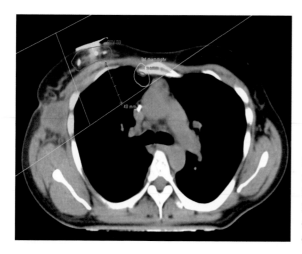

Fig. 4.10 Example of modified wide tangential field technique used in the MA.20 protocol for treatment of the internal mammary chain.

Using either technique, the shape of the patient or tumour location may make one or other technique particularly difficult. For example, in the case of inner quadrant tumours managed by local excision, the field junction between anterior mixed photon/electron field to the IMC and tangential fields to the breast may pass close to, or through, the tumour bed. One option is to adopt the wide tangential field approach. Alternatively, the EORTC technique can be modified by moving the anterior field, including field junction, medially by 1 cm, and angling the anterior mixed photon/ electron field medially by several degrees. In these cases, IMRT methods can be used— either with static beams or using rotational methods. Although the high dose to the lung and heart may be substantially reduced, the trade-off may be an increase in the volume of normal tissue receiving a 'low-dose bath'. The long-term effort of this is unknown. An example of a dose distribution obtained using a rotational technique is shown in Fig. 4.11.

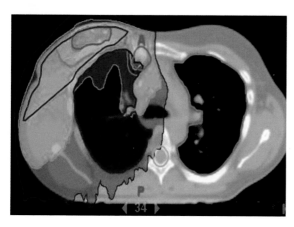

Fig. 4.11 Example of a dose distribution from a rotational method (tomotherapy) showing the good high-dose coverage, and the low-dose bath.

4.5.4 **Matching tangential and lymphatic fields**

Where lymphatic irradiation is required, matching the superior edge of the tangential fields to the inferior edge of the nodal fields is important to avoid the problem of over/underdosing in this region. Beam matching at a specific depth, which leaves a gap at the skin surface, is a possible solution, but there are a number of other methods that enable matching at all depths. Typical examples are given in Tables 4.4 and 4.5.

While all of the techniques indicated produce a satisfactory match plane alignment, the choice may depend on equipment limitations such as CT scanner bore size, potential collisions between portal imaging systems and the couch support system, or an asymmetric field length constraint. The magnitude of couch and collimator rotation on the tangential fields to obtain a match plane is much greater when patients lie flat or at a shallow angle, than where the sternum is horizontal[108]. Restrictions on the CT scanner and collision risk on the linear accelerator may mean a method using asymmetric collimation is preferred. Similarly, a technique with the isocentre on the match plane requires the maximum setting of the jaws with a wedge to be sufficient to cover the majority of breast sizes. It may be necessary to accept a small degree of overlap at the match plane in order to position the beam isocentre inferior to the match plane by 3–4 cm for patients requiring field lengths greater than half the maximum allowable jaw settings.

Table 4.4 Methods for matching tangential and nodal fields when the patient is positioned supine on angled plane with her sternum horizontal

Isocentre position	Tangential fields	Nodal fields
On central plane of breast	Collimator and couch rotated to give vertical superior border	Direct anterior or directly opposed anterior/posterior with half beam blocking on inferior border
At match plane level	Collimator and couch at cardinal angle	Direct anterior or directly opposed anterior/posterior with half beam blocking on inferior border

Table 4.5 Methods for matching tangential and nodal fields when the patient is positioned supine on a flat couch

Isocentre position	Tangential fields	Nodal fields
On central plane of breast	Collimator and couch rotated to give vertical superior border	Direct anterior or directly opposed anterior/posterior with half beam blocking on inferior border
At match plane level	Collimator rotated to track chest wall	Couch rotated to ± 90° and gantry angled to match tilted plan
At match plane level	Collimator and couch at cardinal angles with MLC to track chest wall	Direct anterior or directly opposed anterior/posterior with half beam blocking on inferior border

If a patient is unlikely to require radiotherapy to the supraclavicular and axillary nodes in later years, the divergence of the tangential photon fields at their superior end may be considered unimportant, and a technique with a divergent superior border may be preferred. A gap of appropriate distance must be used if there is a subsequent need for axillary irradiation.

4.5.5 Breast boost

The beam geometry for tumour bed boost treatments with electron beams is usually decided by clinical mark up. It is advised that CT information be used to guide the beam placement, for example, using the information on clip position and the use of surface rendered views from a treatment planning system.

In some cases, mini-tangent fields may be used to deliver the tumour bed boost dose. The treatment geometry is similar to that for the whole breast, with reduced field borders in the superior–inferior direction. Where tighter conformity may be required to the tumour bed PTV, a 3D conformal plan with three to five static beams, or an arc, between the angles of the main opposed fields can be used. The most appropriate applications of these methods are for simultaneous integrated boost treatments to deliver escalated doses to the tumour bed in patients with higher than average recurrence risk[109,110]. Careful patient selection, detailed protocols, and constraints on target volumes and OARs are required, given the increased exposure to radiation of the latter from more complex beam arrangements.

4.6 Dosimetry

4.6.1 General

Ideally, the dose distribution should meet ICRU 50 criteria (−5% to + 7%) for dose variation throughout the whole volume[111]. The majority of whole breast plans require the use of some form of compensation to achieve this. Inhomogeneity corrections must be used when planning on CT datasets to account for the presence of lung correctly. If an external contour only is used, there are several means for adjusting the planning to account for the effect of the lung on the distribution. It is important to make an adjustment for the presence of lung, as the dose to the medial and lateral aspects of the breast, including the underlying ribcage, increases by 2–4% when lung is present.

Choice of photon beam energy

Usually breast patients are treated with a megavoltage photon beam of 4–6 MV. In patients with central slice separation of > 20 cm, it may be preferable to treat with higher energies, e.g. 10 MV. However, the skin sparing distance becomes greater and a balance between the reduction of hot spots and sufficient target volume coverage to 5 mm of the skin surface should be considered. It is likely that over the whole treatment volume, the higher energy does not reduce hot areas as much as is implied by the central contour.

Calculation algorithms

Treatment planning systems vary in how they deal with side scatter, and it may be necessary to make adjustments to account for inaccuracies in the calculation.

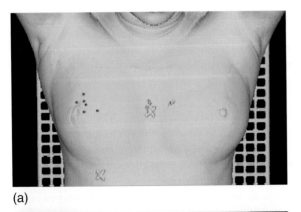

(a)

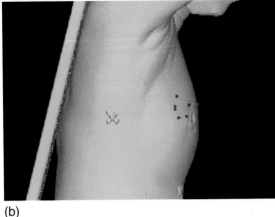

(b)

Fig. 4.12 Example of surface rendered views with surgical clips visible to aid electron beam treatment set up; (a) anterior view and (b) lateral view.

Measurement-based and pencil beam models do not always correctly calculate dose distribution in situations of missing tissue, so that the apparent dose displayed at the breast or chest wall anterior region may be artificially high; the same applies to the dose at the normalization point. The magnitude of the effect is in the order of 2–3%. This is particularly relevant to the calculation of doses in chest wall irradiation, due to the large proportion of the field including either lung or air.

Reference points in the breast, chest wall, and lymphatic pathways

Whole breast and chest wall
The reference point receiving 100% of the prescribed dose should, by definition, be located in the centre of the breast, typically on a point midway between the skin surface and the underlying chest wall and midway between the upper and lower borders of the tangential fields.

Breast boost
If the boost is given by electrons, the dose is prescribed to the 100% isodose. If given by mini-tangential fields, the dose is prescribed at the midplane in the centre of the volume.

Axilla and supraclavicular fossa

The dose is prescribed to the D_{max}. If a posterior axillary field is added, depending on the separation of the axilla, a very small daily applied dose is added in order to bring the axillary midplane dose to 85%. This arrangement of fields and reference points was developed for the UK Standardization of Breast Radiotherapy Fractionation (START) trial[112]. Previously, it was common practice to prescribe to the axillary midplane, but this has disadvantages (see Figs 4.15 and 4.16).

4.6.2 Whole breast dosimetry and compensation

The complex shape of the breast means that optimal dosimetry is not achieved for a significant number of patients with wedge compensation only. Studies of 3D dose distributions in breast plans show that dose gradients > 10% may exist in the superior and inferior aspects of the breast[113,114]. Figure 4.13 (left) demonstrates the typical regions of high dose obtained when only wedge compensation is used.

These can only be removed with some form of compensation in 3D, and Fig. 4.13 (right) shows an example of the results of using full dose compensation. IMRT enables the radiation fluence to vary across the beam, and allows 3D dose compensation. This has the potential to reduce unplanned dose inhomogeneity in tangential field breast irradiation. Donovan et al. assessed and summarized a range of published modulation methods for 3D dose compensation for whole breast radiotherapy[115]. All of the reported modulation methods achieve improvements in the dose variation over the whole breast. Both inverse and forward planned methods are used, although the latter

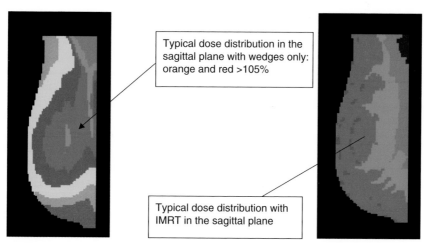

Typical dose distribution in the sagittal plane with wedges only: orange and red >105%

Typical dose distribution with IMRT in the sagittal plane

Fig. 4.13 Dose distributions for whole breast using two different methods of compensation based on wedge only (left) and full dose compensation using MLC (right). Regions coloured red or orange indicate doses > 110% or 105%, respectively, of the prescribed dose. Regions in blue or green indicate areas where the dose is > 95% but < 105% of prescribed dose. (Reprinted from *Radiotherapy and Oncology*, Volume **82**, Ellen Donovan *et al.* Randomised trial of standard 2D radiotherapy (RT) versus intensity modulated radiotherapy (IMRT) in patients prescribed breast radiotherapy, pp. 254–64. Copyright (2007), with permission from Elsevier.)

have the advantage of being simple, and efficient for whole breast treatments as only a few segments are needed to achieve improved dose homogeneity.

Inverse planned solutions

The main feature of the inverse planning algorithms described is the input of a set of criteria that define the ideal solution, and are used to drive the optimization routine. The criteria used in the published work are very similar[116–120]. All authors state that the algorithms attempt to maximize the dose uniformity of the planning target volume while applying dose limits on the nearby critical structures (the heart and lung). In practice, the input is often in the form of points on a dose–volume histogram (DVH) curve and the optimization routine attempts to drive the solution towards the required DVH. This is important, because a DVH loses information on the spatial distribution of dose, and gaining an optimum dose distribution requires a spatially-based input. While inverse planning has been applied to the whole breast, it is likely to be more suitable for complex cases such as bilateral breast irradiation, patients with pectus excavatum, or where several nodal groups need to be irradiated, and more normal tissue sparing is required.

Forward planned solutions

Plane compensation, tissue compensation, and equivalent path length compensation

These three techniques are classified together, as for the near parallel pair situation of a tangential breast treatment, they provide similar solutions. The basic feature of all the methods is the production of a plane of uniform dose in an axis perpendicular to that of the treatment beam. This plane is most likely, but not exclusively, through the isocentre of the beams. It has been implemented on some commercial treatment planning systems, and referred to as electronic compensation.

Beam's eye view isodose contouring

This method uses the visualization tools of a treatment planning system to design beam segments by selecting sets of isodose levels and fitting either MLC leaves or collimator jaws to a best approximation of the shape. The methods described have been used with or without a physical wedge, and with a variety of software to assist the task[121–124]. An example is shown in Fig. 4.14. The shape of the modulation produced in the anterior–posterior direction essentially becomes a step function, where the step height is dependent on the isodose levels chosen for blocking. This method is often referred to as field-in-field.

In practice, the choice of method is likely to be governed by the treatment planning software and linear accelerator hardware available locally.

4.6.3 **Chest wall**

Good dosimetry in this site is difficult to achieve. The tangential fields are narrow with a significant proportion irradiating either lung or air. The target volume for mastectomy (chest wall) treatments includes the skin, and bolus is often applied to achieve an adequate dose. The type and thickness of bolus should be included in the treatment plan.

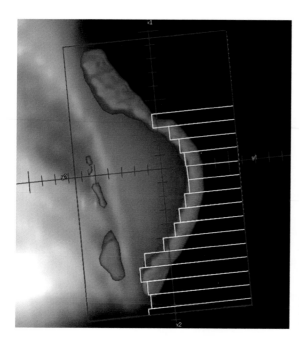

Fig. 4.14 Example of a beam's eye view (BEV) isodose contouring approach to IMRT for the intact breast. The 105% isodose surface in a BEV window is shown in red with some of the MLC leaves positioned over part of this area.

Bolus ensures that the skin receives 100% of the prescribed dose, a basic requirement for any target tissue. Without bolus, skin (the site of nearly all local recurrences) receives roughly 70% of the prescribed dose, whereas the underlying non-target muscles and ribs receive 100%. Depending on the dose regimen, late atrophy with telangiectasia can be marked. For that reason, some oncologists do not use bolus, or use it for half of the number of fractions. In principle, full bolus should be given for the whole course of treatment in all patients. If this exceeds tolerance, in principle, total dose should be reduced and full bolus retained.

Electron dosimetry has its own problems particularly as Monte Carlo algorithms for electron planning are not routinely implemented at most centres. These would allow accurate dosimetry information to be obtained. The appropriate beam energy must be selected to ensure that lung and heart are protected. Ultrasound or CT scanning accurately defines the distance between skin surface and deep fascia. Stand-off and stand-in need to be taken into account when estimating dosimetry. Stand-off markedly changes the percentage dose received at the edges of electron fields in some patients.

4.6.4 Axilla and supraclavicular fossa

The supraclavicular nodes are treated with an applied anterior, or anterior oblique, field with focus skin distance, output factor, and off axis factor corrections as appropriate. When a posterior boost field is required to raise the dose to the deep nodes the dose distribution is very inhomogeneous as shown in Fig. 4.15. If the prescribed dose is delivered to the mid-axillary plane, an overdose ≈ 10% occurs in the region 2–3 cm

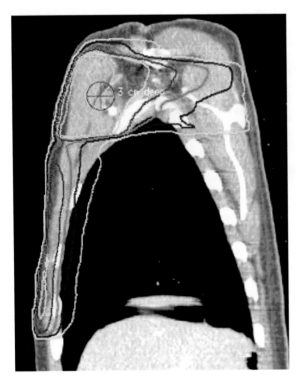

Fig. 4.15 Sagittal view of a four-field treatment for a postmastectomy patient showing that the relative dose at 3 cm deep in the axilla is 113% if the beams are weighted to deliver 100% of the prescribed dose to mid-axillary line. The relative weighting of anterior to posterior fields is 4:1. The isodose contours are represented as: 113% dull yellow, 105% pale pink, 100% red, 95% green, 90% blue, and 80% yellow.

below the anterior skin surface, close to the brachial plexus. To limit the percentage depth dose to the brachial plexus, a list of anterior-to-posterior field monitor unit ratios can be applied to maintain adequate doses to the nodes and limit the maximum dose, as shown in Fig. 4.16.

4.6.5 Tumour bed boost

Electron

If dosimetry for a tumour bed boost using electron beams cannot be calculated on a treatment planning system, then central axis depth dose data, applicator and end-frame output factors, and isodose plots are used to plan the treatment. The use of the CT scan data and internal markers, such as surgical clips, are recommended to ensure the correct electron beam energy and size of applicator are chosen to treat the tumour bed and to minimize the dose to underlying ribcage, lung, and heart. It may not be possible to use the whole breast CT scan if the patient requires a different treatment position for the boost treatment to that for the whole breast, and where there are no internal markers to guide beam placement. All information should be used to reduce the possibility of an inappropriate choice of energy and applicator. It should be noted that the position of the surgical scar is not a valid guide to the position of the tumour bed with modern surgical techniques, which aim to maximum good cosmetic results after surgery.

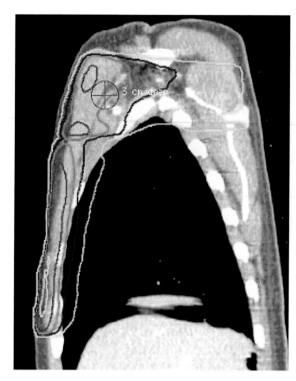

Fig. 4.16 Sagittal view of a four-field treatment for a postmastectomy patient showing that the relative dose at 3 cm deep in the axilla is 95% if the beams are weighted to deliver 100% of the prescribed dose to 1.5 cm deep (D_{max} of a 6-MV beam). The dose at mid-axillary line is 90%. The relative field weightings of anterior to posterior fields are 10:1. (The relative field weightings are taken from the START trial recommendations and are dependent on the axilla separation.) The isodose contours are represented as: 113% dull yellow, 105% pale pink, 100% red, 95% green, 90% blue, and 80% yellow.

Photon

The standard requirement of dose variation within 95–107% of the prescribed dose applies to mini-tangents. A modest amount of dose compensation may be required to achieve this. A field-in-field approach is likely to be appropriate.

If the photon boost is to be delivered with a 3D conformal plan method then specific constraints on PTV coverage and OAR doses are required. In the case of a sequential boost, it is recommended that a composite plan comprising both whole breast and boost plans is produced to assess the total organ doses for the complete treatment. Conformal shaping to the tumour bed PTV using MLC, and simple dose compensation, are required to achieve suitable shaping of the 95% isodose to the PTV. A similar approach can be used for an integrated boost plan. An example dose distribution is given in Fig. 4.17 for a forward planned, co-planar method. It is possible to use rotational delivery methods to deliver conformal photon boosts (both sequential and integrated), however, these may increase mean organ doses, e.g. mean ipsilateral lung dose, to levels several Gy above that from a tangent pair method, and may only be appropriate for specific high risk groups.

4.6.6 **Organs at risk**

For tangential field irradiation, the OARs are heart, lung, muscles, and ribs. When nodal fields are used, they include brachial plexus, major blood vessels, and lung.

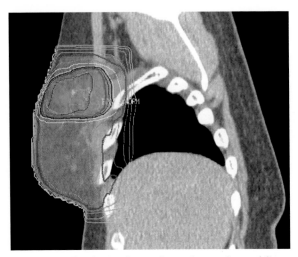

Fig. 4.17 Example of a dose distribution for co-planar photon plan to deliver a tumour bed boost dose. The figure shows the composite distribution for the whole breast and the boost, although the two are treated sequentially. The orange volume is the whole breast PTV; the blue volume is the boost PTV. Dose to the whole breast was 40 Gy in 15 fractions. The dose to the boost volume was 16 Gy in 2-Gy fractions. The bright green line indicates the 95% iosdose of the combined treatment.

In conventional CT planning, exposure of lung and heart can be minimized by maintaining lung and heart depth on the beam's eye view to ≤ 2 cm. Where the brachial plexus is concerned, the priority is to avoid overlap between tangential and nodal fields by ensuring that the patient position is not changed between fields and that the appropriate matching techniques are planned and delivered on a daily basis. The dose prescription and scheduling of cytotoxics are also relevant to brachial plexopathy risk[46,125].

The current approach to minimizing irradiation to OARs is to use volume-based assessment with DVHs. Many of these dose constraints are developed through randomized clinical trials. An example is the IMPORT HIGH trial, which requires the use of more complex radiotherapy techniques, including inverse and forward planned IMRT. In addition, the photon boost requires several beams which will pass through normal tissue. Therefore, it is inadequate to state only the 'traditional' simple normal tissue dose–volume restrictions, such as the maximum heart/lung distance within the treatment field. Dose constraints to the OARs were determined from evidence from the literature and DVH analysis in the IMPORT planning study. Three major normal tissue structures were considered and discussed next and summarized in Table 4.6.

Contralateral breast

The 2007 recommendations from the International Commission on Radiological Protection (ICRP) estimated the risk of radiation-induced breast cancer as follows[126]. Given that there is a 5% risk of cancer per sievert (Sv), and the weighting factor for

Table 4.6 Dose constraints for organs at risk for IMPORT HIGH trial

Dose per fraction	Keep ipsilateral lung volume ≤ 15% (≤ 30% if SCF)	Keep contralateral lung volume ≤ 15%	Keep heart volume ≤ 10%	Keep contralateral mean breast dose
2 Gy	20 Gy	3 Gy	15 Gy	Aim ≤ 0.5 Gy
2.67 Gy	18 Gy	2.5 Gy	13 Gy	Accept ≤ 1.5 Gy

breast is 0.12 (0.06 per breast), then a mean dose to the breast of 1 Gy = 0.06 Sv, which gives a < 0.5% risk of cancer. The IMPORT HIGH planning study showed that a mean dose of < 1 Gy could be achieved in the majority of cases, but occasionally this was exceeded, particularly with very medial tumour beds (Fig. 4.18). Subsequent work has shown 50% of plans submitted to the QA team, were achieving a mean dose of 0.5 Gy or less. Therefore, the current IMPORT HIGH recommendations are to aim for a contralateral breast dose of < 0.5 Gy, but a mean dose of < 1.5 Gy would be acceptable for trial entry.

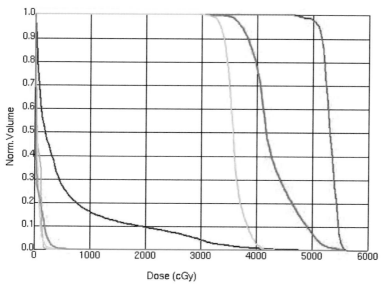

Fig. 4.18 This is a typical dose–volume histogram for the IMPORT HIGH trial test arm 2(102). The red, green and yellow volumes show the 3 distinct PTV targets for the breast: tumour bed, quadrant and whole breast. Heart—the pink line shows the heart DVH and illustrates that a very small volume of heart is irradiated. Lung—the dark blue shows the ipsilateral lung and the light blue shows the contralateral lung. The purple line represents the contralateral breast. (Reprinted from *International Journal of Radiation Oncology, Biology, Physics*, Volume **79**, Ellen M. Donovan *et al.* Planning with intensity-modulated radiotherapy and tomotherapy to modulate dose across breast to reflect recurrence risk (IMPORT High Trial), pp. 1064–72. Copyright (2011), with permission from Elsevier.)

Lung

There is good evidence in non-small cell lung cancer that if V20 (volume receiving 20 Gy) is < 22% for *total* lung, there is no risk of pneumonitis. A breast radiotherapy study showed that V20 < 30% for the *ipsilateral* lung resulted in very few cases of pneumonitis[127]. IMPORT HIGH dose constraints for the lung are set at a lower threshold, to take into account possible increased toxicity with systemic therapy and account for the slightly higher dose per fraction. Therefore, the IMPORT HIGH recommendations are for no more than 15% of the ipsilateral lung to receive 18 Gy, and no more than 15% of the contralateral lung to receive 2.5 Gy.

Heart

Gagliardi et al. reported that only the dose level above 30 Gy to the heart is important to the calculated risk of cardiac toxicity, whereas the curve is almost constant below 30 Gy is almost constant[128]. This is also supported by the results from long-term survivors of Hodgkin's disease receiving RT[129]. IMPORT HIGH dose constraints for the heart are set at a lower threshold, to take into account possible increased toxicity with systemic therapy and account for the slightly higher dose per fraction. Therefore, the IMPORT HIGH recommendations are for no more than 10% of the heart to receive 13 Gy. Cardiac irradiation is discussed in more detail in section 4.6.7.

For lower-risk patients not requiring a boost, the dose constraints can be less stringent. Table 4.7 gives an example from the FAST Forward trial.

4.6.7 **Cardiac irradiation**

The significant excess of non-breast cancer deaths in women randomized to radiotherapy reported by the Early Breast Cancer Trialists' Collaborative Group covers a time period from 1961–1990[1]. This includes trials, which used long-outdated techniques exposing large volumes of cardiac tissue. Improved imaging and planning methods mean that the amount of heart included in the treatment fields is now much reduced, however, Taylor et al. reported that approximately half of all patients with left-side disease received > 20 Gy to some volume of heart with a modern treatment method[130]. Some authors report no increased risk for patients with left breast disease treated with techniques used since about 1970[104,131], while others still claim a doubling of the risk of fatal myocardial infarction for left-sided treatment compared with right-sided[132].

A retrospective study of myocardial perfusion defects compared a cohort of patient treated for left-sided tumours with at least 1 cm heart depth in the treatment field with a matched cohort of patients treated for right-sided tumours[133]. All patients had

Table 4.7 Dose constraints for organs at risk for FAST Forward trial

Dose per fraction	Keep 30% of dose to < 15% of ipsilateral lung volume	Keep 25% of dose to < 5% of heart volume	Keep 5% of dose to < 30% of heart volume
2.67 Gy	12 Gy	10 Gy	2 Gy
5.2/5.4 Gy	8 Gy	7 Gy	1.5 Gy

been treated at least 5 years prior to the reported work. All patients underwent a SPECT (single photon emission CT) scan to assess both myocardial perfusion and function. Perfusion tracer uptake was abnormal in 71% of the left breast patients (median thickness of irradiated heart 1.6 cm range 1.4–2.05) and 17% of the right breast patients. Although the abnormalities were not clinically significant at 5 years postradiotherapy, the study indicated that even with modern radiotherapy methods, damage to the heart still ensues. Estimation of DVH for the techniques used in the Oslo and Stockholm postmastectomy radiotherapy suggests no association between risk of myocardial infarction and volume of heart exposed, although a marked dependency on total dose is seen[128]. If so, cardiac mortality may reflect a risk of radiation-induced atheroma in the anterior descending coronary artery that varies with total dose rather than length of artery exposed, analogous to another serial structure, the spinal cord. This evidence indicates the need to minimize cardiac exposure by using all aspects of technology in planning and treatment delivery.

Methods for reducing cardiac doses

Cardiac shielding blocks and multileaf collimator

The most straightforward method of reducing the dose to the heart is to use shielding blocks or MLC on one or both treatment fields. An example of this is shown in Fig. 4.19. The disadvantage of this approach is the potential for shielding a part of the breast at risk of recurrence. However, this is only a real issue in patients with lower quadrant tumours, where the heart shielding can sometimes encroach on the tumour bed. Quite often, cardiac shielding spares heart and overlying ribcage without encroaching on the breast at all. After mastectomy, it is more difficult to place shielding, as local recurrence risk is highest along the scar line and does not reflect the location of primary tumour.

Patient position

Just by raising the arms above the head, a greater reduction in the amount of irradiated heart can be achieved than by complex planning techniques[134]. When breast patients are CT scanned, the requirement for a treatment position that is either flat or low-angled with the arms drawn over the head, appears to reduce the amount of heart included in the radiotherapy fields. It may be that this development alone will have a long-term positive benefit. The evidence for the benefit of the prone treatment position is ambiguous, however, Kirby et al.[79] show that women with whole breast CTV > 1000 cc receive a lower cardiac dose so it may be possible to chose a cohort of patients for whom this position would be suitable. One disadvantage is the practicalities for patients with impaired mobility.

Breathing manoeuvres

When the ribcage expands on deep inspiration, the chest wall and breast move forward away from the heart. A CT planning study using voluntary breath holding in inspiration has demonstrated a reduction of the amount of heart in the fields[135]. Other authors have confirmed this effect using assisted breathing control with patients undergoing CT scanning for radiotherapy planning[69,136,137]. A robust voluntary breath hold technique could provide a low technology, efficient method to decrease cardiac irradiation in the majority of patients with left-side disease.

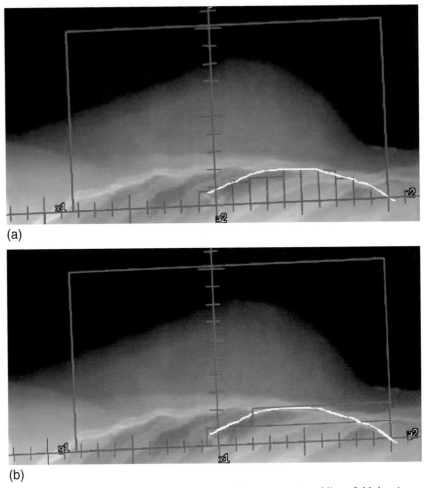

(a)

(b)

Fig. 4.19 Digitally reconstructed radiograph (DRR) from an anterior oblique field showing MLC shielding of the heart (anterior surface indicated in white) from the anterior–posterior direction (a) and the superior–inferior direction (b).

Intensity-modulated radiotherapy

Standard rectangular conformal blocked beams and IMRT fields, using standard tangential beam geometry[116] and with arcs of four and six beams around the patient[138] have been compared. While it is possible to show a marked reduction in cardiac sparing when using IMRT in these cases, there has not necessarily been a gain in dose homogeneity in the breast and, for the arc geometry, there is an inevitable increase in the dose to the contralateral breast (potentially up to 20% of the volume receiving 20% of the prescribed dose) and the ipsilateral lung. Thus some form of partial shielding of the heart in the standard tangential beam arrangement is preferable to a set of beams in an arc geometry with IMRT. The use of IMRT techniques alone does not reduce both unplanned dose inhomogeneity and radiation dose to the heart.

The combination of IMRT methods with patient position modification and respiratory control may enable decreased dose to the heart with improved dosimetry in the breast.

4.6.8 Second cancer risk

The Early Breast Cancer Trialists' Collaborative Group report of 2005[1] presented data showing contralateral breast cancer was increased in the radiotherapy population compared to the control. Darby et al. show an increase in ipsilateral lung cancer compared to contralateral lung cancer in a large population from the SEER database[139]. Smoking is also an important risk factor and substantially increases the risk of developing lung cancer in those patients treated with breast radiotherapy[140].

These data are from women irradiated with old techniques and hence, it is assumed, higher dose. For example, Darby et al. estimate mean lung dose in their population as 5 Gy for contralateral lung and 11 Gy for ipsilateral lung. A modern radiotherapy technique to whole breast only is likely to expose the contralateral lung to a mean dose of 1 Gy and the ipsilateral lung to around 5 Gy mean dose—values which may increase by about 1 Gy if nodal irradiation is added. It would be reasonable to anticipate lower risk with a lower dose compared to the study population. Modern radiotherapy techniques for breast cancer are becoming more sophisticated, however, requiring different beam geometries, rotational techniques, and more intensive imaging at the time of treatment. All of these may contribute to increasing the dose to non-target regions above that from the traditional tangential field arrangement. There is some evidence that recurrence rates for early breast cancer have fallen[141]. In these circumstances, very long-term effects such as a risk of a radiotherapy-induced second malignancy become of greater importance. This should not prohibit the use of more complex methods, or concomitant imaging for verification, but careful selection of the most appropriate patient cohorts is essential.

4.7 Treatment delivery

The delivery of breast radiotherapy will in part depend on the characteristics of the treatment machine. These will govern, for example, whether asymmetric collimation is used as part of the field geometry, and will dictate the level of sophistication of the delivery of IMRT, from physical compensators to MLC on its own, or a combination of wedge and MLC segments. Other factors, such as the use of a universal wedge that linear accelerator control software must insert and check, or the necessity to rotate the couch to achieve a good match plane between tangential and nodal fields, all impact on the overall time for a fraction of treatment.

4.7.1 Treatment machine characteristics

The nature of the field geometry used for breast treatments and the delivery method for 3D compensation will impact on the tolerance and action levels for specific treatment machine parameters. For example, if asymmetric collimation is used to define the match plane of tangential with nodal fields, the tolerances on collimator jaw positions at the central axis may be tighter than when angulation of collimator and couch is used.

When a MLC is used to deliver 3D compensation, whether static or dynamic, low monitor unit, small field characteristics should be measured. These include the dose linearity, and beam stability, energy, flatness, and symmetry[142]. A dynamic MLC delivery requires the checking of MLC leaf position and speed, variations in which can generate significant dose differences.

4.8 Treatment verification

4.8.1 Whole breast

The initial patient set-up is carried out using room lasers pointing at skin marks placed on stable anatomy. A suggested daily set-up tolerance is ± 5 mm for skin marks with respect to the light field. Medial, and possibly lateral, focus-to-skin distance (FSD) measurements may also be used to monitor daily set-up variation. A recent report by The Royal College of Radiologists, Institute of Physics and Engineering in Medicine, and the College of Radiographers has published strategies for the verification of a number of sites in radiotherapy[67]. The approach to the verification is recommended for whole breast and the nodal fields where PTV margin is 10 mm. If smaller margins are required then verification protocols and tolerances should be modified accordingly. For example, the use of on-line imaging, and off-line protocols such the No Action Level (NAL)[143] and extended No Action Level (eNAL) of de Boer et al.[144] may reduce population systematic errors to 1–2 mm with a consequent reduction in the size of the PTV margin[71–73].

For treatments with 3D compensation, portal imaging may be used in order to confirm segment shapes if these are few in number, to sum a number of fields in order to check a fluence profile for a dynamic delivery, or to confirm a multiple static technique with many segments. Portal imaging is the main tool for treatment verification, but there are a number of other systems, which can be used to monitor movement or shape changes during a course of breast radiotherapy. These include external contour outlining systems using lasers in the treatment room, or separate external lasers, for interfraction motion. Tracking systems based on skin marks or external markers may allow monitoring of intrafraction organ motion. Real-time monitoring using internal markers has not been reported for breast radiotherapy treatments at present.

4.8.2 Tumour bed boost

Electron treatments

Initial set-up may be guided by the use of surface rendered views and gantry angles determined from virtual simulation, if CT data with the boost treatment position and internal markers are available. If not, all relevant information should be used to determine appropriate gantry, couch, and collimator angles for treatment. When all treatment parameters have been set then the use of the light field, and the record and verify system for gantry, couch, and collimator angles may be used on a daily basis to ensure a good set-up. Photographs from the initial treatment set-up may assist also.

Mini-tangential fields

Verification is similar to that for whole breast tangential fields and imaging on the first few fractions is sufficient to evaluate any systematic error which can be corrected for the remaining course of treatment. A whole breast treatment field may be useful to check light field against skin tattoos if the mini-tangent fields do not give sufficient information. A larger field may be used for portal imaging for the same reason, and modern imaging systems produce imaging of sufficient quality with 1–2 MU for a tangential view so additional dose is low. The use of kV planar imaging will give good quality images with very low additional dose.

For advanced treatment planning, such as that which may be required for partial breast, conformal boost, or simultaneous boost treatments, internal markers provide a means of verification for the position of the tumour bed. Image-guided techniques using kV or MV imaging are necessary to ensure the quality of treatment, particularly when margins are reduced to 5–6 mm. Off-line verification protocols such as those of de Boer et al.[143,144] provide methods to reduce the population systematic error whilst limiting the concomitant dose burden and imaging workload.

Dose from verification imaging

Harrison et al. have investigated the relative contribution, to non-target organs, of the dose from radiotherapy imaging and that from scattered and leakage radiation from the radiotherapy treatment[145]. For intensive imaging protocols (which were not optimized for breast imaging, e.g. the use of a 360° CT scan and high-dose portal image) they found that imaging contributed no more than 30% of the total organ dose for a range of organs including contralateral breast. Harrison further suggested a practical limit for the ratio of imaging dose to total organ dose of 0.5[146]. Well-designed protocols for the verification of complex breast treatments should be able to meet both this parameter, and reduce population systematic errors, and hence PTV margins.

4.8.3 **Effect of patient movement on advanced radiotherapy treatments**

The availability of CT scanning for breast cancer patients and simple forward planned methods for 3D dose compensation in breast radiotherapy are starting to define a new standard for breast radiotherapy. It is, therefore, important to understand the impact of patient movement on these techniques. Dosimetric consequences of patient movement for wedge-only and a 3D compensated treatment with simple IMRT have been reported[66,68,147]. The results from these studies demonstrate that 3D compensation with IMRT retains a dosimetric advantage over wedge-only compensators, which is not negated by effects of patient movement or delivery technique. More recent work has confirmed this advantage for an IMRT technique combining whole breast fields with a few shaped segments for < 20% of the delivery[148].

4.9 **Dose prescription**

4.9.1 **Whole breast, chest wall, and regional nodes**

It is usual to apply the same dose regimen to reference points in the whole breast, postmastectomy chest wall, and the regional lymphatics. The commonest dose used internationally is 50 Gy in 25 fractions over 5 weeks. There is now a large body of mature outcome data from UK and Canadian randomized trials comparing regimens using 2.67-Gy fractions against 50 Gy in 2-Gy fractions[149–155]. The hypofractionated regimens are:

- 40 Gy in 15 fractions over 3 weeks (UK)
- 42.5 Gy in 16 fractions over 3 weeks + 1 day (Canada).

No significant differences in local control or late adverse effects have been reported between randomized groups. As a consequence, the 15- and 16-fraction regimens have been adopted as standard practice in the UK and Canada.

4.9.2 Breast boost

Conventional boost regimens after external beam doses as defined in the previous section include:

- 16 Gy in eight fractions over 1 week + 3 days
- 10 Gy in five fractions over 1 week.

The dose is defined to the 100% isodose for electrons or intersection point/reference point for photon beam plans. Standard regimens for boost dose after complete tumour excision are strongly influenced by the EORTC trial ($n = 5318$), which tested 16 Gy in eight fractions against no boost after complete microscopic excision of primary tumour[156]. This is usually delivered by electrons, but mini-tangential fields are sometimes needed in patients with large breasts. It is entirely reasonable to hypofractionate the boost schedule, for example, a five-fraction regimen of 2.67 Gy is equivalent to 14 Gy in 2.0-Gy equivalents assuming an alpha/beta value of 3.0 Gy. If excision is incomplete, and re-excision is not possible for some reason, electron doses can be increased to 24 Gy in 2.0-Gy equivalents.

4.9.3 Risk-adapted dose regimens

Individualization of dose intensity based on tumour recurrence risk and/or patient tolerance is currently applied in a very limited way, due to the difficulties of reliably identifying different risk groups for local recurrence or adverse radiotherapy effects. The use of a breast boost based on age and/or tumour-derived prognostic factors is one of the few examples of risk-adapted radiotherapy regimens in the breast conservation setting.

Postmastectomy breast reconstruction may be adversely affected by radiotherapy, so it is important to explain to women that radiotherapy may impact on functional outcome and satisfaction with breast reconstruction. However, there appears to be no reliable evidence to justify routine dose reduction in these patients.

Most authorities recommend avoiding radiotherapy in patients with active collagen vascular disease, especially systemic sclerosis[157].

4.10 Palliative radiotherapy

The same principles apply in breast cancer as elsewhere in oncology. Palliative radiotherapy, by definition, aims to relieve symptoms, not to eradicate cancer. However, the long natural history of disease in a proportion of patients presenting with advanced locoregional disease or skeletal metastases can complicate decision-making.

In a patient whose life expectancy is judged to be years rather than months, low-dose palliation (single fraction of 8 Gy) for uncomplicated skeletal pain is well tested, and reviewed elsewhere in this book. Lifetime palliation of fungating primary tumours or regional lymph nodes often requires more than 20 Gy in five fractions in the absence of life-threatening metastatic disease. Without randomized evidence on which to base reliable treatment recommendations, a reasonable rule is to plan a potentially curative dose to a PTV encompassing all known disease (GTV) plus a 1-cm margin. A curative dose is not precisely defined, but implies a dose in excess of 60 Gy in 2.0-Gy equivalents. Although there are few published data on hypofractionated external beam therapy in this setting, a once-weekly fraction of 6.0 Gy to a total dose of 36 Gy in 6 weeks using a direct electron field (with bolus if needed) will palliate symptoms arising from well-circumscribed lesions in the breast or chest wall if the patient is frail by virtue of comorbidity or extreme age[158]. This schedule often does not cause moist desquamation, a particularly valuable feature in the palliative context. This advantage of extreme hypofractionation cannot be exploited in the lymphatic pathways, due to the fractionation sensitivity of the brachial plexus.

4.11 Partial breast irradiation

Current randomized trials of partial breast radiotherapy are shown in Table 4.8, but until mature outcome data are presented in large patient cohorts, these techniques

Table 4.8 Recently reported and current randomized trials of partial breast radiotherapy

Trial	Target accrual	Test arm: treatment modality	Reported	Median follow-up (months)
Hungarian	258	Multisource brachytherapy	Yes	66
Targit	2200	Intraoperative KV photons	Yes	24
ELIOT	2000	Intraoperative electrons	No	
IMPORT LOW	2000	External beam radiotherapy	No	
GEC-ESTRO	1170	Multisource brachytherapy	No	
NSABP-30	9000	External beam radiotherapy or single/multisource brachytherapy	No	
RAPID	2000	External beam radiotherapy	No	
IRMA	3300	External beam radiotherapy	No	
Danish Breast Cancer Co-operative Group	630	External beam radiotherapy	No	

Reprinted from *The American Journal of Surgery*, Volume **94**, Issue 3, Anna M. Kirby, Charlotte E. Coles and John R. Yarnold. Target volume definition for external beam partial breast radiotherapy: Clinical, pathological and technical studies informing current approaches, pp. 255–63. Copyright (2010), with permission from Elsevier.

remain investigational. Guidelines published by European and American societies are permissive despite the paucity of mature randomized data [159,160].

One of the controversial planning issues of partial breast irradiation is determining the size of the CTV around the tumour bed. The evidence to date is reviewed in by Kirby et al.[161], but more knowledge will become available over the years as the randomized controlled trials mature and report the number and pattern of local recurrences.

4.12 Conclusions

After decades of localizing standardized fields, the definition and localization of tumour CTV and PTV on 3D datasets are driving the planning process. The routine insertion of internal markers for accurate tumour bed localization is an important element of this transition, and the use of other imaging modalities, such as open MR scanners and 3D ultrasound, in radiotherapy planning is likely to improve further the anatomical localization of target tissues in future. Technological advances in treatment delivery not only enable full dose compensation to be offered routinely, but also offer scope for testing the clinical benefits of matching total dose more closely to local recurrence risk. In the context of the preserved breast, the biological advantages of hypofractionation combined with intensity modulation are worth testing in prospective clinical trials. In low recurrence risk patients, systematic evaluation of partial breast radiotherapy is underway using a range of advanced radiotherapy technologies. It is anticipated that the evaluation and implementation of advanced technologies is likely to achieve measurable benefits in cancer cure and quality of life in a cost-effective manner. The increased speed and flexibility of current hardware and software systems appear to accommodate the added complexity with ease.

References

1. Clarke M, Collins R, Darby S, et al. Early Breast Cancer Trialists' Collaborative Group (EBCTCG). Effects of radiotherapy and of differences in the extent of surgery for early breast cancer on local recurrence and 15-year survival: an overview of the randomised trials *Lancet* 2005; **366**: 2087–106.

1a. Early Breast Cancer Trialists' Collaborative Group (EBCTCG), Darby S, McGale P, et al. Effect of radiotherapy after breast-conserving surgery on 10-year recurrence and 15-year breast cancer death: meta-analysis of individual patient data for 10,801 women in 17 randomised trials *Lancet* 2011; **378**(9804): 1707–16.

2. Whelan TJ, Julian J, Wright J, Jadad AR, Levine ML. Does locoregional radiation therapy improve survival in breast cancer? A meta-analysis. Journal of Clinical Oncology 2000; 18(6): 1220–9.

3. Antonini N, Jones H, Horiot JC, et al. Effect of age and radiation dose on local control after breast conserving treatment: EORTC trial 22881–10882. Radiotherapy and Oncology 2007; 82(3): 265–71.

4. Vinh-Hung V, Verschraegen C. Breast-conserving surgery with or without radiotherapy: pooled-analysis for risks of ipsilateral breast tumor recurrence and mortality. Journal of the National Cancer Institute 2004; 96(2): 115–21.

5. Mariani L, Salvadori B, Marubini E, et al. Ten year results of a randomised trial comparing two conservative treatment strategies for small size breast cancer. *European Journal of Cancer* 1998; **34**(8): 1156–62.

6. Veronesi U, Marubini E, Mariani L, *et al.* Radiotherapy after breast-conserving surgery in small breast carcinoma: long-term results of a randomized trial. *Annals of Oncology* 2001; **12**(7): 997–1003.

7. Fisher B, Dignam J, Mamounas EP, *et al.* Sequential methotrexate and fluorouracil for the treatment of node-negative breast cancer patients with estrogen receptor-negative tumors: eight-year results from National Surgical Adjuvant Breast and Bowel Project (NSABP) B-13 and first report of findings from NSABP B-19 comparing methotrexate and fluorouracil with conventional cyclophosphamide, methotrexate, and fluorouracil. *Journal of Clinical Oncology* 1996; **14**(7): 1982–92.

8. Fisher B, Anderson S, Tan-Chiu E, *et al.* Tamoxifen and chemotherapy for axillary node-negative, estrogen receptor-negative breast cancer: findings from National Surgical Adjuvant Breast and Bowel Project B-23. *Journal of Clinical Oncology* 2001; **19**(4): 931–42.

9. Hughes KS, Schnaper LA, Berry D, *et al.* Lumpectomy plus tamoxifen with or without irradiation in women 70 years of age or older with early breast cancer. *New England Journal of Medicine* 2004; **351**(10): 971–7.

10. Fyles AW, McCready DR, Manchul LA, *et al.* Tamoxifen with or without breast irradiation in women 50 years of age or older with early breast cancer. *New England Journal of Medicine* 2004; **351**(10): 963–70.

11. Pötter R, Gnant M, Kwasny W, *et al.* Austrian Breast and Colorectal Cancer Study Group. Lumpectomy plus tamoxifen or anastrozole with or without whole breast irradiation in women with favorable early breast cancer. *International Journal of Radiation Oncology, Biology, Physics* 2007 **68**(2): 334–40.

12. Fisher B, Bryant J, Dignam JJ, *et al.* Tamoxifen, radiation therapy, or both for prevention of ipsilateral breast tumor recurrence after lumpectomy in women with invasive breast cancers of one centimeter or less. *Journal of Clinical Oncology* 2002; **20**(20): 4141–9.

13. Blamey RW, Chetty U, Mitchell A. The BASO II trial of adjuvant radiotherapy. V. None and tamoxifen. V. None in small, node negative, grade I tumours. *European Journal of Cancer* 2002; **38**(Suppl. 3): S149.

14. Liljegren G, Holmberg L, Bergh J, *et al.* 10-year results after sector resection with or without postoperative radiotherapy for stage I breast cancer: a randomized trial. *Journal of Clinical Oncology* 1999; **17**(8): 2326–33.

15. Clark RM, Whelan T, Levine M, *et al.* Randomized clinical trial of breast irradiation following lumpectomy and axillary dissection for node-negative breast cancer: an update. Ontario Clinical Oncology Group. *Journal of the National Cancer Institute* 1996; **88**(22): 1659–64.

16. Lim M, Nixon AJ, Gelman R. A prospective study of conservative surgery (CS) alone without radiotherapy (RT) in selected patients with stage I breast cancer. *Breast Cancer Research and Treatment* 1999; **57**: 34.

17. Bartelink H, Horiot JC, Poortmans PM, *et al.* Impact of a higher radiation dose on local control and survival in breast-conserving therapy of early breast cancer: 10-year results of the randomized boost versus no boost EORTC 22881–10882 trial. *Journal of Clinical Oncology* 2007 **25**(22): 3259–65.

18. Bartelink H, Horiot JC, Poortmans P, *et al.* Recurrence rates after treatment of breast cancer with standard radiotherapy with or without additional radiation. *New England Journal of Medicine* 2001; **345**(19): 1378–87.

19. Liljegren G, Lindgren A, Bergh J, *et al.* Risk factors for local recurrence after conservative treatment in stage I breast cancer. Definition of a subgroup not requiring radiotherapy. *Annals of Oncology* 1997; **8**(3): 235–41.

20. Fisher B, Wickerham DL, Deutsch M, Anderson S, Redmond C, Fisher ER. Breast tumor recurrence following lumpectomy with and without breast irradiation: an overview of recent NSABP findings. *Seminars in Surgical Oncology* 1992; **8**(3): 153–60.

21. Veronesi U, Luini A, Del Vecchio M, *et al.* Radiotherapy after breast-preserving surgery in women with localized cancer of the breast. *New England Journal of Medicine* 1993; **328**(22): 1587–91.

22. Wallgren A, Bonetti M, Gelber RD, *et al.* Risk factors for locoregional recurrence among breast cancer patients: results from International Breast Cancer Study Group Trials I through VII. *Journal of Clinical Oncology* 2003; **21**(7): 1205–13.

23. Rowell NP. Are mastectomy resection margins of clinical relevance? A systematic review. *Breast* 2010; **19**(1): 14–22.

24. Overgaard M, Nielsen HM, Overgaard J. Is the benefit of postmastectomy irradiation limited to patients with four or more positive nodes, as recommended in international consensus reports? A subgroup analysis of the DBCG 82 b&c randomized trials. *Radiotherapy and Oncology* 2007; **82**(3): 247–53.

25. Kyndi M, Overgaard M, Nielsen HM, *et al.* High local recurrence risk is not associated with large survival reduction after postmastectomy radiotherapy in high-risk breast cancer: a subgroup analysis of DBCG 82 b&c. *Radiotherapy and Oncology* 2009; **90**(1): 74–9.

26. Kunkler IH, Canney P, van Tienhoven G, Russell NS; MRC/EORTC (BIG 2–04) SUPREMO Trial Management Group. Elucidating the role of chest wall irradiation in 'intermediate-risk' breast cancer: the MRC/EORTC SUPREMO trial. *Clinical Oncology* 2008; **20**(1): 31–4.

27. Garg AK, Oh JL, Oswald MJ, *et al.* Effect of postmastectomy radiotherapy in patients <35 years old with stage II-III breast cancer treated with doxorubicin-based neoadjuvant chemotherapy and mastectomy. *International Journal of Radiation Oncology, Biology, Physics* 2007; **69**(5): 1478–83.

28. McGuire SE, Gonzalez-Angulo AM, Huang EH, *et al.* Postmastectomy radiation improves the outcome of patients with locally advanced breast cancer who achieve a pathologic complete response to neoadjuvant chemotherapy. *International Journal of Radiation Oncology, Biology, Physics* 2007; **68**(4): 1004–9.

29. Woodward WA, Buchholz TA. The role of locoregional therapy in inflammatory breast cancer. *Seminars in Oncology* 2008; **35**(1): 78–86.

30. Singletary SE. Surgical management of inflammatory breast cancer. *Seminars in Oncology* 2008; **35**(1): 72–7.

31. Mignano JE, Zahurak ML, Chakravarthy A, *et al.* Significance of axillary lymph node extranodal soft tissue extension and indications for postmastectomy irradiation. *Cancer* 1999; **86**: 1258–62.

32. Hetelekidis S, Schnitt SJ, Silver B, *et al.* The significance of extracapsular extension of axillary lymph node metastases in early-stage breast cancer. *International Journal of Radiation Oncology, Biology, Physics* 2000; **46**: 31–4.

33. Grills IS, Kestin LL, Goldstein N, *et al.* Risk factors for regional nodal failure after breast-conserving therapy: regional nodal irradiation reduces rate of axillary failure in patients with four or more positive lymph nodes. *International Journal of Radiation Oncology, Biology, Physics* 2003; **56**(3): 658–70.

34. EBCTCG. Effects of radiotherapy and surgery in early breast cancer. An overview of the randomized trials. Early Breast Cancer Trialists' Collaborative Group. *New England Journal of Medicine* 1995; **333**(22): 1444–55.

35. McKinna F, Gothard L, Ashley S, Ebbs SR, Yarnold JR. Lymphatic relapse in women with early breast cancer: a difficult management problem. *European Journal of Cancer* 1999; **35**(7): 1065–9.

36. Yarnold JR. Selective avoidance of lymphatic irradiation in the conservative management of breast cancer. *Radiotherapy and Oncology* 1984; **2**(2): 79–92.

37. Yates L, Kirby A, Crichton S, *et al.* Risk factors for regional nodal relapse in breast cancer patients with one to three positive axillary nodes. *International Journal of Radiation Oncology, Biology, Physics* 2011; Apr 27. [Epub ahead of print]

38. Handley RS. Carcinoma of the breast. *Annals of the Royal College of Surgeons of England* 1975; **57**(2): 59–66.

39. Krag D, Weaver D, Ashikaga T, *et al.* The sentinel node in breast cancer—a multicenter validation study. *New England Journal of Medicine* 1998; **339**(14): 941–6.

40. Chen RC, Lin NU, Golshan M, *et al.* Internal mammary nodes in breast cancer: diagnosis and implications for patient management—a systematic review. *Journal of Clinical Oncology* 2008 **26**(30): 4981–9.

41. Overgaard M, Hansen PS, Overgaard J, *et al.* Postoperative radiotherapy in high-risk premenopausal women with breast cancer who receive adjuvant chemotherapy. Danish Breast Cancer Cooperative Group 82b Trial. *New England Journal of Medicine* 1997; **337**(14): 949–55.

42. Overgaard M, Jensen MB, Hansen PS, *et al.* Postoperative radiotherapy in high-risk postmenopausal women breast-cancer patients given adjuvant tamoxifen: Danish Breast Cancer Cooperative Group DBCG 82c randomised trial. *Lancet* 1999; **353**: 1641–48.

43. Ragaz J, Jackson SM, Le N, *et al.* Adjuvant radiotherapy and chemotherapy in node-positive premenopausal women with breast cancer. *New England Journal of Medicine* 1997; **337**(14): 956–62.

44. Bijker N, Meijnen P, Peterse JL, *et al.* EORTC Breast Cancer Cooperative Group; EORTC Radiotherapy Group Breast-conserving treatment with or without radiotherapy in ductal carcinoma-in-situ: ten-year results of European Organisation for Research and Treatment of Cancer randomized phase III trial 10853—a study by the EORTC Breast Cancer Cooperative Group and EORTC Radiotherapy Group. *Journal of Clinical Oncology* 2006; **24**(21): 3381–7.

45. Wapnir IL, Dignam JJ, Fisher B, *et al.* Long-term outcomes of invasive ipsilateral breast tumor recurrences after lumpectomy in NSABP B-17 and B-24 randomized clinical trials for DCIS. *Journal of the National Cancer Institute* 2011; **103**(6): 478–88.

46. Cuzick J, Sestak I, Pinder SE, *et al.* Effect of tamoxifen and radiotherapy in women with locally excised ductal carcinoma in situ: long-term results from the UK/ANZ DCIS trial. *Lancet Oncology* 2011; **12**(1): 21–9.

47. Emdin SO, Granstrand B, Ringberg A, *et al.* Swedish Breast Cancer Group. SweDCIS: Radiotherapy after sector resection for ductal carcinoma in situ of the breast. Results of a randomised trial in a population offered mammography screening. *Acta Oncologica* 2006; **45**(5): 536–43.

48. Early Breast Cancer Trialists' Collaborative Group (EBCTCG), Correa C, McGale P, Taylor C, *et al.* Overview of the randomized trials of radiotherapy in ductal carcinoma in situ of the breast. *Journal of the National Cancer Institute Monographs* 2010; **2010**(41): 162–77.

49. Silverstein MJ. The University of Southern California/Van Nuys prognostic index for ductal carcinoma in situ of the breast. *American Journal of Surgery* 2003; **186**(4): 337–43.

50. Goodwin A, Parker S, Ghersi D, Wilcken N. Post-operative radiotherapy for ductal carcinoma in situ of the breast. *Cochrane Database of Systematic Reviews* 2009; 4: CD000563.

51. Omlin A, Amichetti M, Azria D, *et al.* Boost radiotherapy in young women with ductal carcinoma in situ: a multicentre, retrospective study of the Rare Cancer Network. *Lancet Oncology* 2006; **7**(8): 652–6.

52. Early Breast Cancer Trialists' Collaborative Group (EBCTCG). Effects of radiotherapy and of differences in the extent of surgery for early breast cancer on local recurrence and 15-year survival: an overview of the randomised trials. *Lancet* 2005; **366**: 2087–106.

53. Darby SC, Cutter DJ, Boerma M, *et al.* Radiation-related heart disease: current knowledge and future prospects. *International Journal of Radiation Oncology, Biology, Physics* 2010; **76**(3): 656–65.

54. Dewar JA, Benhamou S, Benhamou E, *et al.* Cosmetic results following lumpectomy, axillary dissection and radiotherapy for small breast cancers. *Radiotherapy and Oncology* 1988; **12**(4): 273–80.

55. Pierce SM, Recht A, Lingos TI, *et al.* Long-term radiation complications following conservative surgery (CS) and radiation therapy (RT) in patients with early stage breast cancer. *International Journal of Radiation Oncology, Biology, Physics* 1992; **23**(5): 915–23.

56. Coles CE, Wilson CB, Cumming J, *et al.* Titanium clip placement to allow accurate tumour bed localisation following breast conserving surgery: audit on behalf of the IMPORT Trial Management Group. *European Journal of Surgical Oncology* 2009; **35**(6): 578–82.

57. Lee CE, Prabhu V, Slevin NJ. Collagen vascular diseases and enhanced radiotherapy-induced normal tissue effects—a case report and a review of published studies. *Clinical Oncology* 2011; **23**(2): 73–8.

58. Schnitt SJ, Abner A, Gelman R, *et al.* The relationship between microscopic margins of resection and the risk of local recurrence in patients with breast cancer treated with breast-conserving surgery and radiation therapy. *Cancer* 1994; **74**(6): 1746–51.

59. Schnitt SJ, Connolly JL, Khettry U, *et al.* Pathologic findings on re-excision of the primary site in breast cancer patients considered for treatment by primary radiation therapy. *Cancer* 1987; **59**(4): 675–81.

60. Park CC, Mitsumori M, Nixon A, *et al.* Outcome at 8 years after breast-conserving surgery and radiation therapy for invasive breast cancer: influence of margin status and systemic therapy on local recurrence. *Journal of Clinical Oncology* 2000; **18**(8): 1668–75.

61. Bentel G, Marks LB, Hardenbergh P, Prosnitz L. Variability of the location of internal mammary vessels and glandular breast tissue in breast cancer patients undergoing routine CT-based treatment planning. *International Journal of Radiation Oncology, Biology, Physics* 1999; **44**(5): 1017–25.

62. Nielsen HM, Christensen JJ, Aagaard T *et al.* A simple method to test if the internal mammary lymph nodes are covered by the wide tangent technique in radiotherapy for high-risk breast cancer. *Clinical Oncology* 2003; **15**: 17–24.

63. van Tienhoven G, Lanson JH, Crabeels D, Heukelom S, Mijnheer BJ. Accuracy in tangential breast treatment set-up: a portal imaging study. *Radiotherapy and Oncology* 1991; **22**(4): 317–22.

64. Lirette A, Pouliot J, Aubin M, Larochelle M. The role of electronic portal imaging in tangential breast irradiation: a prospective study. *Radiotherapy and Oncology* 1995; **37**(3): 241–5.

65. Fein DA, Fowble BL, Hanlon AL, Hoffman JP, Sigurdson ER, Eisenberg BL. Does the placement of surgical clips within the excision cavity influence local control for patients treated with breast-conserving surgery and irradiation. *International Journal of Radiation Oncology, Biology, Physics* 1996; **34**(5): 1009–17.

66. Hector CL, Webb S, Evans PM. The dosimetric consequences of inter-fractional patient movement on conventional and intensity-modulated breast radiotherapy treatments. *Radiotherapy and Oncology* 2000; **54**: 57–64.

67. The Royal College of Radiologists, Society and College of Radiographers, Institute of Physics and Engineering in Medicine. *On Target: Ensuring geometric accuracy in radiotherapy.* London: The Royal College of Radiologists, 2008.

68. Hector CL, Evans PM, Webb S. The dosimetric consequences of inter-fractional patient movement on three classes of intensity-modulated delivery techniques in breast radiotherapy. *Radiotherapy and Oncology* 2001; **59**(3): 281–91.

69. Remouchamps VM, Vicini FA, Sharpe MB, Kestin LL, Martinez AA, Wong JW. Significant reductions in heart and lung doses using deep inspiration breath hold with active breathing control and intensity-modulated radiation therapy for patients treated with locoregional breast irradiation. *International Journal of Radiation Oncology, Biology, Physics* 2003; **55**(2): 392–406.

70. Baglan KL, Sharpe MB, Jaffray D, *et al.* Accelerated partial breast irradiation using 3D conformal radiation therapy (3D-CRT). *International Journal of Radiation Oncology, Biology, Physics* 2003; **55**(2): 302–11.

71. Penninkhof J, Quint S, Baaijens M, *et al.* Practical use of the extended no action elvel (eNAL) correction protocol for breast cancer patients with implanted surgical clips. *International Journal of Radiation Oncology, Biology, Physics* 2012; **82**(2): 1031–7.

72. Coles CE, Harris EJ, Donovan EM, *et al.* Evaluation of implanted gold seeds for breast radiotherapy planning and on treatment verification: A feasibility study on behalf of the IMPORT trialists. *Radiotherapy and Oncology* 2011; 100(2): 276–81.

73. Leonard CE, Tallhamer M, Johnson T, *et al.* Clinical experience with image-guided radiotherapy in an accelerated partial breast intensity-modulated radiotherapy protocol. *International Journal of Radiation Oncology, Biology, Physics* 2010; **76**(2): 528–34.

74. Dobbs J, Greener T, Driver D. *Geometric Uncertainties in Radiotherapy of the Breast.* London: British Institute of Radiology, 2003.

75. Thilmann C, Adamietz IA, Saran F, Mose S, Kostka A, Bottcher HD. The use of a standardized positioning support cushion during daily routine of breast irradiation. *International Journal of Radiation Oncology, Biology, Physics* 1998; **41**(2): 459–63.

76. Carter DL, Marks LB, Bentel GC. Impact of setup variability on incidental lung irradiation during tangential breast treatment. *International Journal of Radiation Oncology, Biology, Physics* 1997; **38**: 109–15.

77. Merchant TE, McCormick B. Prone position breast irradiation. *International Journal of Radiation Oncology, Biology, Physics* 1994; **30**: 197–203.

78. Grann A, McCormick B, Chabner ES, *et al.* Prone breast radiotherapy in early-stage breast cancer: a preliminary analysis. *International Journal of Radiation Oncology, Biology, Physics* 2000; **47**(2): 319–25.

79. Kirby AM, Evans PM, Donovan EM, *et al.* Prone versus supine positioning for whole and partial-breast radiotherapy:. A comparison of non-target tissue dosimetry *Radiotherapy and Oncology* 2010; **96**: 178–84.

80. Varga Z, Hideghety K, Mezo T, *et al.* Individual positioning: a comparative study of adjuvant breast radiotherapy in the prone versus supine position. *International Journal of Radiation Oncology, Biology, Physics* 2009; **75**: 94–100.

81. Morrow NV, Stepaniak C, White J, *et al.* Intra- and interfractional variations for prone breast irradiation: an indication for image-guided radiotherapy. *International Journal of Radiation Oncology, Biology, Physics* 2007; **69**: 910–7.

82. Kirby AM, Evans PM, Helyer SJ, *et al.* A randomised trial of Supine versus Prone breast radiotherapy (SuPr study): Comparing set-up errors and respiratory motion. *Radiotherapy and Oncology* 2011; **100**(2): 221–6.

83. Hurkmans CW, Borger JH, Pieters BR, *et al.* Variability in target volume delineation on CT scans of the breast. *International Journal of Radiation Oncology, Biology, Physics* 2001; **50**(5): 1366–72.

84. Bentel G, Marks LB, Hardenbergh P, Prosnitz L. Variability of the location of internal mammary vessels and glandular breast tissue in breast cancer patients undergoing routine CT-based treatment planning. *International Journal of Radiation Oncology, Biology, Physics* 1999; **44**(5): 1017–25.

85. Denham JW, Sillar RW, Clarke D. Boost dosage to the excision site following conservative surgery for breast cancer: it's easy to miss! *Clinical Oncology* 1991; **3**(5): 257–61.

86. Bedwinek J. Breast conserving surgery and irradiation: the importance of demarcating the excision cavity with surgical clips. *International Journal of Radiation Oncology, Biology, Physics* 1993; **26**(4): 675–9.

87. Machtay M, Lanciano R, Hoffman J, Hanks GE. Inaccuracies in using the lumpectomy scar for planning electron boosts in primary breast carcinoma. *International Journal of Radiation Oncology, Biology, Physics* 1994; **30**: 43–8.

88. Harrington KJ, Harrison M, Bayle P, *et al.* Surgical clips in planning the electron boost in breast cancer: a qualitative and quantitative evaluation. *International Journal of Radiation Oncology, Biology, Physics* 1996; **34**(3): 579–84.

89. Hunter MA, McFall TA, Hehr KA. Breast-conserving surgery for primary breast cancer: necessity for surgical clips to define the tumor bed for radiation planning. *Radiology* 1996; **200**: 281–2.

90. Krawczyk JJ, Engel B. The importance of surgical clips for adequate tangential beam planning in breast conserving surgery and irradiation. *International Journal of Radiation Oncology, Biology, Physics* 1999; **43**(2): 347–50.

91. Kovner F, Agay R, Merimsky O, Stadler J, Klausner J, Inbar M. Clips and scar as the guidelines for breast radiation boost after lumpectomy. *European Journal of Surgical Oncology* 1999; **25**(5): 483–6.

92. Regine WF, Ayyangar KM, Komarnicky LT, Bhandare N, Mansfield CM. Computer-CT planning of the electron boost in definitive breast irradiation. *International Journal of Radiation Oncology, Biology, Physics* 1991; **20**: 121–5.

93. Florell K, Halvorsen P. An accurate method for localization of the boost volume in breast radiotherapy. *Medical Dosimetry* 1997; **22**(4): 283–91.

94. Surgical guidelines for the management of breast cancer. *European Journal of Surgical Oncology* 2009; **35**(Suppl 1): 1–22.

95. DeBiose DA, Horwitz EM, Martinez AA, *et al.* The use of ultrasonography in the localization of the lumpectomy cavity for interstitial brachytherapy of the breast. *International Journal of Radiation Oncology, Biology, Physics* 1997; **38**(4): 755–9.

96. Haba Y, Britton P, Sinnatamby R, Moody A, Sycamore C, Wilson C. Can ultrasound improve the accuracy of delivery of electron boost treatment following breast conserving surgery? *European Journal of Cancer* 2001; **37**(37): 38.

97. Vicini FA, Jaffray DA, Horwitz EM, *et al.* Implementation of 3D-virtual brachytherapy in the management of breast cancer: a description of a new method of interstitial brachytherapy. *International Journal of Radiation Oncology, Biology, Physics* 1998; **40**(3): 629–35.

98. Coles CE, Cash CJC, Treece GM, *et al.* High definition three-dimensional ultrasound to localise the tumour bed: A breast radiotherapy planning study. *Radiotherapy Oncology* 2007; **84**: 233–41.

99. Whipp EC, Halliwell M. Magnetic resonance imaging appearances in the postoperative breast: The clinical target volume–tumor and its relationship to the chest wall. *International Journal of Radiation Oncology, Biology, Physics* 2008; **72**: 49–57.

100. Kirby AM, de Souza NM, Evans PM, *et al.* MRI delineation of tumour bed for partial breast irradiation: fusion/comparison with CT/Titanium clip method. *Clinical Oncology* 2009; **21**(3): 251–4.

101. Castro Pena P, Kirova YM, Campana F, *et al.* Anatomical, clinical and radiological delineation of target volumes in breast cancer radiotherapy planning: individual variability, questions and answers. *British Journal of Radiology* 2009; **82**: 595–9.

102. Kirova YM, Castro Pena P, Dendale R, *et al.* Simplified rules for everyday delineation of lymph node areas for breast cancer radiotherapy. *British Journal of Radiology* 2010: **83**(992): 683–6.

103. Whipp EC, Hartley-Davies R, Carroll J, *et al.* The inaccuracies of conventional breast radiotherapy planning defined by MR imaging. *ASCO* 2003; 25.

104. Hojris I, Overgaard M, Christensen JJ, Overgaard J. Morbidity and mortality of ischaemic heart disease in high-risk breast-cancer patients after adjuvant postmastectomy systemic treatment with or without radiotherapy: analysis of DBCG 82b and 82c randomised trials. Radiotherapy Committee of the Danish Breast Cancer Cooperative Group. *Lancet* 1999; **354**: 1425–30.

105. Goodman RL, Grann A, Saracco P, Needham MF. The relationship between radiation fields and regional lymph nodes in carcinoma of the breast. *International Journal of Radiation Oncology, Biology, Physics* 2001; **50**: 99–105.

106. Nielsen HM, Christensen JJ, Aagaard T, *et al.* A simple method to test if the internal mammary lymph nodes are covered by the wide tangent technique in radiotherapy for high-risk breast cancer. *Clinical Oncology* 2003; **15**: 17–24.

107. Lievens Y, Poortmans P, Van den Bogaert W. A glance on quality assurance in EORTC study 22922 evaluating techniques for internal mammary and medial supraclavicular lymph node chain irradiation in breast cancer. *Radiotherapy and Oncology* 2001; **60**(3): 257–65.

108. Casebow MP. Matching of adjacent radiation beams for isocentric radiotherapy. *British Journal Radiology* 1984; **57**(680): 735–40.

109. Hurkmans CW, Meijer GJ, van Vliet-Vroegindeweij C, *et al.* High dose simultaneously integrated breast boost using intensity-modulated radiotherapy and inverse optimization. *International Journal of Radiation Oncology, Biology, Physics* 2006; **66**(3): 923–30.

110. Donovan EM, Ciurlionis L, Fairfoul J, *et al.* Planning with intensity-modulated radiotherapy and tomotherapy to modulate dose across breast to reflect recurrence risk (IMPORT HIGH trial). *International Journal of Radiation Oncology, Biology, Physics* 2011; **79**(4): 1064–72.

111. ICRU. *Prescribing, Recording and Reporting Photon Beam Therapy.* Bethesda, MD: International Commission on Radiation Units and Measurements, 1993.

112. Winfield E, Deighton A, Venables K, *et al.* Survey of UK breast radiotherapy techniques: background prior to the introduction of the quality assurance programme for the Standardisation of Radiotherapy (START) trial in breast cancer. *Clinical Oncology* 2002; **14**(4): 267–71.

113. Moody AM, Mayles WP, Bliss JM, *et al.* The influence of breast size on late radiation effects and association with radiotherapy dose inhomogeneity. *Radiotherapy and Oncology* 1994; **33**(2): 106–12.

114. Neal AJ, Torr M, Helyer S, Yarnold JR. Correlation of breast dose heterogeneity with breast size using 3D CT planning and dose-volume histograms. *Radiotherapy and Oncology* 1995; **34**(3): 210–18.

115. Donovan EM, Yarnold JR, Adams EJ, *et al.* An investigation into methods of IMRT planning applied to breast radiotherapy. *British Journal of Radiology* 2008; **81**: 311–22.

116. Hurkmans CW, Cho BC, Damen E, *et al.* Reduction of cardiac and lung complication probabilities after breast irradiation using conformal radiotherapy with or without intensity modulation. *Radiotherapy and Oncology* 2002; **62**(2): 163–71.

117. Hong L, Hunt M, Chui C, *et al.* Intensity-modulated tangential beam irradiation of the intact breast. *International Journal of Radiation Oncology, Biology, Physics* 1999; **44**(5): 1155–64.

118. Chang SX, Deschesne KM, Cullip TJ, *et al.* A comparison of different intensity modulation treatment techniques for tangential breast irradiation. *International Journal of Radiation Oncology, Biology, Physics* 1999; **45**(5): 1305–14.

119. Carruthers LJ, Redpath AT, Kunkler IH. The use of compensators to optimise the three dimensional dose distribution in radiotherapy of the intact breast. *Radiotherapy and Oncology* 1999; **50**(3): 291–300.

120. Aref A, Thornton D, Youssef E, *et al.* Dosimetric improvements following 3D planning of tangential breast irradiation. *International Journal of Radiation Oncology, Biology, Physics* 2000; **48**(5): 1569–74.

121. van Asselen B, Raaijmakers CP, Hofman P, Lagendijk JJ. An improved breast irradiation technique using three-dimensional geometrical information and intensity modulation. *Radiotherapy and Oncology* 2001; **58**(3): 341–7.

122. Donovan EM, Johnson U, Shentall G, *et al.* Evaluation of compensation in breast radiotherapy: a planning study using multiple static fields. *International Journal of Radiation Oncology, Biology, Physics* 2000; **46**(3): 671–9.

123. Kestin LL, Sharpe MB, Frazier RC, *et al.* Intensity modulation to improve dose uniformity with tangential breast radiotherapy: initial clinical experience. *International Journal of Radiation Oncology, Biology, Physics* 2000; **48**(5): 1559–68.

124. Zackrisson B, Arevarn M, Karlsson M. Optimized MLC-beam arrangements for tangential breast irradiation. *Radiotherapy and Oncology* 2000; **54**(3): 209–12.

125. Olsen NK, Pfeiffer P, Johannsen L, *et al.* Radiation-induced brachial plexopathy: neurological follow-up in 161 recurrence-free breast cancer patients. *International Journal of Radiation Oncology, Biology, Physics* 1993; **26**: 43–9.

126. The 2007 Recommendations of the International Commission on Radiological Protection. ICRP publication 103. *Annals of the ICRP* 2007; **37**(2–4): 1–332.

127. Blom-Goldman U, Svane G, Wennberg B, *et al.* Quantitative assessment of lung density changes after 3-D radiotherapy for breast cancer. *Acta Oncologica* 2007; **46**: 187–93.

128. Gagliardi G, Lax I, Ottolenghi A, Rutqvist LE. Long-term cardiac mortality after radiotherapy of breast cancer—application of the relative seriality model. *British Journal Radiology* 1996; **69**(825): 839–46.

129. Hancock SL, Tucker MA, Hoppe RT. Factors affecting late mortality from heart disease after treatment of Hodgkin's disease. *Journal of the American Medical Association* 1993; **270**(16): 1949–55.

130. Taylor CW, Povall J, McGale P, *et al.* Cardiac dose from tangential breast cancer radiotherapy in the year 2006. *International Journal of Radiation Oncology, Biology, Physics* 2008; **72**(2): 501–7.

131. Rutqvist LE, Liedberg A, Hammar N, Dalberg K. Myocardial infarction among women with early-stage breast cancer treated with conservative surgery and breast irradiation. *International Journal of Radiation Oncology, Biology, Physics* 1998; **40**(2): 359–63.

132. Paszat LF, Mackillop WJ, Groome PA, Boyd C, Schulze K, Holowaty E. Mortality from myocardial infarction after adjuvant radiotherapy for breast cancer in the surveillance, epidemiology, and end-results cancer registries. *Journal of Clinical Oncology* 1998; **16**(8): 2625–31.

133. Seddon B, Cook A, Gothard L, *et al.* Detection of defects in myocardial perfusion imaging in patients with early breast cancer treated with radiotherapy. *Radiotherapy and Oncology* 2002; **64**: 53–63.

134. Canney PA, Deehan C, Glegg M, Dickson J. Reducing cardiac dose in post-operative irradiation of breast cancer patients: the relative importance of patient positioning and CT scan planning. *British Journal of Radiology* 1999; **72**(862): 986–93.

135. Lu HM, Cash E, Chen MH, *et al.* Reduction of cardiac volume in left-breast treatment fields by respiratory maneuvers: a CT study. *International Journal of Radiation Oncology, Biology, Physics* 2000; **47**(4): 895–904.

136. Sixel KE, Aznar MC, Ung YC. Deep inspiration breath hold to reduce irradiated heart volume in breast cancer patients. *International Journal of Radiation Oncology, Biology, Physics* 2001; **49**: 199–204.

137. Pedersen A, Korreman S, Nystrom H, Specht L. Efficacy of respiratory gated radiotherapy of breast cancer. In Overgaard J (ed) *21st Annual ESTRO Meeting*, 17–21 September 2002. Praha, Czech Republic: Elsevier, 2002, p. S143.

138. Landau D, Adams EJ, Webb S, Ross G. Cardiac avoidance in breast radiotherapy: a comparison of simple shielding techniques with intensity-modulated radiotherapy. *Radiotherapy and Oncology* 2001; **60**(3): 247–55.

139. Darby SC, McGale P, Taylor CW, *et al.* Long-term mortality from heart disease and lung cancer after radiotherapy for early breast cancer; prospective cohort study of about 300 000 women in US SEER cancer registries. *Lancet Oncology* 2005; **6**(8): 557–65.

140. Lorigan P, Califano R, Faivre-Finn C, Howell A, Thatcher N. Lung cancer after treatment for breast cancer. *Lancet Oncology* 2010; **11**(12): 1184–92.

141. Mannino M, Yarnold JR. Local relapse rates are falling after breast conserving surgery and systemic therapy for early breast cancer: Can radiotherapy ever be safely withheld? *Radiotherapy and Oncology* 2009; **90**: 14–22.

142. Hansen VN, Evans PM, Budgell GJ, *et al.* Quality assurance of the dose delivered by small radiation segments. *Physics in Medicine and Biology* 1998; **43**(9): 2665–75.

143. de Boer HCJ, Heijmen BJM. A protocol for the reduction of systematic patient setup errors with minimal portal imaging workload. *Radiotherapy and Oncology* 2001; **50**: 1350–65.

144. de Boer HCJ, Heijmen BJM. eNAL: an extension of the NAL setup correction protocol for effective use of weekly follow-up measurements. *International Journal of Radiation Oncology, Biology, Physics* 2007; **67**: 1586–95.

145. Harrison RM, Wilkinson M, Rawlings DJ, Moore M. Doses to critical organs following radiotherapy and concomitant imaging of the larynx and breast. *British Journal of Radiology* 2007; **80**: 989–95.

146. Harrison RM. Doses to organs and tissues from concomitant imaging in radiotherapy: a suggested framework for clinical justification. *British Journal of Radiology* 2008; **81**: 970–4.

147. Hector C, Webb S, Evans PM. A simulation of the effects of set-up error and changes in breast volume on conventional and intensity-modulated treatments in breast radiotherapy. *Physics in Medicine and Biology* 2001; **46**(5): 1451–71.

148. Mourik AM, van Kranen S, den Hollander S, *et al.* Effects of setup errors and shape changes on breast radiotherapy. *International Journal of Radiation Oncology, Biology, Physics* 2011; **79**(5): 1557–64.

149. Yarnold J, Ashton A, Bliss J, *et al.* Fractionation sensitivity and dose response of late adverse effects in the breast after radiotherapy for early breast cancer: long-term results of a randomised trial. *Radiotherapy and Oncology* 2005; **75**(1): 9–17.

150. Owen JR, Ashton A, Bliss JM, *et al.* Effect of radiotherapy fraction size on tumour control in patients with early-stage breast cancer after local tumour excision: long-term results of a randomised trial. *Lancet Oncology* 2006; **7**(6): 467–71.

151. START Trialists' Group, Bentzen SM, Agrawal RK, Aird EG. The UK Standardisation of Breast Radiotherapy (START) Trial A of radiotherapy hypofractionation for treatment of early breast cancer: a randomised trial. *Lancet Oncology* 2008; **9**(4): 331–41.

152. START Trialists' Group, Bentzen SM, Agrawal RK, *et al.* The UK Standardisation of Breast Radiotherapy (START) Trial B of radiotherapy hypofractionation for treatment of early breast cancer: a randomised trial. *Lancet* 2008; **371**(9618): 1098–107.

153. Hopwood P, Haviland JS, Sumo G, *et al.*; START Trial Management Group. Comparison of patient-reported breast, arm, and shoulder symptoms and body image after radiotherapy for early breast cancer: 5-year follow-up in the randomised Standardisation of Breast Radiotherapy (START) trials. *Lancet Oncology* 2010; **11**(3): 231–40.

154. Whelan T, MacKenzie R, Julian J, *et al.* Randomized trial of breast irradiation schedules after lumpectomy for women with lymph node-negative breast cancer. *Journal of the National Cancer Institute* 2002; **94**(15): 1143–50.

155. Whelan TJ, Pignol JP, Levine MN, *et al.* Long-term results of hypofractionated radiation therapy for breast cancer. *New England Journal of Medicine* 2010; **362**(6): 513–20.

156. Bartelink H, Horiot JC, Poortmans PM. Impact of a higher radiation dose on local control and survival in breast-conserving therapy of early breast cancer: 10-year results of the randomized boost versus no boost EORTC 22881–10882 trial. *Journal of Clinical Oncology* 2007; **25**(22): 3259–65.

157. Hölscher T, Bentzen SM, Baumann M. Influence of connective tissue diseases on the expression of radiation side effects: a systematic review. *Radiotherapy and Oncology* 2006; **78**(2): 123–30.

158. Rostom AY, Pradhan DG, White WF. Once weekly irradiation in breast cancer. *International Journal of Radiation Oncology, Biology, Physics* 1987; **13**(4): 551–5.

159. Polgár C, Van Limbergen E, Pötter R, *et al.*; GEC-ESTRO breast cancer working group. Patient selection for accelerated partial-breast irradiation (APBI) after breast-conserving surgery: recommendations of the Groupe Européen de Curiethérapie-European Society for Therapeutic Radiology and Oncology (GEC-ESTRO) breast cancer working group based on clinical evidence (2009). *Radiotherapy and Oncology* 2010; **94**(3): 264–73.

160. Smith BD, Arthur DW, Buchholz TA, *et al.* Accelerated partial breast irradiation consensus statement from the American Society for Radiation Oncology (ASTRO). *Journal of the American College of Surgeons* 2009; **209**(2): 269–77.

161. Kirby AM, Coles CE, Yarnold JR. Target volume definition for external partial breast radiotherapy: clinical, pathological and technical studies informing current practice approaches. *Radiotherapy and Oncology* 2010; **94**(3): 255–63.

Radiotherapy for thoracic tumours

Sara C Erridge, Elizabeth Toy, Sorcha Campbell

5.1 Lung cancer

5.1.1 Introduction

Lung cancer is the most common cause of cancer death in the UK, accounting for one in five of all deaths. Unfortunately, despite treatment advances survival remains poor with < 10% of patients alive at 5 years. More than 40,000 cases are diagnosed each year in the UK and though the incidence in males is declining, it has only stabilized in females. This is because 90% of lung cancers are related to active or passive smoking and the smoking rate in women peaked in the 1960s, 20 years later than for men. Lung cancer is therefore becoming a disease of the elderly with the incidence 560 per 100,000 for men over the age of 80 (200 per 100,000 for men aged 60–69) and 273 per 100,000 for women over 80 (140 per 100,000 for women aged 60–69). Over 87% of patients diagnosed with lung cancer between 2006 and 2008 were over the age of 60 years (http://info.cancerresearchuk.org/cancerstats/types/lung).

The increasing age and comorbidity of patients with lung cancer means that many are medically unfit for surgery. Consequently, radical radiotherapy is playing an increasing role in the curative management of lung cancer[1].

There are two main subtypes of lung cancer; small cell (SCLC) and non-small cell (NSCLC). However, the management is becoming increasingly similar so the general principles of lung cancer radiotherapy will be discussed first, followed by the specific features of management of the two pathological entities.

5.1.2 Assessment of patients with lung cancer for radical radiotherapy

Two key factors are important in the assessment of the suitability of patients for radical radiotherapy; comorbid illnesses and the stage/configuration of the tumour and nodes.

Fitness for treatment

Age per se is not a contraindication for radical radiotherapy. When assessing a patient it is important to look at their general well-being and make an estimate of their life expectancy; a fit 80-year-old may live another 10–15 years whereas a 60-year-old with significant ischaemic heart disease and chronic obstructive pulmonary disease (COPD) may have a short life-expectancy making the treatment of their lung cancer futile.

The key comorbid diseases that can preclude the use of radical doses of radiotherapy are COPD and pulmonary fibrosis. Relative contraindications include systemic lupus erythematosus and similar conditions. Classically FEV_1 (volume exhaled in 1 second), FVC (total volume exhaled), and DLCO (transfer factor) have been used to assess suitability for radical treatment but the lower threshold has not been formally established and studies examining the association of these parameters with toxicity have demonstrated conflicting results[2]. However this being said, most clinical trials recommend for a lower limit of FEV_1 and DLCO of 40% predicted and treating with pulmonary function below this threshold requires careful discussion of the risks with the patient.

When assessing a patient's suitability for radical radiotherapy other factors that should be taken into account include weight loss and decline in performance status, both of which are associated with an inferior prognosis.

Disease stage and configuration

All patients undergoing radical treatment should be staged with a contrast-enhanced CT scan of chest and upper abdomen, and those with NSCLC, a PET-CT scan. For NSCLC CT scans have a sensitivity of 60% and specificity of 80% for nodal involvement, compared with 84% and 89%, respectively for PET. PET also has a higher sensitivity (93%) and specificity (96%) for distant disease[3]. However, it performs poorly for the detection of brain metastases (sensitivity 60%) so a CT or MRI brain should be considered, especially for patients with mediastinal lymphadenopathy[3]. If the PET-CT scan demonstrates positive lymph nodes(s) it is good practice to confirm this pathologically with EBUS/EUS (endobronchial/endoluminal ultrasound) and/or mediastinoscopy. Consideration should be given to biopsy of solitary metastatic deposits identified on PET-CT scan.

The location of the primary and involved nodes is often the principal determinant of the suitability for radical radiotherapy. For example, a right upper lobe lesion with a right paratracheal (R4) node may be easily encompassed within a suitable volume, whereas a left lower lobe tumour with an identical node is not, due to the organs at risk.

5.1.3 Patient positioning for radical radiotherapy for lung cancer

- The patient should be planned and treated in the same position; supine with arms above their head. A variety of immobilization devices are available, but none has proved superior to a bar for the hands to hold and support under the elbows (e.g. T-bar and wing-board). For comfort, a knee roll should be used.

- When patients are to be treated with stereotactic body radiotherapy (SBRT) the duration of each treatment is considerably longer than that for conventional treatment. There is clear evidence that patient comfort and minimizing treatment times (< 30 minutes) are the most important factors in minimizing intrafraction movement. Some advocate specialized devices, for example, a body vacuum mould, whereas others suggest standard systems with time taken to ensure patient comfort are equally reproducible[4]. If on assessment the tumour movements exceed 1 cm then abdominal compression should be considered to reduce diaphragmatic movement.

♦ For superior sulcus tumours, the patient's arms should be placed by their sides, and an immobilization shell covering neck and shoulders used to maintain a consistent shoulder position.

♦ The patient should have a CT scan performed on a flat couch top with 3-mm slice thickness covering the entire lungs (cricoid to L2 vertebra). IV contrast should be used to aid target delineation in node positive disease.

♦ Ideally, the treatment isocentre should be fixed and tattooed at the time of the planning CT scan. This minimizes systematic errors which result from isocentre shifts. However, estimating the isocentre position can be difficult if there is disease progression between the diagnostic and planning scans.

♦ If fixing the isocentre at the time of scanning is not possible, radio-opaque markers should be placed on reference tattoos and an isocentre shift calculated at time of planning.

♦ Lateral tattoos should be placed on stable areas of the lower chest to minimize errors caused by lateral rotation.

5.1.4 Tumour motion

Tumour movement, resulting in a geographic miss, is a major concern in the treatment of lung cancers, particularly for lesions close to the diaphragm, which can move as much as 5 cm in the cranio–caudal direction. Several techniques to try and reduce the effect of this have been developed:

1. Deep inspiration breath hold (DIBH): patient is coached to hold their breath in maximum inspiration during treatment.

2. Active breathing control (ABC): the patient is connected to a breathing apparatus with a control valve that immobilizes breathing motion at a fixed level.

3. Gating: the beam is turned on and off, synchronized with respiratory movement to deliver treatment during only part of the respiratory cycle. Gating is challenging in patients with poor pulmonary function as they have unpredictable breathing patterns.

4. Slow CT scanning: the planning CT is acquired at a much slower rotation than usual (4 seconds per revolution) to produce a GTV, which includes all tumour positions during the respiratory cycle. This produces a blurred GTV.

5. Four-dimensional scanning: a multislice CT scan is acquired with respiratory monitoring and the images sorted according to the respiratory cycle to produce a series of GTVs that demonstrate the tumour position at different times.

Four-dimensional-CT is now considered the gold standard for thoracic radiotherapy, particularly for Stage I tumours, with many commercial 'packages' available. The patient is scanned during 6–10 breathing cycles and then the GTV contoured on: (a) the maximum intensity projection (MIP) of the 4D dataset, (b) the MIP (see Fig. 5.1a), the maximum inspiratory and expiratory datasets, or (c) all 10 phases of the 4D CT scan[5].

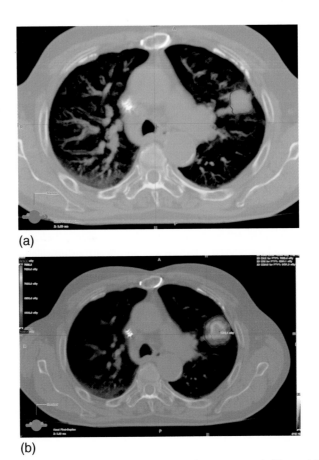

(a)

(b)

Fig. 5.1 (a) Four-dimensional target delineation: ITV drawn on MIP in blue and PTV (ITV + 5 mm) in red. (b) Stereotactic plan using two intensity modulated arcs. (Image courtesy of Corrine Faire-Finn and Gareth Webster.)

5.1.5 Target delineation

The delineation of the GTV, CTV, and PTV varies with the pathological subtypes and location of the tumour and is described in the appropriate later sections.

5.1.6 Organs at risk

Lungs

The lungs should be contoured from apex to bases (usually using automatic contouring sequence) and then combined.

1. V20: this is the volume of normal lung receiving more than 20 Gy. In the original publication by Graham et al. the volume of normal lung within the PTV was removed from the calculation[6], though this is not essential (makes 2–3% difference), provided each centre has a consistent approach and audits their pneumonitis rates. The differing pneumonitis scoring systems and methods used in studies make

drawing conclusions about the optimal threshold challenging, but most recommend that the V20 should be kept $\leq 35\%$[7] to keep the risk of clinically significant pneumonitis to $< 20\%$. Although these thresholds are frequently also applied to the UK fraction schedules (e.g. 55 Gy, 20 fractions), it should be appreciated the data sources for these recommendations used 2 Gy per fraction, and without chemotherapy.

Other dose–volume thresholds have been explored, for example, V13, V10, and V5 and high values of these may be associated with increased pneumonitis and/or late fibrosis, however, a definitive threshold has not been established.

2. Mean lung dose (MLD): this may be a better predictor of the risk of symptomatic pneumonitis than V20[7]. A standard threshold of 20 Gy is recommended[4].

3. When treating with SBRT, the dose recommendations from the ROSEL trial are V20 of $< 10\%$ and the V12.5 $< 15\%$.

Other factors for pulmonary toxicity:

♦ The tumour location appears important with lower lobe lesions having a greater risk of pneumonitis.

♦ Concurrent chemotherapy with platinum and vinca alkaloids or etoposide does not appear to increase the risk of pneumonitis[8].

♦ The presence of idiopathic pulmonary fibrosis is anecdotally linked with increased risk of toxicity but there are no published series from which to calculate the extent of the risk.

Spinal cord

The spinal cord should be defined throughout the thoracic region and for a few centimetres above and below if non-coplanar techniques are employed. A set up margin, similar to that applied to the CTV (3–5 mm), should be added to produce a planning residual volume.

Sometimes the location of the tumour, especially if there is invasion of the vertebral body and into the foramina, can preclude a radical radiotherapy dose. It is estimated that using 2 Gy per fraction a total dose of 50 Gy is associated with a 0.2% risk of myelopathy[9]. However, this dose should be reduced with the use of concurrent chemotherapy (usually by 10% to 46 Gy) or hypofractionated (44 Gy at 2.75 Gy per fraction) or accelerated hyperfractionated schedules where repair maybe incomplete[10].

When treating with SBRT the maximum acceptable point dose to the spinal cord is 18 Gy when giving the primary 54 Gy in three fractions and 25 Gy for the 55 Gy in five-fraction regimens[5].

Oesophagus

The oesophagus should be contoured, including all mucosal and muscular layers, from cricoid to oesophago-gastric junction. For patients receiving radiotherapy alone the risk of clinically significant oesophagitis is low ($< 5\%$) but with the addition of concurrent chemotherapy this is quadrupled[8]. As most cases of oesophagitis settle with supportive measures it is rarely a dose limiting toxicity but establishing the risk is important for obtaining informed consent. Some patients do, however, go on to develop a stricture. There is no single dosimetric parameter on which to base predictions, though a V70 > 20%, V50 > 40%, and a V35 > 50% appear to predict for grade 2+ oesophagitis[11].

When treating with SBRT a maximum acceptable cumulative dose to a volume of 1 cc of the oesophagus is 24 Gy when giving the primary 54 Gy in three fractions and 27 Gy for 55 Gy in five fractions[5].

Heart

The heart should be contoured from the superior aspect of the left atrium and extended to the apex where it touches the diaphragm. Radiation-induced cardiac damage can either be acute (pericarditis) or long-term (ischaemic heart disease, myocardial dysfunction, and valvular disease). The majority of the cardiac tolerance data comes from fit patients with either Hodgkin's lymphoma or breast cancer when radiotherapy is often given in conjunction with anthracyclines. The challenge for patients with lung cancer is they often have pre-existing cardiovascular disease due to smoking and there is little good data on this specific group of patients. The CONVERT trial protocol recommends V66 < 30% and D50 of < 33 Gy.

For SBRT the maximum dose to 1 cc of the heart is 24 Gy and 27 Gy for the three- and five-fraction schedules, respectively[5].

When treating with SBRT additional OARs need to be contoured[5]

Brachial plexus: all major trunks of the ipsilateral plexus should be contoured using the subclavian and axillary vessels as a surrogate for the position of the plexus, extending from the bifurcation of the brachiocephalic trunk to where the neurovascular structures cross the second rib. The maximum dose to a volume of 1 cc is 24 Gy for the three-fraction regimen and 27 Gy for the five-fraction regimen.

Proximal trachea: contours should begin 10 cm superior to the superior extent of the tumour or 5 cm above the carina, whichever is more superior. It should continue to the superior aspect of the proximal bronchial tree. The maximum dose to 1 cc of the proximal trachea is 30 Gy and 32 Gy respectively for the three- and five-fraction regimens.

Proximal bronchial tree: this will include the inferior 2 cm of the trachea, the carina, right and left main bronchi, right and left upper lobe bronchi, right middle lobe bronchus, lingular bronchus, and right and left lower lobe bronchi. The dose constraints for the proximal bronchial tree are identical to those for the trachea.

Others: if non-coplanar beams are used, other organs such as the liver and small bowel may be irradiated and, therefore, will need to be taken into consideration. The body contour should always be outlined and efforts made to ensure beam entry points on the skin do not overlap resulting in unacceptable skin toxicity.

5.1.7 Planning

- ◆ The treatment is usually planned using photons of ≤ 10 MV (usually 6 or 8 MV); beam energies greater than this result in a larger penumbra and, consequently, greater normal tissue irradiation.

- As lungs have a density of around 30% of that of normal soft tissue, the treatment should be planned with a tissue inhomogeneity correction. A planning algorithm which takes into account lateral electron transport should be used (so-called type B models)[4].

- To minimize normal tissue irradiation, conformal shaping should be used on all fields.

- When delivering conventionally fractionated radiotherapy a simple three-field technique with an anterior and two ipsilateral oblique fields will often suffice, but a more complex field arrangement may be required for large, central lesions, where achieving a dose distribution in the ICRU range of 95–107% can be difficult.

- In these cases, additional coplanar beams, or non-coplanar beams exiting into mediastinal structures, may be required. An alternative for difficult cases is to use a two-phase technique starting with parallel-opposed anterior and posterior fields and then a second phase with a three-field technique. However, this technique increases the dose delivered to the oesophagus and spinal cord, which may increase toxicity, especially if delivered with concurrent chemotherapy.

- IMRT has been shown, in a number of planning studies and single institution series, to reduce the doses to OARs (particularly lung and spinal cord) enabling large target volumes (e.g. Stage III with contralateral nodes) to be treated within OAR dose constraints and smaller tumours to have the treatment dose escalated (see Fig. 5.2). However, there are potential concerns about the 'low-dose bath' delivered to the lungs (e.g. V5, V10, and V13) and the impact on late pulmonary toxicity, but the recommended thresholds have not formally been established. Additionally, there is concern about how to take into account tumour and medias-tinal movement during IMRT. Some centres have advocated 4D planning to ensure coverage but this remains experimental.

When planning SBRT in order to achieve adequate target coverage with high conformity and a rapid isotropic dose fall-off beyond the PTV, at least seven beams are required. Alternatively, intensity-modulated arc based therapy can be used (see Fig. 5.1b)[12]. These may be coplanar or non-coplanar. For SBRT it is recommended that 95% of the PTV should receive the prescribed dose, with 99% of PTV receiving at least 90% of this dose. Due to the nature of the planning there is inevitably less homo-geneity in the dose distribution. Therefore, it is recommended that the dose maxi-mum within the PTV should be a minimum of 110% and a maximum of 140% of the prescribed dose[5].

5.1.8 Implementation on the treatment machine

If the treatment has not been planned around a fixed isocentre an isocentre shift must be performed, either at time of a plan check or on the machine prior to treatment. This shift should be verified by comparison of the portal images with digitally reconstructed radiographs (DRRs) derived from the planning CT scan.

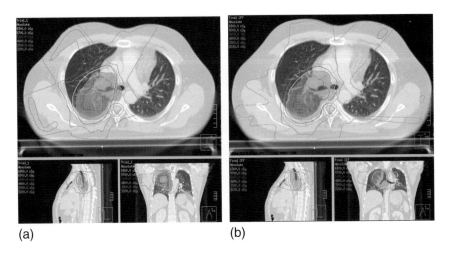

(a)   (b)

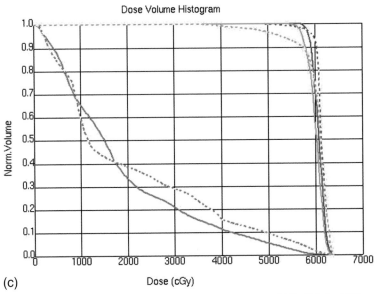

(c)

Fig. 5.2 (a) T4N3M0 NSCLC conventional 3D CRT V20 = 38.5%. (b) IMRT plan V20 = 33.3%. (c) Comparison DVH: improved PTV coverage with IMRT. Key: solid line = IMRT, dotted line = 3DCRT, pale blue = PTV, green = lungs. (Image courtesy of Corrine Faire-Finn and Gareth Webster.)

5.1.9 **Treatment verification**

- ◆ The treatment position should be verified against the original planning CT scan—either comparing the CT set directly with a cone-beam CT (CBCT) or DRRs with orthogonal images.

- For conventionally fractionated schedules imaging should be acquired on a daily basis for at least the first 3 days to ensure no systematic errors have occurred and then at least weekly to ensure no changes in the patient position have occurred.

- Traditionally imaging comparison has looked at the position of the carina, vertebral bodies, sternum, and chest wall but with the advent of CBCT it is feasible to compare the position of the target lesion.

- If daily imaging is performed the cumulative radiation dose should be considered.

- For SBRT daily pretreatment imaging and patient position correction should be performed. CBCT is the most commonly used system, though KV tracking systems (e.g. EXAC TRAC) are also used when available. The images are matched initially to bony anatomy to identify major inaccuracies and to assess rotation. A soft tissue match is then performed in which the average tumour image on the CBCT is manually registered to ensure the tumour localizes well within the internal target volume (ITV)/PTV defined on the planning CT. If the match is outside a 3 mm tolerance a shift is applied. Further imaging during and at the end of treatment may also be performed to ensure patient stability and provide further information on the adequacy of treatment margins.

5.1.10 Radical treatment of non-small cell lung cancer

Stage I and II NSCLC (see Tables 5.1 and 5.2)

Radical radiotherapy is indicated when the patient is medically inoperable or declines surgery. There is currently no evidence for the use of chemoradiation in patients with

Table 5.1 Staging of lung cancer (TNM 7)

T1	≤ 3 cm surrounded by lung or visceral pleura, no evidence of invasion of proximal bronchus
T1a	≤ 2 cm
T1b	>2 cm ≤ 3 cm
T2	> 3 cm but ≤ 7 cm, or invading visceral pleura or causing atelectasis of the lobe
T2a	>3 cm ≤ 5 cm
T2b	>5 cm ≤ 7 cm
T3	Tumour > 7 cm or invading chest wall, diaphragm, mediastinal pleura or pericardium, or lesion < 2 cm from carina or causing atelectasis of whole lung or separate nodules in the same lobe
T4	Invading mediastinum or involving heart, great vessels, trachea, oesophagus, vertebral body, or satellite nodules within same lung
N1	Metastases in peribronchial or hilar nodes
N2	Metastases in subcarinal or ipsilateral mediastinal nodes
N3	Metastases in contralateral mediastinal or hilar nodes, any supraclavicular nodes
M1a	Malignant pleural or pericardial effusion or separate nodules in the contralateral lung or tumour with pleural nodules
M1b	Distant metastases

Source: Data from Edge SB, Byrd DR, Compton CC, eds. *AJCC Cancer Staging Manual. 7th ed.* New York, NY.: Springer, 2010.

Table 5.2 Summary of stages of lung cancer

	N0	N1	N2	N3
T1 T1a	IA	IIA	IIIA	IIIB
T1b	IA	IIA	IIIA	IIIB
T2 T2a	IB	IIA	IIIA	IIIB
T2b	IIA	IIB	IIIA	IIIB
T3	IIB	IIIA	IIIA	IIIB
T4	IIIB	IIIB	IIIB	IIIB
M1a/b	IV	IV	IV	IV

Source: Data from Edge SB, Byrd DR, Compton CC, eds. *AJCC Cancer Staging Manual. 7th ed.* New York, NY.: Springer, 2010.

Stage I–II NSCLC. Patients can be treated using conventional techniques or SBRT. Patients with T1N0M0 or T2/3N0M0 disease where the tumour is < 5 cm may be candidates for SBRT. The lesion must be outside a 2-cm radius from the main airways and proximal bronchial tree. This is to minimize the risk of stenosis or perforation of the main airways.

SBRT aims to improve clinical outcomes in early stage lung cancer by delivering higher biologically effective doses (BED > 100 Gy) than are possible with conventional radiotherapy. It does this by using multiple beams to deliver high doses to a carefully defined target with steep dose gradients beyond the target. There is now a considerable body of evidence (but no randomized clinical trials) supporting SBRT as superior to conventional radiotherapy in terms of local control and survival in patients with inoperable disease[13]. A current international phase III trial (ROSEL) is comparing surgery and SBRT in those patients who have operable disease. Therefore, whilst the results of the ROSEL and other studies are awaited, SBRT should only be offered to patients who not medically operable or who refuse surgery.

Stage III NSCLC

Radical radiotherapy, often combined with concurrent or sequential chemotherapy, is the principal potentially curative treatment for Stage III NSCLC. A recent meta-analysis has confirmed an improved outcome (absolute survival benefit 4.5% at 5 years) with concurrent chemoradiation, but at the cost of increased toxicity, particularly oesophageal[8].

The routine use of preoperative chemoradiation for Stage III NSCLC is controversial; though advocated by some, it has not been shown to improve overall survival when compared with chemoradiation alone. Analysis of the data from the International Association for the Study of Lung Cancer (IASLC) database has suggested that patients with single-station N2 disease have similar survival outcomes to patients with multiple N1 nodal involvement. This has resulted in a growing argument that there may be a subgroup of patients with single level N2 disease, resectable with lobectomy, that might benefit from combined modality treatment with surgery plus chemotherapy ± radiotherapy. There is, however, no randomized data to support this at present.

There are however, some phase II data, which demonstrated good local control results from the use of preoperative chemoradiation in superior sulcus lesions.

Stage IV NSCLC due to a solitary brain metastasis

Some patients with a solitary brain metastasis (confirmed on MRI as solitary) can be cured if the metastases is either resected or treated with radiosurgery. The treatment of the thoracic disease will depend on the configuration of the chest disease. The 5-year overall survival rates quoted in a number of small single-centre series range from 8–20%.

Treatment volume and definition

- The treatment volume consists of the primary lesion and all involved nodes; by limiting the treatment volume to known disease, a higher radiation dose can be delivered with minimal toxicity[4].

- Historically elective nodal irradiation (ENI) was recommended, but the dose that can be delivered safely is limited to around 40 Gy and several studies using involved field radiotherapy have demonstrated that few patients ($< 5\%$) have an isolated mediastinal relapse.

- The GTV consists of the primary lesion and all nodes > 10 mm in the short axis or with central necrosis. All PET avid areas should be included.

- The primary lesion should be contoured on lung windows (W1600, L-600) and the nodes on mediastinal windows (W400, L20)[4].

- The CTV includes a margin of 6 mm added to the GTV for squamous cell lesions and 8 mm for adenocarcinomas.

- The definition of the ITV will depend on the scanning technique used. If conventional scanning is used then a margin needs to be added to account for tumour movement. This is usually of the order 1–2 cm, more in the cranio–caudal direction. If 4D scanning is used then the ITV will be generated using the MIP, end inspiratory and expiratory positions, or composite of the multiple positions throughout the breathing cycle[5].

- To produce the PTV, a margin should be added to the ITV to account for set-up variability. This margin will vary between departments, but is usually around 5 mm, but can be reduced to 3 mm if daily pretreatment imaging is performed (cone-beam CT or orthogonal images).

- For SBRT, the GTV is defined on the 4D CT scan using the MIP dataset, the maximum inspiratory and expiratory datasets, or all 10 phases of the 4D CT scan. By convention, there is no CTV expansion (i.e. microscopic spread is not taken into account). This approach is supported by the high levels of local control in the published series. The ITV is, therefore, derived from the union of the GTV delineations. To form the PTV the ITV is uniformly expanded by 3 mm.

Dose prescription

- There is evidence from several studies of a dose response in NSCLC, but the optimal schedule has yet to be established. Doses in excess of 90 Gy have been safely used in single centre phase I studies and phase III randomized controlled trials are currently being conducted to assess their efficacy and safety.

- Though total dose is important in the radical treatment of NSCLC, so is overall treatment time due to accelerated repopulation.
- Currently recommended conventionally fractionated schedules for Stage I and II patients not treated with SBRT, or Stage III patients unfit for concurrent chemoradiation include:
  - 60–66 Gy to isocentre in 30–33 daily fractions over 6–7 weeks.
  - 55 Gy to isocentre in 20 daily fractions over 4 weeks.
  - 54 Gy to isocentre in 36 thrice-daily fractions (minimum 6-hour gap) over 12 days (CHART).
- Patients with Stage I lesions ≤ 5 cm in diameter and > 2 cm from main carina who are medically inoperable, may benefit from SBRT. The optimal schedule has not been established with some authors suggesting that the doses used in the early studies maybe unnecessarily high. Also, the dose prescription depends on the type of planning algorithm used. For example the ROSEL trial protocol recommends[5]:
- Type A planning system e.g. pencil beam:
  - (i) 60 Gy in three fractions on alternate days over 5–8 days (minimum interval 40 hours) with 95% of the PTV receiving this dose and 99% receiving > 18 Gy per fraction.
  - (ii) Lesions which are in contact with the thoracic wall, mediastinum or heart: 60 Gy in five fractions on alternate days over 10–14 days (minimum interval 40 hours) with 95% of the PTV receiving this dose and 99% receiving > 18 Gy per fraction.
- Type B planning system e.g. collapsed cone:
  - (i) 54 Gy in three fractions on alternate days over 5–8 days (minimum interval 40 hours) with 95% of the PTV receiving this dose and 99% receiving > 48.6 Gy.
  - (ii) Lesions which are in contact with the thoracic wall, mediastinum or heart: 55 Gy in five fractions on alternate days over 10–14 days (minimum interval 40 hours) with 95% of the PTV receiving this dose and 99% receiving > 49.5 Gy.
- For fit patients (WHO PFS 0–1) with Stage III disease concurrent chemoradiation is considered the standard of care:
  - 60–66 Gy to isocentre in 30–33 daily fractions over 6–7 weeks with four cycles of cisplatin-based chemotherapy (usually with vinorelbine or etoposide), starting with the first or second cycle (depending primarily on logistics).
  - Some patients not quite fit enough for concurrent treatment or large volumes may benefit from two to four cycles of neo-adjuvant platinum based chemotherapy followed by standard radiotherapy (as described previously).

Postoperative radiotherapy

A number of meta-analyses of the trials of postoperative radiotherapy (PORT) have been published demonstrating that the routine use of PORT may have a detrimental effect on survival, particularly those with early stage (N0 or N1) disease[14]. Whether in the modern 3D conformal radiotherapy era, PORT would be equally harmful is unknown. Those patients with mediastinal nodes (N2 or N3) had improved local control but no change in survival. Therefore, current recommendations are to consider PORT for patients with either an incomplete resection and/or mediastinal nodal involvement when there has been incomplete surgical staging. The current LUNGART trial is re-examining the role of PORT in patients with N2 disease following R0 resection and complete staging of the mediastinum.

Pulmonary function tests should be repeated postoperatively to ensure the patient has sufficient reserve to tolerate PORT.

Treatment volume and definitions

♦ The treatment volume should be defined after discussion with the operating surgeon to accurately establish the area at increased risk of recurrence. The option of a second operation to complete the resection should also be discussed.

♦ Ideally areas of uncertain surgical clearance should be marked with surgical clips. If a nodal group has been found to be involved, this region should be encompassed within the target volume.

♦ There should not be, by definition, a GTV unless there is gross residual disease. The CTV therefore consists of the bronchial stump, the region of the involved nodes, and the ipsilateral hilar, subcarinal (level 7), and paratracheal (level 4) nodes.

♦ A margin of around 5 mm should be added to account for mediastinal movement and then a further margin applied for set-up inaccuracy according to departmental guidelines, typically 5 mm.

Dose prescription

♦ There is no proven benefit from the use of concurrent chemotherapy with PORT.

♦ The role of adjuvant chemotherapy, and how it should be scheduled with PORT, remains controversial but usually chemotherapy precedes the radiotherapy.

♦ The radiotherapy dose depends on the nature of the residual disease; macroscopic disease should be treated with full dose radical radiotherapy, but the dose may be reduced for microscopic only disease.

♦ Gross residual disease:
  • 64–70 Gy to isocentre in 32–35 daily fractions over 6–7 weeks.
  • 55 Gy to isocentre in 20 fractions over 4 weeks.

♦ Microscopic residual disease:
  • 50–54 Gy to isocentre in 25–30 daily fractions over 5 weeks using 6–8-MV photons.
  • 50 Gy to isocentre in 20 daily fractions over 4 weeks using 6–8-MV photons.

5.1.11 **Radical treatment for small cell lung cancer**

Small cell tumours account for only around 15% of all lung cancers and the proportion is dropping. Though the current recommendation is to use the standard TNM staging, the terminology 'limited stage' (can be encompassed with a radical radiotherapy volume) and 'extensive stage' is still widely used as it describes the potential for cure. Only about a third of patients present with limited stage disease.

Stages I–III (limited stage) SCLC

High-dose thoracic radiotherapy (either delivered concurrently or sequentially with chemotherapy) has been demonstrated to improve the 3-year survival of limited stage SCLC by 5%.

- Early concurrent radiotherapy delivered with the first or second cycle of cisplatin-based chemotherapy (cycle 3 at the latest[15]) appears to have a small survival advantage when compared to radiotherapy delivered later, provided that both the chemotherapy and radiotherapy are delivered with the optimal dose intensity. Modifications of either negate the benefit.

- If early chemoradiation, which has enhanced toxicity, is to be used then it is imperative that the patient should be fully staged to ensure the disease is indeed truly 'limited'. Patients should have CT scan of chest and upper abdomen and a CT (or MRI) brain. The usefulness of PET and bone scans remains unproven.

Treatment volume and definition

- The original studies of thoracic radiotherapy in LSCLC treated the primary, ipsilateral hilum, bilateral mediastinal and often supraclavicular fossa nodes. This was achieved by using parallel-opposed anterior-posterior fields with the dose limited to 40 Gy in 15 fractions[16].

- With the introduction of 3D conformal radiation therapy (3DCRT) there has been a trend to limit the treatment volume to just the primary and involved nodes as this enables dose escalation. Which technique is superior is unknown, as they have never been compared in a clinical trial.

- If the patient has received chemotherapy, a small-randomized trial suggested that the post-chemotherapy volume should be used [17].

- The GTV should be defined using mediastinal and lung windows and consists of the primary tumour and all enlarged (>1 cm short axis) lymph nodes. If prior chemotherapy has been delivered evidence from a small randomized clinical trial suggests that the GTV may be limited to just the residual volume[17].

- The CTV consists of a margin for microscopic spread (usually around 5 mm) around the GTV and the mediastinal nodal groups deemed to be at risk of containing microscopic disease.

- The ITV should be individualized for the tumour movement and include 5 mm for mediastinal movement.

- To produce the PTV a margin for setup error should be added according to departmental guidelines, typically 5 mm.

Dose prescription

- The optimal dose has not been established though there are data to support both a dose response and the importance of overall treatment time.
- The best published survival rates were achieved with hyperfractioned radiotherapy though as with CHART, for logistical reasons, this has not been widely adopted[18].
- An on-going international trial (CONVERT) is comparing 66 Gy in 33 fractions with 45 Gy in 30 twice-daily fractions.

Current recommended dose schedules include:

- 45 Gy to isocentre in 30 twice-daily fractions (minimum 6-hour gap) over 15 days[18].
- 45–50 Gy to isocentre in 20–25 daily fractions over 4–5 weeks.
- 60–66 Gy to isocentre in 30–33 daily fractions of 6–6½ weeks.
- 40 Gy to mid-plane in 15 daily fractions over 3 weeks using anterior–posterior fields with a spinal cord block to limit the cord dose to 35Gy[16] (generally only used for large tumours with bilateral nodes that cannot be treated with 3DCRT).

Prophylactic cranial irradiation (PCI)

- PCI has also shown a 5% improvement in 3-year survival and is recommended for all patients with LSCLC who have achieved a complete response.
- In order to reduce the risk of neurotoxicity, PCI should be delivered after the completion of all chemotherapy.
- There is little data on the use of PCI in patients over 70 years of age. The initial trials excluded older patients (due to concerns regarding potential increased neurocognitive toxicity) and the most recent dose finding study [19] had a median age of only 60 years.
- Most centres use thermoplastic shells for PCI to improve localization and avoid skin marks, which can be easily washed off.
- The treatment is planned using two lateral portals to encompass the whole of the brain, paying particular attention to the temporal lobe reflection.
- MLC may be used to reduce skin dose at the skull vertex and consequently, long-term alopecia.
- The international EULINT trial compared 25 Gy in 10 fractions with 36 Gy in either 18 daily or 24 twice-daily fractions. The intracerebral recurrence rate was similar with 26% of patients developing brain metastases by 2 years and the overall survival was marginally worse in the higher dose arm (42% 2 years in 25 Gy vs. 37% in 36 Gy p = 0.05)[19]. By 3 years both groups were noted to have mild memory and intellectual impairment (5% Grade 2 or more) so patients need to be informed of these at time of consent[20].

The most widely used fractionation schedule is therefore:

- 25 Gy mid-plane dose in 10 fractions over 2 weeks.

Postoperative radiotherapy

- Occasionally patients who have had a lobectomy or pneumonectomy are found, unexpectedly, to have SCLC.
- The outcome may be better as they have more localized disease, but the standard recommendation is that they should receive postoperative treatment with chemotherapy, mediastinal radiotherapy (unless fully explored negative mediastinal nodes), and PCI.

5.1.12 **Palliative thoracic treatment for lung cancer**

Palliative radiotherapy should be considered for all patients who have incurable lung cancer and local symptoms. Radiotherapy is good at palliating haemoptysis (80% of patients benefit), chest pain (60%), and cough (40%), but is less good at helping dyspnoea (30%) and fatigue (20%).

For poor performance status patients with NSCLC who are asymptomatic, palliative radiotherapy can be deferred until symptoms develop. Data from a randomized controlled trial demonstrated that 60% of such patients never required radiotherapy[21]. However, patients with locally advanced NSCLC and a good performance status (WHO 0 or 1) may have a survival benefit from a higher dose of radiotherapy. In an MRC study, patients receiving 39 Gy in 13 fractions had a median survival of 2 months longer than those receiving 17 Gy in two fractions[22]. A similar result was also observed in a Canadian study of 20 Gy in five fractions versus 10 Gy in one fraction[23]. Consequently, high-dose palliative radiotherapy is frequently offered to fitter patients as part of the initial management, regardless of symptoms, either before or following chemotherapy, or for those unsuitable for, or who decline chemotherapy.

Chemotherapy is the mainstay of treatment for both limited (Stages I–III) and extensive (Stage IV) SCLC therefore thoracic palliative radiotherapy is used for patients who:

1. are not fit enough/decline to receive chemotherapy;
2. have progressed during chemotherapy;
3. have a partial response to chemotherapy with persistent local symptoms;
4. relapse shortly after chemotherapy.

The use of consolidation thoracic radiotherapy in patients with Stage IV SCLC remains controversial. One study suggests that patients with low-volume metastatic SCLC, who have had a complete response to their extrathoracic disease, might gain a survival benefit from consolidation thoracic irradiation[24]. A phase III randomized trial (REST) is currently investigating this further (chemotherapy followed by 30 Gy in 10 fractions to residual thoracic disease or observation).

Patient positioning for palliative radiotherapy

Today most patients having palliative thoracic radiotherapy are planned using virtual simulation. This makes assessment of the current disease configuration and design of the treatment portals much easier. However, if required, the fields can be defined using a conventional simulator.

On the CT scanner the patient should be scanned in a comfortable position, usually on a mattress, with their arms by their sides. A scan with 3–5-mm slice thickness should be obtained whilst the patient breaths gently. The centre of the treatment field should be tattooed before the patient moves (if necessary an approximate isocentre can be established swiftly and then the treatment finalized using asymmetric jaws).

Volume/field localization

If using virtual simulation software there are two approaches that can be used depending on clinician's preference; some clinicians define the GTV including the primary disease, enlarged nodes, and any that might be involved, then use this target to design the treatment portals whereas others directly define the portals without initial contouring. Either approach is reasonable and depends on one's experience.

- A margin should be placed around the gross tumour to account for movement and set-up inaccuracies. It should be remembered that often these patients are frail so set-up maybe less accurate. A margin of 1.5–2 cm is usually added to the field edge (50% isodose).

- Most patients can be treated using parallel anterior–posterior beams though some (especially those receiving higher doses) may benefit from the use of a longitudinal wedge to compensate for the slope of the sternum. If the patient has very poor lung function then a formal 3D volume and calculation may be considered to minimize normal lung irradiation.

- Shielding should be used to reduce the risk of lung toxicity.

- When using higher dose palliation with a spinal cord BED > 100 (39 Gy in 13 fractions or 17 Gy in two fractions) a spinal cord block should be considered[25].

Treatment verification

It is considered good practice to verify all treatments on the machine prior to the first fraction. This ensures no systematic error has occurred.

Dose prescriptions

NSCLC

1. High-dose palliation: 39 Gy or 36 Gy mid-plane dose in 13 or 12 fractions over 2½ weeks using ≤ 10-MV photons (block spinal cord on posterior field for first three or four fractions.

2. Poor performance status patients: 10 Gy mid-plane dose in a single fraction provided area of field < 150 cm^2 using ≤ 10-MV photons.

3. If fields are >150 cm^2 or the patient has stridor or superior vena caval obstruction then: 20 Gy mid-plane dose in five fractions over 1 week using ≤ 10-MV photons.

4. An alternative dose is: 17 or 16 Gy in two fractions 1 week apart using ≤ 10-MV photons (block spinal cord for first fraction if using 17 Gy). In some cases with a large variation in separation a superior-inferior wedge may be beneficial.

SCLC

A variety of schedules are used:

1. For patients with localized SCLC unfit for chemotherapy then: 40 Gy mid-plane in 15 fractions over 3 weeks using ≤ 10-MV photons.

2. For fitter patients (PS 0–1) with Stage IV SCLC and a good response to chemotherapy: 30 Gy mid-plane dose in 10 fractions over 2 weeks using ≤ 10-MV photons.

3. For less fit patients more hypofractionated schedules can be used such as: 20 Gy mid-plane dose in five fractions over 1 week using ≤ 10-MV photons. Or 16 Gy mid-plane dose in two fractions 1 week apart using ≤ 10-MV photons.

4. Frail (PFS 3) patients with respiratory symptoms may gain some benefit from 10 Gy mid-plane dose in a single fraction provided area of field < 150 cm².

PCI in Stage IV (extensive) SCLC

A recent trial has demonstrated an improvement in overall survival (27% at 1 year vs. 13% in control) for patients (up to age 75 and WHO performance status 0–2) with extensive SCLC who have responded to chemotherapy with a halving of the incidence of brain metastases (15% vs. 40%)[26]. Because of the shorter duration of survival, a simpler technique is usually employed using two standard whole brain portals without MLC blocking or immobilization.

Dose prescription

Depending on the patient's performance status and extent of disease:

- 20 Gy mid-plane dose in five fractions over 1 week.
- 25 Gy mid-plane dose in 10 fractions over 2 weeks

5.2 **Mesothelioma**

Mesothelioma is one of the most common occupation-related malignancies with the majority of patients having a history of asbestos exposure. Patients usually present with locally advanced disease causing breathlessness and/or pain. Though a staging system exists (IMIG) it is infrequently used as it rarely affects management decisions.

Pleural malignant mesothelioma is a radiosensitive tumour but the diffuse spread throughout the pleural cavity precludes radical radiotherapy with the lung *in situ*. Consequently, a number of studies have been conducted to look at the role of radiotherapy combined with chemotherapy following extrapleural pneumonectomy (EPP)—so called 'trimodality therapy'. The recently reported MARS study showed a significant detriment in survival compared to standard therapy of best supportive care ± chemotherapy, primarily due to high number of postoperative deaths[27]. This approach continues to be advocated by individual centres but should be considered investigational. The radiotherapy modalities used include IMRT and proton therapy, however, the optimal technique has yet to be defined. Studies have stressed the importance of minimizing the dose to the remaining lung[28].

5.2.1 Current indications for radiotherapy

Port site radiotherapy

With the decline in the use of EPP the management of pleural disease has changed; the diagnosis of malignant mesothelioma is confirmed with a video-assisted thoraco-scopic surgery (VATS) biopsy followed by either talc pleurodesis or insertion of an indwelling pleural catheter. Likewise pleural decortication via a limited thoracotomy scar is increasingly common. The reported incidence of procedure tract metastases (PTM) varies widely from 4–40%. Many of these series predate the routine use of chemotherapy for fit patients. Therefore the exact role of adjuvant radiotherapy to operative port sites in the current era is controversial. Some small historical rand-omized trials suggested a reduction of symptomatic local recurrence[29–31] but pooled analysis failed to confirm this benefit[32].

Palliative radiotherapy

Patients with pleural mesothelioma can benefit from localized palliative radiotherapy to aid pain control.

5.2.2 Patient positioning

The treatment field is clinically marked either with the patient on the treatment table or using virtual simulation. The advantage of the latter is that the target depth can be more accurately defined. The patient should be positioned comfortably and reproducibly with the affected area in a position so it is easily accessible by the machine.

5.2.3 Target delineation

Usually a 2–3-cm margin is added to the surgical bed or nodule. Where a pleural catheter has been placed this should include the entire tract.

5.2.4 Planning

Treatment can either be delivered using 9–18-MeV electrons, 250–300-kV or 4–6-MV photons. The electron energy should be selected to treat the entire chest wall to 90%. Appropriate depth bolus should be used to ensure the skin dose is adequate. When treating with photons, either a single applied field or a tangential parallel opposed pair may be used. The dose to the underlying lung should be considered.

5.2.5 Dose prescription

For port site radiotherapy: 21 Gy in three fractions over 3 working days.

Palliative radiotherapy: 10 Gy in single fraction or 20 Gy in five fractions over 1 week.

5.3 Thymic tumours

Thymomas and thymic carcinomas are rare lesions originating in the anterior medi-astinum. Surgery, often requiring extended thymectomy, is the mainstay of treatment

Table 5.3 Staging system for thymoma (Masaoka–Koga)

Stage 1	Grossly and microscopically completely encapsulated tumour
Stage IIa	Microscopic transcapsular invasion
Stage IIb	Macroscopic invasion into thymic or surrounding fatty tissue, or grossly adherent to but not breaking through the mediastinal pleura or pericardium
Stage III	Macroscopic invasion of neighbouring organ, e.g. great vessel, pericardium, or lung
Stage IVa	Pleural and pericardial metastases
Stage IVb	Lymphogenous or haematogenous metastases

Reprinted from *Pathology International*, Volume **44**, Kenji Koga *et al.* A review of 79 thymomas: Modification of staging system and reappraisal of conventional division into invasive and non-invasive thymoma, pp. 359–67. Copyright (1994), with permission from John Wiley & Sons, Inc.

for the majority of lesions and progression-free and overall survival is strongly associated with the completeness of resection.

The most commonly used staging system is that described by Masaoke with the Koga modification, refined by ITMIG (see Table 5.3)[33].

Careful preoperative staging with CT and MRI is required to confirm that resection is likely to be successful and result in a macroscopic clearance of the disease.

5.3.1 Current indications for radiotherapy

- If resection is likely to leave gross residual disease then preoperative treatment with chemotherapy alone or in combination with radiotherapy should be considered, as shrinkage may allow a complete resection.

- Patients who are medically unfit for surgery or who are surgically inoperable due to great vessel involvement may be considered for definitive radiotherapy or chemoradiation.

- Postoperative radiotherapy is recommended for patients with an R1 or R2 resection or who have Stage III or IVa disease. Occasionally postoperative radiotherapy may be offered for Stage II disease who have histological features predicting an aggressive clinical course although the evidence for this is limited.

5.3.2 Radical treatment

Patient positioning

The treatment position depends on the site of the lesion. If the lesion is located superiorly, close to the neck, the patient may be treated with their arms by their sides, sometimes in a mask. Usually, however, they will be immobilized on a lung board, with their arms above their heads and neck slightly extended as for NSCLC. Whilst breathing gently, the patient should, have a contrast-enhanced CT scan performed with 3-mm slice spacing covering the entire lungs.

Treatment volume and definitions

Primary radical/preoperative

◆ The GTV consists of the primary lesion and any enlarged nodes identified on the staging CT and MRI scans. Mediastinal windows (W400, L20) should be used.

◆ The CTV includes a margin of 5–10 mm added for microscopic spread.

◆ The definition of the ITV will depend on the scanning technique used. If conventional scanning is used then a 10–15-mm margin needs to be added. It has been reported that this margin may be reduced to 5–10 mm if 4D scanning is used.

◆ To produce the PTV, a margin should be added to the ITV to account for set-up variability. This margin will vary between departments, but is usually around 5 mm, but can be reduced to 3 mm if daily pretreatment imaging is performed (CBCT or orthogonal images).

Postoperative

◆ GTV is the surgical bed of the tumour defined using the staging CT and MRI scans. Any areas of macroscopic residual disease should be marked by clips at time of surgery and should be boosted.

◆ CTV accounts of microscopic spread and areas that might have been affected by the surgery.

◆ ITV and PTV are the same as for primary radiotherapy.

Dose prescription

The treatment should be planned in the same manner as that described for the radical treatment of NSCLC. A variety of field arrangements have been advocated. Traditionally, anterior field and two anterior oblique or posterior oblique fields were used although increasingly IMRT is utilized to minimize the dose to normal lung[34].

The OAR dose constraints are the same as those used in the treatment of lung cancer.

◆ When radiotherapy is the sole modality of treatment and surgery is not planned the current recommendation is 54 Gy to isocentre in 28–30 fractions using 6–10-MV photons with areas of bulky disease escalated to 60–66 Gy either with a synchronous or sequential boost.

◆ When the treatment is being delivered preoperatively with a view to downstaging there is an increasing trend to use combined chemoradiation. No single regimen can be recommended, but with increasing experience in the preoperative setting in NSCLC, one regimen being increasingly adopted is 45 Gy in 25 fractions over 5 weeks in combination with cisplatin and etoposide. A dose below 40 Gy or above 64 Gy would be deemed inappropriate in this setting.

◆ When delivered postoperatively usually a dose of 45–54 Gy in 25–30 fractions using 6–10-MV photons is recommended. Areas of gross residual disease should be boosted to 64–66 Gy.

Treatment verification

This is identical to the method used for radical treatment of NSCLC.

5.3.3 Palliative radiotherapy

Occasionally, a patient may present with advanced disease who is not fit for either chemotherapy or radical radiotherapy yet has local symptoms or who develops symptomatic recurrences. In these cases the treatment is similar to that described for the palliative treatment of NSCLC.

Acknowledgements

The authors would like to thank their colleagues for their comments in preparation of this chapter and in particular Corrine Faire-Finn and Gareth Webster from the Christie Hospital Manchester for supplying the IMRT images.

References

1. Erridge S, Murray B, Price A, *et al.* Improved treatment and survival for lung cancer patients in South-East Scotland. *Journal of Thoracic Oncology* 2008; **3**(5): 491–8.

2. Brunelli A, Charloux A, Bolliger CT, *et al.* ERS/ESTS clinical guidelines on fitness for radical therapy in lung cancer patients (surgery and chemo-radiotherapy). *European Respiratory Journal* 2009; **34**(1): 17–41.

3. Lim E, Baldwin D, Beckles M, *et al.* Guidelines on the radical management of patients with lung cancer. *Thorax* 2010; **65**(Suppl 3): iii1–27.

4. De Ruysscher D, Faivre-Finn C, Nestle U, *et al.* European Organisation for Research and Treatment of Cancer recommendations for planning and delivery of high-dose, high-precision radiotherapy for lung cancer. *Journal of Clinical Oncology* 2010; **28**(36): 5301–10.

5. Hurkmans CW, Cuijpers JP, Lagerwaard FJ, *et al.* Recommendations for implementing stereotactic radiotherapy in peripheral stage IA non-small cell lung cancer: report from the Quality Assurance Working Party of the randomised phase III ROSEL study. *Radiation Oncology* 2009; **4**: 1.

6. Graham MV, Purdy JA, Emami B, *et al.* Clinical dose-volume histogram analysis for pneumonitis after 3D treatment for non-small cell lung cancer (NSCLC). *International Journal of Radiation Oncology, Biology, Physics* 1999; **45**(2): 323–9.

7. Marks LB, Bentzen SM, Deasy JO, *et al.* Radiation dose-volume effects in the lung. *International Journal of Radiation Oncology, Biology, Physics* 2010; **76**(3 Suppl): S70–6.

8. Aupérin A, Le Péchoux C, Rolland E, *et al.* Meta-analysis of concomitant versus sequential radiochemotherapy in locally advanced non-small-cell lung cancer. *Journal of Clinical Oncology* 2010; **28**(13): 2181–90.

9. Kirkpatrick JP, van der Kogel AJ, Schultheiss TE. Radiation dose-volume effects in the spinal cord. *International Journal of Radiation Oncology, Biology, Physics* 2010; **76**(3 Suppl): S42–9.

10. Dische S, Saunders MI. Continuous, hyperfractionated, accelerated radiotherapy (CHART): an interim report upon late morbidity. *Radiotherapy and Oncology* 1989; **16**(1): 65–72.

11. Werner-Wasik M, Yorke E, Deasy J, Nam J, Marks LB. Radiation dose-volume effects in the esophagus. *International Journal of Radiation Oncology, Biology, Physics* 2010; **76**(3 Suppl): S86–93.

12. Ong CL, Verbakel WF, Cuijpers JP, Slotman BJ, Lagerwaard FJ, Senan S. Stereotactic radiotherapy for peripheral lung tumors: a comparison of volumetric modulated arc therapy with 3 other delivery techniques. *Radiotherapy and Oncology* 2010; **97**(3): 437–42.

13. Palma D, Senan S. Stereotactic radiation therapy: changing treatment paradigms for stage I nonsmall cell lung cancer. *Current Opinion in Oncology* **23**(2): 133–9.

14. PORT Meta-analysis Trialists Group. Postoperative radiotherapy for non-small cell lung cancer. *Cochrane Database of Systematic Reviews* 2005; **2**: CD002142.

15. Fried DB, Morris DE, Poole C, *et al*. Systematic review evaluating the timing of thoracic radiation therapy in combined modality therapy for limited-stage small-cell lung cancer. [Erratum appears in Journal of Clinical Oncology 2005; **23**(1): 248]. *Journal of Clinical Oncology* 2004; **22**(23): 4837–45.

16. Murray N, Coy P, Pater JL, *et al*. Importance of timing for thoracic irradiation in the combined modality treatment of limited-stage SCLC. *Journal of Clinical Oncology* 1993; **11**(2): 336–44.

17. Kies MS, Mira JG, Crowley JJ, *et al*. Multimodal therapy for limited small-cell lung cancer: a randomized study of induction combination chemotherapy with or without thoracic radiation in complete responders; and with wide-field versus reduced-field radiation in partial responders: a Southwest Oncology Group Study. *Journal of Clinical Oncology* 1987; **5**(4): 592–600.

18. Turrisi AT, 3rd, Kim K, Blum R, *et al*. Twice-daily compared with once-daily thoracic radiotherapy in limited small-cell lung cancer treated concurrently with cisplatin and etoposide. *New England Journal of Medicine* 1999; **340**(4): 265–71.

19. Le Péchoux C, Dunant A, Senan S, *et al*. Standard-dose versus higher-dose prophylactic cranial irradiation (PCI) in patients with limited-stage small-cell lung cancer in complete remission after chemotherapy and thoracic radiotherapy (PCI 99–01, EORTC 22003–08004, RTOG 0212, and IFCT 99–01): a randomised clinical trial. *Lancet Oncology* 2009; **10**(5): 467–74.

20. Le Péchoux C, Laplanche A, Faivre-Finn C, *et al*. Clinical neurological outcome and quality of life among patients with limited small-cell cancer treated with two different doses of prophylactic cranial irradiation in the intergroup phase III trial (PCI99–01, EORTC 22003–08004, RTOG 0212 and IFCT 99–01). *Annals of Oncology* 2011; **22**(5): 1154–63.

21. Falk SJ, Girling DJ, White RJ, *et al*. Immediate versus delayed palliative thoracic radiotherapy in patients with unresectable locally advanced non-small cell lung cancer and minimal thoracic symptoms: randomised controlled trial. *British Medical Journal* 2002; **325**(7362): 465.

22. Macbeth FR, Bolger JJ, Hopwood P, *et al*. Randomized trial of palliative two-fraction versus more intensive 13-fraction radiotherapy for patients with inoperable non-small cell lung cancer and good performance status. Medical Research Council Lung Cancer Working Party. *Clinical Oncology* 1996; **8**(3): 167–75.

23. Bezjak A, Dixon P, Brundage M, *et al*. Randomized phase III trial of single versus fractionated thoracic radiation in the palliation of patients with lung cancer (NCIC CTG SC.15). *International Journal of Radiation Oncology, Biology, Physics* 2002; **54**(3): 719–28.

24. Jeremic B, Shibamoto Y, Nikolic N, *et al*. Role of radiation therapy in the combined-modality treatment of patients with extensive disease small-cell lung cancer: A randomized study. *Journal of Clinical Oncology* 1999; **17**(7): 2092–9.

25. Macbeth FR, Wheldon TE, Girling DJ, *et al*. Radiation myelopathy: estimates of risk in 1048 patients in three randomized trials of palliative radiotherapy for non-small cell lung cancer. The Medical Research Council Lung Cancer Working Party. *Clinical Oncology* 1996; **8**(3): 176–81.

26. Slotman B, Faivre-Finn C, Kramer G, *et al*. Prophylactic cranial irradiation in extensive small-cell lung cancer. *New England Journal of Medicine* 2007; **357**(7): 664–72.

27. Treasure T, Lang-Lazdunski L, Waller D, *et al*. Extra-pleural pneumonectomy versus no extra-pleural pneumonectomy for patients with malignant pleural mesothelioma: clinical outcomes of the Mesothelioma and Radical Surgery (MARS) randomised feasibility study. *Lancet Oncology* 2011; **12**(8): 763–72.

28. Chi A, Liao Z, Nguyen NP, *et al*. Intensity-modulated radiotherapy after extrapleural pneumonectomy in the combined-modality treatment of malignant pleural mesothelioma. *Journal of Thoracic Oncology* 2011; **6**(6): 1132–41.

29. Boutin C, Rey F, Viallat J. Prevention of malignant seeding after invasive diagnostic procedures in patients with pleural mesothelioma. A randomized trial of local radiotherapy. *Chest* 1995; **108**: 754–8.

30. Bydder S, Phillips M, Joseph D. A randomised trial of single-dose radiotherapy to prevent procedure tract metastasis by malignant mesothelioma. *British Journal of Cancer* 2004; **91**: 9–10.

31. O'Rourke N, Garcia J, Paul J. A randomised controlled trial of intervention site radiotherapy in malignant pleural mesothelioma. *Radiotherapy and Oncology* 2007; **84**: 18–22.

32. Ung YC, Yu E, Falkson C, *et al*. The role of radiation therapy in malignant pleural mesothelioma: a systematic review. *Radiotherapy and Oncology* 2006; **80**(1): 13–8.

33. Detterbeck F. The Masaoka-Koga Stage classification for thymic malignancies. Clarification and definition of terms. *Journal of Thoracic Oncology* 2011; **6**: S1710–16.

34. Gomez D. Radiation therapy definitions and reporting guidelines for thymic malignancies. *Journal of Thoracic Oncology* 2011; **6**: S1743–46.

6

Upper gastrointestinal tract

Stephen Falk

6.1 Introduction

Cancers of the upper gastrointestinal tract and hepatobiliary system represent a challenge for the practising radiotherapist. The overall outlook for patients with these diseases is poor with survival rates generally < 10% at 5 years worldwide. The majority of patients present with either locally advanced or metastatic disease, typically of poor functional status and unsuitable for aggressive therapies. The technical challenges of these diseases are considerable related to tumour volumes, anatomical situation, and poor normal tissue tolerance particularly of the intra-abdominal contents. More contemporaneous treatment techniques such as IMRT and IGRT have not currently made significant impact in the routine treatment of upper gastrointestinal tumours in the UK.

6.2 Oesophageal cancer

Radiotherapy can be used as a neo-adjuvant prior to surgery; in the adjuvant setting following oesophagectomy; as sole modality (definitive therapy) usually with concomitant chemotherapy; or as palliation. Radiotherapy issues in oesophageal cancer include tumour localization, radiation volumes, dose and fractionation, planning techniques, optimum chemotherapy regimen, and scheduling, and whether or not trimodality or bimodality therapy is appropriate.

6.2.1 Radical primary treatment

In 1966 Pearson published encouraging results for radiotherapy as sole modality with a 23% 5-year survival for middle third tumours, and 17% 5-year survival for lower third tumours. However, a Cochrane meta-analysis indicates a < 10% 5-year survival with radiotherapy alone[1].

Indications

Whether definitive chemoradiotherapy or surgery ± neo-adjuvant chemotherapy is offered in particular to patients with squamous cell carcinoma varies widely largely dependent on available surgical expertise and treatment philosophy. There is no recent randomized trial comparing non-surgical therapies with radical surgery and current views are that either treatment strategy is acceptable in particular for squamous cell carcinoma.

Treatment volume and definition

All patients should receive:

- CT scan of chest, abdomen, and pelvis with intravenous contrast.

- Laparoscopy, biopsy, and peritoneal washings are recommended for full thickness tumours that extend below the diaphragm for evaluation of possible peritoneal spread, which is poorly shown on CT scan.

- 18-FDG PET scans can identify occult metastatic disease in an additional 15% of patients when compared with other imaging modalities, notably CT scans.

- Endoluminal ultrasound (EUS) is the most reliable indicator of the local stage of the disease, the extent of the disease superiorly and inferiorly, being able to visualize disease in the submucosa, and also of the extent of local lymph node spread (Fig. 6.1). The top of the aortic arch and superior and inferior extent of tumour should be recorded to allow accurate planning. In particular, the incisor to carina distance is more variable than previously thought. Classically it is considered to be 25 cm but the range *in vivo* when measured in a clinical study was 20.5–29 cm[2].

- Fibreoptic bronchoscopy should be performed if CT or EUS suggests invasion of the trachea or left main bronchus. Should mucosal involvement of the airway or main bronchi be confirmed histologically then prophylactic placement with a covered metal stent in the airway is recommended to lessen risks of the severe morbidity and mortality associated with tracheo-oesophageal fistula.

The GTV is the gross primary and nodal disease as defined on the planning-CT scan with all available diagnostic information. Elective nodal irradiation is not usually practised in the UK, although surgical series show extensive nodal spread particularly in adenocarcinoma.

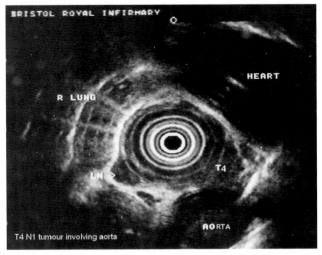

Fig. 6.1 Endoluminal ultrasound showing a locally advanced tumour involving aorta. (From Bristol Royal Infirmary.)

Clinical target definition

The anatomical landmarks of the oesophagus are arbitrary in nature and include the cervical oesophagus which begins at the cricopharyngeal muscle at the level of C7 and extends to the thoracic inlet at T3. The midthoracic segment extends from T4 to T8, and the lower thoracic oesophagus from T8 to T10.

Middle/upper oesophageal cancers:

- The CTV should comprise the GTV plus a margin of up to 1cm laterally, anterior, and posterior; 2cm superiorly and inferiorly (shaped along the contour of the oesophagus).

Lower third tumours, which involve or come within 2 cm of the gastro-oesophageal junction:

- The CTV should comprise the GTV plus a margin of up to 1 cm laterally, anterior, and posterior; 2 cm superiorly (along the line of the oesophagus) and 2 cm inferiorly (this will include the mucosa of the stomach in the direction of the lymph node stations along the lesser curve including the paracardial and left gastric lymph nodes).

The superior and inferior CTV margins are typically extended to 5 or 6 cm and 1–2 cm. laterally, when chemotherapy is contraindicated and radiotherapy is used alone.

Planning target definition

The PTV is created by the addition of the following typical margins:

- Superiorly and inferiorly: 1.0 cm.
- Laterally, anteriorly, and posteriorly: 0.5 cm.

These should be confirmed as appropriate for each individual departmental practice.

The maximum treatment field length is 17 cm, i.e. maximum EUS disease length of primary tumour and lymph nodes is usually 10 cm.

The PTV is, in summary, the EUS defined GTV + affected peritumoral nodal stations with a 1.5-cm lateral, anterior, and posterior margin and 3-cm superior–inferior margin.

For middle and lower third tumours the patient is immobilized in a vacuum fixed polystyrene bag and treated supine with arms above head held on an arm pole, and the shoulders supported for stability by the vacuum bag to allow for the lateral fields. A device similar to a 'knee-fix' is also encouraged. For the treatment of the upper third of oesophagus, a supine cast is made to immobilize the neck and jaw. The patient is positioned with the cervical spine straight and parallel to the couch top.

Dose distribution and field modifications

Radiation techniques for carcinomas of the mid and lower oesophagus comprise a conformally planned single-phase technique usually requiring a multifield five-field plan comprising two posterior oblique, two anterior oblique, and an anterior field (Fig. 6.2). This reduces spinal cord doses and results in relative sparing of the lung fields. Reducing lung volumes is particularly significant if either subsequent planned or salvage surgery is considered. The greatest morbidity and mortality for trimodality patients is pulmonary toxicity.

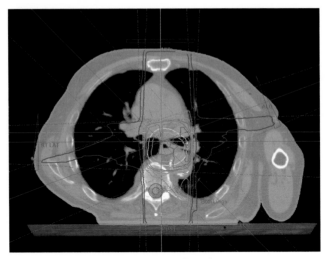

Fig. 6.2 Five-field plan for radical therapy of oesophageal cancer.

For all multifield arrangements account needs to be made of the variation in body contour and the changing position of the oesophagus throughout its length. There will commonly be an inclined volume to the treatment couch, requiring the use of angulation of the treatment head and couch. Traditionally three transverse outlines are taken at the upper, middle, and lower levels of the planned treatment volume. The midline as defined by the central tattoo is marked on each outline as is the interplanar distance and position of the spinal cord is also marked on each outline.

The treatment volume is usually planned on the mid-level outline. The gantry angle of the lateral/oblique fields is usually 90–120° dependent on the spinal cord dose from the parallel opposed pair arrangement. Larger gantry angles decrease the lung dose but increase the dose to the spinal cord. Variations in body contour or depth can be resolved by the use of an individual alloy compensator or a longitudinal wedge, commonly 15° placed in the anterior field with the thick end superiorly (Fig. 6.3). For the lateral/oblique fields a head twist is applied to allow for the volume being angled away from the horizontal.

The cervical oesophagus can usually be treated by a single plan, comprising two anterior placed oblique and wedged fields. The wedge angle is usually 30° when 4–8-MV photons are used. Sometimes the distribution is improved by the addition of a straight anterior field (Fig. 6.4). Given the curvature of the oesophagus within the neck and oesophagus, compensation in the superior–inferior plane with either a tissue compensator, or a longitudinal wedge in the superior–inferior plane is commonly required. The gantry is angled at 50–60° for the lateral fields to allow for changes in depth of the oesophagus, an angle > 65° indicates that the beam would pass through the head of the humerus. The small volumes of lung in the radiation field in the superior portion of the chest allows for treatment with a single phase two- or three-field plan throughout.

Tissue constraints as shown in Table 6.1 should be observed and recorded for each patient plan.

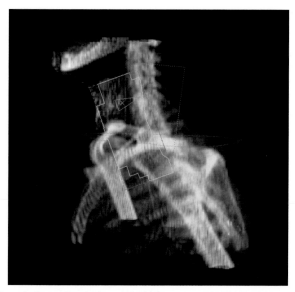

Fig. 6.3 Conformal plan with spinal cord shield for carcinoma of oesophagus.

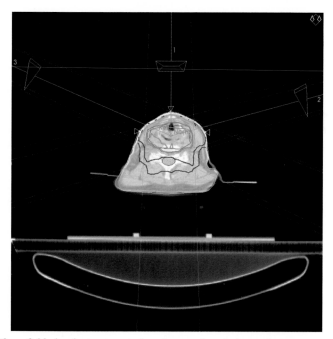

Fig. 6.4 Three-field plan for treatment of carcinoma of cervical oesophagus.

Table 6.1 Tissue constraints

Region of interest/organ at risk	Dose constraint
PTV	V95% (47.5 Gy) > 99.0%
PTV	PTV min > 93% (46.5 Gy)
DMAX	< 107% (53.5 Gy)
GTV	GTV min > 100% (50 Gy)
Spinal cord PRV	Cord max <80% (40 Gy)
Combined lungs	V40% (V20 Gy) < 25%
Heart	V80% (V40 Gy) < 30%
Liver	V60% (V30 Gy) < 60%
Individual kidneys	V40% (V20 Gy) < 25%

Conformal radiotherapy can significantly reduce lung volumes. There is little, however, in the literature concerning IMRT. However, it does not seem that IMRT results in significant improvements in either PTV dose homogeneity or treated lung volumes when compared with conformal techniques. Only limited evidence from a single centre in Hong Kong currently suggests an advantage in tumour coverage resulting from the use of IMRT[3].

The patient is aligned longitudinally using anterior and lateral lasers. The anterior field centre is either marked on the cast for upper third tumours or found in relation to either anterior or lateral tattoos, and sometimes both. All fields are treated isocentrically every day.

Each phase of the plan should be verified in the simulator prior to starting treatment and at least one portal image taken in the first week of each phase on the linear accelerator. The most helpful verification is localization of the anterior field whether by simulator check film or portal imaging. Many centres will verify lateral fields but at a gantry angle of 90°. Anatomical localization becomes more prone to operator error when lateral oblique films are considered.

A Cochrane collaboration meta-analysis indicates that the addition of chemotherapy to radiotherapy improves both local control and survival[1]. Thirteen randomized trials were included in the analysis. There were eight concomitant and five sequential radiotherapy and chemotherapy (RTCT) studies. Concomitant RTCT provided significant overall reduction in mortality at 1 and 2 years of 7% (95% CI: 1–15%). The mortality in the control arms being 62% and 83% respectively. The local recurrence rate for the control arms was in the order of 68%. Combined RTCT provided an absolute reduction of local recurrence rate of 12% (95% CI: 3–22%). There was, however, a significant increase of severe and life threatening toxicities.

Appropriate prescription options include:

- Neo-adjuvant *chemotherapy* is often given for example: cisplatin 60 mg/m² day 1 with capecitabine given continuously at a dose of 625 mg/m² twice daily.

Continuous infusion 5-FU 225 mg/m²/day is substituted for the capecitabine when patients cannot swallow capsules. Four courses of treatment are given repeated every 21 days.

Table 6.2 Patterns of failure of oesophageal cancer

Modality	No. (%)	Local and marginal (%)	Neck (%)	Mediastinum (%)	Local + distant (%)	Distant (%)
RT	517	25–84	25	10–43		23–65
Surgery	266	21–50	44	33		17–65
RT/Surg	2078	22–87	53		20	17–43
RT/Che	254	15–39			5–25	6–25
Tri mod	150	2–36			5–38	16–29

5 years postsurgery 25% will have a second primary of the aero-digestive tract.

From Aisner J, Forastiere A, Aroney R (1983) Patterns of recurrence for cancer of the lung and esophagus. In Wittes RE (ed) *Cancer Treatment Symposia: Proceedings of the Workshop on Patterns of Failure After Cancer Treatment*, Vol **2**, p. 87. US Department of Health and Human Services, Washington DC.

- ◆ *Radiotherapy:* 50 Gy in 25 daily treatments 5 days per week concomitant with chemotherapy, commences at the start of the third chemotherapy cycle.
  When radiotherapy alone is used recommended doses include:
- ◆ 55 Gy in 20 fractions in 4 weeks 5 days per week.
- ◆ 66 Gy in 33 fractions in 6 weeks, 5 days per week.

Treatment is prescribed to the ICRU PTV reference point.

When concomitant chemoradiotherapy is used there is currently no good evidence that radiation doses > 50.4 Gy are beneficial. No survival benefit, yet a significant increase in toxicity, was seen when a concomitant chemoradiation schedule employing 64 Gy was randomized against 50.4 Gy total dose[4]. In spite of this, patterns of failure continue to be predominantly local. Table 6.2 shows patterns of failure in oesophageal cancer associated with modality of therapy[5]. One important current question is whether dose escalation using techniques such as IMRT, associated with improved localization techniques including EUS and PET, may improve outcome when compared with clinical studies performed in the 1980s.

6.2.2 Neo-adjuvant radiotherapy prior to surgery

Practice varies hugely around the world largely according to custom and practice. The following remain acceptable therapies for squamous cell and adenocarcinoma of the oesophagus:

- ◆ Surgery alone (favoured in much of the world).
- ◆ Neo-adjuvant chemotherapy and surgery (favoured in the UK).
- ◆ Neo-adjuvant chemoradiotherapy favoured in Europe and the US.

Preoperative radiotherapy

There is no evidence that sole modality preoperative radiotherapy improves the survival of patients with potentially resectable oesophageal cancer. A Cochrane group quantitative meta-analysis of preoperative radiotherapy using updated data from 1147 patients in five randomized trials has been performed[6]. With a median follow-up of

9 years, in a group of patients with mostly squamous carcinomas, the hazard ratio (HR) of 0.89 (95% CI: 0.78–1.01) suggests an overall reduction in the risk of death of 11% and an absolute survival benefit of 3% at 2 years and 4% at 5 years. This result is not statistically significant (p = 0.062).

Preoperative chemotherapy

British practice has been dominated by the 802-patient MRC OE02 study[7] which reported a significant prolongation of the 2-year survival (43% vs. 34%) and median survival (16.8 vs. 13.3 months) in favour of the chemotherapy arm. In a recent trial update, the survival benefit was maintained at 5 years with 23% overall survival in the chemotherapy arm compared to 17% in surgery alone arm, which was independent of histological subtype (p = 0.03). Gebski et al. (2007) reported a review of eight randomized trials of surgery alone or chemotherapy followed by surgery in patients with potentially operable oesophageal cancer (n = 1724)[8]. The hazard ratio for all-cause survival at 2 years favoured the use of chemotherapy followed by surgery (HR = 0.90 (0.81 to 1.0)) with an absolute survival benefit of 7%. However, no survival advantage was observed in patients with squamous cell carcinoma compared to patients with adenocarcinomas.

Preoperative chemoradiotherapy

Numerous trimodality phase II studies utilizing preoperative chemoradiation (CRT) therapy have suggested prolonged and improved survival compared with surgical reports. Studies are generally flawed by the inclusion of small numbers of patients and heterogeneous treatment schedules, in particular with an inconsistent approach to the use of sequential and/or concomitant radiation with chemotherapy. Treatment-related toxicity can however be significant. In particular most centres report an increase in postoperative mortality from 2–3% for surgery alone to up to 8–10% with trimodality therapy. A recent systematic overview of 3840 patients from 38 trials reported an average R0 resection and pathological complete response (pCR) rate of 88.4%, and 25.8%, respectively with the use of CRT. The postoperative mortality was reported as 5.2% and the 5-year survival rates ranged from 16–59% for all patients and from 34–62% in those with pCR[9].

More promising results have recently been reported by the Dutch Trials Group where 363 patients were randomized to neo-adjuvant CRT with 5 weekly cycles of paclitaxel (50 mg/m^2) and carboplatin (AUC2) combined with concurrent RT (41.4 Gy in 23 fractions) or surgery alone[10]. Most cancers were adenocarcinomas (n = 273) and the reported R0 resection rate was 92.3% in the CRT arm versus 64.9% in the surgery alone arm, and the PCR rate was 32.6%. Importantly postoperative mortality was 3.7% in the surgery alone arm compared to 3.8% in the CRT arm. The OS was significantly better in the group of patients treated with CRT (HR = 0.67 (0.50–0.92)) and the MS was 49 months in the CRT arm versus 26 months in the surgery alone arm[10].

The controversy regarding optimal policy of preoperative therapy as well as regimens remains needing to be answered by further well-designed randomized collaborative studies. However, given R1 resection rates of up to 30% following surgery, this author recommends consideration of trimodality therapy for bulky T3/4 tumours.

Indications

Patients with potentially operable adenocarcinoma or squamous cell carcinoma usually T3–4 N0 or N1 disease of the middle and lower thirds are eligible for trimodality therapy when there is no evidence of disease outside the standard operative field. To be eligible for therapy:

- In general patients will be biologically < 70 years of age without significant comorbidities.
- Have good performance status (WHO 0 or 1), adequate haematological and renal function to tolerate all treatments including cisplatin-based chemotherapy.
- Have adequate respiratory function: a resting pO_2 of > 10 kPa and FEV_1 > 1.5 litres is recommended; transfer factor > 50%.
- Have adequate cardiac function as defined by a cardiac ejection fraction of > 50% and normal echocardiography.
- Patients with distant metastases, including involvement of the supraclavicular lymph nodes are excluded.
- Patients with clear evidence of disease outside a planned radiation volume are excluded. In particular, endoscopically visible involvement of the gastric lesser curve, any evidence of extension to the rest of the stomach (endoscopy or imaging), large-volume involvement of gastrohepatic ligament/coeliac axis lymph nodes). In practice > 2 cm of submucosal spread into the stomach shown on EUS is likely to render this treatment impractical.

Planning for radiotherapy takes place during the induction phase of chemotherapy and starts week 7 concomitantly with the third cycle of cisplatin.

Planning techniques are as described for radical primary treatment.

Dose prescription

Treatment includes:

- Capecitabine 625 mg/m^2 twice a day orally or if there is significant dysphagia 11 weeks of continuous infusional 5-FU initially at a dose of 250 mg/m^2/day reduced to 200–225 mg/m^2/day during radiotherapy.
- Four courses of cisplatin 60 mg/m^2 given at three-weekly intervals with pre- and posthydration and antiemetics.

The specified radiation treatment should be 45 Gy in 25 fractions with the dose prescribed at the isocentre of a planned target volume.

- Patients are treated 5 days per week with all fields treated daily.
- The total spinal cord maximum dose should be 40 Gy.

6.2.3 Postoperative adjuvant therapy

Given the major morbidity of oesophageal cancer surgical procedures, patients are significantly debilitated following surgery and often unfit for postoperative therapy to be given within a reasonable time-frame following operation. Furthermore, attempts at radiation therapy are compromised by the need to irradiate the gastric pull-up or occasionally intestinal interposition. Traditionally postoperative radiation has been reserved for patients with microscopic evidence of residual disease. Postoperative radiation reduces local recurrence rates but with the rare exception of lymph-node negative patients has no clear impact on survival. Techniques and doses are as described in the neo-adjuvant section with particular care to reduce the length of gastric mucosa irradiated to an absolute minimum (maximum advised field length 12 cm).

6.2.4 Palliative therapy

The majority of patients with oesophageal cancer are elderly, and often significantly debilitated following rapid weight loss due to dysphagia. Many have, often asymptomatic, metastatic disease at presentation, discovered during staging procedures as described above. Such patients are suitable for palliative therapies only. The options are

- Palliative radiotherapy including brachytherapy.
- Stent placement.
- Laser treatment or ablation.
- Palliative chemotherapy.

Palliative radiation can improve symptoms of pain and dysphagia in up to 80% of patients. Placement of an expanding metal stent is, however, favoured if there is:

- Complete dysphagia or very tight stricturing at endoscopy (e.g. the gastroscope will not pass through the tumour).
- Evidence either clinically or radiologically of fistulation into the airway.

The CTV is usually defined by barium swallow performed on the simulator. Virtual simulation using CT may also be used.

Planning technique

The patient lies supine with arms by side. Immobilization devices are not usually employed.

The tumour is localized by barium. Typically treatment fields extend 3 cm inferior and superior from the tumour with a field width of 9 cm, which may be increased as necessary if CT scan suggests extra-oesophageal tumour extension. A simple anterior–posterior parallel pair is employed, longitudinal wedges or tissue compensators are not used.

Typical dose prescriptions include 30 Gy in 10 daily fractions treating 5 days per week or 20 Gy in five fractions treating daily over one week.

6.3 Carcinoma of the stomach

6.3.1 Radical primary treatment

Anecdotal evidence suggests that radiotherapy has the ability to salvage a small number of patients who present with adenocarcinoma of the stomach and locally advanced unresectable disease. Probably no more than 5–10% of patients can be cured by such therapy. Standard therapy in the UK is primary chemotherapy with consideration of surgery when there is a favourable response.

The only commonly used indication for radical radiotherapy to the stomach is as consolidation radiotherapy following chemotherapy for high-grade B-cell non-Hodgkin's lymphoma localized to the stomach, as a substitute for surgery; or as primary therapy for MALT lymphomas and low-grade B-cell lymphomas, again localized to the stomach (see Chapter 10).

Typically the radiation volume encompasses the tumour and major draining nodal chains. This includes the lesser and greater curvature of the stomach, coeliac axis including the pancreaticoduodenal, splenic, suprapancreatic, and porta hepatis. Para-aortic lymph nodes to the level of mid L3 or mid L4 are included and paraoesophageal nodes for more proximal lesions encompassing the lower oesophagus.

The treatment volume can be defined by CT scan or less frequently with orthogonal films following barium contrast.

The patient is treated in the supine position. Since a portion of both kidneys will be in the treatment volume, care must be taken to not irradiate at least two-thirds of one kidney. For proximal gastric lesions, 50% or more of the left kidney is inevitably within the treatment volume and the right kidney must be shielded. For more distal lesions the right kidney is commonly within the treatment volume and the left kidney must be spared. For lesions at the gastro-oesophageal junction a CTV margin of 3–5 cm of the distal oesophagus is indicated.

The most efficient way to obtain dose distribution through the often large volume and simultaneously to be able to spare the kidneys is by the use of anterior and posterior opposed fields.

Due to poor tolerance of gastric mucosa the usual dose is 45–50 Gy in 1.8–2.0-Gy daily fractions. It is normal practice to reduce the volume to known disease only after 45 Gy. At this point it may be appropriate and or practical to employ multiple planned beams. A maximum dose of 55 Gy is recommended.

6.3.2 Postoperative adjuvant therapy

Locoregional recurrence in the gastric or tumour bed, at the anastomosis, or regional lymph nodes occurs in 40–65% of patients after gastric resection. The frequency of such relapses makes regional radiation attractive as adjuvant therapy. Randomized trials have addressed the extent of surgical resection (D1 vs. D2 resection). Whilst in Japanese hands extended lymph node dissection has improved outcome, these results have not been replicated in European studies, at least in part due to enhanced morbidity and mortality observed in the groups treated with extended surgery. Radiation therapy by itself has no significant role following R0 resection. Indeed, level A evidence of benefit of any postoperative treatment is limited. Sequential British Stomach Cancer Group (BSCG) trials suggested no survival benefit from postoperative chemotherapy and/or radiotherapy. However, interpretation of these studies is limited by 29% of patients randomized having macroscopic residual disease and 18% having positive resection margins[11]. Such patients would not now be considered eligible for trials of adjuvant therapies. More recent data from MacDonald et al.[12] using a postoperative regimen of 5-FU/leucovorin and radiation demonstrated an improvement in median overall survival from 27 months, in the surgery-only group, to 36 months in the chemoradiotherapy group. This trial has, however, been criticized in Europe for inadequate surgical therapy. Of 552 patients, only 54 (10%) had undergone a formal D2 dissection. A D1 dissection (removal of all invaded (N1) lymph nodes) had been performed in 199 patients (36%), but most patients (54%) had undergone a D0 dissection, which is less than a complete dissection of the N1 nodes. Furthermore survival outcomes in the combined modality arm of the MacDonald study are equivalent to European surgical only studies where D1 and/or D2 dissection has been performed. Postoperative chemoradiotherapy thus remains controversial in Europe and further studies are underway.

If postoperative treatment is planned treatment given should reflect the protocol published by MacDonald et al.[12]. The CTV should encompass the tumour bed, regional nodes, and extend 2 cm beyond the proximal and distal margins of resection. The presence of proximal T3 lesions will necessitate treatment of the medial left hemidiaphragm. Perigastric, coeliac, local para-aortic, splenic, hepatoduodenal or hepatic-portal, and pancreaticoduodenal lymph nodes may be included in the radiation fields. In patients with tumours of the gastroesophageal junction, paracardial and paraesophageal lymph nodes are included in the radiation fields. Exclusion of the splenic nodes is recommended in patients with antral lesions to spare the left kidney.

The CTV is defined by preoperative CT imaging, barium swallow if necessary, and in some instances, if available surgical clips.

The usual way to obtain dose distribution through the often large volume and simultaneously to be able to spare the kidneys is by the use of anterior and posterior opposed fields with the patient prone in an immobilization device such as a vacuum bag with arms extended and holding an arm pole.

More recently a 'split-field' technique has been described in which the PTV is divided into two abutting sections, with each section treated using a separate, optimized field arrangement[13]. This conformal technique provides better coverage of the target volume with 99% of the PTV) receiving 95% of the prescribed dose, compared to 93% using AP–PA fields. Comparative DVHs for the right kidney, left kidney, and spinal cord demonstrate lower radiation doses using the conformal technique, and although the liver dose is higher, it is still well below liver tolerance. The upper half of the PTV which includes the tumour bed, anastomosis, and splenic hilar nodes is treated using a three-field arrangement comprising anterior and posterior fields, and a left lateral field angled to avoid the spinal cord. The lower half of the PTV which includes the subpyloric, pancreaticoduodenal, and local paraaortic nodes is treated using a three-field arrangement comprising right and left lateral fields and an anterior field angled to minimize the dose to the kidneys. The coeliac, suprapancreatic and porta hepatis nodes are included in either the upper or lower half of the PTV, depending upon individual patient anatomy, as well as the level at which the PTV is split. This technique involves the use of a single isocentre placed at the level of the split and asymmetric collimator jaws in the superior–inferior direction to achieve an effective match-line.

Check films on the simulator with intravenous contrast to image the kidneys and portal imaging within the first 3 days of treatment are recommended. Quality assurance for radiotherapy was an essential component of the Macdonald/Intergroup trial. At review, 35% of the treatment plans were found to contain major or minor deviations from the protocol, most of which were corrected before the start of radiotherapy. Given the complexity and potential toxicity of therapy it seems appropriate that there should be at least internal institutional review of all such treated cases when a centre starts these treatments.

Chemotherapy (5-FU, 425 mg/m^2 of body-surface area per day, and folinic acid, 20 mg/m^2 per day, for 5 days) is given on day 1 and followed by chemoradiotherapy beginning 28 days after the start of the initial cycle of chemotherapy.

Chemoradiotherapy consists of 45 Gy radiation delivered in 25 fractions, 1.8 Gy per day, treating 5 days per week for 5 weeks, with 5-FU (400 mg/m^2 per day) and folinic acid (20 mg/m^2 per day) on the first 4 and the last 3 days of radiotherapy.

One month after the completion of radiotherapy, two 5 cycles of 5-FU (425 mg/m^2 per day) plus folinic acid (20 mg/m^2 per day) are given 1 month apart. The dose of 5-FU is reduced in patients who have documented grade 3 or 4 toxicity.

Recent trials examining the substitution of single agent 5-FU with combinations including cisplatin have not demonstrated a survival advantage.

Radiation doses should be restricted so that:

+ Less than 60% of the hepatic volume is exposed to > 30 Gy.

+ The equivalent of at least two-thirds of one kidney is spared from the field of radiation.

+ No portion of the heart representing 30% of the cardiac volume receives > 40 Gy of radiation.

6.3.3 Palliative treatment

The majority of patients with gastric cancer present with advanced or metastatic disease. Chemotherapy has an established role in providing palliation for patients with gastric cancer with evidence of prolongation of quality and quantity of life. There are no randomized published series employing palliative radiotherapy for gastric cancer. The literature contains series describing 50–75% patients obtaining palliation for symptoms by the use of radiotherapy with or without chemotherapy. The median duration of palliation varies from 4–18 months. Recognized symptoms that may be palliated include bleeding and pain due to local tumour infiltration. Whilst the use of radiation in obstruction is described in the literature such therapy has now been superseded by the radiological placement of stents.

The volume is determined by simulation with barium or virtual simulation with CT and includes known tumour, which can be assessed by endoscopic description aided by CT films. Often the whole stomach requires irradiation.

Palliative radiotherapy is given by anterior and posterior opposing fields with the patient in the supine position and arms adducted. Field margins are marked on the patients skin and the central point marked by tattoo.

Bleeding can often be stopped by relatively modest doses of irradiation. Commonly applied treatments comprise 20 Gy in five fractions in 1 week. In very unwell patients a single fraction of 8–10 Gy can often stop bleeding within a few days. Some authors advocate higher doses (such as 40 Gy in 2-Gy daily fractions for pain relief).

6.4 **Carcinoma of pancreas**

Pancreatic cancer has a poor prognosis with an overall survival rate of < 5%. Surgery is considered the only potentially curative treatment for pancreas cancer but no more than 10–20% of patients are able to undergo resection. Fifty to 60% of patients present with metastatic disease and approximately 30–40% have locally advanced disease, usually defined in terms of vascular encasement not amenable to surgical resection and/or vascular reconstruction. The use of radiotherapy in all stages of pancreatic cancer is controversial and the data is of generally poor quality.

6.4.1 **Indications**

Radiotherapy usually with chemotherapy may be considered:

- As *neo-adjuvant therapy*, in particular when the disease is considered borderline resectable.

- In the postoperative setting (*adjuvant*), particularly when the surgical margins are involved.

- As primary *palliative* therapy for clearly unresectable localized disease with reconsideration of surgery in good responders.

Neoadjuvant therapy

A meta analysis of 111 retrospective studies concluded that approximately one-third of initially staged non-resectable tumour patients could be expected to have resectable tumours following neoadjuvant therapy, with similar survival but higher resection-associated morbidity and mortality rates when compared with patients having initially resectable tumour. There is, however, no randomized data to support this finding[14].

Adjuvant therapy

A meta-analysis[15] included five randomized controlled trials of adjuvant chemoradiation and adjuvant chemotherapy. Adjuvant chemotherapy improved survival in patients with R0 resections. The 2- and 5-year survival rates were estimated at 38% and 19%, respectively, with chemotherapy and 28% and 12% without. This benefit was not seen with adjuvant chemoradiation (HR = 1.09, 95% CI: 0.89, 1.32, P = 0.43). The group concluded that adjuvant chemoradiation is only more effective than chemotherapy alone after R1-resections.

Palliative therapy

Studies from the GITSG group in the 1980s suggested a role for chemoradiotherapy (CRT). More recent studies have been inconsistent and there remains no high-quality phase III study randomizing patients between chemotherapy alone and chemoradiotherapy. Indeed one study[16] showed a survival detriment for the use of CRT compared to single agent gemcitabine. However, this trial used an unconventional CRT regimen with large fields encompassing uninvolved regional nodes, a high radiation dose of 60 Gy concurrent with infusional 5-FU (300 mg/m^2/day) and cisplatin.

Several studies have now shown that selected patients who attain stable or partially responsive disease after 3–4 months of induction chemotherapy may benefit from consolidation CRT. In a retrospective analysis of 318 patients treated at the MD Anderson Cancer Centre between 1993 and 2005, the overall survival time to local and distant progression were significantly better in participants receiving neo-adjuvant chemotherapy[17]. Thus standard practice is to give three to four cycles of gemcitabine-based chemotherapy prior to radiotherapy thereby excluding the up to 35% patients with initially rapid progressive disease that is treatment refractory.

6.4.2 Treatment volume and definition

There is a wide variation in the practice of pancreatic RT in international literature with no consensus on definition of CTV, radiation dose, or acceptable dose to OARs.

MRI scans have not shown clear advantages over CT scanning for volume localization in surgical series[18]. FDG-PET may be useful with unexpected distant metastasis detected by whole-body PET in about 40% of the cases. However, active and chronic pancreatitis and autoimmune pancreatitis may sometimes show high FDG accumulation and may mimic pancreatic cancer[19].

The GTV includes the primary tumour as visualized on a contrast-enhanced planning CT scan along with lymph nodes > 1 cm diameter identified on pretreatment scans.

The PTV is defined as the GTV with 15-mm margin in the anterior–posterior and lateral directions and 20-mm margin in superior–inferior direction.

6.4.3 Treatment techniques

Patients should be planned and treated in the supine position with their arms above their heads and immobilized, ideally with the use of a chest-board and knee-fixation. For the CT planning scan, three horizontally aligned tattoos are marked at the right, anterior, and left hand surfaces. Typical *conformally planned* conventional field arrangements include an anterior open field and two lateral wedged fields usually at an angle of 90°. This angle may need to be modified to reduce renal dose (Fig. 6.5). The lateral fields may be weighted in order to reduce the exit dose to the spinal cord from the anterior field. If the volume lies to the right of the midline, anterior and right lateral wedged fields may be used to spare the small bowel and left kidney. An isocentric technique is employed on a linear accelerator; 6–8-MV photons are used.

Verification can take place on the simulator and by using on treatment portal views. Anterior fields are relatively easy to verify but images of the lateral or oblique fields are less easy to interpret.

The use of *intensity-modulated radiotherapy* (IMRT) is attractive to simultaneously allow dose-escalation and further enhance locoregional control while limiting doses in particular to the small bowel and kidneys. Landry et al.[20] showed that IMRT plans were more conformal than 3D-CRT plans. The median volume of small bowel exceeding 50 Gy was $19.2 \pm 11.2\%$ (range 3–45%) compared to 31.4 ± 21.3 (range 7–70%) for 3D-CRT (p = 0.048). The potential roles of daily pretreatment CT scans, implanted fiducial markers in the tumour, and respiratory gating are all being explored. However, it remains unknown if these radiation planning studies will translate into decreased toxicity in the clinical setting.

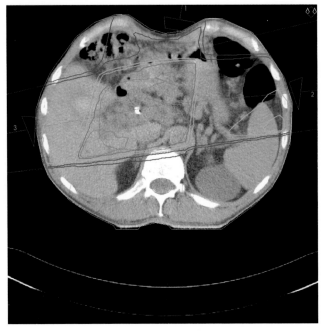

Fig. 6.5 Three-field plan for treatment of carcinoma of pancreas.

Intraoperative radiotherapy (IORT) allows dose limiting normal structures such as the bowel to be physically moved out of the radiation field. A randomized trial by the National Cancer Institute in just 24 patients[21] suggested an improvement in local control with the use of 20-Gy IORT following surgical resection compared to standard therapy but no improvement in survival has been shown in any setting of pancreas cancer.

Stereotactic body radiotherapy (SBRT). The duodenum is the primary dose-limiting normal tissue for radiotherapy and pancreas cancer. One major concern with SBRT is thus small bowel ulceration, perforation, or obstruction. Doses used include 30 Gy in

Table 6.3 Dose–volume constraints

Region of interest/organ at risk	Dose constraint
PTV	V95% > 99.0%
D_{max}	< 107%
Spinal cord PRV	V40 Gy < 0%
Liver	V30 Gy < 40%
Kidney receiving the higher dose	V20 Gy < 40%
Combined kidney	V20 Gy < 30%

three fractions. The Stanford group using a 25-Gy single-dose SBRT have reported an overall rate of freedom from local progression at 6 months and 12 months of 91% and 84%, respectively accompanied by grades ≥ 2 late largely gastro-intestinal toxicity of 11% and 25%.[22]. The addition of an SBRT boost to 45 Gy conventional fractionated RT has also been investigated by the Stanford group, yielding a 94% local control rate and a 12.5% rate of late duodenal ulcers. Although local control rates have been encouraging, median survival times in these trials have not been substantially different compared to historical controls (6–11 months), primarily due to the development of distant disease[23]. Well-conducted phase III trials are required given the not inconsiderable late toxicity rates.

6.4.4 **Dose prescription**

The dose is limited by critical structures. When the dose exceeds 50 Gy with chemotherapy there is a 10% risk of small bowel damage, most commonly seen in the duodenal loop and manifested by bleeding.

◆ Typical prescriptions are 50 Gy in 1.8–2-Gy daily fractions single phase.

◆ Concomitant chemotherapy where applicable comprises:

- Gemcitabine 300mg/m² weekly (intravenous).
- Or 5-FU 200 mg/m² daily by continuous infusion throughout treatment.
- Or capecitabine 830 mg/m² twice daily (orally), Monday–Friday.

6.5 **Gall bladder and biliary tree**

Carcinomas of the gall bladder and extrahepatic biliary tree are uncommon and account for about 4% of all gastrointestinal tumours. The principal therapeutic measure for these tumours remains surgical, whether by excision of the primary tumour, or, where necessary, palliative stenting and bypass procedures to relieve obstructive jaundice.

Given the relative rarity of these tumours the role of radiotherapy is uncertain with the literature containing small, often single-institution, studies describing patients with both gallbladder cancer and cholangiocarcinoma and no randomized data. There is currently no primary role for external beam radiotherapy as curative treatment, and its use is limited to either the postoperative adjuvant setting or for palliation.

6.5.1 **Postoperative adjuvant treatment**

Indications

Even after apparently curative resection locoregional failure is common for both gall bladder and extrahepatic bile duct lesions. In bile duct cancer, proximal and distal margins are often narrow, as are the circumferential margins if lesions extend through the entire duct wall. For ductal lesions local failure is the commonest cause of death. Overall, after simple or extended cholecystectomy for gall bladder cancer, 75–85% of patients with early recurrence will die with or because of local failure. There is therefore a rationale for employing postoperative radiation therapy.

Treatment volume and definition

A major deterrent to the use of radiation is the limited tolerance of surrounding structures, in particular liver, duodenum, jejunum, and stomach. The presence of surgical clips eases difficulties in identifying residual disease. It is helpful to plan treatment with the referring surgeon present to clarify both the postoperative anatomy and areas at risk of relapse.

The patient is treated supine arms above head and holding an arm pole, with the trunk immobilized in an individualized bean-bag. Patient alignment is from lateral tattoos.

Planning technique

The CTV is defined as the tumour bed, any residual tumour, lymph node drainage along the porta hepatis, pancreatico-duodenal system and coeliac axis.

One phase technique using a three-field plan using one anterior and two angled lateral oblique fields. Other techniques employ shrinking fields where phase 1 comprises a large volume parallel opposed pair to this volume then reduced using a three-field plan with one anterior and two angled lateral oblique fields to the tumour bed. The large volume may need to be reduced at 35 Gy to avoid radiation-induced hepatitis.

Verification

Verification can take place on the simulator and/or using linear accelerator portal imaging views. Whilst anterior fields are relatively easy to verify, images of the lateral fields are difficult to interpret and probably non-contributory. Image-guided radiation techniques with soft tissue algorithms in development offer the prospect of significantly improving quality of radiotherapy delivery.

Dose prescription

Tolerance and potential for morbidity dictates dose.

- 50 Gy in 25 fractions in 5 weeks, single phase.
- Concomitant chemotherapy where applicable comprises:
  - 5-FU 500 mg/m^2 days 1–3 with folinic acid 20 mg/m^2 days 1–3, weeks 1 and 5.
  - Or Gemcitabine 300mg/m^2 weekly (intravenous).

An alternative when large volume treatments are employed is 45 Gy in 25 fractions in 5 weeks to the large volume followed by a boost using multiple field, arc techniques to a total dose of 55–60 Gy over 6–7 weeks.

Buskirk et al.[24] showed that with total doses of 55 Gy or less, the risk of severe gastrointestinal complications was 5–10%. At doses > 55 Gy one-third of patients developed severe problems. In rabbits, significant biliary fibrosis is seen at doses > 30 Gy. Temporary biliary tree fibrosis with consequent secondary biliary cirrhosis is recognized usually resolving within 18–24 months of treatment.

6.5.2 **Palliative treatment**

Palliative treatments are rarely employed. Intraluminal brachytherapy may be used for palliation and its use has been facilitated by the development of smaller more flexible catheters for use with high-dose rate machines.

Significant palliation can be achieved using techniques identical to the postoperative adjuvant therapies discussed previously. Permanent local control is uncommon and doses are usually limited to 40–60 Gy over 4½–7 weeks. Fluoropyrimidines or more recently gemcitabine chemotherapy is commonly co-administered with treatment.

References

1. Wong R, Malthaner R. Combined chemotherapy and radiotherapy (without surgery) compared with radiotherapy alone in localized carcinoma of the esophagus (Cochrane Review). In *The Cochrane Library*, Issue 2, Chichester: John Wiley & Sons, Ltd, 2004.

2. Rice PF, Crosby TL, Roberts SA. Variability of the carina-incisor distance as assessed by endoscopic ultrasound. *Clinical Oncology* 2003; **15**(7): 383–5.

3. Wu VW, Kwong DL, Sham JS. Target dose conformity in 3-dimensional conformal radiotherapy and intensity modulated radiotherapy. *Radiotherapy and Oncology* 2004; **71**(2): 201–6.

4. Minsky B, Pajak T, Ginsberg R, *et al.* INT 0123 (Radiation Therapy Oncology Group 94–05) Phase III trial of combined-modality therapy for esophageal cancer: high-dose versus standard-dose radiation therapy. *Journal of Clinical Oncology* 2002; **20**(5): 1167–74.

5. Aisner J, Forastiere A, Aroney R. Patterns of recurrence for cancer of the lung and esophagus. In Wittes RE (ed) *Cancer Treatment Symposia: Proceedings of the Workshop on Patterns of Failure After Cancer Treatment, Vol. 2.* Washington, DC: US Department of Health and Human Services, 1983, p. 87.

6. Arnott SJ, Duncan W, Gignoux M, *et al.* (Oesophageal Cancer Collaborative Group). Preoperative radiotherapy for esophageal carcinoma. *Cochrane Database of Systematic Reviews* 2005; **4**: CD001799.

7. Allum, WH, Stenning, SP, Bancewicz, J, *et al.* (2009). Long-term results of a randomized trial of surgery with or without preoperative chemotherapy in oesophageal cancer. *Journal of Clinical Oncology* 2009; **27**:5062.

8. Gebski V, Burmeister B, Smithers BM, *et al.* Australasian Gastro-Intestinal Trials Group. Survival benefits from neo-adjuvant chemoradiotherapy or chemotherapy in esophageal cancer—a meta-analysis. *Lancet Oncology* 2007; **8**: 226–34

9. Courrech Staal EF, Aleman BM, Boot H, *et al.* Systematic review of the benefits and risks of neoadjuvant chemoradiation for oesophageal cancer. *British Journal of Surgery* 2010; **97**: 1482–96.

10. Gaast AV, van hagen P, Hulshof M, *et al.* Effect of preoperative concurrent chemoradiotherapy on survival of patients with resectable esophageal or esophagogastric junction cancer: Results from a multicenter randomized phase III study. *Journal of Clinical Oncology* 2010; **28**:15s(suppl; abstr) 4004.

11. Hallissey MT, Dunn JA, Ward LC, Allum WH. The second British Stomach Cancer Group trial of adjuvant radiotherapy or chemotherapy in resectable gastric cancer: five-year follow-up. *Lancet* 1994; **343**: 1309–12.

12. Macdonald JS, Smalley SR, Benedetti, J, *et al.* Chemoradiotherapy after surgery compared with surgery alone for adenocarcinoma of the stomach or gastroesophageal junction. *New England Journal of Medicine* 2001; **345**: 725–30.

13. Leong T, Willis D, Joon DL, *et al.* 3D conformal radiotherapy for gastric cancer—results of a comparative planning study. *Radiotherapy and Oncology* 2005; **74**: 301–6.

14. Gillen S, Schuster T, Meyer Zum Büschenfelde C, *et al*. Preoperative/neoadjuvant therapy in pancreatic cancer: a systematic review and meta-analysis of response and resection percentages. *PLoS Medicine* 2010; **7**(4): e1000267.

15. Stocken DD, Büchler MW, Dervenis C, *et al*. Meta-analysis of randomised adjuvant therapy trials for pancreatic cancer. *British Journal of Cancer* 2005; **92**: 1372–81.

16. Chauffert B, Mornex F, Bonnetain F, *et al*. Phase III trial comparing intensive induction chemoradiotherapy (60 Gy, infusional 5-FU and intermittent cisplatin) followed by maintenance gemcitabine with gemcitabine alone for locally advanced unresectable pancreatic cancer. Definitive results of the 2000–01 FFCD/SFRO study. *Annals of Oncology* 2008; **19**: 1592–9.

17. Krishnan S, Rana V, Janjan NA, *et al*. Induction chemotherapy selects patients with locally advanced, unresectable pancreatic cancer for optimal benefit from consolidative chemoradiation therapy. *Cancer* 2007; **110**: 47–55.

18. Pauls S, Sokiranski R, Schwarz M, *et al*. Value of spiral CT and MRI (1.5 T) in preoperative diagnosis of tumors of the head of the pancreas. *Rontgenpraxis* 2003; **55**(1): 3–15.

19. Higashi T, Saga T, Nakamoto Y, *et al*. Diagnosis of pancreatic cancer using fluorine-18 fluorodeoxyglucose positron emission tomography (FDG PET)—usefulness and limitations in 'clinical reality'. *Annals of Nuclear Medicine* 2003; **17**, 261–79.

20. Landry JC, Yang GY, Ting JY, *et al*. Treatment of pancreatic cancer tumors with intensity-modulated radiation therapy (IMRT) using the volume at risk approach (VARA): employing dose-volume histogram (DVH) and normal tissue complication probability (NTCP) to evaluate small bowel toxicity. *Medical Dosimetry* 2002; **27**: 121–9.

21. Sindelar WF, Kinsella TJ. Studies of intraoperative radiotherapy in carcinoma of the pancreas. *Annals of Oncology* 1999; **10**(Suppl. 4): S226–30.

22. Chang DT, Schellenberg D, Shen J. Stereotactic radiotherapy for unresectable adenocarcinoma of the pancreas. *Cancer* 2009; **115**: 665–72.

23. Koong AC, Le QT, Ho A, *et al*. Phase II study to assess the efficacy of conventionally fractionated radiotherapy followed by a stereotactic radiosurgery boost in patients with locally advanced pancreatic cancer. *International Journal of Radiation Oncology, Biology, Physics* 2005; **63**: 320–3.

24. Buskirk SJ, Gunderson LL, Adson MA, *et al*. Analysis of failure following curative irradiation of gallbladder and extrahepatic bile duct carcinoma. *International Journal of Radiation Oncology, Biology, Physics* 1984; **10**: 2013.

7 Lower gastrointestinal tract

7a

Rectal cancer

Rob Glynne-Jones, Mark Harrison,
David Sebag-Montefiore

7.1 Introduction

Approximately 15 000 patients in the UK develop rectal cancer each year. Unlike the colonic portion of the large bowel the majority of the rectum lies below the peritoneal reflection and has no serosa. This allows tumour to penetrate deeply into perirectal fat. An additional issue is that the rectum lies within the bony anatomy of the pelvis. These factors have previously been associated with a high risk of locoregional failure within the pelvis[1]. The lack of a precise definition of the upper rectum influences the apparent local recurrence rate, since the risks of local failure are much higher for cancers in the lower rectum. In the UK the upper limit of the rectum is defined as 15 cm from the anal verge.

The past 20 years have seen a significant evolution in surgical practice, particularly with the technique of meticulous mesorectal dissection, whereby the surgeon removes all of the surrounding mesorectal fat in a neat anatomical package, which is associated with much lower rates of local recurrence and improved survival[2]. This technique is termed total mesorectal excision (TME). Surgery remains the mainstay of treatment, yet the optimal multimodality management of stages II and III rectal cancer currently remains an increasing challenge.

There is still little worldwide consensus of what constitutes locally advanced rectal cancer. For the purpose of this chapter we have used the definition 'rectal tumours, which would require more than a standard conventional TME as defined by staging MRI'.

Local recurrence after potentially curative resection is partly explained by the work of Quirke and colleagues[3]. The presence of microscopic tumour cells within 1 mm of the radial or circumferential resection margin (CRM) is clearly demonstrated to be associated with a very high rate of local recurrence and poor survival. This view is supported by national audit data from Norway and within the context of phase III trials testing the role of adjuvant radiation[4,5].

This chapter assesses the role of radiation therapy in rectal cancer, with emphasis on preoperative imaging, patient selection for preoperative chemoradiotherapy (CRT) and short-course preoperative radiotherapy (SCPRT), and postoperative chemoradiation. We describe the various available planning techniques. More conformal techniques such as intensity-modulated radiotherapy (IMRT), volume-modulated arc therapy

(VMAT), and brachytherapy are also described. In addition, chemoradiation and radiotherapy as an adjunct to local excision and endoluminal irradiation are also reviewed.

7.1.1 Staging

Rectal cancer is classified according to the TNM system, but it is controversial which TNM version should be used. The 1997 (TNM version 5) definition states tumour deposits should be considered as positive lymph nodes if > 3 mm in size. This definition is reproducible and comparisons with radiological imaging can be performed. The majority of units in the UK continue to recommend TNM version 5, whilst others endorse the current version 7. It is recommended that the multidisciplinary team (MDT) should be aware which version is being routinely used. In addition, TNM staging as recommended for low rectal cancer is probably inappropriate, or at least clinically unhelpful, since as with anal cancer it is based on size rather than depth of penetration.

7.1.2 Imaging

High-resolution pelvic MRI is routine in the UK as a preoperative staging and selection tool for the use of preoperative radiation. MRI can predict the likelihood of involvement of the CRM particularly in the mid-rectum, involvement of the levators in the low rectum, and the extramural depth of invasion[6,7]. This provision of accurate information on primary tumour local extension, precise location, nodal-stage, potential CRM involvement, and extramural venous invasion is essential for defining the optimum treatment strategy on an individual basis. Since CRM can only be defined postoperatively by histological examination of the surgical plane, it may be more appropriate to describe tumour in relation to an anatomical structure, such as the mesorectal fascia (MRF) or the levator musculature[8]. Obviously, this preoperative assessment can identify patients at risk of an R1 resection and can accurately predict ultimate outcome.

It is also important to identify the nodal status not only inside but also outside the MRF. Currently, however, MRI, multislice CT, and ERUS are all equally inadequate for the detection of involved lymph nodes despite specific imaging features such as size ≥ 8 mm/round/heterogenous/irregular in nodal border. FDG-PET has not improved the accuracy, and may lead to nodal upstaging.

Although diagnostic MRI can accurately define the extent of the primary tumour, and may reduce interobserver variation, it is usually performed with the patient supine. Studies are required to evaluate the role of MRI/CT coregistration for the planning process and whether prone pelvic MRI is required. Some recent clinical studies have looked at MRI CT fusion for delineation of the CTV[46]. Others have examined whether greater accuracy can be achieved using MRI and FDG-PET-CT to give additional information to standard pretreatment evaluation and whether it would change the shape and the size of the GTV delineation.

Hence, the integration of PET scanning in combination with conventional morphological imaging is under investigation. Evidence suggests there may be more accurate and reliable definition of the tumour volume leading to changes in the GTV compared to CT and more consistency in planning[47–49].We consider that at present, FDG-PET-CT coregistered with a planning CT requires further validation before routine implementation. It also remains unclear whether this use of different volumes impacts on clinical outcome, unless a boost to higher than conventional doses is planned.

7.1.3 Choice of preoperative treatment— short-course preoperative radiotherapy (5 × 5 Gy) or chemoradiation

There are two preoperative radiation approaches currently in common use in the UK. These are SCPRT and long-course preoperative CRT. Current trials suggest that in resectable cancers, where the preoperative MRI suggests/MRF is not potentially involved, SCPRT and CRT are equivalent in terms of outcomes such as local recurrence, disease-free survival (DFS) and overall survival (OS)[9,10]. For more advanced cases, where the CRM/MRF is breached or threatened according to the MRI, the addition of 5-FU to radiation has favourable effects on relapse-free survival, and cancer-specific survival with a trend to improve OS[11].

7.1.4 Short-course preoperative radiotherapy

SCPRT employs a short intensive course of radiotherapy of 25 Gy over 5 days followed by surgery within 2–5 days, with the predominant aim of reducing the risk of pelvic recurrence. This approach has gained widespread acceptance in many European and UK centres after the publication of the Swedish, Dutch, and UK rectal cancer trials[12–14]. The landmark Swedish rectal cancer trial[12] randomized patients to surgery or SCPRT followed by surgery and demonstrated not only a significant reduction in local recurrence but also a 10% absolute improvement in survival.

The Dutch trial examined the routine use of SCPRT followed by TME, against TME and selective postoperative radiotherapy in the event of histopathological evidence of involvement of the circumferential resection margin. This trial demonstrated local recurrence rates at 43 months of 4.1% after SCPRT compared with 11.5% after initial TME alone[15]. The 10-year local recurrence cumulative incidence was 5% in the group assigned to SCPRT versus 11% in the surgery-alone group (P < 0.001) demonstrating both a sustained long-term improvement in local control as well as an improved 10-year survival from 40–50% in a subgroup with stage III disease and a negative CRM. However, the overall results of the Dutch trial do not show a difference in survival which implies that some groups (? node negative) are disadvantaged in terms of survival by radiotherapy. The deaths from second malignancy were higher in the RT arm than the TME-alone arm (13.7% vs. 9.4%). Given this finding is seen after only 11.6 years of follow-up—this difference is highly likely to widen at 15–25 years.

The MRC CR07 trial[5,14] randomized 1350 rectal cancer patients to either SCPRT (5 × 5 Gy) or selective postoperative chemoradiation (25 × 1.8 Gy with concurrent 5-FU) administered only for patients with histologically involved (≤ 1 mm) resection margins. Overall clinically significant absolute risk reduction in 3-year local recurrence rate of 6.2% was observed, corresponding to a relative risk reduction of 61%. DFS was 6% better in the preoperative radiotherapy group, but there was no improvement in OS. The CR07 trial suggests SCPRT is of benefit in terms lowering local recurrence for all tumour locations, all pathological stages, and good, average or poor quality surgery.

7.1.5 Chemoradiation

Since the early 1980s, 5-FU alone, and more recently combinations of cytotoxic chemotherapy using oxaliplatin or irinotecan, have formed the basis of chemotherapy treatment for patients with metastatic colorectal cancer. Historical postoperative studies in the 1980s examined 5-FU-based chemotherapy and radiotherapy[16] or their combination, and showed a significant benefit for the concurrent chemoradiation[17,18]. With the introduction of improved preoperative imaging (CT, transrectal ultrasound, and MRI) to stage the patient, the strategy of postoperative chemoradiation has been extrapolated to the preoperative setting.

Randomized controlled trials have proved preoperative radiotherapy or chemoradiation[19–23] are more effective compared to postoperative chemoradiation therapy in terms of local recurrence with less acute and late toxicity.

In the German Trial CAO/ARO/AIO—94[22] a total of 823 patients were randomized between preoperative CRT and postoperative CRT (patients received postoperative adjuvant chemotherapy in both arms of this trial). Acute and late toxicity were significantly reduced with the preoperative approach. Locoregional failure was 6% in the preoperative arm versus 13% in the postoperative arm. There was, however, no difference in the distant metastases rate or OS.

The integration of chemotherapy is, therefore, attractive as it may act as both a radiosensitizing agent within the pelvis, with systemic effects to eradicate distant micrometastases.

5-FU-based chemoradiation is effective at downstaging patients with rectal cancer and between 10–15% of patients will have achieved a pathological complete response (pCR) at surgery[23–26]. The converse of this is that many still fail to respond sufficiently. When compared with radiation alone, 5-FU-based chemoradiation has improved locoregional control[25,26], without any improvement in DFS or OS. However, for more advanced unresectable cases or when the preoperative MRI shows a threatened or breached CRM, 5-FU-based chemoradiation has a statistically significant effect on resectability and DFS[11,27].

The NSABP R03 used a similar design but only recruited 267 of its planned patient target number (n = 900), so results should be interpreted with caution. In the preoperative CRT arm 15% of patients achieved a pCR. Five-year locoregional recurrence was 10.7% in each treatment arm (p = 0.693). A significant improvement of 5-year DFS (65% vs. 53%, p = 0.011), and a non-significant improvement in 5-year OS (75% vs. 66%, p = 0.065) were also observed for the preoperative arm. Since local recurrence remained 10.7% in both arms, improvements in local control are unlikely to be responsible.

Interest in intensifying chemotherapy has been stimulated by the success of improved PFS of combination chemotherapy in advanced disease and the improvement in DFS and OS when oxaliplatin is added to a fluoropyrimidine in the adjuvant setting in colon cancer[28,29]. However, to date these combinations in chemoradiation have not clearly improved early clinical or pathological outcomes in resectable rectal cancer[30–32].

Molecularly targeted agents such as cetuximab, panitumumab, and bevacizumab have also been integrated into standard chemotherapy regimens in colorectal cancer, and therefore have been incorporated in phase I/II studies into chemoradiation schedules[33]. However, although the observed increased toxicity is acceptable there is no current evidence to support additive effects. In fact a possible antagonism has been postulated when cetuximab is added to fluoropyrimidine containing CRT[34]. At present combination CRT and the addition of a biological agent remain investigational.

7.1.6 Selection of patients for short-course preoperative radiotherapy or chemoradiotherapy

Currently, in the UK patients are categorized by MRI criteria into three groups—'The good, the bad and the ugly'[35]. This allows definition of three different settings where preoperative neoadjuvant treatment may be required. Early cT1/T2 tumours which are not usually treated with radiotherapy; more advanced T3 tumours in which the patient is at risk of local recurrence[13,14] and who require TME and preoperative radiotherapy; and patients where the MRI suggests the CRM/MRF is potentially involved and hence require preoperative chemoradiotherapy for downstaging (MERCURY). Currently recommended radiotherapy field sizes are the same in the latter circumstances.

Following this lead, the 2011 colorectal guidelines from NICE describe three different risk groups of patients with rectal cancer, defined by their possibility of local recurrence. These groups are defined in Table 7a.1.

The appropriate choice of preoperative regimen is summarized in Fig. 7a.1.

Table 7a.1 Risk groups of patients with rectal cancer

Risk of local recurrence for rectal tumours as predicted by MRI	Characteristics of rectal tumours predicted by MRI
High	A threatened (<1 mm) or breached resection margin **or**
	Low tumours encroaching onto the inter-sphincteric plane or with levator involvement
Moderate	Any cT3b or greater, in which the potential surgical margin is not threatened **or**
	Any suspicious lymph node not threatening the surgical resection margin **or**
	The presence of extramural vascular invasion
Low	cT1 or cT2 or cT3a **and**
	No lymph node involvement

National Institute for Health and Clinical Excellence (2011) CG 131 Colorectal cancer: The diagnosis and management of colorectal cancer. London: NICE. Available from www.nice.org.uk/guidance/CG131. Reproduced with permission.

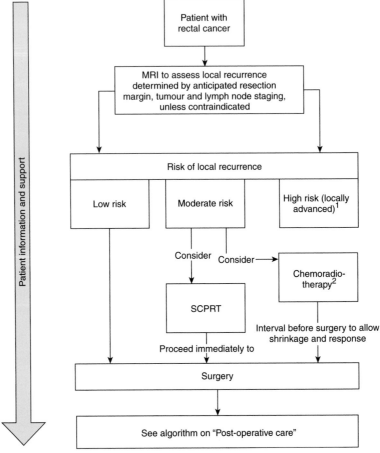

Fig. 7a.1 Management of local disease—patients with rectal cancer.

7.1.7 **Rendering unresectable tumours resectable**

A minority of patients (approximately 20% defined by MRI) will present with a rectal tumour that cannot be completely resected with a standard conventional TME without leaving residual macroscopic or microscopic disease. It is essential that these patients are identified before an (unsuccessful) attempt at surgical resection is made. Once identified preoperative strategies to facilitate tumour shrinkage and an R0 resection can be initiated.

Two small phase III trials[11,27] in unresectable rectal cancer compared radiation alone with CRT, demonstrating improved resectability and local control with the use of CRT.

In those patients with locally advanced rectal cancer the pelvic MRI can demonstrate the relationship of the primary tumour to the surrounding mesorectal fascia and the surrounding organs, such as bladder, vagina, and levators, as well as enlarged lymph nodes outside the mesorectal fascia. Thus patients may be selected for preoperative chemoradiation in this category if there is evidence of:

- Primary tumour involving or extending beyond the mesorectal fascia.
- Primary tumour within 1–2 mm of the mesorectal fascia. Whether this should be 1 or 2 mm is debated.
- Involved lymph nodes outside the mesorectal fascia (usually pelvic side wall or iliac nodes).
- Invasion of other solid pelvic organs, i.e. prostate/uterus.

There is general agreement that this group of patients should receive preoperative CRT. Current approaches use a fluoropyrimidine combined with radiation although a considerable number of phase II and phase III studies of the integration of combination chemotherapy schedules have recently been presented. To date, this approach remains investigational and preliminary results have been disappointing.

The majority of randomized trials in rectal cancer have been performed in patients with resectable rectal cancer where the aim of treatment has been to lower the risk of local recurrence, which has often formed the primary endpoint.

Systematic reviews/meta-analyses have been published on the role of radiotherapy in rectal cancer. A Cochrane review protocol is also examined the role of preoperative radiotherapy in rectal cancer[36] and supports these findings, but did not confirm a significant reduction in mortality (HR 0.93, 95% CI: 0.87–1)(absolute difference 2% if the expected survival rate is 60%),or a sphincter-sparing benefit.

7.1.8 Postoperative chemoradiotherapy

The North American standard of care was to deliver postoperative chemotherapy and CRT to all patients post-operatively with T3/4 or N+ disease (approximately 80% of all resected cancers). This approach was defined within the NIH consensus statement in 1990 based on the results of three USA trials. Based on the results of the German AIO/ARO study[22], Europe and the USA have extrapolated to the preoperative setting and deliver preoperative CRT to patients who are considered T3/4 or N+ on the basis of preoperative transrectal ultrasound.

7.1.9 A blanket approach to short-course preoperative radiotherapy

In the UK, some colorectal MDTs have adopted the non-selective use of routine SCPRT whilst others use pelvic MRI to determine patients whose primary tumour is predicted to be clear of the CRM and in whom initial surgery is performed. If surgery is performed and there is unexpected involvement of the CRM then postoperative CRT is required.

The present authors do not feel comfortable with a blanket approach. Studies have demonstrated that radiotherapy has acute and long-term detrimental effects on quality of life with significant risks of permanent morbidity. About 5–10% of patients will experience such grade 3 or 4 late morbidity[37,38]. Effects on sexual functioning[39], urinary incontinence[40], bowel function[41], and an increase in faecal incontinence[42,43] and of insufficiency fractures in the pelvis[44] have all been documented after SCPRT. These complications depend on the size of the radiation field, shielding, the overall treatment time, the fraction size, and total dose. There may also be unexplained late cardiac effects[40]. At present, late effects have been less well described following long-course preoperative chemoradiation though this is probably a consequence of the shorter follow-up seen in these studies. A subset of French patients within the EORTC 22921 trial suggests that the addition of chemotherapy to radiotherapy adversely affects sexual functioning, diarrhoea and quality of life[45].

The most recent studies suggest that 17–20 patients need to undergo adjuvant radiotherapy to prevent one local recurrence using a blanket approach. For example, the Dutch trial demonstrated an absolute reduction in local recurrence of 6%. Thus, if 100 are irradiated, then six local recurrences are prevented, so the number of patients treated to prevent one local recurrence is 16.7. Current evidence does not therefore support the widespread advocacy for routine adjuvant radiotherapy as used in the treatment arms of recent trials.

There is increasing evidence that patients with very low tumours that require abdominoperineal excision are at higher risk of involvement of the CRM, local recurrence, and inferior survival. Many concur that routine preoperative radiotherapy is indicated for this group of patients. It is a source of considerable debate whether SCPRT may be used in some, CRT is preferred in others and if a wider 'cylindrical' or extralavator surgical technique should be adopted to reduce the rate of an involved CRM.

7.2 Radiation treatment planning

To target the tumour accurately, and get the best functional outcome out of the combination of surgery and radiotherapy, a good collaboration between surgeons, radiologists, and radiation oncologists is essential. In the UK this approach is facilitated by the MDT meeting, which discusses every patient both in the preoperative and postoperative setting. Useful information includes knowledge of the extent and position of the tumour, the likely operation that will be performed, and any relevant previous medical, surgical, and oncological treatment history.

7.2.1 Computed tomography planning techniques

Patients are given instructions to maintain a full bladder as far as possible prior to each simulation/scanning/treatment session in order to push small bowel cranially out of the field as far as possible, and are reminded throughout the treatment period. A CT planning scan with 3-mm thick slices should be performed with the patient prone in the treatment position. Intravenous contrast is ideally required, as the vessels cannot always be easily identified from sequential slices. A radio-opaque marker is useful to demonstrate the position of the anal verge.

Patients are scanned from the superior aspect of L5, to 2 cm beyond the anal marker in order to cover the whole of the pelvis, recto sigmoid, and rectum. The anterior borders can be defined in female patients with the use of a radio-opaque tampon placed in the vagina at CT planning or at simulation. Dilute small bowel contrast may be used at this stage (gastrograffin 20 mL in 1 L of water given 45–60 minutes prior to the planning scan).

The radiation fields used in current practice are based on a number of factors including the fields used in previous and current Phase III trials, the limited information available on the pattern of failure, and differences in the philosophy of field size according to whether certain lymph node regions should or should not be included.

A number of series prior to the TME era mapped recurrence, including the seminal work from Gunderson at the Mayo clinic, amassed from 'second-look' surgery[1] and then made recommendations for CTV contouring based on observed sites of local recurrence[50–53]. Early studies defined the superior border as the junction of L5/S1, but some studies treated up to the origin of the inferior mesenteric artery at L4 and were associated with significant morbidity. Most current radiation fields are still based on patterns of historical failure.

Rectal surgical quality has markedly improved with TME, and more recent studies examining the site of local recurrence show these are often either at the site of the anastomosis, the posterior pelvis and lateral pelvic side-wall, or low down representing inadequate surgery[54]. The risk of involvement of regional lymph node groups is also recognized to depend on whether the primary tumour lies in the upper middle- or lower-third of the rectum. All these sites have a slightly different natural history with different areas of known lymphatic drainage. With this knowledge several groups have recently attempted to redefine planning volumes for rectal cancer[55,56]. The Roels guidelines identified five predominant areas of risk for local recurrence and potential lymph node involvement. In contrast, the RTOG guidelines are a one-size-fits-all consensus of experts[56]. Further recent modifications in CTV are derived from a study where the site of recurrence has been analysed in patients within the Dutch TME study[57]. No recurrences were found cranial to the S2–S3 junction in patients without nodal involvement. In the absence of nodal involvement and with a negative CRM, only one recurrence was found cranial to the S2–S3 junction allowing a major reduction in exposure of small bowel.

Prior to initiating a Phase III multicentre chemoradiotherapy trial in the UK for rectal cancer (Aristotle), we reviewed the literature on the definition of target volumes in rectal cancer, the evidence of the site of lymph node metastases, and potential areas of subclinical disease, the site of recurrence after TME for more locally advanced disease and late morbidity[58]. With the collaboration of the trial management group, we have established recommendations for target volume definition in a simple practical guide for 3D planning within the trial (http://www.ukctg.nihr.ac.uk/trialdetails/ISRCTN09351447).

A summary of this approach is provided as follows.

Gross tumour volume

GTV is defined as all sites of gross tumour seen on the planning CT scan with the help of the information from diagnostic CT and MRI scans clinical examination, endoscopy, barium enema, and PET scan.

This concept is more difficult than in some other disease sites. The discontinuous nature of many rectal cancers with extranodal deposits may require the demarcation of more than one GTV area. Also, if there is a large lymph node separate to the primary tumour then both should be outlined separately. However, loops of unopacified small bowel and perirectal soft tissue densities may be easily confused with lymph node structures and correlation with the diagnostic pelvic MRI is important.

It is useful to document the sites of all areas of GTV at the time of MDT discussion. This includes extra nodal deposits, involved lymph nodes with irregular borders and mixed signal characteristics, and extra mural vascular invasion.

It is recommended that the normal rectal wall is also included if the GTV is not circumferential (see Fig. 7a.2a, left-hand panel). Clearly, following surgery when postoperative radiotherapy is given, it is not possible to define a GTV but preoperative imaging can indicate the site of the primary tumour prior to surgery.

Clinical target volume

The CTV will encompass areas of microscopic spread beyond the defined GTV. The CTV is defined in three separate steps:

1. CTVA: a 1-cm margin is applied in all directions to all sites of GTV (Fig. 7a.2a, right-hand panel).

2. CTVB: includes sites of potential microscopic disease including the mesorectal, presacral, and internal iliac node regions. The limits of this volume are defined as follows:

   - *Superior limit:* is the S2/S3 junction provided that there is a 2-cm margin above the most superior aspect of the GTV (Fig. 7a.2b, left-hand panel, black line). If this is not the case, the superior border is defined as 2 cm superior to the most superior aspect of GTV (Fig. 7a.2b, right-hand panel, white line).

   - *Inferior limit:* is at the level of puborectalis which corresponds with the level that the mesorectal fat is no longer seen on the axial planning CT scans (Fig. 7a.2c, left-hand panel) *providing* that there is at least a 2-cm margin below the most inferior aspect of the GTV. If this is not the case (i.e. a very low tumour) the CTV is placed 2 cm below the inferior aspect of the GTV (Fig. 7a.2c, grey line).

   - *Posterior limit:* is the anterior surface of the sacrum and the coccyx. In the presence of the symptom of nerve infiltration but in the absence of macroscopic tumour the CTVB may be placed 0.5 cm posterior to the anterior border of the sacrum.

   - *Anterior limit:* is determined at different levels of the pelvis:

     - Upper pelvis—a 7-mm margin is applied to the internal iliac arteries. The most anterior of the arteries is used to determine the border (Fig. 7a.2d).

     - Mid pelvis—is 1cm anterior to the anterior mesorectal fascia *or* the anterior limit of the lateral (iliac) pelvic lymph node compartment, whichever is the more anterior. In the example shown, the patient has a very small mesorectum and the border is determined by the lateral nodal compartment.

     - Lower pelvis—is the outer aspect of the sphincter complex.

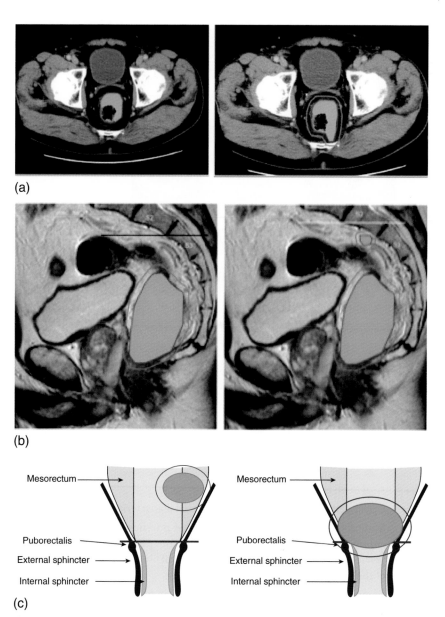

Fig. 7a.2 (a) Left hand panel: GTV. Right hand panel: CTVA and GTV. (b) Superior border CTVB. Left hand panel: S2/3 junction. Right hand panel: 2 cm superior to the most superior limit of GTV. (c) Inferior border CTVB at level of puborectalis (left-hand panel) and 2 cm below most inferior limit of HGTV (right-hand panel).

(*Continued*)

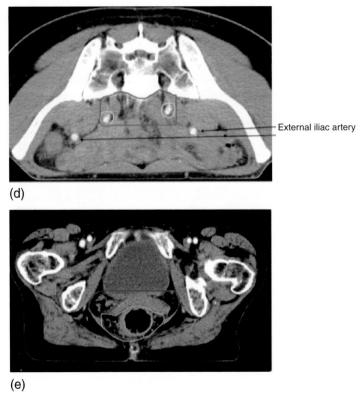

External iliac artery

(d)

(e)

Fig. 7a.2 *Continued* (d) Anterior CTVB border upper pelvic. (e) Anterior CTVB border mid pelvis—this patient has a small mesorectum and the border is determined by the lateral pelvic nodal compartment.

- *Lateral limit:* is also determined at three levels:
  - Upper pelvis—7 mm lateral to the internal iliac arteries.
  - Mid pelvis—the medial aspect of obturator internus in the absence of nodal enlargement outside the mesorectum. In the presence of involved lateral pelvic nodes the border is the lateral border of obturator internus on the side of the involved nodes.
  - Lower pelvis—the outer aspect of the sphincter complex unless there is involvement of the sphincter complex when a 1-cm margin later to the sphincter complex is used.

The ARISTOTLE trial management group did not find any evidence to justify the inclusion of the entire ischiorectal fossa or the external iliac nodes in the CTVB.

3. Final clinical target volume (CTVF)—defined by the union of CTVA and CTVB.

Planning target volume

The PTV is defined by applying a 1-cm direction in all directions to the CTVF.

This volume ensures coverage of the CTV taking into account the systematic and random set-up errors and internal movement the may occur when delivering a radical course of radiation. Including variations in tissue position, size, shape and also variations in patient position (ICRU Report 50)[59]. Few departments have data to determine their systematic and random set up errors.

7.2.2 Organs at risk

Three-dimensional treatment planning allows visualization on non-axial planes with a beam configuration, and will allow a DVH and hence derive optimal and maximal doses for both tumour and normal tissue. Normal critical tissues such as small bowel, femoral heads, ureter, and bladder can be contoured and doses to these organs kept to a minimum. In anal cancer, retrospective studies suggest the incidence of late bone injury is 9% for patients with a femoral head dose of $\geq$ 44 Gy versus 0.7% for patients receiving lesser doses ($p = 0.008$)[56]. It appears that the small bowel is often close to the target volume, and in our units the dose is specified such that not more than 250 mL receives in excess of 45 Gy. Significant small bowel sparing is offered in the future by the increasing use of IMRT in rectal cancer. However, doses to significant volumes of small bowel in excess of 50.4 Gy are not recommended.

7.2.3 Normal tissue tolerances

The advent of 3D computerized planning has led to the concept of tolerance of an organ being replaced by DVHs, which focus on the dose delivered to the percentage of any specific organ. Most normal tissue tolerance algorithms assume a daily dose of 1.8 Gy per fraction. It is also wise to maintain the dose per fraction to small bowel, ureter, and bladder to 2 Gy or less, i.e. avoiding gross inhomogeneities of dose. When adjuvant radiation is given the tolerance doses of most organs is not exceeded with the exception of the small bowel (see section 7.2.4). Although DVHs may be produced for organs/structures at risk, it is currently unclear whether the planned treatment approach should be altered if predefined dose constraints cannot be met, except possibly for the small bowel. It is not possible to generate a true DVH, because currently, the assessment of dose to small bowel is usually based on the volume (cm^3) of small bowel within the high-dose volume in the pelvis rather than the entire small bowel volume within the whole abdominal cavity (which is difficult to measure).

7.2.4 Small bowel

Small bowel, baldder, and femoral heads have usually been considered the main dose-limiting organs when high doses of radiotherapy are delivered to the pelvis. Late effects in small bowel range from malabsorption to strictures, obstruction, and ulceration. The volume of the small bowel within the radiation field is crucial. Late small bowel complications are estimated to occur with an actuarial 5-year risk of between 5–40% incidence following pelvic postoperative radiotherapy to a dose of 45–50 Gy in rectal cancer[60–62]. IMRT may substantially reduce the dose of radiation that the small bowel within the pelvis may receive but at the expense of a larger volume of small bowel receiving a more modest radiation dose[63–65].

7.2.5 **Patient position**

The prone position is thought to displace small bowel out of the treatment volume mainly by anterior displacement. It also allows the accurate placement of a radio-opaque marker on the anal verge and the use of a belly board. It is, however, uncomfortable, particularly when the patient has a stoma, and is often unfeasible because of impaired mobility and comorbidities. Whilst the supine position is considered more reproducible, most clinicians are concerned that this position may increase the small bowel in the treated volume. The validity of this is unclear with studies that show no significant difference in the irradiated small bowel volume between the two treatment positions[66].

7.2.6 **Small bowel exclusion techniques**

Most units employ bowel displacement techniques such as a Styrofoam mould[50] or the belly board (the use of a table cut-out which allows the upper abdominal small bowel contents to fall forwards and upwards, and further reduce the volume of small bowel irradiated[67,68]. There is no consensus as to the optimal standardized position of the belly board opening. However, if the opening is placed near the lumbosacral junction, the reduction in volume of small bowel irradiated appears to be maximal[69].

7.2.7 **Organ motion set-up variation and treatment margins in radical radiotherapy**

Studies in organ motion and set-up uncertainties have usually focused on prostate irradiation because radiotherapy doses are very high. In contrast, radiotherapy in rectal cancer delivers doses which are relatively modest (45–54 Gy) and can be boosted with endoluminal brachytherapy, if required. The rectum does move during a course of radiotherapy[70,71]. However, there is little data to quantify rectal motion and set-up variation. Anatomical considerations suggest the rectum is more fixed at the distal end than proximally.

7.2.8 **Postoperative radiotherapy clinical target volume**

Postoperative radiotherapy is currently delivered for those patients who did not receive preoperative RT and proceeded to surgery, but on histology review are seen to have an involved circumferential margin (≤ 1 mm), large defects in the mesorectum, evidence of perforation, or widespread extranodal deposits. We do not recommend the routine use of postoperative chemoradiation for pT3 N0 or N1 provided there is a good quality of mesorectal excision[5]. The indication of postoperative chemoradiation for patients with pT4 and/or pN2 histology remains a matter of debate.

CTV—after surgery there is no defined GTV. The CTV includes a surrounding safety margin beyond the mesorectum and the nodal groups at risk of presumed subclinical involvement. In patients who have had abdominoperineal resection, the authors do not include the perineal scar. However, there is no consistent approach or agreement as to the CTV.

Displacement of the omentum and attachment within the pelvis, and the application of ligaclips at the time of surgery are helpful procedures in order to localize areas at high risk of local recurrence, and also minimize the late effects of RT.

7.2.9 Total dose

Conventionally, when 1.8 Gy per fraction is used, total doses in the range 45–50.4 Gy have been delivered in the preoperative setting, and 50.4 Gy with the option of a 5.4-Gy boost to the tumour bed in the postoperative setting. It is assumed that the treatment will be delivered 5 days per week, one fraction per day, 1.8 Gy per fraction.

7.2.10 Preoperative dose

Of the SCPRT studies that delivered a biologically equivalent dose > 30 Gy, one used 20 Gy in four fractions[72] but the Swedish Rectal Cancer Study, The Dutch study, and the CR07 study use 25 Gy in five fractions, which most clinicians accept is a dose fractionation that cannot safely be exceeded. However, with the use of 1.8–2.0 Gy per fraction the total dose may range from 45–54 Gy in most reported series. This upper limit is 20% higher than 45 Gy and the balance between improved local control and the increased risk of late toxicity is continually debated.

Early retrospective studies suggest an increase in local control with dose of 50 Gy compared to 40 Gy[73]. There is also more recent data from Phase II studies that higher doses may be associated with lower risks of local relapse[74-77].

A sequential Phase II from Canada has been reported. Three sequential schedules combined radiation with infusional 5-FU, escalated from 40 Gy in 20 fractions to 46 Gy in 23 fractions and finally to 50 Gy in 25 fractions. A statistically significant difference in terms of local control was observed for doses of 46 Gy and above, but no difference between 46 Gy and 50 Gy[78]. The same study also appeared to show a trend to higher pathological complete response rates with increasing radiation dose of 13%, 21% and 31% for 40 Gy, 46 Gy, and 50 Gy respectively.

Currently, there are proponents that a much higher dose (in the region of 60 Gy) can be safely delivered with the high conformality of IMRT or brachytherapy[79] in order to achieve more R0 resections.

7.2.11 Standard dose prescriptions

- Short course preoperatively: 25 Gy in five daily 5-Gy fractions.
- Preoperative CRT: 45–50.4 Gy in 25–28 daily 1.8-Gy fractions.
- Postoperative CRT: 50.4–54 Gy in 28–30 daily 1.8-Gy fractions.

It is conventional to report the dose to the ICRU reference point, the maximum dose to the PTV, and the minimum dose to the PTV. The isocentric treatment plan is usually specified to receive 100% with the 95% isodose line encompassing the PTV. The minimum dose is defined as 99% of the PTV receiving ≥ 95% of the prescribed dose. The maximum dose is defined as < 5% and < 2% of the PTV receiving 105% and 110% of the prescribed dose respectively.

7.2.12 Image-guided radiotherapy

Systematic errors in terms of set-up uncertainties result from the fact that the imaging performed for treatment planning is typically just a snapshot, and the target position determined at that moment may differ from the average target position at any subsequent treatment time. The random error is the day-to-day deviation from the average target position (which depends on internal organ motion and the repeated treatment set-up over 25 treatments). The concept of IGRT achieves tumour and soft tissue imaging in real-time to allow correction for both systematic and random errors on a daily basis. The accurate target localization demanded in IMRT and SBRT and the delivery of boost doses to small volumes identified with functional and biological imaging (e.g. PET glucose metabolism or hypoxia). Hence, IGRT should utilize ideally 4D assessment of the target volume; efficient comparisons of images with reference data and fast automated corrections.

7.2.13 Portal imaging

Portal imaging for verification of isocentre position and treatment fields should be acquired on the first treatment session both for AP and lateral images and compared to simulator films. Electronic portal imaging (EPI) can monitor set-up displacement on a daily basis in the initial phase of treatment[80]. Fields should be moved if they fall outside an agreed tolerance level—usually 5–8 mm for patients who are treated prone. This process also allows radiographers to evaluate the whole set-up, and thus to assess and correct systematic errors. Reverification is recommended on a weekly basis. These EPI images should be audited at the clinician's weekly meeting. The MLC configuration can also be verified for consistency and reproducibility.

7.3 Supportive care during radiotherapy

Expected acute side effects include diarrhoea, proctitis, urinary frequency and dysuria, erythema, and moist desquamation of the perineum in low rectal cancers. Patients should be assessed at least once weekly with regard to toxicity and their overall tolerance to treatment. Toxicity should be scored according to the Common Toxicity Criteria (CTC) v 3.0.

It is important to maximize patient tolerance with the use of simple antiemetics, antidiarrhoeals, analgesia, and nutritional support. When SCPRT is used a small minority of patients may experience an acute sensory neuropathy occurring within a few hours of delivery of the radiation. If this occurs during the treatment course, the superior border should be lowered to the S2/3 junction and usually results in disappearance of symptoms. If this is not the case, the authors currently limit the radiation dose to 15 Gy in Three fractions, as this approach was shown to significantly lower the local recurrence rate in a randomized trial[81].

7.4 Intensity-modulated radiotherapy

Technical advances such as IMRT and more recently VMAT allow greater precision and sparing normal surrounding structures such as small bowel compared to conventional 2D or 3D planning. IMRT can conform radiation dose to the target structure by partitioning the radiation field into smaller beamlets of different dose intensities. During IMRT treatment planning, standard radiation dose constraints can be used to spare critical normal tissues. Hence, IMRT may allow improved compliance, or facilitate dose-escalation, without increasing late morbidity. The benefit of reducing the irradiated small bowel volume is greatest when a very large clinical target is treated. For example, if the CTV extends superiorly to the sacral promontory or the external iliac nodes are electively irradiated. These approaches are not routinely used by the authors.

In the EORTC 22921 study trial, 522 patients retained their sphincters and of these, only 1.4% required surgery for small bowel complications. This low level of late morbidity calls into question the drive to deliver IMRT to unselected patients with the aim of reducing small bowel toxicity, unless dose escalation becomes more commonly employed.

7.5 Endocavitary radiotherapy

Appropriately selected patients with early rectal cancer can be controlled locally, and experience long-term survival with the use of endocavitry or contact radiation therapy. Treatment guidelines and indications for this technique have been developed originally by Papillon. Most clinicians would accept that the use of this radical technique as a sole modality is only appropriate if the tumour is confined to the rectal wall (T1 N0). Very small volumes of tissue are irradiated in this situation. Doses of 100 Gy can be delivered to such a small volume without risk of unacceptable late complications[82,83]. Some clinicians endorse treatment with combined external beam and contact therapy[84].

A 3-cm diameter applicator can be introduced into the rectum using local anaesthesia, and can be performed on an outpatient basis. Standard treatment regimen is four applications of 30 Gy (applied dose) with an interval of at least 2 weeks between each application. The applicator is a short focal contact X-ray unit using 50-kV rays with a dose rate of approximately 10 Gy per minute. Lack of response after two applications of radiotherapy is usually considered a sign of radioresistance and most will then proceed to surgery. High rates of local control are possible with this approach[85–89].

7.6 High dose rate brachytherapy

High dose rate intraluminal brachytherapy (HDR-ILBT) has the advantage of high conformality—i.e. a rapid fall-off of radiation dose, which allows the delivery of a high dose to the tumour while sparing normal surrounding structures such as the adjacent normal rectal mucosa, bladder, and small bowel[90]. This technique has been used at Mount Vernon Cancer Centre for the last 10 years[91] both as a boost alongside external chemoradiation, as a sole treatment for small localized tumours, and for short-term palliation for advanced symptomatic tumours particularly in the very elderly and frail. It is possible to place clips at the superior inferior extent of the tumour although these can only be relied upon to be maintained for 10–12 days. When the tumour is not circumferential, it is possible to use segmental shielding with an applicator which shields 25% or 50% of the rectal circumference. Treatment prescription is defined at 1 cm from the source access. Fractionated courses of 6 Gy up to 36 Gy for radical treatments or 10–12 Gy in one to two fractions can be delivered. In a group of 25 assessable patients treated at the Mount Vernon Cancer Centre, 14 have achieved complete clinical response. Median survival for the entire group was 6 months (range 1–54 months) and for patients treated with radical intent, 25 months (range 1.5–54)[92]. This technique is also being used in the preoperative setting[90]. However, there is limited data available evaluating the advantages of HDR-ILBT with EBRT as compared to EBRT alone. HDR-ILBT for advanced or inoperable tumours of the rectum has been used both in the palliative setting and to dose escalate after chemoradiation for curative treatment[93]. Future multicentre trials are planned.

7.7 Intraoperative radiotherapy

Intraoperative radiotherapy (IORT) allows additional irradiation of the tumour bed (i.e. higher doses) without compromising the surrounding organs at risk. IORT is usually delivered under anaesthetic using electrons, and doses range between 10–20 Gy. However, no randomized trials have evaluated the benefit of IORT in addition to EBRT. Recent pooled results of multimodality treatment of locally advanced rectal cancer in four major treatment centres with particular expertise in IORT suggest an advantage to IORT[94]. The limitations of these analyses are the inclusion of patients treated during a relatively long period, the heterogeneity of external beam treatments, and variations in adjuvant chemotherapy. This technique is extensively used particularly in locally advanced and recurrent rectal cancer in Europe, but as far as we are aware there are no centres in the UK performing IORT.

7.7.1 Locally recurrent rectal cancer

Isolated local recurrence after preoperative radiotherapy and TME surgery is infrequent and rarely isolated[15], and has been almost invariably associated with a fatal outcome[95]. More recent series have documented 50% R0 resections and 40% 5-year survival. In contrast, surgery for central/anastomotic local recurrences resulted in the most favourable outcomes, with 77% R0 resections and 60% 5-year survival[96]. Currently in the UK, the widespread use of preoperative radiotherapy, either short-course or long-course with chemoradiation, has ensured that isolated local pelvic recurrence is a rarity.

When local recurrence occurs when radiotherapy has *not* previously been administered, radiotherapy or chemoradiation can produce good palliation of symptoms, but long-term local control is seldom achieved. The duration of effective palliation is usually short with further progression of symptoms within 3–6 months after irradiation[97], and complete responses are rarely achieved even with high radical doses in the region of 60 Gy[98]. More recently a study of patients with unresectable or locally recurrent rectal cancer treated using 3D planning and prolonged venous infusion of 5-FU to a dose of 45 Gy. There was no difference between unresectable locally advanced cancers and recurrent cancers as regards outcome[99]. IORT has also been advocated in these circumstances.

7.7.2 Re-irradiation

Further re-irradiation in these circumstances remains a controversial issue[100,101]. Some single-centre experience suggests this practice may be safe in the short term, although long-term evidence is sparse[102]. Others have recommended hyperfraction-ated accelerated radiotherapy with doses of 39 Gy (1.5 Gy twice daily) provided previ-ous pelvic radiotherapy with a retreatment interval of > 1 year)[103]. The present authors would not usually recommend doses in excess of 36 Gy in combination with 5-FU based chemotherapy, and should only be considered in circumstances where concerns regarding small bowel toxicity were absent.

7.8 The future

We recommend examining dose escalation of radiotherapy to the primary tumour where MRI predicts a threatened CRM—either with external beam or brachytherapy. Potential strategies of neo-adjuvant, concurrent, consolidation (after chemoradiation and before surgery), and postoperative adjuvant chemotherapy with cytotoxic agents, are promising, but consolidation chemotherapy following chemoradiation in locally advanced disease, and neo-adjuvant chemotherapy in MRI selected patients who do not require radiation appear the most attractive strategies. Improvement in the quality of surgery is also an important future goal. Stereotactic body radiotherapy, whereby a small number of high-dose-per-fraction treatments are given to a small target volume also warrants further attention particularly in the setting of local rectal recurrence and oligometastatic disease.

7.9 Conclusion

Today in the UK, the importance of achieving a R0 (CRM ≥ 1 mm) endpoint, with the emphasis on the circumferential margin, remains paramount.

Carefully controlled studies with data on functional outcome enrolling large numbers of patients will need to focus on cancers 0–10 cm from the anal margin. We will also need much longer follow-up than is presently undertaken if we are to assess accurately the long-term results of neo-adjuvant preoperative chemoradiation.

The future hope is of increasingly accurate methods of selecting patients—potentially by means of even more sophisticated MRI staging and molecular predic-tive markers—who would benefit most and least from preoperative treatment. If we could more accurately predict the risks of local recurrence, we could spare those for whom radiotherapy may be unnecessary. The tumour–node–metastasis stage (obtained either clinically by MRI or histologically at surgery) is the only proven prognostic marker to aid in the identification of patients with aggressive patterns of disease. However, clinical assessment of nodal status remains inaccurate. Current phase III trials therefore have an obligation to collect tissue and perform translational studies.

The recommendations proposed here are limited by current knowledge and subjec-tive biases and should not be considered definitive. We have purposely not addressed patient positioning and more sophisticated technical details such as organ motion and respiratory gating. We will also need to use modern, more precise treatment planning systems and develop conformal and IMRT or VMAT approaches to the treatment of rectal cancer—particularly if we intend to dose escalate above 45 Gy.

References

1. Gunderson LL, Sosin H. Areas of failure found at reoperation (second or symptomatic look) following curative surgery of adenocarcinoma of the rectum. Clinicopathological correlation and implications for adjuvant therapy. *Cancer* 1974; **34**: 1278–92.

2. Heald RJ. Total mesorectal excision is optimal surgery for rectal cancer: a Scandinavian consensus. *British Journal of Surgery* 1995; **82**: 127–9.

3. Quirke P, Durdey P, Dixon MF, Williams NS. Local recurrence of rectal adenocarcinoma due to inadequate surgical resection. Histopathological study of lateral tumour spread and surgical excision. *Lancet* 1986; **2**: 996–9.

4. Nagtegaal ID, Marijnen CA, Kranenbarg EK, *et al.* Circumferential margin involvement is still an important predictor of local recurrence in rectal cancer: not 1 millimetre but 2 millimetres is the limit. *American Journal of Surgery and Pathology* 2002; **26**: 350–7.

5. Quirke P, Steele R, Monson J, *et al.* MRC CR07/NCIC-CTG CO16 Trial Investigators; NCRI Colorectal Cancer Study Group. Effect of the plane of surgery achieved on local recurrence in patients with operable rectal cancer: a prospective study using data from the MRC CR07 and NCIC-CTG CO16 randomised clinical trial. *Lancet* 2009; **373** (9666): 821–8.

6. Brown G, Radcliffe AG, Newcombe RG, Dallimore NS, Bourne MW, Williams GT. Preoperative assessment of prognostic factors in rectal cancer using high-resolution magnetic resonance imaging. *British Journal of Surgery* 2003; **90**: 355–64.

7. Mercury Group. Diagnostic accuracy of preoperative magnetic resonance imaging in predicting curative resection of rectal cancer: prospective observational study. *British Medical Journal* 2006; **333**(7572): 779–84.

8. Glimelius B, Beets-Tan R, Blomqvist L, *et al.* Mesorectal fascia instead of circumferential resection margin in preoperative staging of rectal cancer. *Journal of Clinical Oncology* 2011; **29**(16): 2142–3.

9. Bujko K, Nowacki MP, Nasierowska-Guttmejer A, *et al.* Long-term results of a randomised trial comparing preoperative short-course radiotherapy with preoperative conventionally fractionated chemoradiation for rectal cancer *British Journal of Surgery* 2006; **93**: 1215–23.

10. Ngan S, Fisher R, Goldstein D, *et al.* TROG, AGITG, CSSANZ, and RACS. A randomized trial comparing local recurrence (LR) rates between short-course (SC) and long-course (LC) preoperative radiotherapy (RT) for clinical T3 rectal cancer: an intergroup trial (TROG, AGITG, CSSANZ, RACS) [abstract no. 3509). *Journal of Clinical Oncology* 2010; **28**(15 Suppl.): 3509.

11. Braendengen M, Tveit KM, Berglund A, *et al.* Randomized phase III study comparing preoperative radiotherapy with chemoradiotherapy in nonresectable rectal cancer. *Journal of Clinical Oncology* 2008; **26**(22): 3687–94.

12. Anonymous. Improved survival with preoperative radiotherapy in resectable rectal cancer. Swedish Rectal Cancer Trial. *New England Journal of Medicine* 1997; **336**: 980–7.

13. Kapiteijn E, Marijnen CA, Nagtegaal ID, *et al.* Dutch Colorectal Cancer Group. Preoperative radiotherapy combined with total mesorectal excision for resectable rectal cancer. *New England Journal of Medicine* 2001; **345**: 638–46.

14. Sebag-Montefiore D, Stephens RJ, Steele R, *et al.* Preoperative radiotherapy versus selective postoperative chemoradiotherapy in patients with rectal cancer (MRC CR07 and NCIC-CTG C016): a multicentre, randomised trial. *Lancet* 2009; **373**(9666): 811–20.

15. Van den Brink M, Stigglebout AM, van den Hout WB, *et al.* Clinical nature and prognosis of locally recurrent rectal cancer after total mesorectal excision with or without preoperative radiotherapy. *Journal of Clinical Oncology* 2004; **22**: 3958–64.

16. Fisher B, Wolmark N, Rockette H, *et al.* Postoperative adjuvant chemotherapy or radiation therapy for rectal cancer: Results from NSABP protocol R01. *Journal of the National Cancer Institute* 1988 2; **80**(1): 21–9.

17. Gastrointestinal Tumour Study Group – GiTSG 7175. Prolongation of disease free interval in surgical treated rectal carcinoma. *New England Journal of Medicine* 1985; **312**: 1464–72.

18. Krook JE, Moertel CG, Gunderson LL, *et al.* Effective surgical adjuvant therapy for high-risk rectal carcinoma. *New England Journal of Medicine* 1991; **324**: 709–15.

19. Cedermark B, Johansson H, Rutqvist LE, Wilking N. The Stockholm I trial of preoperative short term radiotherapy in operable rectal carcinoma. A prospective randomized trial. Stockholm Colorectal Cancer Study Group. *Cancer* 1995; **75**: 2269–75.

20. Dahl O, Horn A, Morild I, *et al.* Low-dose preoperative radiation postpones recurrences in operable rectal cancer. Results of a randomized multicenter trial in western Norway. *Cancer* 1990; **66**(11): 2286–94.

21. Pahlman L, Glimelius B. Pre- or postoperative radiotherapy in rectal and rectosigmoid carcinoma. Report from a randomized multicenter trial. *Annals of Surgery* 1990; **211**(2): 187–95.

22. Sauer R, Becker H, Hohenberger W, *et al.* Preoperative versus postoperative chemoradiotherapy for rectal cancer. *New England Journal of Medicine* 2004; **351**: 1731–40.

23. Roh MS, Colangelo LH, O'Connell MJ, *et al.* Preoperative multimodality therapy improves disease-free survival in patients with carcinoma of the rectum: NSABP-R03. *Journal of Clinical Oncology* 2009; **27**: 5124–30.

24. Bujko K, Nowacki MP, Nasierowska-Guttmejer A, *et al.* Sphincter preservation following preoperative radiotherapy for rectal cancer: report of a randomised trial comparing short-term radiotherapy vs. conventionally fractionated radiochemotherapy. *Radiotherapy and Oncology* 2004; **72**: 15–24.

25. Gerard JP, Conroy T, Bonnetain F, *et al.* Preoperative radiotherapy with or without concurrent fluorouracil and leucovorin in T3-T4 rectal cancers: results of FFCD 9203. *Journal of Clinical Oncology* 2006; **24**: 4620–5.

26. Bosset JF, Collette L, Calais G, *et al.*Chemoradiotherapy with preoperative radiotherapy in rectal cancer. *New England Journal of Medicine* 2006; **355**: 1114–23.

27. Frykholm GJ. Pahlman L. Glimelius B. Combined chemo- and radiotherapy vs. radiotherapy alone in the treatment of primary, nonresectable adenocarcinoma of the rectum. *International Journal of Radiation Oncology, Biology, Physics* 2001; **50**: 427–34.

28. Kuebler JP, Wieand HS, O'Connell MJ, *et al.* Oxaliplatin combined with weekly bolus fluorouracil and leucovorin as surgical adjuvant chemotherapy for stage II and III colon cancer: results from NSABP C-07. *Journal of Clinical Oncology* 2007; **25**(16): 2198–204.

29. André T, Boni C, Navarro M, *et al.* Improved overall survival with oxaliplatin, fluorouracil, and leucovorin as adjuvant treatment in stage II or III colon cancer in the MOSAIC trial. *Journal of Clinical Oncology* 2009; **27**(19): 3109–16.

30. Mohiuddin M, Winter K, Mitchell E, *et al.* Radiation Therapy Oncology Group Trial 0012. Randomized phase II study of neoadjuvant combined-modality chemoradiation for distal rectal cancer: Radiation Therapy Oncology Group Trial 0012. *Journal of Clinical Oncology* 2006; **24**(4): 650–5.

31. Gérard JP, Azria D, Gourgou-Bourgade S, *et al.* Comparison of two neoadjuvant chemoradiotherapy regimens for locally advanced rectal cancer: results of the phase III trial ACCORD 12/0405-Prodige 2. *Journal of Clinical Oncology* 2010; **28**(10): 1638–44.

32. Aschele C, Cionini L, Loardi S, *et al*. Primary tumor response to preoperative chemoradiation with or without oxaliplatin in locally advanced rectal cancer: Pathologic results of the STAR-01 randomized phase III trial. *Journal of Clinical Oncology* **29**: 2773–80.

33. Pinto C, Di Fabio F, Maiello E, *et al*. Phase II study of panitumumab, oxaliplatin, 5-fluorouracil, and concurrent radiotherapy as preoperative treatment in high-risk locally advanced rectal cancer patients (StarPan/STAR-02 Study). *Annals of Oncology* 2011; **29**(20): 2773–80.

34. Glynne-Jones R, Mawdsley S, Harrison M. Cetuximab and chemoradiation for rectal cancer—is the water getting muddy? *Acta Oncologica* 2010; **49**(3): 278–86.

35. Smith N, Brown G. Preoperative staging of rectal cancer. *Acta Oncologica* 2008; **47**(1): 20–31.

36. Wong R, Tandan V, De Silva S, Figueredo A. Preoperative radiotherapy and curative surgery for the management of localized rectal cancer. *Cochrane Database of Systematic Reviews* 2007; **2**: CD002102.

37. Dahlberg M, Glimerlius B, Pahlman L. Preoperative radiation effect functional results after surgery for rectal cancer: results from a randomised study. *Diseases of the Colon and Rectum* 1998; **41**: 543–9.

38. Tepper JE, O'Connell M, Niedzwiecki D, *et al*. Adjuvant therapy in rectal cancer: analysis of stage, sex, and local control—final report of intergroup 0114. *Journal of Clinical Oncology* 2002; **20**: 1744–50.

39. Marijnen CA, van de Velde CJ, Putter H, *et al*. Impact of short-term preoperative radiotherapy on health-related quality of life and sexual functioning in primary rectal cancer: report of a multicenter randomized trial. *Journal of Clinical Oncology* 2005; **23**(9): 1847–58.

40. Pollack J, Holm T, Cedermark B, *et al*. Late effects of short-course preoperative radiotherapy in rectal cancer. *British Journal of Surgery* 2006; **93**(12): 1519–25.

41. Peeters KC, van de Velde CJ, Leer JW, *et al*. Late side effects of short-course preoperative radiotherapy combined with total mesorectal excision for rectal cancer: increased bowel dysfunction in irradiated patients—a Dutch colorectal cancer group study. *Journal of Clinical Oncology* 2005; **23**(25): 6199–206.

42. Lange MM, den Dulk M, Bossema ER, *et al*.; Cooperative Clinical Investigators of the Dutch Total Mesorectal Excision Trial. Risk factors for faecal incontinence after rectal cancer treatment. *British Journal of Surgery* 2007; **94**(10): 1278–84.

43. Stephens RJ, Thompson LC, Quirke P, *et al*. Impact of short-course preoperative radiotherapy for rectal cancer on patients' quality of life: data from the Medical Research Council CR07/National Cancer Institute of Canada Clinical Trials Group C016 randomized clinical trial. *Journal of Clinical Oncology* 2010; **28**(27): 4233–9.

44. Herman MP, Kopetz S, Bhosale PR, *et al*. Sacral insufficiency fractures after preoperative chemoradiation for rectal cancer: incidence, risk factors, and clinical course. *International Journal of Radiation Oncology, Biology, Physics* 2009; **74**(3): 818–23.

45. Tiv M, Puyraveau M, Mineur L, *et al*. Long-term quality of life in patients with rectal cancer treated with preoperative (chemo)-radiotherapy within a randomized trial. *Cancer Radiotherapy* 2010; **14**(6–7): 530–4.

46. Tan J, Lim Joon D, Fitt G, *et al*. The utility of multimodality imaging with CT and MRI in defining rectal tumour volumes for radiotherapy treatment planning: a pilot study. *Journal of Medical Imaging and Radiation Oncology* 2010; **54**: 562–8.

47. Anderson C, Koshy M, Staley C, *et al.* PET-CT fusion in radiation management of patients with anorectal tumors. *International Journal of Radiation Oncology, Biology, Physics* 2007; **69**: 155–62.

48. Bassi M, Turri L, Sacchetti G, *et al.* FDG-PET/CT imaging for staging and target volume delineation in preoperative conformal radiotherapy of rectal cancer. *International Journal of Radiation Oncology, Biology, Physics* 2008; **70**: 1423–6.

49. Brændengen M, Hansson K, Radu C, Siegbahn A, Jacobsson H, Glimelius B. Delineation of gross tumor volume (GTV) for radiation treatment planning of locally advanced rectal cancer using information from MRI or FDG-PET/CT: A prospective study. *International Journal of Radiation Oncology, Biology, Physics* 2011; **81**(4): e439–45.

50. Myerson R, Drzymala R. Technical aspects of image-based treatment planning of rectal carcinoma. *Seminars in Radiation Oncology* 2003; **13**: 433–40.

51. Gunderson LL, Haddock MG, Gervaz PA. Clinical target volume in conformal and intensity modulated radiation therapy. A clinical guide to cancer treatment. In Grégoire V, Scalliet P, Ang KK (eds) *Rectal and anal cancer in conformal radiotherapy planning (Selection and delineation of lymph node areas)*. Heidelberg: Springer-Verlag, 2003, pp. 187–97.

52. Portaluri M, Bambace S, Perez C, *et al.* Clinical and anatomical guidelines in pelvic cancer contouring for radiotherapy treatment planning. *Cancer Radiotherapy* 2004; **8**(4): 222.

53. Hocht S, Hammad R, Thiel HJ, *et al.* Recurrent rectal cancer within the pelvis. A multicenter analysis of 123 patients and recommendations for adjuvant radiotherapy. *Strahlentherapie und Onkologie* 2004; **180**: 15–20.

54. Syk E, Torkzad M, Blomqvist L, *et al.* Radiological findings do not support lateral residual tumour as a major cause of local recurrence of rectal cancer. *British Journal of Surgery* 2006; **93** (1): 113–19.

55. Roels S, Duthoy W, Haustermans K, *et al.* Definition and delineation of the clinical target volume for rectal cancer. *International Journal of Radiation Oncology, Biology, Physics* 2006; **65**(4): 1129–42.

56. Myerson RJ, Garofalo Mc, Naqa I, *et al.* Elective clinical target volumes for conformal therapy in anorectal cancer: A radiation therapy oncology group consensus panel contouring atlas. *International Journal of Radiation Oncology, Biology, Physics* 2009 **74**(3): 824–30.

57. Nijkamp J, Kusters M, Beets-Tan RG, *et al.* Three-dimensional analysis of recurrence patterns in rectal cancer: the cranial border in hypofractionated preoperative radiotherapy can be lowered. *International Journal of Radiation Oncology, Biology, Physics* 2011; **80**(1): 103–10.

58. Dulley L, Harrison M, Glynne-Jones R. One size does not fit all. Planning volumes for radiotherapy in rectal cancer: should we tailor radiotherapy fields to stage and risk? *Current Colorectal Cancer Reports* 2010; **7**(1): 89–96.

59. International Commission on Radiation Units and Measurements. *Prescribing, recording and reporting photon beam therapy. Report 50.* Bethesda, MD: ICRU, 1993.

60. Gallagher MJ, Brereton HD, Rostock RA. A prospective study of treatment techniques to minimise the volume of pelvic small bowel with reduction of acute and late effects associated with pelvic irradiation. *International Journal of Radiation Oncology, Biology, Physics* 1986; **12**: 1565–73.

61. Tepper JE, Cohen AM, Wood WC, *et al.* Postoperative radiation therapy of rectal cancer. *International Journal of Radiation Oncology, Biology, Physics* 1987; **13**(1): 5–10.

62. Letschert JGJ, Lebesque JV, Aleman VMP. The volume effect in radiation related late small bowel complications: results of a clinical study of the EORTC Radiotherapy Co-operative Group in patients treated for rectal carcinoma. *Radiotherapy and Oncology* 1994; **32**: 116–23.

63. Robertson JM, Yan D, Girimomnte PE, *et al.* The potential benefit of intensity modulated radiation therapy (IMRT) for rectal cancer. *International Journal of Radiation Oncology, Biology, Physics* 1999; **45S**: 248–9.

64. Nutting CM, Convery DJ, Cosgrove VP, *et al.* Reduction of small and large bowel irradiation using an optimised intensity modulated pelvic radiotherapy technique in patients with prostate cancer. *International Journal of Radiation Oncology, Biology, Physics* 2000; **48**: 649–56.

65. Laub W, Yan D, Robertson J, Martinez A. Intensity modulated radiation therapy (IMRT) in the radiotherapy treatment of colorectal cancer: the influence of profile smoothing on the efficiency of delivery. *Journal of Radiotherapy in Practice* 2002; **2**: 189–98.

66. Drzymala M, Hawkins MA, Henrys AJ, *et al.* The effect of treatment position, prone or supine, on dose-volume histograms for pelvic radiotherapy in patients with rectal cancer. *British Journal of Radiology* 2009; **82**(976): 321–7.

67. Mak AC, Rich TA, Schultheiss TH, *et al.* Late complications of postoperative radiation therapy for cancer of the rectum and rectosigmoid. *International Journal of Radiation Oncology, Biology, Physics* 1994; **28**: 597–603.

68. Rodat V, Flentje M, Engenhart R, *et al.* The belly-board technique for the sparing of small intestine. *Studies on positioning accuracy taking into consideration conformational irradiation techniques. Strahlenther Onkol* 1995; **171**: 437–43.

69. Koelbl O, Vordermark D, Flentje M. The relationship between belly board position and patient anatomy and its influence on dose-volume histogram of small bowel for postoperative radiotherapy of rectal cancer. *Radiotherapy and Oncology* 2003; **67**: 345–9.

70. Roeske JC, Forman JC, Mesina CF, *et al.* Evaluation of changes in the size and location of the prostate, seminal vesicles, bladder and rectum during a course of external beam radiation therapy. *International Journal of Radiation Oncology, Biology, Physics* 1995; **33**: 1321–9.

71. Lebesque JV, Bruce AM, Kroes AP, *et al.* Variation in volumes, dose-volume histograms, and estimated normal tissue complication probabilities of rectum and bladder during conformal radiotherapy of T3 prostate cancer. *International Journal of Radiation Oncology, Biology, Physics* 1995; **33**: 1109–19.

72. Marsh PJ, James RD, Schofield PF. Adjuvant preoperative radiotherapy for locally advanced rectal carcinoma. Results of a prospective, randomized trial. *Diseases of the Colon & Rectum* 1994; **37**: 1205–14.

73. Fortier GA, Krochak RJ, Kim JA, *et al.* Dose response to preoperative irradiation in rectal cancer: implications for local control and complications associated with sphincter sparing surgery and abdominoperineal resection. *International Journal of Radiation Oncology, Biology, Physics* 1987; **12**: 1559–63.

74. Chan AK, Wong AO, Langevin J, *et al.* Preoperative chemotherapy and pelvic radiation for tethered or fixed rectal cancer: a phase II dose escalation study. *International Journal of Radiation Oncology, Biology, Physics* 2000; **48**: 843–56.

75. Ahmed NR, Marks G, Mohiuddin M, *et al.* High dose preoperative radiation for cancer of the rectum: impact of radiation dose on patterns of failure and survival. *International Journal of Radiation Oncology, Biology, Physics* 1993; **27**: 773–8.

76. Mohiuddin M, Regine WF, John WJ, *et al*. Preoperative chemoradiation in fixed distal rectal cancer: dose time factors for pathological complete response. *International Journal of Radiation Oncology, Biology, Physics*. 2000; **46**(4): 883–8.

77. Valentini V, van Stiphout RG, Lammering G, *et al*. Nomograms for predicting local recurrence, distant metastases, and overall survival for patients with locally advanced rectal cancer on the basis of European randomized clinical trials. *Journal of Clinical Oncology* 2011; **29**(23): 3163–72.

78. Wiltshire KL, Ward IG, Swallow C, *et al*. Preoperative radiation with concurrent chemotherapy for resectable rectal cancer: effect of dose escalation on pathologic complete response, local recurrence-free survival, disease-free survival, and overall survival. *International Journal of Radiation Oncology, Biology, Physics* 2006; **64**(3): 709–16.

79. Tinger A, Michalski JM, Bosch WR, Valicenti RK, Low DA, Myerson RJ. An analysis of intratreatment and intertreatment displacements in pelvic radiotherapy using electronic portal imaging. *International Journal of Radiation Oncology, Biology, Physics* 1996; **34**(3): 683–90.

80. Jakobsen AKM, Appelt AL, Lindebjerg J, *et al*. The dose-effect relationship in preoperative chemoradiation of locally advanced rectal cancer: Preliminary results of a phase III trial. *Journal of Clinical Oncology* **29**: 2011 (suppl; abstr 3512).

81. Goldberg PA, Nicholls RJ, Porter NH, Love S, Grimsey JE. Long-term results of a randomised trial of short-course low-dose adjuvant pre-operative radiotherapy for rectal cancer: reduction in local treatment failure. *European Journal of Cancer* 1994; **30A**(11): 1602–6

82. Papillon J. Present status of radiation therapy in the conservation management of rectal cancer. *Radiotherapy and Oncology* 1990; **17**(4): 275–83.

83. Gerard JP, Coquard R, Fric D, *et al*. Curative endocavitary irradiation of small rectal cancers and preoperative radiotherapy in T2 T3 (T4) rectal cancer. A brief overview of the Lyon experience. *European Journal of Surgical Oncology* 1994; **20**(6): 644–7.

84. Birnbaum EH, Ogunbiyi OA, Gagliardi G, *et al*. Selection criteria for treatment of rectal cancer with combined external and endocavitary radiation. *Diseases of the Colon and Rectum* 1999; **42**: 727–35.

85. Papillon J. *Rectal and anal cancers: conservative treatment by irradiation: an alternative to radical surgery*. New York: Springer-Verlag, 1982.

86. Sichy B. The use of endocavity irradiation for selected carcinomas of the rectum 10 year experience. *Radiotherapy and Oncology* 1985; **4**: 97–101.

87. Papillon J, Berard P. Endocavity irradiation in the conservative treatment of adenocarcinoma of the low rectum. *World Journal of Surgery* 1992; **16**(3): 451–7.

88. Gerard JP, Roy P, Coquard R, *et al*. Combined curative radiation therapy alone in (T1) T2–3 rectal adenocarcinoma: a pilot study of 29 patients. *Radiotherapy and Oncology* 1996; **38**(2): 131–7.

89. Rauch P, Bey P, Peiffert D, *et al*. Factors affecting local control and survival after treatment of carcinoma of the rectum by endocavitary radiation: a retrospective study of 97 cases. *International Journal of Radiation Oncology, Biology, Physics* 2001; **49**: 117–24.

90. Vuong T, Beliveau PJ, Michel RP, *et al*. Can formal preoperative endorectal brachytherapy treatment for locally advanced rectal cancer: early results of a Phase I/II study. *Diseases of the Colon and Rectum* 2002; **45**: 1486–5.

91. Corner C, Bryant L, Chapman C, Glynne-Jones R, Hoskin PJ. High-dose-rate afterloading intraluminal brachytherapy for advanced inoperable rectal carcinoma. *Brachytherapy* 2010; **9**: 66–70.

92. Hoskin PJ, de Canha SM, Brownes P, Bryant L, Glynne-Jones R. High dose afterloading intraluminal brachytherapy for advanced inoperable rectal carcinoma. *Radiotherapy and Oncology* In press.

93. Jakobsen A, Mortensen JP, Bisgaard C, Lindebjerg J, Hansen JW, Rafaelsen SR. Preoperative chemoradiation of locally advanced T3 rectal cancer combined with an endorectal boost. *International Journal of Radiation Oncology, Biology, Physics* 2006; **64**: 461–5.

94. Kusters M, Marijnen CA, van de Velde CJ, *et al.* Patterns of local recurrence in rectal cancer: a study of the Dutch TME trial. *European Journal of Surgical Oncology* 2010; **36**(5): 470–6.

95. Holm T, Cedermark B, Rutqqvist LE. Local recurrence of rectal adenocarcinoma after curative surgery, with or without preoperative radiotherapy. *British Journal of Surgery* 1994; **81**: 452–5.

96. Kusters M, Dresen RC, Martijn H, *et al.* Radicality of resection and survival after multimodality treatment is influenced by subsite of locally recurrent rectal cancer. *International Journal of Radiation Oncology, Biology, Physics* 2009; **75**(5): 1444–9.

97. James RD, Johnson RJ, Edelstone B, *et al.* Prognostic factors in locally recurrent rectal carcinoma treated by radiotherapy. *British Journal of Surgery* 1983; **70**: 469–72.

98. Dubrowsky W. Mitomycin C. 5FU in radiation advanced locally advanced rectal cancer. *British Journal of Radiology* 1992; **65**: 143–7.

99. Myerson RJ, Valentini V, Birnbaum E, *et al.* Phase I/II trial of three dimensionally planned concurrent boost radiotherapy and protracted venous infusion of 5FU chemotherapy for locally advanced rectal carcinoma. Response to treatment. *International Journal of Radiation Oncology, Biology, Physics* 2001; **50**: 1299–308.

100. Mohiuddin M, Lingareddy V, Rakinic J, Marks G. Reirradiation for rectal cancer and surgical resection after ultra-high doses. *International Journal of Radiation Oncology, Biology, Physics* 1993; **27**: 1159–63.

101. Glimelius B. Recurrent rectal cancer. The pre-irradiated primary tumour: can more radiotherapy be given? *Colorectal Disease* 2003; **5**: 501–3.

102. Mohiuddin M, Marks G, Marks J. Long-term results of reirradiation for patients with recurrent rectal carcinoma. *Cancer* 2002; **95**: 1144–50.

103. Das P, Delclos ME, Skibber JM, *et al.* Hyperfractionated accelerated radiotherapy for rectal cancer in patients with prior pelvic irradiation. *International Journal of Radiation Oncology, Biology, Physics* 2010; **77**(1): 60–5.

Squamous carcinoma of the anus

David Sebag-Montefiore, Mark Harrison, Rob Glynne-Jones

7.10 Introduction

Squamous cell cancer (SCC) of the anus is a rare cancer whose incidence appears to be increasing. Anal cancers are more common in women than men[1,2] and just under half the patients are over the age of 65. An indolent natural history and a low rate of distant metastases[3,2] determine locoregional control as the primary aim of treatment. The relative 5-year survival rate is 62%[4], and has changed little for patients treated in the last two decades[4].

Human papilloma virus (HPV) infection is closely correlated with squamous cell anal carcinoma[5]. Immunosuppression is an important risk factor; particularly in renal and cardiac transplant recipients. Human immunodeficiency virus (HIV) is also a recognized risk factor, but the precise relationship is difficult to separate from the prevalence of HPV. Anal cancer is strongly associated with cigarette smoking[6].

Tumours of the anal canal are often poorly differentiated SCC, and anal margin well-differentiated but grading is subject to considerable interobserver variability and considerable heterogeneity in larger tumours. Hence, males have an excess of well-differentiated tumours. Although high-grade tumours are generally accepted to have a worse prognosis, this has not been confirmed in multivariate analysis[7].

7.11 Classification

Staging is based on the TNM (tumour, nodes, metastases) classification developed by the American Joint Committee on Cancer (AJCC) and the Union Internationale Contre le Cancer (UICC) is used (Table 7b.1). Because few cancers are resected surgically this classification is based on clinical factors such as tumour size (assessed by clinical examination and imaging studies). Nodal status is based on distance from the primary site rather than the number of nodes involved, as this has more prognostic significance, and it should be noted that the definition is subtly different for cancers in the anal canal and margin (Table 7b.1).

Small early cancers often cause few symptoms, and are sometimes diagnosed serendipitously with the removal of anal tags. More advanced lesions present as non-healing ulcers, perineal pain, sensation of a mass, rectal bleeding, itching, discharge, and faecal incontinence. Tumours may also be diagnosed concomitantly with a benign anal condition such as haemorrhoids, anal fissure, or fistula.

Table 7b.1 TNM classification for anal cancer

Stage	T*	N+	M
0	T_{is}	N0	M0
I	T1	N0	M0
II	T2	N0	M0
	T3	N0	M0
IIIA	T1	N1	M0
	T2	N1	M0
	T3	N1	M0
	T4	N0	M0
IIIB	T4	N1	M0
	Any	N2	M0
	Any	N3	M0
IV	Any	Any	M1

*Tumour stages: Tis, carcinoma *in situ*; T1, < 2 cm; T2, 2–5 cm; T3, > 5 cm; T4, invading adjacent organs but not anal sphincter

+Nodal stages: N0, no regional nodes; N1, perirectal nodes; N2, unilateral internal iliac or inguinal nodes; N3, perirectal and inguinal, or bilateral internal iliac or bilaiteral inguinal nodes.

Reproduced from Edge SB, Byrd DR, Compton CC, eds. *AJCC Cancer Staging Manual, 7th ed.* New York, NY.: Springer 2010. Used with the permission of the American Joint Committee on Cancer (AJCC), Chicago, Illinois. The original source for this material is the AJCC Cancer Staging Manual, Seventh Edition (2010) published by Springer Science and Business Media LLC, www.springer.com.

Primary chemoradiation with concurrent mitomycin C (MMC) and 5-fluorouracil (5-FU) is the current standard management. A small minority of investigators continue to use external beam radiotherapy alone followed by a small volume boost either with photons or electrons (particularly for small tumours). Early studies showed encouraging outcome achieved using radiotherapy combined with chemotherapy in patients with extensive disease preoperatively (or in patients not suitable for surgery). The initial studies of Nigro[8,9] demonstrated high rates of local control with the use of approximately 30 Gy of irradiation with concurrent MMC/5-FU. Cummings[10] reported a series of 190 patients treated with RT or CRT using seven sequential regimens and concluded retrospectively improved local control with CRT and with the addition of MMC to 5-FU.

Interstitial implantation of radioactive sources as a sole modality or as a boost after external beam radiotherapy[11,12] has been and continues to be used in parts of Europe. Scandinavian retrospective series[13] have described the use of high doses in the region of 60-Gy external beam radiotherapy with results that compare favourably with the randomized trials. Very high rates of local control have been observed in a series from Institute Gustave Roussy of 91% for T1, T2 tumours[14] and in a study from San Francisco node negative patients with T1, T2 tumours had a very high 5-year survival rate of 92%[15]. In contrast to this experience in specialist centres, multicentre anal cancer trials[3] demonstrate that even T1, T2 lesions appear to obtain the benefit from the addition of chemotherapy.

The evidence for the advantage of chemoradiation is based on a series of Phase III trials which compared radiotherapy with radiotherapy and concurrent 5-FU and MMC[3,16]. CRT improved disease-free survival (DFS) compared to radiotherapy alone. The RTOG-8704 subsequently confirmed the superiority of 5-FU and MMC over 5-FU alone when combined with radiotherapy[8]. Two of these Phase III trials addressed the fundamental question of combining radiotherapy with concurrent 5-FU and MMC[3,16]. Both used a split-course treatment with a planned gap of 6 weeks at 45 Gy. Both demonstrated that 5-FU and MMC in CRT improved DFS compared to radiotherapy alone. In a third Phase III trial with a median dose of 48 Gy and a boost of 9 Gy to histologically confirmed residual disease, the RTOG-8704 confirmed the superiority of 5-FU and MMC over 5-FU alone when combined with radiotherapy[14].

Further trials support the continued use of 5-FU and MMC. The strategies of utilizing cisplatin either as a radiation-sensitizer or systemically as induction or maintenance treatment have not improved outcome. The RTOG-9811[17] and the ACCORD-03 Phase III trials[18] tested induction chemotherapy with cisplatin and radiation dose-escalation with no benefit either in local control or DFS. The colostomy rate appears higher with cisplatin[12]. The preliminary results of the ACT II trial[19] failed to confirm an advantage in terms of outcome by replacing cisplatin in the CRT, or from additional 5-FU and cisplatin-based maintenance chemotherapy. There was no difference between MMC-based and cisplatin-based chemoradiation in response at 6 months or DFS and overall survival (OS)—although MMC provoked more haematological toxicity.

In this chapter, we examine treatment schedules of chemoradiation in terms of current chemotherapy treatments in all stages of squamous cell anal cancer, radiation dose, field size, fraction size, brachytherapy, IMRT; and define questions for future studies. We do not discuss other biologically distinct anal tumours such as melanoma, neuroendocrine tumours, adenocarcinomas, lymphomas, and GIST tumours. We have also elected not to discuss patients with either low CD4 counts or positive HIV status, partly because experience will be limited to super-specialized centres, and also because with highly active antiretroviral therapy (HAART), HIV-positive patients can expect similar outcomes to HIV-negative patients in terms of local control—although reports differ as to the tolerability and toxicity of treatment.

7.12 Phase III trials that establish chemoradiotherapy as the treatment of choice

The UK trial (ACT I)[20] was the largest trial with 585 patients[3] and demonstrated at a median follow-up of 42 months a 46% reduction (95% CI: 0.42–0.69, $\chi^2 = 24.6$, p < 0.0001) in the risk of local treatment-failure using combined modality treatment (CMT) with 5-FU/MMC and RT, over that achieved by RT alone (61% vs. 39% at 3 years). The UKCCR trial delivered 45 Gy in 20 or 25 fractions using large parallel-opposed fields followed by a boost of 15 Gy in six fractions (external beam) or 25 Gy to the 85% isodose (brachytherapy). There was also a reduced risk of death from anal cancer and a non-significant overall survival advantage. Long-term data from ACT I with a median follow-up of 12 years, showed a 25% difference in time to local failure in favour of CRT over RT alone[21] but no significant difference in survival.

The European Organisation for Research and Treatment of Cancer (EORTC) trial of 110 patients used the same design as the ACT 1 trial but was restricted to T3/T4 and T1–4N+ tumours[22]. This trial also used a single dose of MMC on day 1 in the combined modality arm and closed when the outcome of the UK trial was known[22]. Local control was also improved with chemoradiotherapy (68% vs. 55% at 3 years).

The US trial run by the Radiation Therapy Oncology Group RTOG-8704[23] demonstrated the advantage of adding two courses of MMC at a dose of 10 mg/m^2. With a median dose of 48 Gy and a boost of 9 Gy to histologically confirmed residual disease, the RTOG-8704 confirmed the superiority of 5-FU and MMC over 5-FU alone when combined with radiotherapy[8].

Studies vary in the proportion of early (T1 and T2) tumours between 16% in the EORTC 22861 study, 41% in ACT I, and 57% in the RTOG-8704. The T-staging classification also varies, since early trials used the 1978 UICC, or the 1985 UICC classification based on anatomical extent, and the proportion of the circumference involved by tumour. Hence the outcome from these studies should not be compared.

7.12.1 **RTOG 98–11**

The US Intergroup RTOG 98–11 phase III trial[24] randomly assigned 682 patients with anal canal tumours (35% with T3/4 tumours, and 26% with clinically involved lymph nodes to either neo-adjuvant 5-FU and cisplatin for two cycles prior to concurrent chemoradiation with 5-FU and cisplatin (n = 341), or the standard arm of concurrent chemoradiation with 5-FU and mitomycin (n = 341). The primary endpoint was 5-year DFS.

The neo-adjuvant cisplatin-based chemotherapy arm failed to improve DFS, locoregional control and distant relapse or OS. In fact, trends favoured the control arm of mitomycin. The 5-year DFS rates were 61% in the mitomycin arm versus 54% for the cisplatin arm, and 5-year OS rates were 75% and 70%, respectively ($P = 0.1$). A recent presentation of the trial data suggests the outcome for cisplatin could be significantly inferior[17]. By contrast, the requirement for a colostomy was significantly higher in the cisplatin arm compared with the mitomycin arm (19% versus 10%; $P = 0.02$). Haematological toxicity was worse with mitomycin. Similar compliance to both chemotherapy and radiotherapy in each arm suggests that the differences did not relate to excess toxicity. Both neo-adjuvant and concurrent cisplatin was used in the same experimental arm of the trial, however, making analysis of the individual role of each strategy difficult. In total 10% of patients developed long-term toxicity after chemoradiotherapy, with 5% requiring a colostomy for treatment-related problems.

7.12.2 **ACCORD-03**

The Action Clinique Coordonees en Cancerologie Digestive ACCORD-03 phase III trial[18] tested neo-adjuvant chemotherapy (NACT) with cisplatin and also radiation dose-escalation in a factorial 2 × 2 trial design. The trial compared 45 Gy in 25 daily fractions plus a 15-Gy boost with a higher boost dose, but found no benefit in colostomy-free survival at doses above 59 Gy. Event-free survival for the induction high dose and high dose-alone arms were 78% and 68% compared to the reference arm with 67%. In contrast the colostomy free survival was 85% and 80% arms respectively, compared to

86% in the control arm[18]. These results suggest that the boost could improve control—but at the expense of a higher risk of a colostomy.

7.12.3 ACT II

Preliminary results of The ACT II multi-centre, randomized trial which recruited 940 patients have been presented[19]. Patients received 5-FU (1000 mg/m^2/day on days 1–4, 29–32) and radiotherapy (50.4 Gy in 28 daily fractions), and randomized to receive MMC (12 mg/m^2, day 1; (n = 471) or CDDP (60 mg/m^2 on day 1, 29; (n = 469). A second randomization directed 2 courses of consolidation therapy (n = 448) 5 and 8 weeks after CRT (5-FU/CDDP, i.e. weeks 11, 14), or no consolidation (n = 446). Response was assessed clinically at 11 and 18 weeks, and by CT at 6 months. Preliminary results show almost identical complete response rates (95%) in both arms, with 3-year recurrence-free survival rates of 75% in T1/2 tumours overall, and 68% for more advanced T3/T4 tumours[19]. Hence neither strategy, i.e. CRT with CDDP versus CRT with MMC, nor chemotherapy consolidation with CDDP was more effective for achieving cCR, reducing tumour relapse or cancer- specific deaths than the standard of MMC/CRT. Acute haematological toxicity was more pronounced for MMC, but non-haematological toxicity similar.

In summary, the RTOG-9811, ACCORD-03, and ACT II phase III trials in anal cancer showed no benefit for cisplatin-based induction and maintenance chemotherapy, or radiation dose-escalation above 59 Gy. Neither the RTOG-9208 trial nor the ACCORD-03 trial, which compared radiation dose-escalation from 60 Gy to 65–70 Gy support the view that radiation dose-escalation within a CRT schedule increases local control in anal cancer. Normal tissue effects may be dose-limiting over a given threshold, and outweigh any potential tumoricidal advantages from higher radiation doses.

The present authors feel that 5-FU and MMC (12 mg/m^2 day 1) is the recommended standard. A recent update of the RTOG 98–11 presented at the 2011 Gastrointestinal Cancers Symposium suggests that with more mature follow-up there is a significant advantage in 5-year DFS for 5-FU/MMC over induction cisplatin and CRT with 5-FU/cisplatin (67.7 vs. 57.6% respectively; p = 0.0044) and 5-year OS (78.2 vs. 70.5% respectively; p = 0.02). A minority of patients will not be suitable for primary CRT.

7.13 Radiotherapy dose

The determination of optimal dose fractionation is limited by a lack of data regarding the pattern of failure. No randomized study has described the site(s) of local failure (within, marginal to, or outside of the radiotherapy field). Therefore it remains uncertain whether the majority of locoregional failure is due to inadequate CTVs, insufficient radiation dose, or intrinsic radioresistance.

7.14 Gaps in treatment

The established European philosophy of treatment has usually relied on split-course radiotherapy with high total doses and the use of interstitial implants based on the tradition of Papillon[11] to allow sufficient time for the bulk of the tumour to shrink and hence facilitate the delivery of high doses of radiation using an interstitial implant to the smallest possible volume. This practice also minimizes the risk of necrosis in the high dose area. In addition the long gap thereby allows selection of patients who fail to respond who may be salvaged by surgical resection. Current studies in anal cancer in Europe limit the gap to 2 weeks if clinically feasible. However, this practice defies standard radiotherapy principles regarding overall treatment time, breaks in treatment, and the concept of repopulation. The present authors believe the use of a planned gap is detrimental to local control, and should be avoided[20].

7.15 Brachytherapy

Brachytherapy may potentially increase dose to the primary tumour in T3/T4 tumours[25] but requires skill and expertise to avoid radionecrosis due to a unsatisfactory dose distribution. A low dose-rate iridium interstitial implant was originally advocated as a boost following radiotherapy alone[11] after an interval of 6 weeks, but this technique does not achieve current standards of conformal treatment. The strategy influenced the design of the two early European phase III studies[22], where a brachytherapy boost delivered 25 Gy following CRT after an interval of 6 weeks. Enthusiasm that brachytherapy achieves better outcomes in terms of local control probably reflects patient selection—patients with smaller tumours or good responders to chemoradiation being preferentially selected for brachytherapy boost.

7.16 Radical primary treatment

7.16.1 Indications

Patients who should receive radical CRT

Definitive chemoradiation with curative intent is recommended for patients with an initial diagnosis of epidermoid (squamous) carcinoma of the anal canal and margin in whom metastatic disease has been excluded by a staging CT scan of thorax and abdomen. It is important to emphasize that patients should be assessed for their fitness to receive this relatively intensive treatment approach taking into account performance status and medical comorbidity—particularly renal function. It is assumed that that scattered dose of radiation to the testes is very likely to induce permanent sterility and therefore sperm banking should be discussed with male patients who wish to preserve fertility prior to the commencement of treatment.

Patients in whom radical CRT may be considered but where caution is necessary

There *are* a number of clinical scenarios where definitive chemoradiation may be indicated but where there is an increased risk of treatment-related toxicity or where modifications to the standard treatment approach might be necessary. The scenarios include:

- ◆ HIV infection.
- ◆ Renal transplant.

◆ Inflammatory bowel disease (IBD).

◆ Prior pelvic resectional surgery.

There is an increased incidence of anal cancer in renal transplant patients. This produces challenges due to the site of the transplant kidney in the pelvis. Depending on its position this may require significant reduction in field sizes to avoid significant irradiation to the kidney. Surgery may still therefore have a role in selected cases.

Patients with IBD and those who have undergone prior resectional pelvic surgery are at increased risk of treatment-related toxicity with radical chemoradiation. After resectional pelvic surgery, the volume of small bowel in the pelvis is increased and this may lead to increased risks of early and late small bowel complications. In patients with IBD either the small or large bowel may be involved by this disease process and limit the tolerance of these normal tissues to standard doses of radical chemoradiation.

These difficult cases need discussion within the multidisciplinary team. Although there are increased risks of treatment-related toxicity with radical chemoradiation, there are also commonly increased risks of the alternative approach of radical surgery. In the experience of the authors these difficult cases commonly receive modified CRT and considerable multidisciplinary expertise is required to minimize the risks of treatment related morbidity and mortality.

Patients who should not receive radical chemoradiation

It is preferable to consider local excision of small anal margin tumours < 2 cm in diameter if there is no evidence of disease within the anal canal, there is no evidence of nodal spread clinically or on imaging and if clear lateral and deep margins are likely to be obtained.

Sometimes patients are referred with symptoms and in whom biopsy has shown anal intraepithelial neoplasia, which can be graded from 1 to 3 in severity. The natural history of progression from AIN3 to invasive malignancy is poorly documented but a recent estimate is 10% after 5 years[26]. At present it is recommended that chemoradiation is reserved for the development of established invasive carcinoma.

Radical chemoradiation should be used with great caution in patients who have received prior pelvic radiation. Depending on the treatment previously given, it might be possible to use modified chemoradiation to a total dose of 30–40 Gy (see section 7.17).

7.16.2 **Treatment volume and definition**

Clinical examination is an extremely valuable method of assessment. The clinical oncologist requires different information to a surgical assessment at the time of examination under anaesthetic. The clinical oncologist is interested in accurate measurements of tumour length, degree of circumferential involvement, and the extension of any macroscopic disease in all directions including below the anal verge. Vaginal examination is essential in female patients to assess whether there is any extension into the postvaginal wall or even breaching vaginal mucosa. It is also most important that this examination is documented in the patient record to assist post-treatment assessment of response to CRT.

Examination under anaesthetic allows more detailed palpation of the anorectal and pelvic structures and is used particularly when clinical examination is limited by discomfort or pain.

7.16.3 General points

Patients should be assessed for performance status, renal function, and other medical comorbidity prior to treatment. Patients should be tested for relevant infections, and other malignancies.

Assessment of the cervix, vagina, and vulva is suggested in female patients, and includes screening for vaginal and cervical cancer (and the penis in men), because of the common role of HPV in these tumours.

HIV testing is recommended in any patient with a lifestyle that puts them at risk of contracting HIV infections because of its implications of excess toxicity from radiotherapy and chemotherapy and the development of infections. In HIV-positive patients, the CD4 count, measured viral load, and optimization of highly antiretroviral therapy are all essential to determine the management plan.

Sperm banking should be discussed prior to the commencement of treatment with male patients who wish to preserve fertility. Premenopausal women should be informed that fertility will be lost, and hormone replacement therapy may be appropriate in those in whom an early menopause is induced. A defunctioning colostomy should be considered in patients with transmural vaginal involvement (at risk of development of an anorectal–vaginal fistula), or faecal incontinence.

Staging investigations should include:

◆ Clinical examination.

◆ Whole body CT (chest, abdomen, and pelvis).

◆ High-resolution MRI of the pelvis.

MRI is more accurate in distinguishing and delineating primary tumour and lymph nodes and offers the advantage of coronal sagittal and axial views of the extent of the primary tumour and is of considerable aid to radiotherapy treatment planning.

The optimal method to determine inguino-femoral lymph node status of patients with anal canal cancer remains controversial. Clinical examination supplemented by fine-needle aspiration cytology (FNAC) is the traditional method[27] but may not be sufficiently accurate. Approximately one-third of patients have enlarged inguinal lymph nodes but on biopsy only 50% will confirm metastatic spread. The remainder are caused by secondary infection. Tumours of the anal margin are more likely to involve inguinal nodes than anal canal tumours. In retrospective surgical series 30% of patients will have involvement of their inguinal nodes; however, in early T1/T2 tumours the rate of involvement is approximately only 12%.[28–30] Involvement is usually unilateral and occasionally bilateral but not usually contralateral to the tumour.

Small shotty nodes may be inflammatory whereas palpable nodes > 1 cm are clearly at risk of microscopic involvement. Clinically suspicious nodes should be assessed by biopsy where possible. However, formal biopsy of these nodes significantly delays CRT and leaves a surgical scar with the potential for seeding microscopic tumour cells.

In contrast a fine-needle aspirate or core biopsy is only helpful if cancer cells are detected. A negative sample is compatible with either a sampling error of an involved node or a truly uninvolved node. Where there is suspicion the nodes should be treated as involved.

The pelvic nodes are frequently involved particularly with increasing tumour stage and in poorly differentiated tumours. If the tumour extends up into the rectum, spread may occur via the inferior mesenteric lymph nodes. The overall incidence of pelvic lymph node metastases is in the region of 25–30%[30].

FDG PET may have a role to investigate equivocal lesions on CT or MRI not amenable to biopsy, but some series show that up to 40% of patients with PET/CT avid nodes are false-positive results. Haematogenous spread at presentation is noted in < 5% of cases and predominantly involves lung or liver.

7.16.4 Sentinel lymph node biopsy

Sentinel lymph node biopsy (SLNB) has not fulfilled the initial hopes of this staging modality, and in addition management strategies are not in place to stratify treatment for the findings of macroscopic involvement, microscopic involvement, and the presence of isolated cells. The presence of inguinal node metastases which are not clinically palpable are of uncertain relevance, since a worse survival relates to clinically involved palpable inguinal LN. Some studies suggest that all patients found negative for inguinal lymph node metastases at PET-CT were also negative at SNB, i.e. few false-negative results are observed. Also, neither a negative PET-CT nor a negative SLNB excludes involved nodes, as elective bilateral inguinal node dissection is not normally performed. In addition, the morbidity of irradiating to 30.6 Gy or even 50.4 Gy after + SLNB, is unknown—particularly as some SLNB will require bilateral nodes to be removed.

SLNB is probably most useful in the setting of locoregional recurrence after CRT in deciding whether a radical inguinal dissection should be performed in addition to salvage APER.

7.16.5 Radiotherapy planning technique

The important principles of radical chemoradiation currently used in the UK are that all microscopic disease at risk receives a minimum of 30 Gy of radiation, and secondly that all macroscopic disease receives at least 50 Gy. This is achieved by the use of a two-phase shrinking field technique

Although investigators are interested in the introduction of more complex radiotherapy techniques such as IMRT, it is important to ensure that the introduction of more complex techniques do not compromise the excellent long-term outcomes achieved in the majority of patients. Important issues include immobilization of the patient, the accurate delineation of the CTV, and precise definition of normal tissue constraints and dose fractionation that is used without prolongation of the overall treatment time.

The current approach in the UK is the use of the protocol used for the ACT 2 trial and is described in he following sections. The margins used are generous.

ACT II planning technique

Patient position and immobilization

Patients are treated prone for both phases of treatment. This allows direct visualization of the anal verge, the application of perianal bolus, and assists in the anterior placement of mobile small bowel for the second phase of treatment.

There is no consensus on the use of immobilization techniques. It is common that immobilization devices are not used and an individual centre employs their standard technique to attempt to achieve a reproducible position for treatment. Some centres use belly boards to assist in the anterior displacement of small bowel.

Target volume and field definition

Gross tumour volume

It is recommended that both phases of treatment are planned at the same time and that GTV is delineated for each phase. It is common for macroscopic disease to regress quickly during chemoradiation so that if the second phase if treatment is planned 2 weeks into treatment the marked macroscopic disease may be smaller than the initial extent.

All macroscopic disease in the inguino-femoral and perianal region should be identified by the use of radio-opaque markers at initial simulation. The extent of gross tumour within the anal canal is difficult to determine on a non-contrast CT planning scan. A number of approaches are used. Firstly the measurements from clinical examination are extremely useful with the length of tumour extension up the anal canal measured from the anal verge. Using a radio-opaque marker on the anal verge the superior extent of the GTV can be determined in this way.

High-resolution MRI also provides this measurement as well as measurements of gross tumour extension in the lateral anterior and posterior directions. There is a need to accurately fuse the diagnostic pelvic MRI in the supine position with the planning CT images performed on a flat couch to improve definition of the GTV. Until this is available the clinician must take measurements from the MRI scan and translate this on to the planning CT scan. The use of rectal contrast or a radio-opaque wire in the anal canal may assist this process.

All areas of macroscopic disease should be identified as GTV. In patients with clinically or radiologically significant lymphadenopathy these areas are identified as separate GTVs.

Treatment fields (including the clinical target volume and planning target volume)

There is considerable difficulty in applying the principles of ICRU 62 to the target volume definition for anal cancer. There is little information on the pattern of failure of anal cancer, which is essential when defining the CTV. There is also a need for further studies to determine the extent of organ motion and departmental set up errors for the techniques used in the treatment of anal and rectal cancer.

Phase 1—large parallel-opposed fields

This phase includes all sites of GTV and all microscopic disease at risk in the inguino-femoral and pelvic lymph nodes (Fig. 7b.1) and is used in all patients receiving radical chemoradiation. This approach may be planned using a conventional or virtual simulator.

All borders described below are *field*. By describing field borders this technique does not require the specific delineation of the clinical or planning target volumes. This very simple technique was considered necessary at the start of the ACT II trial to ensure high protocol compliance.

◆ Superior border: 2 cm above inferior aspect of the sacro-iliac joints. The superior field border is standard unless pelvic lymph nodes are seen on CT scan or the primary tumour extends to within 3 cm of this border, in which case the border is recommended to extend 3 cm above the upper limit of macroscopic disease.

◆ Lateral border: to cover fully both inguinal nodal regions—in practice this field border is approximately the midpoint of the femoral neck.

◆ Inferior border: 3 cm below the anal verge (for disease confined to the anal canal only) or 3 cm below most inferior extent of tumour (for anal margin tumours).

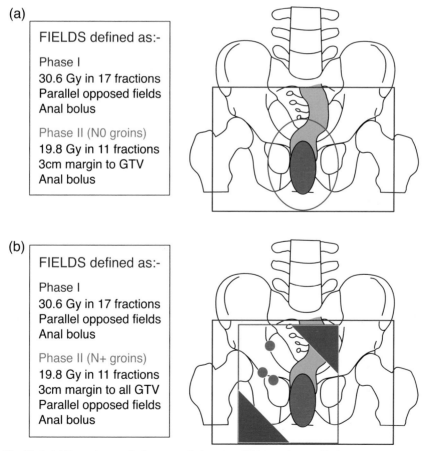

Fig. 7b.1 (a) Two-phase radiotherapy technique N0. (b) Two-phase radiotherapy technique N+.

Parallel-opposed fields are used with equal weighting. Wax bolus is used *placed between the buttocks* for all patients with anal margin tumour and in those patients with anal confined disease that extends down to within 2 cm of the anal verge (the vast majority of patients). A wedge-shaped piece of bolus is more comfortable for the patient than layers of bolus material and also corrects the differences in separation in this region between the buttocks. A longitudinal wedge may improve the homogeneity of the dose distribution in some patients. A minimum energy of 8 MV is used.

Dose prescription
+ Phase 1—30.6 Gy in 17 fractions of 1.8 Gy per fraction.

Phase 2
The technique used is different depending on the presence or absence of clinically or radiologically significant lymphadenopathy and the position of the primary tumour.

+ If there is no evidence of lymphadenopathy then the GTV is treated with a 3-cm margin for anal canal tumours.
+ If the tumour is confined to the anal margin only, a direct photon field may be used with 3-cm lateral superior and inferior margins. Electrons are not recommended.
+ If clinically or radiologically significant lymphadenopathy is present then the nodal and primary GTVs are defined and reduced parallel opposed fields with 3-cm lateral superior and inferior margins are used.

This phase may be planned using orthogonal films, CT planning, or virtual simulation. It is essential that all visible tumour at and around the anal margin is marked using a radio-opaque marker. Rectal contrast is also recommended. If the disease is confined to the anal canal, then a radio-opaque marker placed on the anal verge is essential. All significant inguino-femoral lymphadenopathy is also marked with radio-opaque markers. The GTV is determined using clinical and diagnostic MRI measurements.

Primary tumour without significant lymphadenopathy—anal canal
A 3-cm margin is applied superiorly, laterally, anteriorly, and posteriorly to determine the treatment fields. The inferior 3-cm margin is applied to the anal verge if tumour is confined to the anal canal or 3 cm inferior to the inferior extent of GTV if tumour extends inferior to the anal verge marker.

A three- or four-field arrangement is used. This is usually posterior and two wedged lateral fields. Occasionally an anterior fourth field is required to improve any inhomogeneity.

Primary tumour without significant lymphadenopathy—anal margin only
For anal margin confined tumours a direct photon field (electrons should not be used) is used and the margins (superior, inferior and lateral) are 3cm from the limits of the GTV to define the treatment fields (see Fig. 7b.1). Wax bolus as described for phase 1 is used for phase 2.

Primary tumour with significant lymphadenopathy

All macroscopic disease will be considered as GTV including the primary site and all sites of involved nodal disease. The field borders are defined as:

- Superior field border: 3 cm superior to the most superior extent of GTV.
- Inferior field border: as phase I—3cm below the anal margin (for disease confined to the anal canal only) or 3 cm below most inferior extent of tumour (for anal margin tumours).
- Lateral field border: 3 cm lateral to the most lateral extent of the GTV.

Lead shielding and multileaf collimation may be used to shield normal tissue providing a minimum of a 3-cm margin from all delineated GTV is achieved.

Parallel opposed fields are used with equal weighting.

Dose prescription for phase 2

- 19.8 Gy in 11 fractions at 1.8 Gy per fraction to the ICRU intersection point (for anal canal tumours or when parallel opposed fields are used. When a direct field is used for an anal margin tumour the dose is prescribed to the anal margin.

Implementation and quality assurance

The techniques described require electronic portal imaging[31] to ensure the reproducibility of the treatment set up. *In vivo* dosimetry is not routinely used.

Fields should be moved if they fall outside an agreed tolerance level—usually 5 mm for patients who are treated prone. This process also allows radiographers to evaluate the whole set-up, and thus to assess and correct systematic errors. Reverification is recommended on a weekly basis. These EPI images should be audited at the clinician's weekly meeting.

Implementing complex radiotherapy techniques for anal cancer

The ACT II technique is associated with significant acute toxicity, particularly of the perineal skin and genitalia. IMRT can reduce mean and threshold doses to genitals, perineum, small bowel, and bladder.

The delineation of pelvic nodes radiologically is described in a recent review[32], and other relevant pelvic nodal atlases, but is, to date, insufficiently relevant to anal cancer. With routine use of CT simulation and intravenous contrast, contouring vessels on the CT image can be used as a surrogate for lymph node localization and can offer a more precise, and individualized field delineation compared to that achieved when using conventional pelvic fields. In addition involved nodes can be imaged on MRI[33].

On review of the literature we could find no meta-analyses or systematic reviews in anal cancer to offer guidance on the optimal total dose or target definition. In addition, no dose–response curves have been proposed in the literature either for radiotherapy or chemoradiation. The RTOG have made suggestions based on a consensus of nine experts in radiation oncology[34]. No consensus on proper selection and delineation of subclinical lymph nodal areas in definitive chemoradiation therapy has been suggested either conclusively or proposed for further investigation. Work is ongoing at Mount Vernon Cancer to produce an atlas for treatment of anal cancer.

Future UK studies plan to reduce dose to early tumours, and dose escalate to T3/T4 tumours, using 3D-conformal or IMRT, or VMAT techniques. We have therefore calculated retrospectively the dose received by the PTV in patients planned according to the ACT II protocol, for clinically involved areas (primary tumour/lymph nodes) and determined the dose to relevant uninvolved lymph nodes (nodal CTVs), and OARs to design a pilot conformal 3D-regimen. A draft document of consensus recommendations for CTV definition is being created and will be used to aid in contouring in a future Phase II/III trial.

Intensity-modulated radiotherapy

IMRT allows both precision and sparing normal surrounding structures (perineal skin, external genitalia, and bladder) compared to conventional 3D planning[35–37], which may lead to reduced acute toxicity and less treatment breaks, which may in the past have compromised efficacy. IMRT may also allow radical treatment to be delivered to tumours with para-aortic node involvement[38]. However, in obese patients with non-reproducible external skin contours, or a major component of tumour outside the anal canal, IMRT may prove more problematic.

IMRT is challenging even for experienced clinicians. In a recent multicentre study, even after centres had been approved and accredited, 79% of IMRT plans required field modification of elective nodes after central review[39]. The RTOG-0529 Phase II study confirms that IMRT can reduce the overall treatment time (OTT). When the RTOG-9811 and the RTOG-0529 studies were compared, the median RT duration was 49 days and 42 days respectively. A consensus is forming as to the planning tumour volumes required for IMRT[33]. Thus IMRT in the future offers the promise of less toxicity, shorter OTT, and potentially higher radiotherapy doses.

In order to use IMRT or VMAT, we need to define subclinical areas, which potentially harbour microscopic disease, the optimal dose for macroscopic and microscopic disease. We also need to decide a suitable prescribed planning dose to the PTV and an appropriate PTV margin.

7.16.6 Chemotherapy

The cytotoxic drugs are given during the first and fifth week of radiotherapy using:

- MMC day 1, 12 mg/m^2 bolus day 1 only (max. 20 mg).
- 5-FU 750mg/m^2 in 1 litre N saline over 24 hours days 1–5, 29–33.

An estimated glomerular filtration rate of > 50 mL/min is required for this prescription.

For patients who are elderly and there is concern about a possible increased risk of neutropenic sepsis dose modifications are used:

- MMC day 1, 8 mg/m^2 bolus day 1 only (max. 10 mg).
- 5-FU 750 mg/m^2 in 1 litre N saline over 24 hours days 1–4, 29–32.

7.17 Postoperative adjuvant treatment

7.17.1 Indications

A few patients will be referred for consideration of chemoradiation after local excision of a small tumour either in the anal canal or margin. The local excision may be performed for a polypoid perianal mass that appears like a haemorrhoid but where there is clinical suspicion at the time of removal that it may be malignant. Alternatively a local excision may be performed of a small mass considered likely to be a carcinoma but where clear margins might be obtained. This occurs most commonly with small anal margin tumours. This approach should be avoided in anal canal tumours and the surgical approach restricted to small representative biopsies only. An incomplete local excision of an anal canal squamous cell carcinoma compromises long-term sphincter function from the surgical procedure alone and requires additional post operative chemoradiation which itself will also have an impact on long-term anorectal function. There is no evidence to support a 'debulking' approach prior to chemoradiation.

A few patients may undergo initial abdominoperineal excision as definitive treatment of their anal cancer. Postoperative chemoradiation should be considered when there is evidence of involvement of the circumferential resection margin.

The indications for postoperative adjuvant treatment are therefore:

Absolute indications

+ Incomplete local excision of squamous carcinoma of the anus (deep or lateral resection margins ≤ 1 mm).

+ Involvement of the circumferential resection margin after initial abdominoperineal excision.

Relative indications

+ After local excision:
  - Narrow margin > 1–4 mm, or
  - Primary tumour with clear margins but primary > 2–5 cm (T2).
+ After APER:
  - Node +ve CRM–ve defects in the specimen.

These indications are uncommon, and there is uncertainty as to the best approach with respect to postoperative chemoradiation. It is important that patients are carefully examined, as there may be palpable residual disease after local excision. All patients should undergo staging investigations as described in previous sections.

If initial surgery, comprising either local excision for a ≤ 2 cm tumour or an abdominoperineal excision with involvement of the CRM has occurred and there is no evidence of nodal or distant spread, it is recommended that patients are treated with radical chemoradiation using the technique described for a primary tumour without lymphadenopathy using a two-phase technique. However for the phase 2 component of the planning process the site of the initial (excised) tumour should be used as the 'presumed GTV' even though this now should only harbour microscopic disease at risk. It is reasonable to assume that this area may harbour a greater microscopic disease burden and require the full dose of radiation.

If a small (< 2 cm) tumour has been treated by local excision and there is a close or involved margin with no evidence of nodal or distal spread a number of approaches have been used including brachytherapy alone, a single phase of involved field irradiation[40] either alone or combined with concurrent chemotherapy (similar to the phase 2 approach described previously for radical chemoradiation) or the two-phase shrinking field approach described earlier for radical chemoradiation.

In both of these situations there is concern about the risks of increased acute and late toxicity. This results in discussion about the extent of the radiotherapy fields (whether smaller fields can be treated) and the total dose (whether a lower total dose may be used). There is insufficient evidence to make clear recommendations at this time.

7.18 Supportive care during radiotherapy

Patients should be assessed at least once weekly with regard to skin toxicity and their overall tolerance to treatment. It is advisable to check full blood counts at least once weekly during radiation therapy particularly if MMC is used. It is also important to maximize patient tolerance with the use of simple antiemetics, antifungals, analgesia, skin care, and nutritional support. Expected acute side effects include diarrhoea, proctitis, urinary frequency and dysuria, loss of pubic hair, and erythema and moist desquamation of the skin in the groins and perineum.

Patients should be informed of the negative effect of smoking before chemoradiation starts. Smoking may worsen acute toxicity during treatment and lead to a poorer outcome in terms of DFS and CFS[41].

The post-treatment use of vaginal dilators in sexually active females has been recommended, although the evidence for their effectiveness is weak.

Skin effects rapidly disappear within 2–3 weeks after treatment is completed. Longer-lasting side effects are unusual with external beam alone but within the UKCCR Act I trial were more common after a brachytherapy boost. Long-term morbidity includes ulceration of the anal/rectal area, small bowel obstruction, urethral obstruction, and fistula formation.

7.19 Palliative treatment

A small minority (< 10%) of patients will have synchronous metastases at presentation. Clinical experience with cytotoxic drugs (5-FU, cisplatin, carboplatin, paclitaxel, irinotecan) or the biologicals (bevacizumab, cetuximab) is limited, with few publications regarding the efficacy of chemotherapy alone[42].

If the patient is fit for systemic treatment, we recommend the majority should receive initial chemotherapy—usually with a combination of cisplatin and infusional 5-FU. Responses are rarely complete and usually of short duration. If the volume of metastatic disease is relatively small, then in order to attempt to achieve locoregional control in the pelvis, patients may benefit from a total dose of 30–50 Gy combined with systemic chemotherapy as described in the earlier section on radical chemoradiation. Modification of this approach to an involved field single-phase of treatment of GTV + 3 cm may be appropriate for some patients.

There remains a group of patients who are commonly elderly and of poor performance status where local palliative treatment is required. The most appropriate approach for such patients may range from a single fraction of 8 Gy to 20 Gy in five fractions, again using an involved field approach particularly to attempt to control bleeding, pain, or enlargement of a fungating tumour.

However, if significant tumour regression is to be achieved it is the authors' experience that modified chemoradiotherapy is more effective than palliative radiotherapy alone. The use of 30 Gy in 15 fractions combined with 600 mg/m^2 5-FU by continuous infusion days 1–4 to an involved field (GTV + 3 cm) is associated with major tumour regression and is feasible in elderly frail patients[43].

7.20 Therapeutic assessment and follow-up

Patients who do not respond to CRT are usually treated with abdominoperineal resection hence it is important to assess patients following treatment. MRI is performed most commonly 6–8 weeks following treatment although there is little data of its accuracy or predictive value. One study has suggested MRI appearances such as tumour size reduction and signal intensity change are predictive of a good outcome.

Surveillance is typically performed by clinical examination with the aid of additional proctoscopy. Given that metastatic relapse is uncommon, the scheduling of CT scans for metastatic surveillance outside trials remains controversial.

7.21 Current and future research

The results for T1/T2 anal cancers appear good, but for T3/T4 tumours remain poor. A 'one size fits all' stages approach is no longer appropriate. Probably early T1 tumours are over-treated; more advanced T3/T4 merit treatment escalation. IMRT abolishes the gap and shortens OTT, and looks promising in this setting.

The integration of other potential chemotherapy agents including oxaliplatin, gemcitabine carboplatin, vinorelbine, paclitaxel, etoposide, topotecan, and the biological agents are all potential future strategies. SCC of the anus commonly overexpresses EGFR, and Kras and BRAF mutations appear rare. As yet, there is no data of the efficacy of biologicals combined with chemoradiation although several trials have been performed but not reported (ECOG E3205, AMC045, and a FNLCC trial). Preclinical data show that cetuximab increases radiation-induced apoptosis, and the effect of EGFR inhibition appears greater if administration is extended beyond the end of radiotherapy. Hence, a future clinical strategy could employ consolidation or maintenance treatment inhibiting EGFR after chemoradiation.

References

1. Jemal A, Siegel R, Ward E, *et al*. Cancer statistics, 2008. *CA: A Cancer Journal for Clinicians* 2008; **58**(2): 71–96.

2. Bilimoria KY, Bentrem DJ, Rock CE, *et al*. Outcomes and prognostic factors for squamous-cell carcinoma of the anal canal: analysis of patients from the National Cancer Data Base. *Diseases of the Colon and Rectum* 2009; **52** (4): 624–31.

3. The UKCCCR Anal Cancer Trial Working Party. Epidermoid Anal Cancer: Results from the UKCCCR Randomised Trial of Radiotherapy Alone versus Radiotherapy, 5-Fluorouracil and Mitomycin C. *Lancet* 1996; **348**: 1049–54.

4. Miller BA, Ries LAG, Hankey BF, *et al*. *SEER Cancer Statistics Review 1973–1989. NIH pub No 94–2789*. Bethesda, MD: National Cancer Institute, 1992.

5. Williams IG. Carcinoma of the anus and anal canal. *Clinical Radiology* 1962; **13**: 30–5.

6. Frisch M, Glimelius B, van den Brule AJ, *et al*. Sexually transmitted infection as a cause of anal cancer. *New England Journal of Medicine* 1997; **337**: 1350–8.

7. Shepherd NA, Scoffield JH, Love SB *et al*. Prognostic factors in anal squamous carcinoma: a multi variant analysis of clinical, pathological and flow cytometric perimeters in 235 cases. *Histopathology* 1990; **16**: 545–55.

8. Nigro ND, Vaitkevicus VK, Considine B Jr. Combined therapy for cancer of the anal canal: a preliminary report. *Diseases of the Colon and Rectum* 1974; **17**: 354–35.

9. Leichman L, Nigro N, Vaitkevicius VK, *et al*. Cancer of the anal canal: model for preoperative adjuvant combined modality therapy. *American Journal of Medicine* 1985; **78**(2): 211–21.

10. Cummings BJ, Keane TJ, O'Sullivan B, *et al*. Epidermoid anal cancer: treatment by radiation alone or by radiation and 5-flurouracil with and without mitomycin C. *International Journal of Radiation Oncology, Biology, Physics* 1991; **21**: 1115–25.

11. Papillon J, Montbarbon JF. Epidermoid carcinoma of the anal canal. A series of 276 cases. *Diseases of the Colon and Rectum* 1987; **30**: 324–33.

12. Newman G, Calverley DC, Acker BD, *et al*. Management of carcinoma of the anal canal by external beam radiotherapy, experience in Vancouver 1971–1988. *Radiotherapy and Oncology* 1992; **25**: 196–202.

13. Glimelius B, Pahlman L. Radiation therapy of anal epidermoid carcinoma. *International Journal of Radiation Oncology, Biology, Physics* 1987; **13**(3); 305–12.

14. Eschwege F, Lasser ER, Chavey A, *et al*. Squamous cell carcinoma of the anal canal: treatment by external beam irradiation. *Radiotherapy and Oncology* 1985; **3**: 145–9.

15. Cantril ST, Green JR, Shal GL, Shaupp WC. Primary irradiation therapy in the treatment of anal carcinoma. *International Journal of Radiation Oncology, Biology, Physics* 1983; **9**: 1271–8.

16. Doggett SW, Green JP, Cantril ST. Efficacy of radiation alone for limited squamous cell carcinoma of the anal canal. *International Journal of Radiation Oncology, Biology, Physics* 1986; **15**(5): 1069–72.

17. Gunderson LL, Winter KA, Ajani JA, *et al*. Long-term update of US GI Intergroup RTOG 98–11 phase III trial for anal carcinoma: Comparison of concurrent chemoradiation with 5FU-mitomycin versus 5FU-cisplatin. Gastrointestinal Cancers Symposium 2011. *ASCO 2011*, abstract **367**, p.140.

18. Conroy T, Ducreux M, Lemanski C. Treatment intensification by induction chemotherapy (ICT) and radiation dose-escalation in locally advanced squamous cell anal canal carcinoma (LAAC): Definitive analysis of the intergroup ACCORD-03 trial. *Journal of Clinical Oncology* 2009; **27**:15s (Part I of II):176s (abstract 4033).

19. James R, Wan S, Glynne-Jones R, *et al*. A randomised trial of chemoradiation using Mitomycin or cisplatin, with or without maintenance cisplatin/5FU in squamous cell carcinoma of the anus. *Journal of Clinical Oncology* 2009; (Proc ASCO) **27**: 18S(part II of II); 797s (abstract LBA-4009).

20. Glynne-Jones R, Sebag-Montefiore D, Adams R, *et al*. for the UKCCCR Anal Cancer Trial Working Party.'Mind the Gap' – The Impact of Variations in the Duration of the Treatment Gap and Overall Treatment Time in the First UK Anal Cancer Trial (ACT I). *International Journal of Radiation Oncology, Biology, Physics* 2011; **81**(5): 1488–94.

21. Northover J, Glynne-Jones R, Sebag-Montefiore D, *et al*. Chemoradiation for the treatment of epidermoid anal cancer: 13-year follow-up of the first randomised UKCCCR Anal Cancer Trial (ACT I). *British Journal of Cancer* 2010; **102**(7): 1123–8.

22. Bartelink H, Roelofson F, Eschwege F, *et al*. Concomitant radiotherapy and Chemotherapy is Superior to radiotherapy alone in the treatment of locally Advanced Anal Cancer: Results of a Phase III Randomized Trial of the European Organisation for Research and Treatment of Cancer Radiotherapy and Gastrointestinal groups. *Journal of Clinical Oncology* 1997; **15**: 2040–9.

23. Flam M, John M, Pajak TF, *et al*. Role of mitomycin in combination with fluorouracil and radiation, and of salvage chemoradiation in the definitive nonsurgical treatment of epidermoid carcinoma of the anal canal: results of a phase III randomised intergroup study. *Journal of Clinical Oncology* 1996; **14**: 2527–39.

24. Ajani JA, Winter KA, Gunderson LL, *et al*. Fluorouracil, mitomycin and radiotherapy vs fluorouracil, cisplatin and radiotherapy for carcinoma of the anal canal: a randomised controlled trial. *Journal of the American Medical Association* 2008; **199**: 1914–21.

25. Hwang JM, Rao AR, Cosmatos HA, *et al*. Treatment of T3 and T4 anal carcinoma with combined chemoradiation and interstitial implantation: a 10 year experience. *Brachytherapy* 2004; **3**(2): 95–100.

26. Scholefield JH, Nugent KP. Anal Cancer. Position statement of the Association of Coloproctology of Great Britain and Ireland introduction. 2011 *Colorectal Disease* 2011; **13**, S1: 3–10.

27. Gerard JP, Chapet O, Samiei F, *et al*. Management of inguinal lymph node metastases in patients with carcinoma of the anal canal: experience in a series of 270 patients treated in Lyon and review of the literature. *Cancer* 2001; **92**(1): 77–84.

28. Clark J, Petrelli N, Herrera L, Mittelman A. Epidermoid carcinoma of the anal canal. *Cancer* 1986; **57**(2): 400–6.

29. Pyper PC, Parks TG. The results of surgery for epidermoid carcinoma of the anus. *British Journal of Surgery* 1985; **72**(9): 712–14.

30. Stearns MW, Urmacher C, Sternberg SS. Cancer of the anal canal. *Current Problems in Cancer* 1980; **4**: 1–44.

31. Tinger A, Michalski JM, Bosch WR, Valicenti RK, Low DA, Myerson RJ. An analysis of intratreatment and intertreatment displacements in pelvic radiotherapy using electronic portal imaging. *International Journal of Radiation Oncology, Biology, Physics* 1996; **34**(3): 683–90.

32. Lengelé B, Scalliet P. Anatomical bases for the radiological delineation of lymph node areas. Part III: Pelvis and lower limbs. *Radiotherapy and Oncology* 2009; **92**(1): 22–33.

33. Roach SC, Hulse PA, Moulding FJ, Wilson R, Carrington BM. Magnetic resonance imaging of anal cancer. *Clinical Radiology* 2005; **60**(10): 1111–19.

34. Myerson RJ, Garofalo MC, El Naqa I, *et al*. Elective clinical target volumes for conformal therapy in anorectal cancer: a radiation therapy oncology group consensus panel contouring atlas. *International Journal of Radiation Oncology, Biology, Physics* 2009; **74**(3): 824–30.

35. Milano MT, Jani AB, Farrey KJ, *et al*. Intensity-modulated radiation therapy (IMRT) in the treatment of anal cancer: Toxicity and clinical outcome. *International Journal of Radiation Oncology, Biology, Physics*2005; **63**: 354–61.

36. Mell LK, Schomas DA, Salama JK, *et al*. Association between bone marrow dosimetric parameters and acute haematologic toxicity in anal cancer patients treated with concurrent chemotherapy and intensity-modulated radiotherapy. *International Journal of Radiation Oncology, Biology, Physics* 2008; **71**: 1431–7.

37. Menkarios C, Azria D, Laliberte B, *et al*. Optimal organ-sparing intensity-modulated radiation therapy (IMRT) regimen for the treatment of locally advanced anal canal carcinoma: a comparison of conventional and IMRT plans. *Radation Oncology* 2007; **2**: 41.

38. Hodges JC, Das P, Eng C, *et al*. Intensity-modulated radiation therapy for the treatment of squamous cell anal cancer with para-aortic nodal involvement. *International Journal of Radiation Oncology, Biology, Physics* 2009; **75**(3): 791–4.

39. Kachnic L, Winter K, Myerson R, *et al*. RTOG 0529: A phase II evaluation of dose painted IMRT in combination with 5-fluorouracil and mitomycin C for the reduction of acute morbidity in carcinoma of the anal canal. *International Journal of Radiation Oncology, Biology, Physics* 2010; **78**(3 Suppl. 55).

40. Hatfield P, Cooper R, Sebag-Montefiore D. Involved-field, low-dose chemoradiotherapy for early-stage anal carcinoma. *International Journal of Radiation Oncology, Biology, Physics* 2008; **70**(2): 419–24.

41. Mai SK, Welzel G, Haegele V, Wenz F. The influence of smoking and other risk factors on the outcome after chemoradiotherapy for anal cancer. *Radation Oncology* 2007; **2**: 30.

42. Pathak P, King BT, Ohinata A, *et al*. The treatment of metastatic squamous cell carcinoma of the anal canal: a single institution experience. *ASCO GI Cancers Symposium* 2008; abstract A352.

43. Charnley N, Choudhury A, Chesser P, Cooper RA, Sebag-Montefiore D. Effective treatment of anal cancer in the elderly with low-dose chemoradiotherapy. *British Journal of Cancer* 2005; **92**(7): 1221–5.

8 Urology

8a

External beam radiotherapy for prostate cancer

Shaista Hafeez, Antonia Creak,
David P Dearnaley

8.1 Introduction

Prostate cancer is the most prevalent cancer in men in the United Kingdom with an estimated lifetime risk of one in nine, and the second most common cause of cancer-related death in men[1]. A rise in prostate cancer incidence over the last 20 years, due initially to increased use of transurethral resection of the prostate (TURP) and more recently to greater prostate-specific antigen (PSA) testing, has not been reflected in changes in mortality rates. Histopathological data from postmortem studies demonstrate that prostate cancer is found in approximately half of all men in their 50s and in 80% of men by age 80 but only one in 26 men (3.8%) will die from this disease[2,3]. The challenge is to identify those men who will benefit from radical treatment and in whom it is therefore justifiable to risk the side effects associated with such treatment.

Active surveillance aims to identify low/intermediate-risk prostate cancer and is becoming increasingly popular. It involves regular clinical review, PSA testing, prostate biopsies, in some cases MRI scans, and allows the clinician to select only those men with significant cancers for radical treatment.

Radical treatment options for localized disease include radical prostatectomy, radical external beam radiotherapy usually with hormonal therapy, and interstitial brachytherapy. Contemporary series suggest that outcomes of each of these treatment modalities are similar[4]. There is a paucity of randomized comparisons and long-term follow-up data after brachytherapy is at present less mature.

In recommending whether and how to treat a man with localized prostate cancer, one needs to consider the risk grouping (see Table 8a.1), the life expectancy of the patient, any comorbidities, and his preference for and between treatment options, taking into account their expected side effects. Radiotherapy is the most commonly used curative treatment modality for localized prostate cancer in the UK and about 10 000 men are treated annually with 3D conformal radiotherapy (3DCRT) becoming the standard of care[5,6]. However with technical advances in treatment planning and delivery, intensity-modulated radiotherapy (IMRT)-based techniques are increasingly being used.

Radical treatment reduces prostate cancer death. A randomized Scandinavian trial has shown that radical prostatectomy for localized prostate cancer reduces prostate cancer mortality and risk of metastases compared to watchful waiting[7]. A similar improvement

Table 8a.1 Risk stratification for men with localized prostate cancer

	PSA		Gleason score		Clinical stage
Low risk	< 10 ng/mL	and	≤ 6	and	T1–T2a
Intermediate risk	10–20 ng/mL	or	7	or	T2b–T2c
High risk	> 20 ng/mL	or	8–10	or	T3–T4

in absolute difference in overall survival was also observed but failed to reach statistical significance. The landmark SPCG-7 trial was the first to show an overall survival advantage for radiotherapy in the primary treatment of prostate cancer[8]. This trial compared androgen suppression alone to the addition of radiotherapy to the prostate and seminal vesicles in men with locally advanced disease. Results from an interim analysis of the broadly similarly designed Medical Research Council sponsored PR07 trial[9] support the addition of radiotherapy in men with locally advanced prostate cancer.

8.2 **Indications**

Radiotherapy is employed in six settings for prostate cancer which will be discussed in turn:

1. Radical radiotherapy to prostate ± seminal vesicles.
2. Radical radiotherapy to prostate and pelvic lymph nodes.
3. Radical radiotherapy following prostatectomy.
4. Palliative radiotherapy to prostate ± pelvis
5. Palliative radiotherapy to distant metastases.
6. Breast bud radiotherapy.

8.2.1 **Radical radiotherapy to prostate ± seminal vesicles**

Radical radiotherapy to the prostate is the most commonly used curative treatment in the UK[5]. The seminal vesicles are treated if they are clinically or radiologically involved. They may also be treated if they are at a high risk of being microscopically involved based on the Roach formula[10]:

$$\text{Seminal vesicle risk (\%)} = \text{PSA} + [(\text{Gleason score-6}) \times 10]$$

A systematic review of 26 trials using conventional external beam radiotherapy in the treatment of prostate cancer has shown a 10-year disease-free survival of 100%, 69%, and 57% for T1a, T1b, and T2 disease, respectively[11].

Technological advances over recent years have increased the precision of external beam radiotherapy and have led to improved outcomes. The use of 3DCRT reduces the dose limiting side effects and has allowed for dose escalation to the whole prostate. It is the standard of care for treating localized prostate cancer[12,13].

Radiotherapy is usually given in combination with hormonal therapy. The duration of hormonal therapy depends on the risk grouping. External beam radiotherapy may be given with a short course of neo-adjuvant androgen suppression for low or intermediate risk patients or combined with longer term hormonal therapy for patients with advanced or high grade cancers. See section 8.10.

8.2.2 Radical radiotherapy prostate plus pelvis

Pelvic radiotherapy has been used as standard by many North American and European centres for men with locally advanced disease and has been included in phase III trials for these groups of men[14–17], see Table 8a.2. In the UK, pelvic radiotherapy has been infrequently used due to concerns over bowel toxicity.

Two large phase III trials evaluating the role of whole pelvic radiotherapy in patients with intermediate- and high-risk prostate cancer have been recently published. RTOG trial 94–13[18] included 1323 patients with estimated lymph node risk ≥ 15%. Patients were randomized between prostate only (dose received 70.2 Gy in 1.8-Gy fractions) and whole pelvis radiotherapy (dose to whole pelvis 50.4 Gy in 1.8-Gy fractions), and in the second randomization between neo-adjuvant/concurrent and adjuvant hormonal therapy. Updated results show no statistically significant benefit in biochemical control with pelvic radiotherapy compared with prostate only radiotherapy.

GETUG-01, a smaller French phase III trial[19] of 444 patients also failed to show any difference between whole pelvic (46 Gy in 2-Gy fractions) and prostate only radiotherapy (66–70 Gy in 2-Gy fractions, median dose to the prostate was 68 Gy): the GETUG group used a lower radiotherapy dose to the prostate than RTOG, with a significant cohort being treated to 66 Gy. They used a lower superior border of the pelvic field than in the RTOG trial and > 50% of patients had < 15% risk of lymph node involvement; these factors may have contributed to the lack of an observed effect.

As a result the treatment volume (i.e. prostate or prostate and pelvic lymph nodal regions) in high-risk patients remains an unresolved question and further trials are currently underway[20,21].

Pelvic radiotherapy may be offered to patients with a high predicted risk of microscopic pelvic lymph node involvement (≥ 15–30%) as defined using the equation devised by Roach[10]:

$$\text{Lymph node risk (\%)} = 2/3 \times \text{PSA} + [(\text{Gleason score} - 6) \times 10]$$

High-risk patients (e.g. Gleason 8–10, clinical T3–T4 tumours, or lymph node risk > 30%), should also be considered for 2–3 years of neo-adjuvant and adjuvant hormonal therapy.

8.2.3 Postprostatectomy radiotherapy

The number of radical prostatectomies being performed in the UK is increasing, with laparoscopic techniques including minimally invasive and robot-assisted procedures increasing in popularity over recent years. However, there is still considerable uncertainty over the optimal postoperative oncological management.

Salvage prostate bed irradiation can be offered to patients post prostatectomy with biochemical failure, defined as *either* two consecutive rises in PSA and final PSA > 0.1 ng/mL *or* three consecutive rises in PSA. Adjuvant radiotherapy can be considered in patients with positive surgical margins, and/or pT3/4 disease due to the associated high risk of residual local disease.

Table 8a.2 Summary of randomized trials of prostate cancer radiotherapy

Trial	Eligible patients	Radiotherapy treatment ± randomization	Hormonal treatment ± randomization	Outcome
SPCG7/ SFUO-3[8]	T1b–T2 (grade 2–3) or T3 and N0 M0 (PSA < 70 ng/ mL)	Prostate RT (70 Gy) vs. no RT	MAB 3 months followed by continuous flutamide	Addition of RT resulted in decrease in cancer specific and overall mortality
PR07[9]	T3/T4 N0 M0 or T2 (PSA > 40 ng/mL) or T2 (PSA > 20 ng/mL) and Gleason ≥ 8	No RT vs. RT to prostate (65–69 Gy) ± RT to pelvis (45 Gy)	Life long suppression (orchidectomy or LHRH analogues)	Significant increase in overall survival and disease-specific survival for combined modality
RTOG 75–06[178]	Stage C, or stage A-2–B with pelvic nodal involvement	Pelvis and prostate vs. pelvis prostate and para-aortic RT	No hormonal treatment	No difference A proportion of node-positive patients cured at 10-year follow-up by RT
RTOG 77–06[14]	Stage A and B	Prostate only vs. prostate and pelvis RT	No hormonal treatment	No difference
RMH/ICR conformal trial[13]	Those having radical RT for prostate cancer	64 Gy/32 fractions conventional vs. conformal RT	3 months neo-adjuvant and concurrent LHRH agonist	Lower incidence of late rectal side effects in conformal arm
RTOG 86–10[15]	Bulky T2–T4 N0–1	Pelvic and prostate RT	MAB for 4 months neo-adjuvant and concurrent vs. no hormones	Improvement in all endpoints except overall survival for whole group in hormones arm. Preferential effect of hormones in Gleason ≤ 6 subgroup in which there was a survival advantage
RTOG 85–31[16]	T3 and/or N +	Prostate and pelvic RT	Long-term adjuvant LHRH agonist vs. LHRH agonist at relapse	Improvement in all endpoints including overall survival in adjuvant arm
MD Anderson Dose escalation trial[179]	T1–T3	70 Gy vs. 78 Gy to prostate	No hormonal treatment	Better freedom from failure in high dose arm for intermediate- and high-risk patients. Greater rectal toxicity in high dose arm.

Table 8a.2 *continued*

Trial	Eligible patients	Radiotherapy treatment ± randomization	Hormonal treatment ± randomization	Outcome
EORTC[17]	T3–4 and/or any high grade	Pelvic and prostate RT	3 years concurrent and adjuvant goserelin vs. no adjuvant hormones	Improvement in overall and disease-free survival in adjuvant hormones arm.
RTOG 94-13[18]	Localized disease

Lymph node risk ≥ 15% PSA < 100 ng/mL | Pelvic vs. prostate only RT | 2 months MAB: neoadjuvant and concurrent vs. adjuvant | Superior progression free-survival for whole pelvis RT and neoadjuvant/concurrent MAB compared to other treatment combinations |
RTOG 92-02[150]	T2c–T4	Prostate and pelvic RT	4 months neoadjuvant and concurrent MAB vs. additional 24 months adjuvant MAB	Superior progression free-survival but not overall survival in long-term arm. Survival advantage in subset with Gleason 8–10 histology
Yeoh[180]	T1–T2, PSA < 80 ng/mL	64 Gy/32 vs. 55 Gy/20 to prostate and base SV, non-conformal RT	None	Biochemical relapse-free survival significantly better with the hypofractionated schedule. No difference in overall survival. No difference in gastrointestinal and genitourinary toxicity
Lukka[181]	T1–T2	Prostate RT, 66 Gy/33 vs. 52.5 Gy/20	None	Comparable late toxicity, failure rate higher in 52.5-Gy arm. Results consistent with low alpha/beta ratio for prostate cancer
Canadian[182]	T2 or T3	Prostate only RT	RT alone vs. 3 months neoadjuvant androgen suppression vs. 10 months neoadjuvant, concomitant and adjuvant	Superior biochemical control in arms including hormones compared to RT alone. No difference between two durations of hormonal treatment
Bolla[26]	Post prostatectomy. Positive surgical margins and/or capsular invasion and/or seminal vesicle invasion	60 Gy/30 immediately vs. at time of palpable local recurrence	None	Improved biochemical and clinical PFS in immediate treatment arm

Table 8a.2 *continued*

Trial	Eligible patients	Radiotherapy treatment ± randomization	Hormonal treatment ± randomization	Outcome
D'Amico[145]	PSA 10–40 ng/mL, Gleason > 6 or extracapsular disease on imaging	70 Gy to prostate in two phases	6 months MAB starting 2 months pre-radiotherapy vs. no hormonal therapy	Improved overall survival in hormonally treated arm
ICR/RMH dose escalation trial[68]	T1b–T3b N0 M0	74 Gy vs. 64 Gy, 1.5-cm vs. 1-cm margin	3 months neoadjuvant and concurrent LHRH agonist	Suggestion of improved freedom from failure in high dose arm, greater rectal toxicity. Reduced margin reduces toxicity without compromising freedom from failure
RT01[133]	T1b–T3b N0 M0	64 Gy/32 vs. 74 Gy/34	3–6 months neo-adjuvant and concurrent LHRH agonist	Improved bPFS, PFS, and decrease use of salvage androgen suppression
TROG 96.01[144]	T2b, T2c, T3 and T4 N0 M0	RT alone (66 Gy/33) vs. RT and hormones.	3 months vs. 6 months hormones; (neoadjuvant, 1 month concurrent); goserelin with flutamide 250mg tds.	Decreased distant progression, prostate cancer mortality and all cause mortality with 6 months hormones and RT
RTOG 94-08[183]	T1b–T2b, PSA < 20 ng/mL	RT alone vs. RT and hormones	4 months total androgen suppression (neoadjuvant and concurrent)	Decreased disease specific mortality and improved overall survival with hormones and RT

Retrospective reviews suggest that results are better for patients with pre-radiotherapy PSA < 0.5ng/mL, PSA doubling time ≥ 9 months, and positive surgical margins[22]. It remains contentious whether radiotherapy is best given immediately following surgery (adjuvant) for patients at high risk of local recurrence, or reserved for those with a documented PSA relapse (salvage)[23–25].

There are three published randomized controlled trials addressing this question. Firstly, the EORTC 22911 trial[26] which recruited patients with pT3 pN0 disease and/or positive surgical margins post prostatectomy, who were randomized between observation and adjuvant radiotherapy (60 Gy in 30 fractions). A statistically significant advantage was seen for adjuvant radiotherapy in terms of biochemical progression-free survival and clinical progression-free survival. There was no evidence of a difference in overall survival. It should be noted that the indication for treatment in the deferred radiotherapy group was defined as palpable local recurrence as opposed to PSA failure. It is therefore not certain whether immediate adjuvant radiotherapy is preferable to early salvage for recurrence defined by PSA testing, particularly using super-sensitive assays.

The second trial, SWOG 8794 (NCIC CTG PR-2)[27] had a similar design and has longer follow-up. Once again, adjuvant radiotherapy was associated with a statistically significant increase in biochemical control and importantly both metastasis-free and overall survival were improved with HR 0.71 and 0.72 respectively (p = 0.02). Toxicity was more common in the adjuvant group, with statistically significant increases in proctitis, urethral stricture, and urinary incontinence.

Thirdly, the smaller ARO 96–02 trial[28] randomized men with pT3 disease with a postoperative undetectable PSA, to either observation or adjuvant radiotherapy. Again adjuvant radiotherapy was associated with improved biochemical control. AR0 96–02 was not sufficiently powered to address the effect of adjuvant treatment on clinical outcomes such as survival.

The rationale for limiting radiotherapy to patients developing PSA failure is that it would reduce the numbers of patients requiring radiotherapy and thus spare them the associated toxicity.

A recent survey of 49 UK Clinical Oncologists[29] found opinion divided between those recommending early adjuvant radiotherapy for pT3 margin positive cases with an undetectable postoperative PSA, and those who preferred to offer salvage radiotherapy. For an isolated rising PSA post prostatectomy, with no evidence of metastatic disease and no prior adjuvant radiotherapy, most of those surveyed would recommend salvage radiotherapy, although the PSA threshold for instituting such treatment varied between clinicians. Most of those recommending salvage radiotherapy used androgen deprivation in combination with radiotherapy. In a second survey of 188 UK oncologists and urologists there was widespread uncertainty regarding the use of both adjuvant radiotherapy and the mode, timing, and duration of hormone therapy[30]. A randomized clinical trial[24] is underway to address these questions.

8.2.4 Palliative radiotherapy to prostate ± pelvis

Patients with locally advanced disease, with or without evidence of distant spread, may be offered radiotherapy to the prostate and pelvic nodes if present, as a means of palliation. A minority of patients without distant spread may achieve long-term symptomatic and PSA control.

8.2.5 Palliative radiotherapy to metastases

External beam radiotherapy is commonly used for palliation of bone pain from metastatic disease. A single fraction of 8 Gy is appropriate for symptom relief[31,32]. This may be repeated safely if required. A higher, fractionated dose, e.g. 20 Gy in five fractions, is traditionally used for overt or subclinical spinal cord compression, or if there is concern over risk of pathological fracture. Symptomatic nodal disease and rarely visceral metastases may also benefit from a short course of palliative radiotherapy.

8.2.6 Breast bud radiotherapy for the prevention or treatment of gynaecomastia

Patients treated with long-term oestrogens or anti-androgens may develop gynaecomastia which can be painful and distressing. Radiotherapy to the breast buds can reduce the incidence of gynaecomastia if used prophylactically, or provide relief of established symptoms[33].

Target volume is the glandular tissue of the breast, with a typical field size being an 8-cm diameter circle centred on the nipple. A single fraction of 8 Gy is used for prophylactic treatment and 12 Gy in two fractions for established gynaecomastia. Applied doses are prescribed in the case of orthovoltage treatment and to the 90% isodose if electrons are used. For orthovoltage treatment 160-kv photons are employed if breast thickness is < 2.5 cm, above which 300 kv is used. Electron energy is typically 6 or 9 MeV.

Tamoxifen is a valuable and effective alternative to radiotherapy to prevent breast related side effects induced by bicalutamide monotherapy[34,35].

8.3 Radical radiotherapy planning

8.3.1 Patient position and immobilization

The patient lies in the treatment position, supine with arms across the chest as shown in Fig. 8a.1.

A variety of immobilization devices are available to ensure patient comfort and to maintain daily position reproducibility. These include head pads, knee pads, ankle stocks, and various types of immobilization casts. Immobilization of the legs, and specifically the pelvis, has been shown by some to improve set-up accuracy for prostate cancer radiotherapy[36,37]. Available evidence, however, suggests that ankle stocks combined with a foam head-pad and knee supports provide a high degree of accuracy in patient positioning, making pelvic immobilization unnecessary[38,39].

Principal causes for prostate displacement are a very full bladder or a distended rectum[40]. Therefore many centres attempt to minimize prostate movement by standardizing bladder and rectal filling.

The bladder should be comfortably full prior to scanning. This also serves to reduce the proportion within the PTV. The patient therefore drinks approximately 350 ml during the hour prior to scanning. This protocol is modified as needed for patients with urinary symptoms.

The rectum should ideally be empty of both faeces and flatus. However, changes in rectal filling occur both during and between radiotherapy treatments, and this in turn

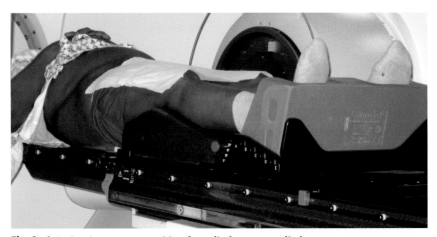

Fig. 8a.1 Patient in treatment position for radical prostate radiotherapy.

influences the position of the prostate[41–43]. Patients with a distended rectum at planning have been reported to have increased biochemical and local failure after radiotherapy[44,45].

In an attempt to achieve a consistent degree of rectal filling during planning and treatment, some advocate the routine use of laxatives or enemas, prior to each radiotherapy fraction[46]. The effectiveness of this approach in terms of reducing prostate movement however has not been fully evaluated. Others have investigated the use of endorectal balloons or endorectal obturators during planning and treatment[47–51]. Such devices may reduce prostatic movement in the anterior to posterior direction, but issues remain over patient acceptability.

Intraprostatic radio-opaque markers placed in the prostate before treatment planning can be used to allow visualization of the prostate during treatment and to correct errors accordingly. These are discussed further in section 8.7.

8.3.2 CT scanning

The patient is positioned on the CT scanner in the treatment position. The skin surface is marked at points anteriorly in the midline of the pubic symphysis, and laterally over each hip. Radio-opaque reference markers are placed over the skin marks.

Scout views are performed followed by a short axial scan starting superiorly at the level of the pubic symphysis. The axial images are reviewed prior to proceeding to the full helical scan. The rectal diameter is measured at the level of the prostate. If the rectal diameter is > 4 cm in the anterior–posterior plane the scan is usually aborted[52]. The patient is assessed and re-scanned after implementing appropriate intervention to aid rectal voiding (i.e. laxatives or suppositories as appropriate). If the rectal diameter is < 4 cm the full helical scan is acquired. The planning CT scan is generated using a slice interval of 5 mm or less from the L3/4 interspace to 2 cm below the ischial tuberosities. For those having pelvic radiotherapy, scans are taken from the bottom of the L1/L2 vertebral space. Inclusion of the whole bladder and rectum is required to ensure that the DVH dose constraints can be calculated and safety standards maintained. Following the CT scan, the skin marks will be made with permanent tattoos. The CT data is then transferred to a radiotherapy planning computer for outlining.

8.4 Target volume definition

8.4.1 Radical radiotherapy: prostate ± seminal vesicles

Standard nomenclature for target volume definition[53] does not strictly apply to conformal prostate radiotherapy. The GTV is often difficult to define precisely by clinical examination and conventional imaging. Current practice is to define the entire prostate gland and all or part of the seminal vesicles, using the diagnostic MRI scan if available. MRI provides better soft tissue contrast compared to CT, making it particularly useful in identifying the prostatic apex and distinguishing between bladder base and anterior rectal wall[54]. On T2-weighted MRI images tumour within the prostate gland can be seen as a focus of low signal intensity. The use of an endorectal coil (probe) at the time of diagnostic MRI scanning also increases the detection of cancer from 67% to 77%[55].

Several studies have shown that the prostate, as defined on CT, is often overestimated in size[56–59]. MRI for radiotherapy planning is complicated by the geometric distortion and chemical shift artefact associated with this imaging modality, as well as lack of direct electron density data arising from it. Solutions, however, are being developed. Newer imaging techniques such as dynamic contrast enhanced MRI (dMRI)[60] and magnetic resonance spectroscopic imaging (MRSI)[61] are being utilized to clarify intraprostatic lesions, differentiating malignant disease from potential benign prostatic tissue and making it possible to deliver a radiotherapy boost to the lesion.

Coregistration of the planning CT scan with the diagnostic MRI scan can also be achieved, as is common with other tumour types, however consideration has to be given to scanning the patient in the same anatomical position[56,62].

The prostate is first outlined on the central slice where the gland is clearly demarcated, and then on successive caudal slices (see Fig. 8a.2). Differentiating the prostate becomes more difficult near the apex and tracing the urethral bulb which is sited below the pelvic floor aids identification of the inferior limit of the prostatic apex. If available, a coronal MRI image may help to define the position of the apical extent of the gland which, with reference to structures such as the femoral heads, can be related to the planning CT. The prostate is then outlined on slices cranial to the central slice. The pelvic sling muscles should be excluded from the prostate outline, and often a fat plane exists between these structures and the prostate. Care is needed to define the prostate in the more cranial slices because the prostatic base may bulge into the bladder.

The base of the seminal vesicles is outlined for all patients. The extent of outlining of the remainder of the seminal vesicles depends on the patient's individual risk of involvement and anatomy. If the seminal vesicles extend predominantly laterally, then they can be included in their entirety without significantly affecting rectal dose. If there is significant posterior extension of the seminal vesicles and they are closely

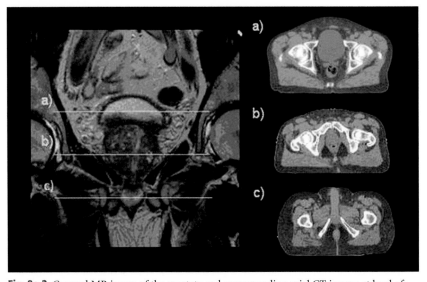

Fig. 8a.2 Coronal MR image of the prostate and corresponding axial CT images at level of: (a) the seminal vesicles, (b) the mid prostate, (c) the penile bulb.

applied to the rectal wall, then rectal dose constraints may be exceeded unless the tips of the seminal vesicles are excluded from the volume. Unless the seminal vesicles are specifically outlined, they are not reliably included in the high dose volume[63].

Pathological data[64] has suggested three patterns of seminal vesicle involvement: (1) tumour spread along ejaculatory ducts (35% of cases); (2) direct extension through capsule (61% of cases); and (3) the presence of isolated tumour deposits (12% of cases). One series[65] showed that the median distance from the prostate to seminal vesicle invasion was 1 cm, and that in 90% of cases involvement was limited to the proximal 2 cm. They advocate including the proximal 2–2.5 cm in the CTV when it is desired to treat the seminal vesicles. Davis et al.[66], however, found that tumour was evident within 0.5 cm of the tip of the seminal vesicles in 40% of patients with known seminal vesicle involvement, and advocate treating the seminal vesicles in their entirety. The entire seminal vesicle is included for stage T3 tumours, or if the risk of seminal vesicle involvement is > 15%, provided that the predicted dose to the rectum is acceptable. If the rectal dose is unacceptable then the tips of the seminal vesicles are excluded, but the proximal 2 cm is treated. If the predicted risk of microscopic seminal vesicle involvement is < 15%, the base of the seminal vesicles (proximal 1–2 cm) is outlined.

The target volume is expanded with an adequate margin to allow for microscopic spread. In a surgical series of patients with clinically localized prostate cancer[67], the median distance of extra capsular spread measured radially from the capsule was 2 mm (range 0.5–12 mm). The PTV also includes an additional margin to allow for patient and prostate movement and variations in treatment set-up. In practice the prostate and some or all of the seminal vesicles are outlined. Margins are then grown to form the PTV. In a small randomized trial[68] comparing a margin of 1.0 cm with 1.5 cm, there was no difference in tumour control but an increase in rectal and bladder side effects with the larger margin. In practice margins will usually be 1.0 cm but may be non-uniform, with tighter margins (5 mm) posteriorly to spare the posterior wall of the rectum[69].

Table 8a.3 shows the treatment protocols for inclusion of the seminal vesicles in recent and ongoing major clinical trials. Note that two or three phases of treatment are used in some studies, with reduced margins for higher dose levels.

Normal tissue structures are also contoured in order to generate DVHs to evaluate dose to OARs (rectum, bladder, femoral heads, bowel and urethral bulb). They are outlined as solid organs by defining the outer wall. The bladder is outlined from the base to dome. The rectum is outlined from the anus (or 1 cm below the lower margin of the PTV whichever is more inferior) to the rectosigmoid junction, frequently the position is best appreciated on the sagittal CT reconstruction as the level at which the rectum comes anteriorly. Additional bowel within the PTV is defined separately.

8.4.2 Radical radiotherapy: pelvis and prostate

Normal sized lymph nodes are not readily identifiable on the planning CT scans. Vascular and bony anatomy is therefore used to define the lymph node CTV. Conventional whole pelvic nodal fields used bony landmarks to localize pelvic nodal groups, however, data suggests the pelvic vasculature is a more accurate surrogate for nodal position[70–73].

Table 8a.3 Summary of contemporary radiotherapy trials in prostate cancer showing details of radiotherapy technique

Trial	SV involvement risk* (%)	Target + margin (mm)** Phase I	Phase II	Phase III	Dose*** (Gy) Ph I	Ph II	Ph III	Total
Dutch	Low (< 10)	P + 10	P + 5/0	–	68	0,10	–	68, 78
	Mod (10–25)	P + SV + 10	P + 10	P + 5/0	50	18	0,10	68, 78
	High (> 25)	P + SV + 10	P + 5/0	–	68	0,10	–	68, 78
	involved	P + SV + 10	P + SV + 5/0	–	68	0,10	–	68, 78
Protect	Low (< 15)	P + bSV + 10/5	P + 0	–	56	18	–	74
	High (≥ 15)	P + SV	P + 0	–	56	18	–	74
RT 01	Low (<15)	P + bSV + 10	P + 0	–	64	0,10	–	64, 74
	High (≥15)	P + SV	P + 0	–	64	0,10	–	64, 74
EORTC 22991	All	P + SV + 10 ± nodes	P + bSV + 10	P + 5/0	46	24	0, 4, 8	70, 74, 78 (not randomized)
RTOG P-0126	All	P + SV + 10/5	P + 10/5	–	58	15, 24	–	73, 82
RTOG 9406	Low (< 15)	P + 5–10	P + 5–10	–	68	0, 6, 11	–	68, 74, 79
	High ≥ 15	P + SV + 5–10	P + 5–10	–	56	12, 8, 23	–	68, 74, 79
	Involved	P + SV + 5–10	–	–	68, 74, 79	–	–	68, 74, 79
MSKCC	All	P + SV + 10/6	P + SV + 10/6 Rectal block	–	65, 70, 76, 76	0, 0, 0, 5	–	65, 70, 76, 81
CHHIP (conventionally fractionated arm)	Low (< 15)	P + bSV + 10	P + 10/5	P + 5/0	54	16	4	74
	High(≥ 15)	P + SV + 10	P + 10/5	P + 5/0	59	12	3	74

* Risk of SV based on Roach formula and Partin tables. ** '/' indicates margin: all around prostate/prostate–rectum interface. *** Multiple figures in each risk group indicate different dose arms of trials. bSV, base of seminal vesicle; P: prostate; SV, seminal vesicle.

The choice of lymph node groups to be included in whole pelvic radiotherapy is guided mainly by surgical literature reporting pathological lymph node involvement following extended pelvic lymph node dissections[72,74–77].

Lymphangiographic techniques poorly visualize certain lymph node groups, particularly pre-sacral and internal iliac regions. The use of ultra-small super paramagnetic iron oxide particles (USPIO) have also been reported to be effective in discriminating between normal and pathological lymph nodes[78,79]. Their use has aided pelvic lymph node mapping[80,81]. As well as identifying metastases in non-enlarged lymph nodes (which appear as a filling defects by virtue of poor contrast take up by malignant tissue)[82], these agents may allow better targeting of lymphatic tissue with radiotherapy, particularly in conjunction with IMRT techniques, which allow better conformity to concave target volumes. Unfortunately these agents have not yet been approved for general use.

In practice, the target volume includes the distal common iliac, internal iliac (hypogastric) lymph nodes, including the presciatic nodes, the external iliac lymph nodes, including the obturator lymph nodes, and the presacral lymph nodes anterior to S1–S3.

Before outlining it is advisable to identify the bowel and pelvic vasculature. Administration of intravenous contrast aids identification of the pelvic vessels. Bowel can be identified with more confidence by following its course over several slices. The use of oral contrast can help define small bowel.

There is no data to help define precisely the superior extent of the target volume, although planning studies have demonstrated feasibility with low toxicity[83]. One randomized controlled trial has shown no advantage for para-aortic radiotherapy in addition to pelvic radiotherapy in individuals with locally advanced tumours or involved pelvic lymph nodes[84].

The Radiation Therapy Oncology Group (RTOG)[80] has recently published a consensus document on guidelines for lymph node delineation. The current randomized phase II PIVOTAL trial utilizes a modified version of this guideline known as the 'ICR-modified RTOG guidelines'[21]. These documents should be referred to for a detailed description and atlas to aid lymph node delineation.

In summary, the pelvic vessels are outlined on each axial CT slice in continuity. The volume should encompass the iliac vessels which high in the pelvis lie on its posterior wall, encompass the internal iliac/presciatic and external iliac vessels in continuity, and cover the upper pre-sacral area anterior to the upper three sacral segments.

The RTOG guidelines suggest a 7-mm radial margin around the pelvic blood vessels to cover the involved lymph nodes (the CTV), whilst excluding adjacent bowel, bladder, bone and pelvic muscles. Volumes begin at the L5/S1 interspace and end at the superior aspect of the pubic bone. The PTV is then created by applying a 5-mm margin in all directions.

8.4.3 Radical radiotherapy: following prostatectomy

For patients undergoing adjuvant or salvage radiotherapy post prostatectomy, the CTV consists of the prostate bed, i.e. the estimated location of the preoperative prostate volume (including sites of possible microscopic tumour extension), plus the extent of the surgical bed, and should normally include any surgical clips provided that the normal-tissue dose-constraints are within tolerance. The seminal vesicles should usually be included in the CTV. In some high-risk cases, the pelvic lymph node regions may also be included.

The volume is localized using CT (as previously described). Preoperative imaging (pelvic CT/MRI), operative notes, and histopathological details from the prostatectomy specimen, including prostate size and tumour extent to specific boundaries of the surgical resection, as well as the anatomy seen on the postoperative planning CT scan, can all help to define the CTV.

Since the prostate is located between the rectum posteriorly, the pubis anteriorly, and the pelvic sling muscles laterally, the volume of the prostate bed can be defined in relation to these structures. Superiorly, since the bladder will fill part of the space previously occupied by the prostate between the pubis and the rectum, the volume will out of necessity include the inferior part of the bladder. The inferior extent of the CTV should lie at the level of the pelvic floor above the penile bulb.

A useful guide for outlining taken from a current trial protocol follows[24]:

- Inferior border: 5mm cranial to the superior border of the penile bulb.
- Anterior border:
  - Caudal (< 2 cm above anastomosis)—posterior aspect of symphysis pubis.
  - Cranial (> 2 cm above anastomosis)—posterior one-third of bladder wall.
- Posterior border: anterior rectal wall.
- Lateral border: medial border of obturator internus and levator ani muscles.
- Superior border:
  - If SV involvement is low risk and pathologically uninvolved: include base of SV.
  - If SV involvement is high risk or are pathologically involved: aim to include tips of SV.
  - If SV absent, the superior border should be determined with reference to the estimated position of the preoperative SV.
- Prostate bed PTV: add 1.0cm in all directions, for day-to-day variation in set-up and for CTV motion.

In a survey of UK practice, Morris et al.[29] found that most UK oncologists used preoperative imaging if available to assist with localization, but the location of surgical clips and bony landmarks were also commonly used as planning aids. As localization is less precise than with the prostate *in situ*, target volumes may be larger than for radical prostate treatments. However, full rectal sparing can be achieved and standard dose of 66 Gy in 2-Gy fractions is generally well tolerated.

8.4.4 Palliative radiotherapy to the prostate/pelvis

The same principles for localization apply here as for radical treatment. The size of the PTV and field arrangement will impact on the dose that can be safely prescribed.

8.5 Dose distribution

8.5.1 3D Conformal radiotherapy

A treatment planning computer which can accurately account for tissue inhomogeneity is used to generate a 3D conformal plan. A three- or four-coplanar field arrangement is usually chosen for treatment. Although some centres use six-field coplanar beams for high-dose treatments, a comprehensive review concluded that three-field arrangements gave equivalent or improved dose distributions[85]. A three-field technique would normally consist of an anterior field and two wedged opposed lateral fields as shown in Fig. 8a.6. In patients with asymmetry in the posterior extent of the PTV, substituting a posterior oblique field for one lateral field may improve target coverage.

An anterior and two wedged lateral fields are also normally used to treat the pelvis, although a posterior field may be added to improve the dose distribution in larger individuals.

Acceptability of a particular plan is assessed by inspection of target and OARs DVHs. Dose is prescribed to the isocentre (100%).

In assessing the acceptability of the plan, the clinician should:

◆ Ensure target volume covers prostate + seminal vesicles with adequate margins in all directions.

◆ Check 95% isodose cover of PTV.

◆ Confirm that no unacceptable 'hot spots' occur within or outside the PTV.

◆ Confirm volume of rectum irradiated, particularly avoiding circumferential exposure.

◆ Assess dose to rectum, bladder and femoral heads, using DVH data (see Table 8a.4).

◆ Verify field sizes (approximately 8 × 8 cm).

◆ Check weighting of different fields.

DVHs are used to develop local guidelines to help produce acceptable planning constraints. We use literature derived 'cut-offs' for rectum[86–88], bowel[89], bladder[40 90], and femoral heads[91]. Current constraints are based on treatment using conformal techniques including IMRT and dose escalation studies. Recent review of the dose volume dependence of external beam radiotherapy toxicity has been carried out[92]. Those for rectum have been derived from studies comparing DVH data in patients who have or have not developed rectal morbidity. The volume of rectum receiving 60 Gy is associated with the risk of Grade 2 rectal toxicity or rectal bleeding. Conservative constraints for 3D conformal planning are V50 ≤ 50%, V60 < 35%, V70 < 15%, and V75 < 3%[86,88]. The NTCP models predict that using these constraints will limit Grade 2 late rectal toxicity to < 15% and the probability of Grade 3 late rectal toxicity to < 10% for prescriptions up to 79.2 Gy in standard fractionation[88]. Factors which may be associated with increased toxicity are diabetes mellitus[93–96], haemorrhoids[97], inflammatory bowel disease[98], and prior rectal or abdominal surgery[93]. Bladder constraints are derived from acute side effect data only. There is only limited evidence available in the literature regarding the risk of small bowel toxicity. When delineating contours of bowel loops, the volume of small bowel receiving 15 Gy or more should be < 120 cc where possible to minimize severe acute toxicity[99]. Studies have found an association between the dose to the urethral bulb and erectile dysfunction[100,101]. Although the PTV coverage should not be compromised, some recommend the mean dose to 95% of the urethral bulb volume to be < 50 Gy[102]. Suggested optimal and mandatory dose constraints can be found in Table 8a.3.

Table 8a.4 Suggested dose–volume constraints for organs at risk

Organ at risk	Dose (Gy) at 2-Gy per fraction to 100%	Dose volume constraint (cc)	
		Optimal	Mandatory
Rectum	30	80	–
	40	65	–
	50	50	60
	60	35	50
	70	15	15
	75	3	5
Bladder	50	50	–
	60	25	–
	65	–	50
	70	5	35
Sigmoid, small & large bowel	45	78	158
	50	17	110
	55	14	28
	60	0.5	6
	65	0	0
Urethral bulb	50	50	–
	60	10	–
Femoral heads	50	5	25

8.5.2 Intensity-modulated radiotherapy

In the UK, IMRT is being developed as the standard delivery technique for radical external beam prostate radiotherapy[103]. Forward planned techniques are commonly used and inverse planned methods are evolving[104].

IMRT to the prostate and seminal vesicles provides highly conformal dose distributions compared with 3DCRT (see Fig. 8a.6) and has produced encouraging results, with regard to both improved tumour control and sparing of OARs[105–107]. The latter may allow further dose escalation.

Inverse planned IMRT is increasingly being used for pelvic lymph node irradiation. The superior conformality allows better sparing of bowel and bladder, as shown in preclinical studies[108] (see Fig. 8a.7). By reducing the dose to OAR, IMRT reduces treatment-related complications, allowing dose escalation to high-risk lymph node areas[20,109–111]. The PIVOTAL trial is underway to assess the use of dose escalation with IMRT in this setting[21].

There are disadvantages to IMRT, namely the higher number of monitor units (MUs) required and the resulting longer treatment times per fraction when compared to unmodulated 3DCRT. There are also concerns about a potential increased risk of secondary cancers with IMRT, due to increased scatter dose and increased spread of a 'low dose bath' around the pelvis[112].

8.6 Implementation

Before the first treatment, set up accuracy is assessed by comparing the portal images with digitally reconstructed radiographs (DRRs). This has the advantage of omitting the simulator visit for the patient and reduces the risk of introducing systematic errors in positioning.

The patient is treated with a linear accelerator of appropriate energy (normally 6 MV or greater) using an isocentric technique. The same patient conditions as for planning are sought, i.e. with the patient supine, using the same immobilization technique and similar rectal and bladder filling. Lasers are aligned with the reference tattoos such that the approximate treatment isocentre is positioned at that of the machine. The appropriate couch moves are then made according to the treatment plan to position the true treatment isocentre correctly.

An allowance can also be made for 'couch sag' on an individual patient basis, which is measured by reading the couch height with the lateral (coronal) laser aligned with the couch top. Skin marks are made corresponding to the shifts from the approximate to true treatment isocentre which should correspond to the position of the lasers on the skin. These can be observed remotely during treatment to ensure that the treatment position is maintained. A measurement of focus-to-skin distance (FSD) is made from the treatment machine, which is compared to the expected measurement calculated from the treatment plan. Using an isocentric technique, the sum of anterior FSD and the distance from the anterior skin surface to the isocentre, as per treatment plan, should equal 100 cm. Day-to-day variations in bladder filling may cause variation in FSD and discrepancies of up to 1 cm can be accepted. Larger discrepancies indicate a set up error or a more significant change in the patient outline. All fields are treated daily with the appropriate shielding with MLCs.

8.7 Verification

Accurate positioning of the target volume in relation to shaped radiation fields is essential to maximize the effectiveness of treatment.

Two types of positioning error have been identified[113,114]

- Systematic errors can occur as a result of incorrect target outlining and data transfer between the treatment planning stage and actual treatment set-up. It may also be caused by incorrect design, marking or positioning of treatment accessories such as shielding blocks and immobilization devices.

- Random errors include daily variations in patient set-up and anatomical changes between treatments due to bladder and rectal filling, tumour growth or shrinkage, respiratory movement, and human error.

The practical effect of a systematic error is to shift a dose distribution with reference to its PTV, and that of a series of random errors is to 'blur' its edges, creating a smaller high-dose volume and a broader penumbra. Algorithms have been derived allowing the calculated treatment margins to account for the effects of both random and systematic geometrical deviation with known probability[113].

The verification process is designed to detect all positioning errors. For each daily treatment, the position of bony structures obtained by a linear accelerator portal image is compared with their position in reference images obtained during treatment planning. The reference images are usually DRRs. Electronic portal imaging devices (EPIDs), which use electronic radiation sensors instead of conventional film based imaging, allow multiple images to be obtained rapidly during treatment. The images may be enhanced electronically and allow in-treatment motion to be assessed if required[114–117]. The EPID system is then compared with the reference image and discrepancies between the position of landmarks or field in the two images, which indicate a positioning error, are corrected.

Image contrast using megavoltage radiographs is poor compared with orthovoltage X-rays. Additional contrast may be provided by radio-opaque markers, including in-tissue markers. These are placed prior to treatment planning and are therefore present on both reference and portal images, allowing discrepancies to be detected more easily[115,117]. The placement of radio-opaque markers within the prostate allows errors arising from prostatic movement to be identified, and incorporated into the overall assessment and correction of errors (see Figs 8a.3 and 8a.4).

Gold seeds as fiducial markers within the prostate have been investigated[40]. Studies have suggested that insertion, although invasive, is tolerable and seed migration is rare[118]. These studies confirm motion of the prostate occurs most frequently in the anterior-posterior and superior-inferior direction as a result of variable bladder and rectal filling. With daily on-line verification using gold seeds to aid set up, there is evidence that the CTV to PTV margin could be further reduced[46, 119–122].

Beacons or transponders, although larger, can also be implanted in a similar way to gold seeds and their position during treatment is monitored by an in room electro-magnetic Calypso localization system. This allows real time tracking data of prostate motion to be obtained[122]. These transponders can cause artefact with MRI scanning and, although they are usually inserted after diagnostic imaging has been performed, they can make subsequent imaging to assess disease progression difficult[123].

Other, less invasive methods of localizing the prostate during radiotherapy, specifically ultrasound and CT have been evaluated. The B-mode acquisition and targeting (BAT) ultrasound system provides a daily visualization of prostate in three orthogonal planes in the radiotherapy treatment room. Software within the device compares the estimated position of the prostate as defined by the operator using the device, with the volume defined on the planning CT, to produce the estimated shifts in prostate position from that initially defined. Although having the advantage of being non-invasive, limitations of the device include interoperator variability, possible shifts in the prostate caused by the pressure of the ultrasound probe itself on the bladder during the scan, and difficulties in making comparisons between imaging modalities[124–129].

Cone beam CT incorporates a diagnostic CT device within the linear accelerator gantry to produce kV images. These CT images can be used to localize the prostate

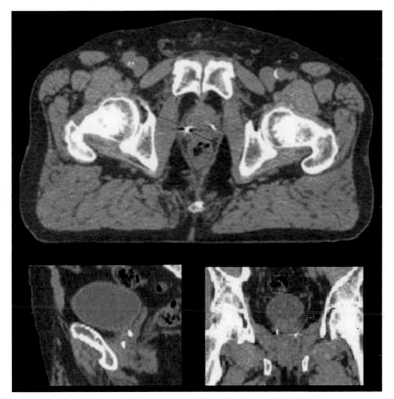

Fig. 8a.3 Planning CT scan with markers to aid localization and verification.

during a course of treatment and errors due to organ movement are corrected accordingly (see Fig. 8a.5). Future advances using combined cone beam CT information and automatic methods for target identification may allow rapid and accurate corrections to be made prior to each radiotherapy treatment.

Stereoscopic KV allows both 3D and 4D information to be obtained using two X-ray sources and detectors which takes images simultaneously. They can be used to track movement of both internal and external markers in real time and are used in treatment systems such as EXACTRAC and CyberKnife.

If discrepancies are detected they are corrected before treatment begins as appropriate. Gross inaccuracies may be corrected by repositioning the patient in real time[116]. For small inaccuracies the radiotherapy field position may also be adjusted, but must take account of random variations[130]. At the Royal Marsden, the field isocentre is adjusted if a systematic displacement of > 3 mm is detected in any direction. With conventional EPID imaging we recommend taking at least three images in the first week of treatment in order to judge random and systematic components of set up uncertainty. If accuracy is within 3 mm then subsequent weekly checks are adequate. If errors ≥ 5 mm are found then accuracy should be ensured before the delivery of the

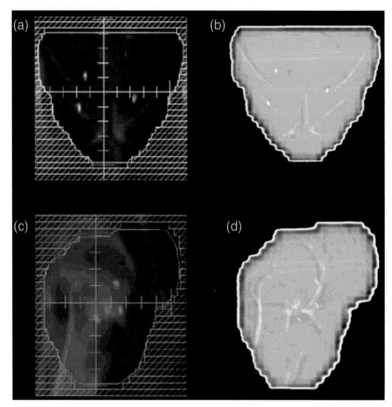

Fig. 8a.4 Gold markers as seen on verification imaging (a) anterior DRR (b) anterior electronic portal image (c) lateral DRR (d) lateral electronic portal image.

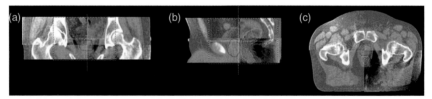

Fig. 8a.5 Cone-beam CT (Elekta Synergy) registered with the planning CT and viewed using a cut plane view (a) coronal (b) sagittal (c) axial views.

next treatment fraction. If adjustments are required, then three sequential measurements are again made before moving to weekly measurements.

8.8 Dose prescription

With the introduction of 3DCRT techniques, it was first possible to more closely match the high dose volume to the tumour target, whilst reducing the radiation to dose limiting normal tissues[131]. This resulted in advantages including a potential reduction in radiation related side effects, as well as a potential improvement in tumour control. It also allowed the possibility of dose escalation.

Dose escalation in prostate cancer is now the 'standard of care' in the UK[12], supported by evidence from several Phase III trials showing a consistent benefit in tumour control with dose escalation[132–135], although no statistically significant overall survival benefit has yet been shown. The MD Anderson trial[135] compared 70 and 78 Gy in high and intermediate risk patients, without neoadjuvant androgen deprivation. In the 78-Gy group, there was a statistically significant benefit in biochemical control, as well as a reduction in nodal and distant failure, which was restricted to men with a presenting PSA > 10 micrograms/L or high-risk disease.

An update of the PROG 95–09 trial[134], which used a proton beam boost to compare doses of 70.2 Gy equivalent and 79.2 Gy equivalent (without androgen deprivation) in mostly low- or intermediate-risk patients, has also reported a benefit in terms of biochemical control and a reduction in local failure.

The Dutch trial[132], compared 78 Gy to 68 Gy (with androgen deprivation used in 21% of patients only) in predominantly intermediate and high risk patients. They also found a benefit in biochemical control for the higher dose. However, no difference in freedom from clinical failure was shown.

The MRC RT01 trial[133] compared a standard dose of 64 Gy in 32 fractions to an escalated dose of 74 Gy in 37 fractions. It is the only dose escalation trial to use neo-adjuvant androgen suppression in all patients. It also confirmed a benefit in biochemical control, at a cost of increased late bowel toxicity.

A meta-analysis of dose escalated radiotherapy in localized prostate cancer, including seven randomized controlled trials with a total patient population of 2812, confirmed the reduction in biochemical failure, in all risk groups, with high dose radiotherapy. A significant increase in Grade > 2 late bowel toxicity was also noted. No survival benefit has yet been shown[136].

A further issue is that of fractionation; prostate cancer may have a lower alpha/beta ratio (according to the linear quadratic model) than other tumours, and consequently larger fraction sizes may be associated with improvements in tumour control for a given level of radiation related side effects. If this is the case, then such schedules should become the standard approach to treatment as they would be more convenient for patients and make better use of radiotherapy resources. Presently there are no long-term data using higher dose hypofractionated radiotherapy and work is underway to test this in the clinical setting[137], although some centres have traditionally used fraction sizes of > 2 Gy with apparently equivalent results[138].

8.8.1 Recommended doses

All doses are prescribed to the ICRU reference point.

Conformal radiotherapy (two-phase)/IMRT radical treatment to prostate ± seminal vesicles (SVs) with short-course neo-adjuvant androgen deprivation:

◆ Prostate: 74 Gy in 37 fractions over 7½ weeks.

◆ SVs: (equivalent to) 54–56Gy.

Radical post prostatectomy radiotherapy:

♦ Prostate bed: 66 Gy in 33 fractions over 6½ weeks.

Radical pelvis and prostate radiotherapy in combination with long-term androgen deprivation:

♦ Prostate and pelvic 3DCRT:
  • Prostate and SVs: 74 Gy in 37 fractions over 7½ weeks.
  • Pelvic lymph nodes: 46 Gy in 23 fractions over 4½ weeks.
♦ Prostate and pelvic IMRT:
  • Prostate: 74 Gy in 37 fractions over 7½ weeks.
  • Pelvic lymph nodes and SVs: 55 Gy in 37 fractions over 7½ weeks.

Palliative treatment:

♦ Prostate ± SVs:
  • 30 Gy in 10 fractions over 2 weeks.
  • 36 Gy in weekly fractions over 6 weeks (three or four-field plan).

8.9 Toxicity and care during treatment

8.9.1 Acute side effects

These occur during and up to 6 months after the start of treatment. The RT01 trial[133] treated patients using 3D-conformal techniques to doses of 74 Gy (in the dose escalated group). The reported acute toxicity is included as follows:

Gastrointestinal toxicity

Symptoms of acute proctitis include tenesmus, rectal bleeding, pain, and mucous discharge. There may be accompanying perianal reaction and soreness. In RT01, cumulative acute RTOG Grade 2 bowel toxicity was 33%. The majority of men (65%) had RTOG Grade 0–1 bowel toxicity during weeks 1–18. During weeks 8–18, 21% of men had Grade 2 bowel side effects[139].

It is important that stools remain soft, and bulking agents (bran, Normacol, Fybogel) may be helpful. If bowel frequency is a problem, loperamide, initially at low dosage, is effective. Proctitis/anal soreness are helped by steroid/local anaesthetic suppositories or cream (e.g. Scheriproct). Diarrhoea with flatulence can indicate small bowel toxicity, particularly in patients who are receiving radiotherapy to the whole pelvis. This can be treated with loperamide, taking care to avoid constipation.

Genitourinary toxicity

Urinary symptoms result from a combination of obstructive and irritative prostatic symptoms, often on a background of pre-existing prostatism. Symptoms include frequency, urgency, poor stream, and rarely haematuria In RT01 trial, cumulative acute RTOG Grade 2 bladder toxicity was 39%. The majority of patients (61%) had Grade 0–1 bladder toxicity during weeks 1–18. At weeks 8–18 30% of patients had RTOG Grade 2 or more toxicity.

Patients are usually advised to remain well hydrated to decrease bladder irritation and prevent infection although increased fluid intake may contribute to increased frequency.

Urinary tract infection should be ruled out by analysis of a mid-stream urine sample. Urinary frequency and poor stream secondary to worsening prostatic obstruction during treatment may be improved by an alpha adrenergic blocking drug, such as tamsulosin 400 micrograms daily, increasing to 800 micrograms if necessary[140]. Simple analgesics and antispasmodics may alleviate dysuria and irritative symptoms. Rarely acute urinary obstruction occurs, necessitating short-term catheterization (2% of men in RT01 trial)[139]. Suprapubic catheters may be better tolerated than urethral catheters. Acute urinary and bowel symptoms are expected to settle in 12–18 weeks or less.

Other acute toxicities

These include skin erythema or dry desquamation, which can be treated with an emollient or occasionally moist desquamation which may require a hydrogel preparation to assist healing. Fatigue, lethargy, and pubic hair loss are also commonly observed during treatment.

8.9.2 **Late side effects**

The RT01 trial reported late toxicity for the escalated dose group (74 Gy) as follows[133]:

Late genitourinary toxicity

This includes cystitis, haematuria, and urethral stricture. At 5 years, 11% had reported RTOG Grade 2 or higher genitourinary toxicity. Urinary symptoms can be evaluated with urodynamic studies. Obstructive symptoms can be treated with alpha adrenergic antagonist drugs, or urethral dilatation in the case of stricture formation. Irritative bladder symptoms can be treated with anticholinergic agents. Overall, however, bladder symptoms improve after prostate radiotherapy, presumably due to prostate shrinkage and reduction in obstructive symptoms[68,139].

Late gastrointestinal toxicity[141]

After 5 years, 33% of men had reported RTOG grade ≥ 2 bowel toxicity. The peak prevalence of bowel dysfunction after radiotherapy was reported 6–24 months after treatment. The most frequently reported grade ≥ 2 rectal toxicities on the LENT/SOM scales were tenesmus (Grade ≥ 2 = 17% cumulative incidence at 5 years for the 74-Gy group), rectal bleeding (Grade ≥ 2 = 32% cumulative incidence at 5 years, with incidence peaking around 3 years after radiotherapy) and mucosal loss (Grade ≥ 2 = 8% cumulative incidence at 5 years). Cumulative incidence of RTOG grade ≥ 2 diarrhoea was 15% of patients at 5 years. RTOG grade ≥ 2 proctitis was present in 25% of patients by 5 years and peaked at 18–36 months. Serious bowel injury including bowel obstruction, rectal–anal stricture, and ulceration were all rare. At 5 years, the cumulative incidence of patient-reported severe bowel problems was 8% and severe distress was 5%[141].

If late rectal/bowel symptoms occur it is important to rule out comorbid conditions. Rectal bleeding should be investigated by flexible sigmoidoscopy. In approximately 30% of patients, gastrointestinal symptoms will be as a result of conditions unrelated to radiotherapy[142].

If treatment is required for proctitis/bleeding, simple outpatient treatments such as corticosteroid containing enemas are commonly used, but evidence of benefit is weak.

Sucralfate enemas may be beneficial, but some patients find them difficult to administer. Stool softeners, e.g. Normacol, are helpful. Pelvic floor exercises and bowel 'training' may help symptoms of rectal leakage. Bowel frequency may respond to low dose loperamide. Troublesome symptoms refractory to such measures may be treated with topical formalin application, endoscopic argon laser coagulation, or hyperbaric oxygen therapy. Refractory late rectal toxicity may occasionally necessitate major surgery such as defunctioning colostomy, although this is very uncommon.

Erectile dysfunction

Erectile dysfunction is prevalent in the first 6 months, mainly due to neo-adjuvant androgen suppression. However, in many men it continues long term. At 5 years, the proportion of men in the best-functioning group before treatment halved, while that in the worst-functioning group doubled[133]. There appears to be a dose–volume effect on the penile bulb and erectile dysfunction[100,102].

Some men benefit from drug treatment such as an oral phosphodiesterase type-5 inhibitor[143]. Of this class of drugs sildenafil (Viagra) is commonly used initially. If treatment is unsuccessful with sildenafil, vardenafil (Levitra) may still prove helpful. Tadalafil (Cialis) results in a longer duration of erectile ability after administration than either sildenafil or vardenafil, and consequently can allow a greater degree of spontaneity of sexual activity. Other drugs include prostaglandin E1 administered either intra-urethrally (MUSE), or by intra-cavernosal injection (Caverject). A vacuum device is an alternative strategy for the management of erectile dysfunction.

8.10 The role of hormonal therapy in combination with radical radiotherapy

Radiotherapy alone with conventional dose is not a satisfactory approach for the treatment of localized prostate cancer[68]. Based on the results of Phase III trials[16,144,145] it is common practice to use a short course of androgen suppression prior to and during radiotherapy treatment. These studies were carried out in patients of clinical stage T2 and above, and some employed whole pelvic radiotherapy. There is little evidence as yet to support the use of hormonal therapy in addition to radiotherapy in patients with clinical stage T1 disease in terms of improved treatment efficacy but is widely used.

8.10.1 Adjuvant androgen suppression with radiotherapy

Meta-analysis of randomized control trials demonstrate that adjuvant hormone therapy following radiotherapy for localized and locally advanced prostate cancer has significant clinical benefit, with improved overall survival, disease-specific survival and disease-free survival up to 10 years with no additional toxicity[146,147].

The first compelling evidence for this came from EORTC trial 22863—415 patients with T2–4 and/or high-grade prostate cancer were randomly assigned to receive radiotherapy to the prostate and pelvis alone or radiotherapy with LHRH agonists given during and after external been radiotherapy for 3 years. The 5-year overall survival was significantly better in the combined treatment arm (79%, CI: 72–86%) than the radiotherapy alone arm (62%, CI: 52–72%). The statistical advantage in both disease-free and overall survival was maintained at 10 years[17]. These results are consistent with RTOG 85–31 which randomized men with T3 and/or N1 disease to radiotherapy alone, or with the addition of long-term adjuvant hormonal therapy. Initial results showed a statistically significant cause specific and overall survival advantage in favour of the combined treatment arm in the subgroup of patients with Gleason 8–10 tumours. Updated results show this benefit extends to the whole study population (53% vs. 38%, $p < 0.0043$)[16].

8.10.2 Neo-adjuvant and concomitant androgen suppression with radiotherapy

Neo-adjuvant hormonal therapy reduces the size of the target volume in such patients and therefore may permit smaller treatment fields which encompass less normal tissue[148]. The RTOG 94–13 suggested that in patients with estimated lymph node risk ≥ 15% and receiving whole pelvis and prostate radiotherapy, a total of 4 months of neo-adjuvant and concurrent hormonal therapy is superior to the same duration of hormonal therapy given after radiotherapy[18]. In men with locally advanced prostate cancer (T2b–T4 N0), 6 months of neo-adjuvant androgen deprivation followed by radiotherapy to the prostate and seminal vesicles significantly reduced distant progression, prostate cancer specific and all cause mortality compared to those receiving 0 or 3 months neo-adjuvant hormones[144].

8.10.3 Short-term versus long-term androgen suppression

The combination of radiotherapy and long-course hormone treatment for 2–3 years should be regarded as the standard of care in for patients with high-risk disease[149,150]. The RTOG 92–02 study, demonstrated that in locally advanced (T2c–T4) high-grade disease (Gleason 8–10) radical radiotherapy to the prostate and pelvis with 4 months of neo-adjuvant androgen suppression followed by an additional 2 years of adjuvant hormones significantly improved 5-year overall survival (80% vs. 69%) compared to those who did not receive adjuvant long-term hormones[150].

8.11 Treatment outcome

Prognostic groups can be defined using combinations of PSA level, Gleason score, and clinical stage[151–153]. Table 8a.5 is derived from Shipley et al.[151].

After completion of radiotherapy and hormonal treatment, testosterone recovery usually occurs. This can cause some PSA elevation that is related to normal prostate tissue recovery and not reflective of disease recurrence.

Rising PSA after radical radiotherapy may be a sign of local failure, metastatic disease or both. Local failure is more likely in patients with low- to intermediate-risk prostate cancer at diagnosis who have a slow rising PSA, with PSA failure occurring some time after radiotherapy.

The definition of PSA relapse after radiotherapy varies. In 1996 the American Society for Radiation Oncology (ASTRO) defined biochemical failure as three consecutive PSA rises following a nadir[154]. In 2005 this was revised as the 'Phoenix criteria' nadir plus 2 ng/mL[155].

Table 8a.5 Stratification of localized prostate cancer into 4 risk groups with corresponding 5-year biochemical failure free survival after radiotherapy. Median prostate dose was 69.4 Gy and no hormonal manipulation was used with radiotherapy[151]

Characteristics	5-year biochemical failure-free survival (%)
PSA < 10 ng/mL, any Gleason score	81
PSA 10–20 ng/mL, any Gleason score	69
PSA > 20 ng/mL, Gleason score 2–6	47
PSA > 20 ng/mL, Gleason score ≥ 7	29

Patients who have local failure only could be considered for salvage therapy. Such therapies include radical prostatectomy, cryotherapy and high-intensity focused ultrasound (HIFU).

The role of salvage radical prostatectomy, cryotherapy and brachytherapy for disease recurrence following definitive radiotherapy has been reviewed[156]. Salvage radical prostatectomy was associated with 5-year biochemical disease free survival rate of 55–69%. There was significant incidence of complications including anastomotic stricture, urinary incontinence and rectal injury. The outcome of salvage cryotherapy was shown to be related to the definition of biochemical failure; generally ranging from 40% when failure was defined as PSA 2 plus nadir[157] to 73% when failure was defined as PSA> 4ng/mL[158]. Salvage cryotherapy was complicated by erectile dysfunction, pain (in the pelvis, rectum or perineum), rectourethral fistula, bladder outlet obstruction, and urethral stricture[156]. Salvage brachytherapy following external beam radiation requires further evaluation; however, in one series 14% of patients experienced bladder outlet obstruction requiring TURP. Haematuria was seen in 4% of patients; 6% developed uretheral stricture; 4% rectal ulcers and 2% experienced rectal bleeding requiring colostomy[156,159].

Salvage HIFU for local recurrence of prostate cancer after external beam radiotherapy demonstrates definitive local control. In one series 73% of patients achieved a negative biopsy result with a 3-year progression-free survival of 53%, 42%, and 25% in the low-, intermediate-, and high-risk patient groups respectively. Although no rectal toxicity is reported, individuals may suffer from urinary incontinence[160].

8.12 **Future developments**

Future technical developments in prostate radiotherapy are aimed at improving the dose delivered to the tumour while reducing irradiation to the organs at risk. Advances can be divided into three general categories:

- Improved target definition.
- Improved radiotherapy delivery.
- Image guided therapy.

These three complementary aspects all contribute to improving the therapeutic ratio of prostate cancer radiotherapy and allows safe dose escalation.

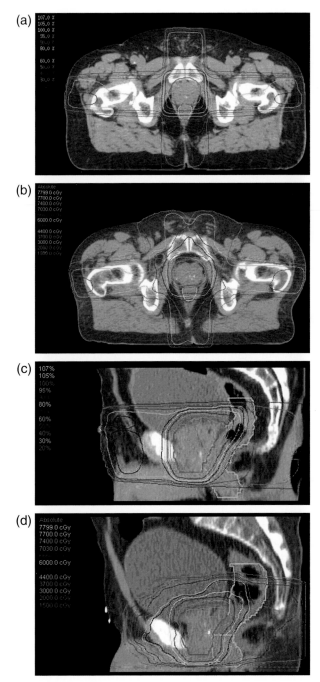

Fig. 8a.6 Axial dose distribution showing prostate conventional 3D plan (a) and IMRT plan (b) (PTV magenta; rectum orange); Sagittal dose distribution showing prostate conventional 3D plan (c) and IMRT plan (d) (PTV magenta; rectum orange).

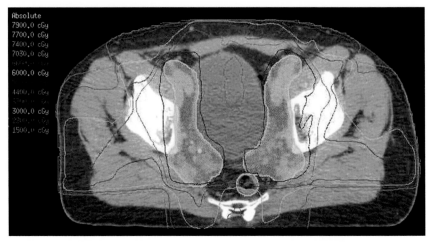

Fig. 8a.7 Axial dose distribution for pelvic nodal treatment volume using 5 field IMRT technique (lymph node PTV yellow; rectum orange).

8.12.1 Improved target definition

The greater degree of dose heterogeneity achievable with IMRT makes it possible to deliver boost doses of radiotherapy to discrete areas of intraprostatic tumour[161]. Standard imaging techniques have not reliably localized such dominant intraprostatic tumour nodules; however, functional imaging techniques are promising.

Functional MRI techniques exploit the increased vascularization and angiogenesis occurring within the tumour to aid delineation. Diffusion-weighted MRI, magnetic resonance spectroscopy and dynamic contrast-enhanced MRI have all shown to be effective in differentiating between normal tissue and malignant prostate tissue[162–166].

PET allows biological imaging of tumours. 18-FDG highlights proliferating areas in many tumour types[167]. However, this commonly used tracer is ineffective in delineating malignant prostate tissue. Its limitation is due predominantly due to the low glucose metabolism of the tumour and the urinary excretion of 18-FDG[168]. Alternative tracers such as [11C] choline, which are not excreted by this means, have been investigated[169] and may be valuable in defining disease recurrence after prostatectomy, lymph node involvement and in identifying intraprostatic nodules.

8.12.2 Improved radiotherapy delivery

The following techniques enable high doses of conformal radiotherapy to be delivered, optimizing tumour control while sparing normal tissue.

Volumetric-modulated arc therapy

Volumetric-modulated arc therapy (VMAT) is a novel form of IMRT optimization that allows the radiation dose to be delivered in a single gantry rotation of up to 360°. Several planning studies have been published on this subject[170–173]. The results show reduced treatment delivery time (which in turn potentially reduces the risk of intra-fraction organ motion), reduced number of MUs, and better OAR sparing with comparable target dose coverage to IMRT. VMAT achieves this by allowing more degrees of freedom such as variations in gantry-speed, dose rate and collimator angle in addition to dynamically changing MLC shaped fields. The potential disadvantages include the theoretical risk of increased secondary malignancy. There is less scatter dose from the reduced MU but a larger proportion of non-target tissue volume receives 5–40 Gy with VMAT techniques (see Fig. 8a.8a), which may well increase the number of secondary cancers[174].

Tomotherapy

Treatment delivery is based on a CT scanner where the diagnostic X-ray tube has been replaced with a 6-MV linac. The patient is treated in slices by a narrow photon beam. As the image acquisition is integrated in to the treatment system (same delivery device), MV CT images are generated. This has an advantage over kV CT imaging in that the MV CT is less likely to introduce significant artefacts because of the dominance of Compton scatter at 6 MV which has a linear relationship between electron density and Hounsfield units[175].

Stereotactic radiotherapy

Stereotactic body radiotherapy delivers highly conformal, dose sculpting radiotherapy treatment in large fraction sizes. It can be delivered a using linear accelerator or CyberKnife. CyberKnife is a linear accelerator mounted on a robotic arm which allows delivery of 6-MV photons in multiple non-coplanar beam directions without movement of the patient (see Fig. 8a.8b). Results from a number of trials using extreme hypofractionation including patients with low- or intermediate-risk prostate cancer suggest that the technique is feasible[176, 177]. Longer follow-up is required is assess efficacy.

Proton beam therapy

Charged particles such as protons deposit their energy within a small area known as the Bragg peak, the radiation dose beyond this rapidly falls to zero. The advantage is that the volume of normal tissue receiving low-dose radiation is reduced. There is limited data of its effective use in localized prostate cancer[177–178]. The significant financial resources required to implement proton therapy at present restricts its widespread use.

8.12.3 Imaged-guided therapy

As newer radiotherapy techniques provide greater 3D conformity around the PTV, it is becoming increasingly necessary to identify prostate motion during and between treatments to avoid geometric uncertainty or miss. Although a number of methods of tracking prostate motion are available, its widespread use in routine clinical practice is yet to be seen. Such measures may permit further dose escalation, and allow the use of novel fractionation schedules.

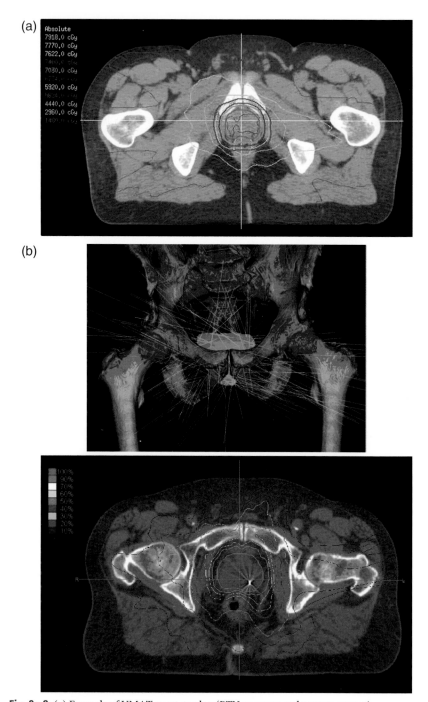

Fig. 8a.8 (a) Example of VMAT prostate plan (PTV magenta and rectum orange).
(b) Example of CyberKnife prostate plan (demonstrating non-coplanar beams and isodose distribution).

References

1. Cancer Research UK. *CancerStats – cancer Statistics for the UK*. http://info. cancerresearchuk.org/cancerstats/ On-line, accessed 21 January 2011.

2. Sakr WA, Grignon DJ, Haas GP, Heilbrun LK, Pontes JE, Crissman JD. Age and racial distribution of prostatic intraepithelial neoplasia. *European Urology* 1996; **30**(2): 138–44.

3. Burford DC KM, Austoker J. *Prostate Cancer Risk Management Programme information for Primary Care; PSA testing for asymptomatic men*. London: NHS Cancer Screening Programmes, 2008.

4. Kupelian PA, Potters L, Khuntia D, *et al*. Radical prostatectomy, external beam radiotherapy <72 Gy, external beam radiotherapy > or = 72 Gy, permanent seed implantation, or combined seeds/external beam radiotherapy for stage T1-T2 prostate cancer. *International Journal of Radiation Oncology, Biology, Physics* 2004; **58**(1): 25–33.

5. Mayles WP. Survey of the availability and use of advanced radiotherapy technology in the UK. *Clinical Oncology* 2010; **22**(8): 636–42.

6. NICE. *Improving Outcomes in Urological Cancer*. London: NICE, 2002.

7. Bill-Axelson A, Holmberg L, Ruutu M, *et al*. Radical prostatectomy versus watchful waiting in early prostate cancer. *New England Journal of Medicine* 2011; **364**(18): 1708–17.

8. Widmark A, Klepp O, Solberg A, *et al*. Endocrine treatment, with or without radiotherapy, in locally advanced prostate cancer (SPCG-7/SFUO-3): an open randomised phase III trial. *Lancet* 2009; **373**(9660): 301–8.

9. Warde PR, Mason MR, Sydes MR, *et al*. Intergroup randomized phase III study of androgen deprivation therapy (ADT) plus radiation therapy (RT) in locally advanced prostate cancer (CaP) (NCIC-CTG, SWOG, MRC-UK, INT: T94–0110; NCT00002633). 2010 ASCO Annual meeting, Oral Abstract Session, Genitourinary Cancer (Prostate) 2010.

10. Roach M, 3rd. Re: The use of prostate specific antigen, clinical stage and Gleason score to predict pathological stage in men with localized prostate cancer. *Journal of Urology* 1993; **150**(6): 1923–4.

11. Nilsson S, Norlen BJ, Widmark A. A systematic overview of radiation therapy effects in prostate cancer. *Acta Oncologica* 2004; **43**(4): 316–81.

12. NICE. *Clinical Guidelines on Prostate Cancer (CG58)*. London: NICE, 2008.

13. Dearnaley DP, Khoo VS, Norman AR, *et al*. Comparison of radiation side-effects of conformal and conventional radiotherapy in prostate cancer: a randomised trial. *Lancet* 1999; **353**(9149): 267–72.

14. Asbell SO, Krall JM, Pilepich MV, *et al*. Elective pelvic irradiation in stage A2, B carcinoma of the prostate: analysis of RTOG 77–06. *International Journal of Radiation Oncology, Biology, Physics* 1988; **15**(6): 1307–16.

15. Roach M, 3rd, Bae K, Speight J, *et al*. Short-term neoadjuvant androgen deprivation therapy and external-beam radiotherapy for locally advanced prostate cancer: long-term results of RTOG 8610. *Journal of Clinical Oncology* 2008; **26**(4): 585–91.

16. Pilepich MV, Winter K, Lawton CA, *et al*. Androgen suppression adjuvant to definitive radiotherapy in prostate carcinoma—long-term results of phase III RTOG 85–31. *International Journal of Radiation Oncology, Biology, Physics* 2005; **61**(5): 1285–90.

17. Bolla M, Van Tienhoven G, Warde P, *et al*. External irradiation with or without long-term androgen suppression for prostate cancer with high metastatic risk: 10-year results of an EORTC randomised study. *Lancet Oncology* 2010; **11**(11): 1066–73.

18. Lawton CA, DeSilvio M, Roach M, 3rd, *et al.* An update of the phase III trial comparing whole pelvic to prostate only radiotherapy and neoadjuvant to adjuvant total androgen suppression: updated analysis of RTOG 94–13, with emphasis on unexpected hormone/radiation interactions. *International Journal of Radiation Oncology, Biology, Physics* 2007; **69**(3): 646–55.

19. Pommier P, Chabaud S, Lagrange JL, *et al.* Is there a role for pelvic irradiation in localized prostate adenocarcinoma? Preliminary results of GETUG-01. *Journal of Clinical Oncology* 2007; **25**(34): 5366–73.

20. Guerrero Urbano T, Khoo V, *et al.* Intensity-modulated radiotherapy allows escalation of the radiation dose to the pelvic lymph nodes in patients with locally advanced prostate cancer: preliminary results of a phase I dose escalation study. *Clinical Oncology* 2010; **22**(3): 236–44.

21. Portfolio NIfHRCRNT. Pivotal. A randomised phase II trial of prostate and Pelvis versus prostate alone treatment for locally advanced prostate cancer. The Institute of Cancer Research. CRUK/10/022; 2011.

22. Stephenson AJ, Scardino PT, Kattan MW, *et al.* Predicting the outcome of salvage radiation therapy for recurrent prostate cancer after radical prostatectomy. *Journal of Clinical Oncology* 2007; **25**(15): 2035–41.

23. Catton C, Parker C, Saad F, Sydes M. Prostate radiotherapy after radical prostatectomy: sooner or later? *BJU International* 2010; **106**(7): 946–8.

24. Parker C, Sydes MR, Catton C, *et al.* Radiotherapy and androgen deprivation in combination after local surgery (RADICALS): a new Medical Research Council/National Cancer Institute of Canada phase III trial of adjuvant treatment after radical prostatectomy. *BJU International* 2007; **99**(6): 1376–9.

25. Dearnaley DP. Additional treatment for pT3 prostate cancer: now, later or never. *BJU International* 2007; **100**(5): 977–9.

26. Bolla M, Van Poppel H, Collette L. [Preliminary results for EORTC trial 22911: radical prostatectomy followed by postoperative radiotherapy in prostate cancers with a high risk of progression]. *Cancer Radiotherapy* 2007; **11**(6–7): 363–9.

27. Thompson IM, Tangen CM, Paradelo J, *et al.* Adjuvant radiotherapy for pathological T3N0M0 prostate cancer significantly reduces risk of metastases and improves survival: long-term followup of a randomized clinical trial. *Journal of Urology* 2009; **181**(3): 956–62.

28. Wiegel T, Bottke D, Steiner U, *et al.* Phase III postoperative adjuvant radiotherapy after radical prostatectomy compared with radical prostatectomy alone in pT3 prostate cancer with postoperative undetectable prostate-specific antigen: ARO 96–02/AUO AP 09/95. *Journal of Clinical Oncology* 2009; **27**(18): 2924–30.

29. Morris SL, Parker C, Huddart R, Horwich A, Dearnaley D. Current opinion on adjuvant and salvage treatment after radical prostatectomy. *Clinical Oncology* 2004; **16**(4): 277–82.

30. Lee LW, Clarke NW, Ramani VA, Cowan RA, Wylie JP, Logue JP. Adjuvant and salvage treatment after radical prostatectomy: current practice in the UK. *Prostate Cancer and Prostatic Diseases* 2005; **8**(3): 229–34.

31. van der Linden YM, Lok JJ, Steenland E, Martijn H, van Houwelingen H, Marijnen CA, *et al.* Single fraction radiotherapy is efficacious: a further analysis of the Dutch Bone Metastasis Study controlling for the influence of retreatment. *International Journal of Radiation Oncology, Biology, Physics* 2004; **59**(2): 528–37.

32. Price P, Hoskin PJ, Easton D, Austin D, Palmer SG, Yarnold JR. Prospective randomised trial of single and multifraction radiotherapy schedules in the treatment of painful bony metastases. *Radiotherapy and Oncology* 1986; **6**(4): 247–55.

33. Dicker AP. The safety and tolerability of low-dose irradiation for the management of gynaecomastia caused by antiandrogen monotherapy. *Lancet Oncology* 2003; **4**(1): 30–6.

34. Bedognetti D, Rubagotti A, Conti G, *et al.* An open, randomised, multicentre, phase 3 trial comparing the efficacy of two tamoxifen schedules in preventing gynaecomastia induced by bicalutamide monotherapy in prostate cancer patients. *European Urology* 2010; **57**(2): 238–45.

35. Fradet Y, Egerdie B, Andersen M, Tammela TL, Nachabe M, Armstrong J, *et al.* Tamoxifen as prophylaxis for prevention of gynaecomastia and breast pain associated with bicalutamide 150 mg monotherapy in patients with prostate cancer: a randomised, placebo-controlled, dose-response study. *European Urology* 2007; **52**(1): 106–14.

36. Fiorino C, Reni M, Bolognesi A, Bonini A, Cattaneo GM, Calandrino R. Set-up error in supine-positioned patients immobilized with two different modalities during conformal radiotherapy of prostate cancer. *Radiotherapy and Oncology* 1998; **49**(2): 133–41.

37. Soffen EM, Hanks GE, Hwang CC, Chu JC. Conformal static field therapy for low volume low grade prostate cancer with rigid immobilization. *International Journal of Radiation Oncology, Biology, Physics* 1991; **20**(1): 141–6.

38. Weber DC, Nouet P, Rouzaud M, Miralbell R. Patient positioning in prostate radiotherapy: is prone better than supine? *International Journal of Radiation Oncology, Biology, Physics* 2000; **47**(2): 365–71.

39. Nutting CM, Khoo VS, Walker V, *et al.* A randomized study of the use of a customized immobilization system in the treatment of prostate cancer with conformal radiotherapy. *Radiotherapy and Oncology* 2000; **54**(1): 1–9.

40. Crook JM, Raymond Y, Salhani D, Yang H, Esche B. Prostate motion during standard radiotherapy as assessed by fiducial markers. *Radiotherapy and Oncology* 1995; **37**(1): 35–42.

41. Adamson J, Wu Q. Inferences about prostate intrafraction motion from pre- and posttreatment volumetric imaging. *International Journal of Radiation Oncology, Biology, Physics* 2009; **75**(1): 260–7.

42. Padhani AR, Khoo VS, Suckling J, Husband JE, Leach MO, Dearnaley DP. Evaluating the effect of rectal distension and rectal movement on prostate gland position using cine MRI. *International Journal of Radiation Oncology, Biology, Physics* 1999; **44**(3): 525–33.

43. van Herk M, Bruce A, Kroes AP, Shouman T, Touw A, Lebesque JV. Quantification of organ motion during conformal radiotherapy of the prostate by three dimensional image registration. *International Journal of Radiation Oncology, Biology, Physics* 1995; **33**(5): 1311–20.

44. Heemsbergen WD, Hoogeman MS, Witte MG, Peeters ST, Incrocci L, Lebesque JV. Increased risk of biochemical and clinical failure for prostate patients with a large rectum at radiotherapy planning: results from the Dutch trial of 68 GY versus 78 Gy. *International Journal of Radiation Oncology, Biology, Physics* 2007; **67**(5): 1418–24.

45. de Crevoisier R, Tucker SL, Dong L, *et al.* Increased risk of biochemical and local failure in patients with distended rectum on the planning CT for prostate cancer radiotherapy. *International Journal of Radiation Oncology, Biology, Physics* 2005; **62**(4): 965–73.

46. Wu J, Haycocks T, Alasti H, *et al.* Positioning errors and prostate motion during conformal prostate radiotherapy using on-line isocentre set-up verification and implanted prostate markers. *Radiotherapy and Oncology* 2001; **61**(2): 127–33.

47. Wachter S, Gerstner N, Dorner D, *et al.* The influence of a rectal balloon tube as internal immobilization device on variations of volumes and dose-volume histograms during treatment course of conformal radiotherapy for prostate cancer. *International Journal of Radiation Oncology, Biology, Physics* 2002; **52**(1): 91–100.

48. D'Amico AV, Manola J, Loffredo M, *et al.* A practical method to achieve prostate gland immobilization and target verification for daily treatment. *International Journal of Radiation Oncology, Biology, Physics* 2001; **51**(5): 1431–6.

49. McGary JE, Teh BS, Butler EB, Grant W, 3rd. Prostate immobilization using a rectal balloon. *Journal of Applied Clinical Medical Physics* 2002; **3**(1): 6–11.

50. Teh BS, McGary JE, Dong L, *et al.* The use of rectal balloon during the delivery of intensity modulated radiotherapy (IMRT) for prostate cancer: more than just a prostate gland immobilization device? *Cancer Journal* 2002; **8**(6): 476–83.

51. van Lin EN, van der Vight LP, Witjes JA, Huisman HJ, Leer JW, Visser AG. The effect of an endorectal balloon and off-line correction on the interfraction systematic and random prostate position variations: a comparative study. *International Journal of Radiation Oncology, Biology, Physics* 2005; **61**(1): 278–88.

52. Stillie AL, Kron T, Fox C, *et al.* Rectal filling at planning does not predict stability of the prostate gland during a course of radical radiotherapy if patients with large rectal filling are re-imaged. *Clinical Oncology* 2009; **21**(10): 760–7.

53. Measurements ICoRUa. *ICRU Report 62, Prescribing, Recording and Reporting Photon Beam Therapy (Supplement to ICRU Report 50)*. Bethesda, MD: Nuclear Technology Publishing; 1999.

54. Khoo VS, Padhani AR, Tanner SF, Finnigan DJ, Leach MO, Dearnaley DP. Comparison of MRI with CT for the radiotherapy planning of prostate cancer: a feasibility study. *British Journal of Radiology* 1999; **72**(858): 590–7.

55. Hricak H, White S, Vigneron D, Kurhanewicz J, Kosco A, Levin D, *et al.* Carcinoma of the prostate gland: MR imaging with pelvic phased-array coils versus integrated endorectal— pelvic phased-array coils. *Radiology* 1994; **193**(3): 703–9.

56. Rasch C, Barillot I, Remeijer P, Touw A, van Herk M, Lebesque JV. Definition of the prostate in CT and MRI: a multi-observer study. *International Journal of Radiation Oncology, Biology, Physics* 1999; **43**(1): 57–66.

57. Roach M, 3rd, Faillace-Akazawa P, Malfatti C, Holland J, Hricak H. Prostate volumes defined by magnetic resonance imaging and computerized tomographic scans for three-dimensional conformal radiotherapy. *International Journal of Radiation Oncology, Biology, Physics* 1996; **35**(5): 1011–18.

58. Kagawa K, Lee WR, Schultheiss TE, Hunt MA, Shaer AH, Hanks GE. Initial clinical assessment of CT-MRI image fusion software in localization of the prostate for 3D conformal radiation therapy. *International Journal of Radiation Oncology, Biology, Physics* 1997; **38**(2): 319–25.

59. Sannazzari GL, Ragona R, Ruo Redda MG, Giglioli FR, Isolato G, Guarneri A. CT-MRI image fusion for delineation of volumes in three-dimensional conformal radiation therapy in the treatment of localized prostate cancer. *British Journal of Radiology* 2002; **75**(895): 603–7.

60. Padhani AR. Dynamic contrast-enhanced MRI in clinical oncology: current status and future directions. *Journal of Magnetic Resonance Imaging* 2002; **16**(4): 407–22.

61. Kurhanewicz J, Swanson MG, Nelson SJ, Vigneron DB. Combined magnetic resonance imaging and spectroscopic imaging approach to molecular imaging of prostate cancer. *Journal of Magnetic Resonance Imaging* 2002; **16**(4): 451–63.

62. Kessler ML. Image registration and data fusion in radiation therapy. *British Journal of Radiology* 2006; **79** Spec No 1: S99–108.

63. Parker C, Haycocks T, Bayley A, Alasti H, Warde P, Catton C. A dose-volume histogram analysis of the seminal vesicles in men treated with conformal radiotherapy to 'prostate alone'. *Clinical Oncology* 2002; **14**(4): 298–302.

64. Ohori M, Scardino PT, Lapin SL, Seale-Hawkins C, Link J, Wheeler TM. The mechanisms and prognostic significance of seminal vesicle involvement by prostate cancer. *American Journal of Surgical Pathology* 1993; **17**(12): 1252–61.

65. Kestin L, Goldstein N, Vicini F, Yan D, Korman H, Martinez A. Treatment of prostate cancer with radiotherapy: should the entire seminal vesicles be included in the clinical target volume? *International Journal of Radiation Oncology, Biology, Physics* 2002; **54**(3): 686–97.

66. Davis BJ, Cheville JC, Wilson TM, Slezak JM, Pisansky TM. Histopathological Characterisation of Seminal Vesicle Invasion In Prostate Cancer: Implications for Radiotherapeutic Management. *International Journal of Radiation Oncology, Biology, Physics* 2001; **51**(3 S1): 140–1.

67. Teh BS, Bastasch MD, Mai WY, Butler EB, Wheeler TM. Predictors of extracapsular extension and its radial distance in prostate cancer: implications for prostate IMRT, brachytherapy, and surgery. *Cancer Journal* 2003; **9**(6): 454–60.

68. Dearnaley DP, Hall E, Lawrence D, *et al.* Phase III pilot study of dose escalation using conformal radiotherapy in prostate cancer: PSA control and side effects. *Br J Cancer* 2005; **92**(3): 488–98.

69. Pickett B, Roach M, 3rd, Verhey L, *et al.* The value of nonuniform margins for six-field conformal irradiation of localized prostate cancer. *International Journal of Radiation Oncology, Biology, Physics* 1995; **32**(1): 211–8.

70. Bonin SR, Lanciano RM, Corn BW, Hogan WM, Hartz WH, Hanks GE. Bony landmarks are not an adequate substitute for lymphangiography in defining pelvic lymph node location for the treatment of cervical cancer with radiotherapy. *International Journal of Radiation Oncology, Biology, Physics* 1996; **34**(1): 167–72.

71. Shih HA, Harisinghani M, Zietman AL, Wolfgang JA, Saksena M, Weissleder R. Mapping of nodal disease in locally advanced prostate cancer: rethinking the clinical target volume for pelvic nodal irradiation based on vascular rather than bony anatomy. *International Journal of Radiation Oncology, Biology, Physics* 2005; **63**(4): 1262–9.

72. Heidenreich A, Ohlmann CH, Polyakov S. Anatomical extent of pelvic lymphadenectomy in patients undergoing radical prostatectomy. *European Urology* 2007; **52**(1): 29–37.

73. Staffurth J. Adams E BS, Sohaib A, Huddart RA, Dearnaley DP. Target volume definition for intensity modulated whole pelvic radiotherapy (IMWPRT). *ASCO Prostate Cancer Symposium; 17/18 February; Orlando, Florida* 2005; Abstract No 185.

74. Golimbu M, Morales P, Al-Askari S, Brown J. Extended pelvic lymphadenectomy for prostatic cancer. *Journal of Urology* 1979; **121**(5): 617–20.

75. Wawroschek F, Vogt H, Weckermann D, Wagner T, Hamm M, Harzmann R. Radioisotope guided pelvic lymph node dissection for prostate cancer. *Journal of Urology* 2001; **166**(5): 1715–9.

76. Heidenreich A, Varga Z, Von Knobloch R. Extended pelvic lymphadenectomy in patients undergoing radical prostatectomy: high incidence of lymph node metastasis. *Journal of Urology* 2002; **167**(4): 1681–6.

77. Fowler JE, Jr., Whitmore WF, Jr. The incidence and extent of pelvic lymph node metastases in apparently localized prostatic cancer. *Cancer* 1981; **47**(12): 2941–5.

78. Khoo VS, Joon DL. New developments in MRI for target volume delineation in radiotherapy. *British Journal of Radiology* 2006; **79** Spec No 1: S2–15.

79. Heesakkers RA, Hovels AM, Jager GJ, *et al.* MRI with a lymph-node-specific contrast agent as an alternative to CT scan and lymph-node dissection in patients with prostate cancer: a prospective multicohort study. *Lancet Oncology* 2008; **9**(9): 850–6.

80. Lawton CA, Michalski J, El-Naqa I, *et al.* RTOG GU Radiation oncology specialists reach consensus on pelvic lymph node volumes for high-risk prostate cancer. *International Journal of Radiation Oncology, Biology, Physics* 2009; **74**(2): 383–7.

81. Taylor A, Rockall AG, Reznek RH, Powell ME. Mapping pelvic lymph nodes: guidelines for delineation in intensity-modulated radiotherapy. *International Journal of Radiation Oncology, Biology, Physics* 2005; **63**(5): 1604–12.

82. Harisinghani MG, Barentsz J, Hahn PF, *et al.* Noninvasive detection of clinically occult lymph-node metastases in prostate cancer. *New England Journal of Medicine* 2003; **348**(25): 2491–9.

83. Fonteyne V, De Gersem W, De Neve W, *et al.* Hypofractionated intensity-modulated arc therapy for lymph node metastasized prostate cancer. *International Journal of Radiation Oncology, Biology, Physics* 2009; **75**(4): 1013–20.

84. Pilepich MV, Krall JM, Johnson RJ, *et al.* Extended field (periaortic) irradiation in carcinoma of the prostate—analysis of RTOG 75–06. *International Journal of Radiation Oncology, Biology, Physics* 1986; **12**(3): 345–51.

85. Khoo VS, Bedford JL, Webb S, Dearnaley DP. Class solutions for conformal external beam prostate radiotherapy. *International Journal of Radiation Oncology, Biology, Physics* 2003; **55**(4): 1109–20.

86. Gulliford SL, Foo K, Morgan RC, *et al.* Dose-volume constraints to reduce rectal side effects from prostate radiotherapy: evidence from MRC RT01 Trial ISRCTN 47772397. *International Journal of Radiation Oncology, Biology, Physics* 2010; **76**(3): 747–54.

87. Fiorino C, Fellin G, Rancati T, *et al.* Clinical and dosimetric predictors of late rectal syndrome after 3D-CRT for localized prostate cancer: preliminary results of a multicenter prospective study. *International Journal of Radiation Oncology, Biology, Physics* 2008; **70**(4): 1130–7.

88. Michalski JM, Gay H, Jackson A, Tucker SL, Deasy JO. Quantec: Organ Specific Paper: Radiation dose-volume effects in radiation-induced rectal injury. *International Journal of Radiation Oncology, Biology, Physics* 2010; **76**(3 Suppl): S123–9.

89. Gallagher MJ, Brereton HD, Rostock RA, *et al.* A prospective study of treatment techniques to minimize the volume of pelvic small bowel with reduction of acute and late effects associated with pelvic irradiation. *International Journal of Radiation Oncology, Biology, Physics* 1986; **12**(9): 1565–73.

90. Storey MR, Pollack A, Zagars G, Smith L, Antolak J, Rosen I. Complications from radiotherapy dose escalation in prostate cancer: preliminary results of a randomized trial. *International Journal of Radiation Oncology, Biology, Physics* 2000; **48**(3): 635–42.

91. Emami B, Lyman J, Brown A, *et al.* Tolerance of normal tissue to therapeutic irradiation. *International Journal of Radiation Oncology, Biology, Physics* 1991; **21**(1): 109–22.

92. Jackson A, Marks LB, Bentzen SM, *et al.* The lessons of QUANTEC: recommendations for reporting and gathering data on dose-volume dependencies of treatment outcome. *International Journal of Radiation Oncology, Biology, Physics* 2010 Mar 1; **76**(3 Suppl): S155–60.

93. Peeters ST, Lebesque JV, Heemsbergen WD, *et al.* Localized volume effects for late rectal and anal toxicity after radiotherapy for prostate cancer. *International Journal of Radiation Oncology, Biology, Physics* 2006; **64**(4): 1151–61.

94. Akimoto T, Muramatsu H, Takahashi M, *et al.* Rectal bleeding after hypofractionated radiotherapy for prostate cancer: correlation between clinical and dosimetric parameters and the incidence of grade 2 or worse rectal bleeding. *International Journal of Radiation Oncology, Biology, Physics* 2004; **60**(4): 1033–9.

95. Herold DM, Hanlon AL, Hanks GE. Diabetes mellitus: a predictor for late radiation morbidity. *International Journal of Radiation Oncology, Biology, Physics* 1999; **43**(3): 475–9.

96. Skwarchuk MW, Jackson A, Zelefsky MJ, *et al.* Late rectal toxicity after conformal radiotherapy of prostate cancer (I): multivariate analysis and dose-response. *International Journal of Radiation Oncology, Biology, Physics* 2000; **47**(1): 103–13.

97. Cheung R, Tucker SL, Ye JS, *et al.* Characterization of rectal normal tissue complication probability after high-dose external beam radiotherapy for prostate cancer. *International Journal of Radiation Oncology, Biology, Physics* 2004; **58**(5): 1513–9.

98. Willett CG, Ooi CJ, Zietman AL, *et al.* Acute and late toxicity of patients with inflammatory bowel disease undergoing irradiation for abdominal and pelvic neoplasms. *International Journal of Radiation Oncology, Biology, Physics* 2000; **46**(4): 995–8.

99. Kavanagh BD, Pan CC, Dawson LA, *et al.* Quantec: Organ Specific Paper: Radiation dose-volume effects in the stomach and small bowel. *International Journal of Radiation Oncology, Biology, Physics* 2010; **76**(3 Suppl): S101–7.

100. Mangar SA, Sydes MR, Tucker HL, *et al.* Evaluating the relationship between erectile dysfunction and dose received by the penile bulb: using data from a randomised controlled trial of conformal radiotherapy in prostate cancer (MRC RT01, ISRCTN47772397). *Radiotherapy and Oncology* 2006; **80**(3): 355–62.

101. Fisch BM, Pickett B, Weinberg V, Roach M. Dose of radiation received by the bulb of the penis correlates with risk of impotence after three-dimensional conformal radiotherapy for prostate cancer. *Urology* 2001; **57**(5): 955–9.

102. Roach M, 3rd, Nam J, Gagliardi G, El Naqa I, Deasy JO, Marks LB. Quantec: Organ Specific Paper: Radiation dose-volume effects and the penile bulb. *International Journal of Radiation Oncology, Biology, Physics* 2010; **76**(3 Suppl): S130–4.

103. Department of Health. Radiotherapy: developing a world class service for England. (May 2007, on-line accessed 19 July 2011). http://wwwdhgovuk/en/Publicationsandstatistics/Publications/PublicationsPolicyAndGuidance/DH_074575.

104. South CP, Khoo VS, Naismith O, Norman A, Dearnaley DP. A comparison of treatment planning techniques used in two randomised UK external beam radiotherapy trials for localised prostate cancer. *Clinical Oncology* 2008; **20**(1): 15–21.

105. Zelefsky MJ, Fuks Z, Hunt M, *et al.* High-dose intensity modulated radiation therapy for prostate cancer: early toxicity and biochemical outcome in 772 patients. *International Journal of Radiation Oncology, Biology, Physics* 2002; **53**(5): 1111–6.

106. Kupelian PA, Reddy CA, Carlson TP, Altsman KA, Willoughby TR. Preliminary observations on biochemical relapse-free survival rates after short-course intensity-modulated radiotherapy (70 Gy at 2.5 Gy/fraction) for localized prostate cancer. *International Journal of Radiation Oncology, Biology, Physics* 2002; **53**(4): 904–12.

107. Corletto D, Iori M, Paiusco M, *et al.* Inverse and forward optimization of one- and two-dimensional intensity-modulated radiation therapy-based treatment of concave-shaped planning target volumes: the case of prostate cancer. *Radiotherapy and Oncology* 2003; **66**(2): 185–95.

108. Nutting CM, Convery DJ, Cosgrove VP, *et al.* Reduction of small and large bowel irradiation using an optimized intensity-modulated pelvic radiotherapy technique in patients with prostate cancer. *International Journal of Radiation Oncology, Biology, Physics* 2000; **48**(3): 649–56.

109. McVey GP, Thomas K, *et al.* Intensity modulated radiotherapy (IMRT) can safely deliver 60Gy to the pelvic lymph node regions in patients with prostate cancer: report of a

Phase 1 dose escalation study. *International Journal of Radiation Oncology Biology & Physics* 2009; **75**(Supl 1): S48.

110. Adams EJ, Convery DJ, Cosgrove VP, *et al.* Clinical implementation of dynamic and step-and-shoot IMRT to treat prostate cancer with high risk of pelvic lymph node involvement. *Radiotherapy and Oncology* 2004; **70**(1): 1–10.

111. Clark CH, Mubata CD, Meehan CA, *et al.* IMRT clinical implementation: prostate and pelvic node irradiation using Helios and a 120-leaf multileaf collimator. *Journal of Applied Clinical Medical Physics* 2002; **3**(4): 273–84.

112. Ruben JD, Davis S, Evans C, *et al.* The effect of intensity-modulated radiotherapy on radiation-induced second malignancies. *International Journal of Radiation Oncology, Biology, Physics* 2008; **70**(5): 1530–6.

113. van Herk M, Remeijer P, Rasch C, Lebesque JV. The probability of correct target dosage: dose-population histograms for deriving treatment margins in radiotherapy. *International Journal of Radiation Oncology, Biology, Physics* 2000; **47**(4): 1121–35.

114. Webb S. *The physics of three dimensional radiation therapy.* Bristol and Philadelphia Institue of Physics Publishing. 1993.

115. Vigneault E, Pouliot J, Laverdiere J, Roy J, Dorion M. Electronic portal imaging device detection of radioopaque markers for the evaluation of prostate position during megavoltage irradiation: a clinical study. *International Journal of Radiation Oncology, Biology, Physics* 1997; **37**(1): 205–12.

116. Gildersleve J, Dearnaley DP, Evans PM, Law M, Rawlings C, Swindell W. A randomised trial of patient repositioning during radiotherapy using a megavoltage imaging system. *Radiotherapy and Oncology* 1994 May; **31**(2): 161–8.

117. Mubata CD, Bidmead AM, Ellingham LM, Thompson V, Dearnaley DP. Portal imaging protocol for radical dose-escalated radiotherapy treatment of prostate cancer. *International Journal of Radiation Oncology, Biology, Physics* 1998; **40**(1): 221–31.

118. Moman MR, van der Heide UA, *et al.* Long-term experience with transrectal and transperineal implantations of fiducial gold markers in the prostate for position verification in external beam radiotherapy; feasibility, toxicity and quality of life. *Radiotherapy and Oncology* 2010; **96**(1): 38–42.

119. Graf R, Wust P, Budach V, Boehmer D. Potentials of on-line repositioning based on implanted fiducial markers and electronic portal imaging in prostate cancer radiotherapy. *Radiation Oncology* 2009; **4**: 13.

120. Kotte AN, Hofman P, Lagendijk JJ, van Vulpen M, van der Heide UA. Intrafraction motion of the prostate during external-beam radiation therapy: analysis of 427 patients with implanted fiducial markers. *International Journal of Radiation Oncology, Biology, Physics* 2007; **69**(2): 419–25.

121. Greer PB, Dahl K, Ebert MA, Wratten C, White M, Denham JW. Comparison of prostate set-up accuracy and margins with off-line bony anatomy corrections and online implanted fiducial-based corrections. *Journal of Medical Imaging and Radiation Oncology* 2008; **52**(5): 511–16.

122. McNair HA, Hansen VN, Parker CC, *et al.* A comparison of the use of bony anatomy and internal markers for offline verification and an evaluation of the potential benefit of online and offline verification protocols for prostate radiotherapy. *International Journal of Radiation Oncology, Biology, Physics* 2008; **71**(1): 41–50.

123. Zhu X, Bourland JD, Yuan Y, Zhuang T, O'Daniel J, Thongphiew D, *et al.* Tradeoffs of integrating real-time tracking into IGRT for prostate cancer treatment. *Physics in Medicine and Biology* 2009; **54**(17): N393–401.

124. Little DJ, Dong L, Levy LB, Chandra A, Kuban DA. Use of portal images and BAT ultrasonography to measure setup error and organ motion for prostate IMRT: implications for treatment margins. *International Journal of Radiation Oncology, Biology, Physics* 2003; **56**(5): 1218–24.

125. Huang E, Dong L, Chandra A, *et al.* Intrafraction prostate motion during IMRT for prostate cancer. *International Journal of Radiation Oncology, Biology, Physics* 2002; **53**(2): 261–8.

126. Morr J, DiPetrillo T, Tsai JS, Engler M, Wazer DE. Implementation and utility of a daily ultrasound-based localization system with intensity-modulated radiotherapy for prostate cancer. *International Journal of Radiation Oncology, Biology, Physics* 2002; **53**(5): 1124–9.

127. Serago CF, Chungbin SJ, Buskirk SJ, Ezzell GA, Collie AC, Vora SA. Initial experience with ultrasound localization for positioning prostate cancer patients for external beam radiotherapy. *International Journal of Radiation Oncology, Biology, Physics* 2002; **53**(5): 1130–8.

128. Lattanzi J, McNeeley S, Pinover W, Horwitz E, Das I, Schultheiss TE, *et al.* A comparison of daily CT localization to a daily ultrasound-based system in prostate cancer. *International Journal of Radiation Oncology, Biology, Physics* 1999; **43**(4): 719–25.

129. Trichter F, Ennis RD. Prostate localization using transabdominal ultrasound imaging. *International Journal of Radiation Oncology, Biology, Physics* 2003; **56**(5): 1225–33.

130. Denham JW, Dally MJ, Hunter K, Wheat J, Fahey PP, Hamilton CS. Objective decision-making following a portal film: the results of a pilot study. *International Journal of Radiation Oncology, Biology, Physics* 1993; **26**(5): 869–76.

131. Fuks Z, Horwich A. Clinical and technical aspects of conformal therapy. *Radiotherapy and Oncology* 1993; **29**(2): 219.

132. Al-Mamgani A, van Putten WL, Heemsbergen WD, *et al.* Update of Dutch multicenter dose-escalation trial of radiotherapy for localized prostate cancer. *International Journal of Radiation Oncology, Biology, Physics* 2008; **72**(4): 980–8.

133. Dearnaley DP, Sydes MR, Graham JD, *et al.* Escalated-dose versus standard-dose conformal radiotherapy in prostate cancer: first results from the MRC RT01 randomised controlled trial. *Lancet Oncology* 2007; **8**(6): 475–87.

134. Zietman AL, Bae K, Slater JD, *et al.* Randomized trial comparing conventional-dose with high-dose conformal radiation therapy in early-stage adenocarcinoma of the prostate: long-term results from proton radiation oncology group/american college of radiology 95–09. *Journal of Clinical Oncology* 2010; **28**(7): 1106–11.

135. Kuban DA, Levy LB, Cheung MR, *et al.* Long-term failure patterns and survival in a randomized dose-escalation trial for prostate cancer. *Who dies of disease? International Journal of Radiation Oncology, Biology, Physics* 2011; **79**(5): 1310–7.

136. Viani GA, Stefano EJ, Afonso SL. Higher-than-conventional radiation doses in localized prostate cancer treatment: a meta-analysis of randomized, controlled trials. *International Journal of Radiation Oncology, Biology, Physics* 2009; **74**(5): 1405–18.

137. Khoo VS, Dearnaley DP. Question of dose, fractionation and technique: ingredients for testing hypofractionation in prostate cancer—the CHHiP trial. *Clinical Oncology* 2008; **20**(1): 12–4.

138. Logue JP, Cowan RA, Hendry JH. Hypofractionation for prostate cancer. *International Journal of Radiation Oncology, Biology, Physics* 2001; **49**(5): 1522–3.

139. Dearnaley DP, Sydes MR, Langley RE, *et al.* The early toxicity of escalated versus standard dose conformal radiotherapy with neo-adjuvant androgen suppression for patients with localised prostate cancer: results from the MRC RT01 trial (ISRCTN47772397). *Radiotherapy and Oncology* 2007; **83**(1): 31–41.

140. Prosnitz RG, Schneider L, Manola J, *et al.* Tamsulosin palliates radiation-induced urethritis in patients with prostate cancer: results of a pilot study. *International Journal of Radiation Oncology, Biology, Physics* 1999; **45**(3): 563–6.

141. Syndikus I, Morgan RC, Sydes MR, Graham JD, Dearnaley DP. Late gastrointestinal toxicity after dose-escalated conformal radiotherapy for early prostate cancer: results from the UK Medical Research Council RT01 trial (ISRCTN47772397). *International Journal of Radiation Oncology, Biology, Physics* 2010; **77**(3): 773–83.

142. Andreyev HJ, Vlavianos P, Blake P, Dearnaley D, Norman AR, Tait D. Gastrointestinal symptoms after pelvic radiotherapy: role for the gastroenterologist? *International Journal of Radiation Oncology, Biology, Physics* 2005; **62**(5): 1464–71.

143. Raina R, Agarwal A, Goyal KK, *et al.* Long-term potency after iodine-125 radiotherapy for prostate cancer and role of sildenafil citrate. *Urology* 2003; **62**(6): 1103–8.

144. Denham JW, Steigler A, Lamb DS, Joseph D, Turner S, Matthews J, *et al.* Short-term neoadjuvant androgen deprivation and radiotherapy for locally advanced prostate cancer: 10-year data from the TROG 96.01 randomised trial. *Lancet Oncology* 2011; **12**(5): 451–9.

145. D'Amico AV, Chen MH, Renshaw AA, Loffredo M, Kantoff PW. Androgen suppression and radiation vs radiation alone for prostate cancer: a randomized trial. *Journal of the American Medical Association* 2008; **299**(3): 289–95.

146. Bria E, Cuppone F, Giannarelli D, *et al.* Does hormone treatment added to radiotherapy improve outcome in locally advanced prostate cancer?: meta-analysis of randomized trials. *Cancer* 2009; **115**(15): 3446–56.

147. Shelley MD, Kumar S, Coles B, Wilt T, Staffurth J, Mason MD. Adjuvant hormone therapy for localised and locally advanced prostate carcinoma: a systematic review and meta-analysis of randomised trials. *Cancer Treatment Reviews* 2009; **35**(7): 540–6.

148. Fiorino C, Sanguineti G, Cozzarini C, *et al.* Rectal dose-volume constraints in high-dose radiotherapy of localized prostate cancer. *International Journal of Radiation Oncology, Biology, Physics* 2003; **57**(4): 953–62.

149. Bolla M, de Reijke TM, Van Tienhoven G, *et al.* Duration of androgen suppression in the treatment of prostate cancer. *New England Journal of Medicine* 2009; **360**(24): 2516–27.

150. Horwitz EM, Bae K, Hanks GE, *et al.* Ten-year follow-up of radiation therapy oncology group protocol 92–02: a phase III trial of the duration of elective androgen deprivation in locally advanced prostate cancer. *Journal of Clinical Oncology* 2008; **26**(15): 2497–504.

151. Shipley WU, Thames HD, Sandler HM, *et al.* Radiation therapy for clinically localised prostate cancer: a multi-institutional pooled analysis. *Journal of the American Medical Association* 1999; **281**(17): 1598–604.

152. Roach M, Lu J, Pilepich MV, *et al.* Four prognostic groups predict long-term survival from prostate cancer following radiotherapy alone on Radiation Therapy Oncology Group clinical trials. *International Journal of Radiation Oncology, Biology, Physics* 2000; **47**(3): 609–15.

153. Roach M, 3rd, Chen A, Song J, Diaz A, Presti J, Jr., Carroll P. Pretreatment prostate-specific antigen and Gleason score predict the risk of extracapsular extension and the risk of failure following radiotherapy in patients with clinically localized prostate cancer. *Seminars in Urology and Oncology* 2000; **18**(2): 108–14.

154. Consensus statement: guidelines for PSA following radiation therapy. American Society for Therapeutic Radiology and Oncology Consensus Panel. *International Journal of Radiation Oncology, Biology, Physics* 1997; **37**(5): 1035–41.

155. Roach M, 3rd, Hanks G, Thames H, **Jr.**, *et al.* Defining biochemical failure following radiotherapy with or without hormonal therapy in men with clinically localized prostate cancer: recommendations of the RTOG-ASTRO Phoenix Consensus Conference. *International Journal of Radiation Oncology, Biology, Physics* 2006; **65**(4): 965–74.

156. Touma NJ, Izawa JI, Chin JL. Current status of local salvage therapies following radiation failure for prostate cancer. *Journal of Urology* 2005; **173**(2): 373–9.

157. Izawa JI, Madsen LT, Scott SM, *et al.* Salvage cryotherapy for recurrent prostate cancer after radiotherapy: variables affecting patient outcome. *Journal of Clinical Oncology* 2002; **20**(11): 2664–71.

158. Chin JL, Pautler SE, Mouraviev V, Touma N, Moore K, Downey DB. Results of salvage cryoablation of the prostate after radiation: identifying predictors of treatment failure and complications. *Journal of Urology* 2001; **165**(6 Pt 1): 1937–41; discussion 41–2.

159. Grado GL, Collins JM, Kriegshauser JS, Balch CS, Grado MM, Swanson GP, *et al.* Salvage brachytherapy for localized prostate cancer after radiotherapy failure. *Urology* 1999; **53**(1): 2–10.

160. Murat FJ, Poissonnier L, Rabilloud M, *et al.* Mid-term results demonstrate salvage high-intensity focused ultrasound (HIFU) as an effective and acceptably morbid salvage treatment option for locally radiorecurrent prostate cancer. *European Urology* 2009; **55**(3): 640–7.

161. Nutting CM, Corbishley CM, Sanchez-Nieto B, Cosgrove VP, Webb S, Dearnaley DP. Potential improvements in the therapeutic ratio of prostate cancer irradiation: dose escalation of pathologically identified tumour nodules using intensity modulated radiotherapy. *British Journal of Radiology* 2002; **75**(890): 151–61.

162. desouza NM, Reinsberg SA, Scurr ED, Brewster JM, Payne GS. Magnetic resonance imaging in prostate cancer: the value of apparent diffusion coefficients for identifying malignant nodules. *British Journal of Radiology* 2007; **80**(950): 90–5.

163. Morgan VA, Kyriazi S, Ashley SE, DeSouza NM. Evaluation of the potential of diffusion-weighted imaging in prostate cancer detection. *Acta Radiologica* 2007 Jul; **48**(6): 695–703.

164. Kozlowski P, Chang SD, Jones EC, Berean KW, Chen H, Goldenberg SL. Combined diffusion-weighted and dynamic contrast-enhanced MRI for prostate cancer diagnosis—correlation with biopsy and histopathology. *Journal of Magnetic Resonance Imaging* 2006; **24**(1): 108–13.

165. Ogura K, Maekawa S, Okubo K, Aoki Y, Okada T, Oda K, *et al.* Dynamic endorectal magnetic resonance imaging for local staging and detection of neurovascular bundle involvement of prostate cancer: correlation with histopathologic results. *Urology* 2001; **57**(4): 721–6.

166. Riches SF, Payne GS, Morgan VA, Sandhu S, Fisher C, Germuska M, *et al.* MRI in the detection of prostate cancer: combined apparent diffusion coefficient, metabolite ratio, and vascular parameters. *AJR. American Journal of Roentgenology* 2009; **193**(6): 1583–91.

167. Senft A, de Bree R, Hoekstra OS, *et al.* Screening for distant metastases in head and neck cancer patients by chest CT or whole body FDG-PET: a prospective multicenter trial. *Radiotherapy and Oncology* 2008; **87**(2): 221–9.

168. Shreve PD, Grossman HB, Gross MD, Wahl RL. Metastatic prostate cancer: initial findings of PET with 2-deoxy-2-[F-18]fluoro-D-glucose. *Radiology* 1996; **199**(3): 751–6.

169. Giovacchini G, Picchio M, Coradeschi E, Scattoni V, Bettinardi V, Cozzarini C, *et al.* [(11) C]choline uptake with PET/CT for the initial diagnosis of prostate cancer: relation to PSA levels, tumour stage and anti-androgenic therapy. *European Journal of Nuclear Medicine and Molecular Imaging* 2008; **35**(6): 1065–73.

170. Palma D, Vollans E, James K, Nakano S, Moiseenko V, Shaffer R, *et al.* Volumetric modulated arc therapy for delivery of prostate radiotherapy: comparison with intensity-modulated radiotherapy and three-dimensional conformal radiotherapy. *International Journal of Radiation Oncology, Biology, Physics* 2008; **72**(4): 996–1001.

171. Zhang P, Happersett L, Hunt M, Jackson A, Zelefsky M, Mageras G. Volumetric modulated arc therapy: planning and evaluation for prostate cancer cases. *International Journal of Radiation Oncology, Biology, Physics* 2010; **76**(5): 1456–62.

172. Kjaer-Kristoffersen F, Ohlhues L, Medin J, Korreman S. RapidArc volumetric modulated therapy planning for prostate cancer patients. *Acta Oncologica* 2009; **48**(2): 227–32.

173. Jouyaux F, De Crevoisier R, Manens JP, Bellec J, Cazoulat G, Haigron P, *et al.* [High dose for prostate irradiation with image guided radiotherapy: contribution of intensity modulation arctherapy]. *Cancer Radiotherapy* 2010; **14**(8): 679–89.

174. Hall EJ. Intensity-modulated radiation therapy, protons, and the risk of second cancers. *International Journal of Radiation Oncology, Biology, Physics* 2006; **65**(1): 1–7.

175. Korreman S, Rasch C, McNair H, *et al.* The European Society of Therapeutic Radiology and Oncology-European Institute of Radiotherapy (ESTRO-EIR) report on 3D CT-based in-room image guidance systems: a practical and technical review and guide. *Radiotherapy and Oncology* 2010; **94**(2): 129–44.

176. Townsend NC, Huth BJ, Ding W, *et al.* Acute toxicity after CyberKnife-delivered hypofractionated radiotherapy for treatment of prostate cancer. *American Journal of Clinical Oncology* 2011; **34**(1): 6–10.

177. Buyyounouski MK, Price RA, Jr., Harris EE, *et al.* Stereotactic body radiotherapy for primary management of early-stage, low- to intermediate-risk prostate cancer: report of the American Society for Therapeutic Radiology and Oncology Emerging Technology Committee. *International Journal of Radiation Oncology, Biology, Physics* 2010; **76**(5): 1297–304.

178. Hanks GE, Buzydlowski J, Sause WT, *et al.* Ten-year outcomes for pathologic node-positive patients treated in RTOG 75–06. *International Journal of Radiation Oncology, Biology, Physics* 1998; **40**(4): 765–8.

179. Pollack A, Zagars GK, Starkschall G, *et al.* Prostate cancer radiation dose response: results of the M. D. Anderson phase III randomized trial. *International Journal of Radiation Oncology, Biology, Physics* 2002; **53**(5): 1097–105.

180. Yeoh EE, Botton RJ, Butters J, *et al.* Hypofractionated versus conventionally fractionated radiation therapy for prostate carcinoma: Final results of a phase III randomized trial. *International Journal of Radiation Oncology, Biology, Physics* 2011; **81**(5): 1271–78.

181. Lukka H, Hayter C, Warde P, *et al.* A randomized trial comparing two fractionation schedules for patients with localized prostate cancer. *International Journal of Radiation Oncology, Biology, Physics* 2003; **57**(2 Suppl): S126.

182. Laverdiere J, Nabid A, De Bedoya LD, *et al.* The efficacy and sequencing of a short course of androgen suppression on freedom from biochemical failure when administered with radiation therapy for T2-T3 prostate cancer. *Journal of Urology* 2004; **171**(3): 1137–40.

183. Jones CU, Hunt D, McGowan DG, Amin MB, Chetner MP, Bruner DW, *et al.* Radiotherapy and short-term androgen deprivation for localized prostate cancer. *N Engl J Med* 2011; **365**(2): 107–18.

Bladder cancer

Nicholas James, Elizabeth Southgate,
Anjali Zarkar

8.13 Indications

There is considerable controversy as to the optimal management of localized, muscle invasive bladder cancer. Surgical removal of the bladder is considered the 'gold standard' in many countries but there is no randomized trial to underpin this statement. A recent attempt in the UK, the SPARE trial, failed to recruit and the question is unlikely to ever be definitively addressed. Many patients are unfit for surgery, for example, the median age in two large published surgical series is in the mid-60s[1,2], whereas median age at diagnosis is around a decade higher (CRUK Cancerstats: http:// info.cancerresearchuk.org/cancerstats/).

Furthermore, reported 5-year survival rates with radiotherapy are remarkably similar to surgical series. For example, Stein et al. reported results from a large series of 1054 surgically treated patients, obtaining 5- and 10-year survivals of 60% and 43%[1]. However, when the surgical results are confined to those patients with muscle invasive disease, overall 5-year survival drops to 47%. This is very similar to the 5-year survival observed in the surgery-only arm of the SWOG 8710 trial of chemotherapy + surgery versus surgery alone. In contrast, Rödel et al.[3] reported results with endoscopic resection and radiotherapy. When patients with inoperable disease are removed from the radiotherapy series to allow direct comparison, 5-year survival is reported as 45%. A population-based study from Ontario looking at bladder cancer outcomes could find no link between treatment modality and survival which was solely determined by tumour related factors such as stage and grade[4]. This lack of data supporting a survival advantage for surgery does not stop its proponents presenting it as the gold standard[1,5,6]. It is, however, more likely that survival in bladder cancer is driven by the presence or absence of distant spread at the time of local therapy and will not be affected by the means adopted for local control.

Furthermore, all patients undergoing surgery will need either reconstructive bladder surgery or an ileal diversion. Thus even if surgery is genuinely better than radiotherapy for patients fit for both approaches, there are many patients for whom radical surgery is simply not suitable and hence bladder-preserving techniques are appropriate. Despite this, use of radiotherapy varies enormously worldwide with possibly a majority receiving radiotherapy in the UK[7], around 25% in Scandinavia[8], but only around 10% in the USA[9]. Radiotherapy alone suffers from a relatively high rate of incomplete response or local recurrence (up to 50% or more) possibly due to the effects of case selection with many poor-risk patients unsuitable for surgery being referred. A report of long-term follow-up from our institution reported a salvage cystectomy rate of around 24% with a median time to cystectomy of 12–18 months[10,11].

8.1.3.1 Indications by stage

(See Table 8b.1 for TNM staging.)

- ◆ CIS, Ta, T1: no role for radiotherapy.
- ◆ T2–T4a N0 M0: potential role for radiotherapy, combined with synchronous chemotherapy if patient sufficiently fit.
- ◆ Tany N1–3 M0 or Tany Nany M1: there is no role for radical radiotherapy as sole treatment for Stage IV disease. It may be worth considering, however, as part of a package of 'radical' palliation in concert with systemic chemotherapy. No randomized data on the use of radiotherapy in this setting beyond studies of fractionation.

8.14 Radical primary treatment

8.14.1 Treatment volume and definition

The CTV consists of the bladder including the primary lesion.

The question of whether whole bladder irradiation is necessary is partially unresolved. Generally, the whole bladder has been included in the radiotherapy fields as it has been thought that the development of bladder cancer is associated with a mucosal field change. However, the need to the treat the whole bladder rather than the tumour alone has not been clearly established. The BC2001 trial compared radiotherapy to the whole bladder to a reduced dose of 85% to uninvolved bladder in a 2 × 2 trial, which also compared radiotherapy with chemo-radiotherapy with 5-FU/mitomycin C (see section 8.14.5). The reduced dose group showed no improvement in either acute or late toxicity, nor was there any impact on locoregional control. One interpretation of these results is that dose to uninvolved bladder was not reduced sufficiently and that with more sophisticated planning and delivery systems, dose to tumour could be escalated and dose to uninvolved bladder reduced. For the time being, the whole bladder should be considered as constituting the CTV.

The normal tissues of concern when treating pelvic lesions are the rectum, small bowel, and to a lesser extent prostatic urethra if there is no involvement of the urethra with tumour. Female patients who are sexually active should be counselled about the risk of vaginal dryness and stenosis. They should be routinely offered treatment with vaginal dilators, as for women receiving treatment for cervical, vaginal or endometrial cancers. Male patients are at risk of erectile dysfunction and should be counselled about this. It should be noted, however, that the long-term risk of sexual dysfunction may be higher in those undergoing radical surgery[12].

Table 8b.1 (a) Staging of bladder cancer (TNM 2009)

Tx	Primary tumour can not be assessed
T0	No evidence of primary tumour
Ta	Non-invasive papillary carcinoma
Tis	Carcinoma *in situ*: 'flat tumour'
T1	Tumour invades subepithelial connective tissue
T2	Tumour invades muscle
T2a	Superficial muscle (inner half)
T2b	Deep muscle (outer half)
T3	Tumour invades perivesical tissue:
T3a	Microscopically
T3b	Macroscopically (extravesical mass)
T4	Tumour invades any of the following: prostate stroma, seminal vesicles, uterus, vagina, pelvic wall, abdominal wall
T4a	Tumour invades prostate stroma, seminal vesicles, uterus or vagina
T4b	Tumour invades pelvis or abdominal wall
N—regional lymph nodes	Defined as nodes of the true pelvis below the bifurcation of the common iliac arteries. Laterality does not affect the N classification
Nx	Nodes cannot be assessed
N0	No lymph node metastasis
N1	Metastasis in a single lymph node in the true pelvis (hypogastric, obturator, external iliac, or presacral)
N2	Metastasis in multiple lymph nodes in the true pelvis (hypogastric, obturator, external iliac, or presacral)
N3	Metastasis in a common iliac lymph node(s)
M—distant metastasis	
Mx	Cannot be assessed
M0	No distant metastasis
M1	Distant metastasis present
G—histopathological grading	
Gx	Grade of differentiation can not be assessed
G1	Well differentiated
G2	Moderately differentiated
G3–4	Poorly differentiated/undifferentiated

Table 8b.1 *continued* (b) Stage grouping

Stage 0a	Ta	N0	M0
Stage 0is	Tis	N0	M0
Stage I	T1	N0	M0
Stage II	T2a,b	N0	M0
Stage III	T3a,b	N0	M0
	T4a	N0	M0
Stage IV	T4b	N0	M0
	Any T	N1, 2, 3	M0
	Any T	Any N	M1

From Edge SB, Byrd DR, Compton CC, eds. *AJCC Cancer Staging Manual. 7th ed.* New York, NY.: Springer, 2010. Used with the permission of the American Joint Committee on Cancer (AJCC), Chicago, Illinois. The original source for this material is the AJCC Cancer Staging Manual, Seventh Edition (2010) published by Springer Science and Business Media LLC, www.springer.com.

8.14.2 Planning technique

Patient position and immobilization

◆ The patient should be planned and treated in the same position; supine with arms on their chest. Knee and ankle immobilization should be used to ensure patient positioning is reproducible.

◆ The rectum should be empty of flatus and faeces. The use of daily micro-enemas may be considered.

◆ Patients will be asked to empty their bladder 15 minutes prior to scan.

◆ Whilst breathing normally, the patient should have a CT scan performed with 3–5-mm slice spacing. Patients are scanned from bottom of ischial tuberosities to 3 cm above the dome of the bladder or bottom of L5 (whichever is higher). A flat top CT scanner should be used.

◆ Neither IV nor oral contrast is thought to be of benefit in this instance.

◆ Reference tattoos should be made at the base of the abdomen and over each hip. The location of the tattoos should be marked on the planning scan by the use of radio-opaque markers to allow cross-referencing of planning scan and set-up instructions.

Volume/field localization

◆ The GTV can be difficult to define and should integrate information from the staging CT or MRI as well as the diagnostic transurethral resection of the tumour (TURBT). MRI/CT fusion may be helpful, where available.

◆ The use of fiducial markers or contrast medium such as lipidiol at the time of TURBT has been explored and may help identify tumour for image guided adaptive radiotherapy.

◆ There are little in the way of data on the optimal radiotherapy volume (Fig. 8b.1). A standard approach is to define the PTV as the whole bladder identified by its non-involved outer bladder wall with a 1.5-cm margin plus extravesical extent of tumour with a 2-cm margin (Fig. 8b.2)

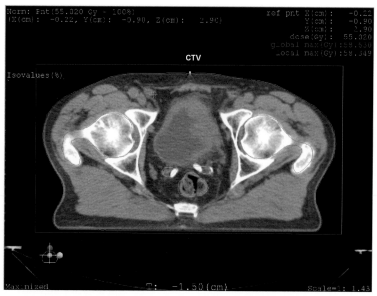

Fig. 8b.1 CTV has been outlined in red.

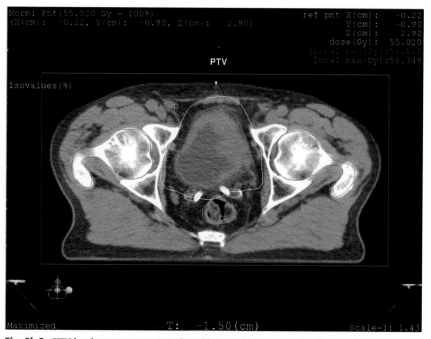

Fig. 8b.2 CTV has been grown to PTV by addition of 1.5 cm margin all around PTV.

- All planning and treatment should be carried out with the bladder empty to minimize the risk of geographical miss and to keep the treated volumes as small as possible. Patients with significant residual volumes post voiding should be considered for planning and treatment with a catheter *in situ*, although this is likely to increase urinary toxicity.

- There are no data to support the routine irradiation of radiologically negative lymph nodes. The nodal relapse rate in the BC2001 trial, with PTV and CTV defined as previously, was only 4% in the chemoradiotherapy arm and 6% with radiotherapy only.

Dose distribution, fields, and dose constraints

- The treatment is planned using photons of > 8 MV, beam energies less than this result in higher superficial doses and greater normal tissue irradiation (depending of course on patient size, larger patients will require photon energies in the 10–15-MV range).

- PTV should be covered to encompass the PTV in the 95% isodose (Figs 8b.3 and 8b.4).

- Usually conformal technique using either three- (an anterior and two lateral/ posterior oblique fields) or four-field plan is created.

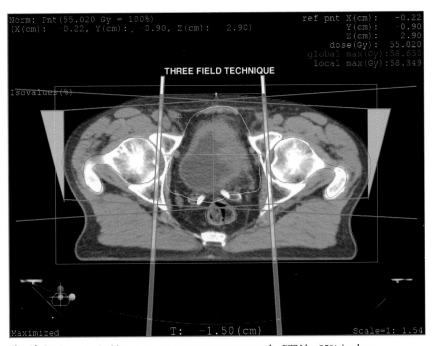

Fig. 8b.3 Shows typical beam arrangement to encompass the PTV by 95% isodose.

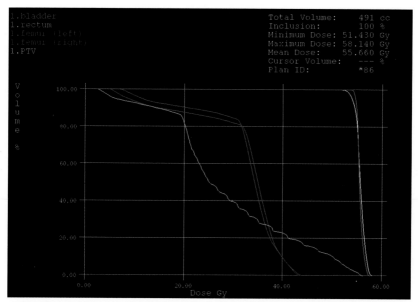

Fig. 8b.4 Dose–volume histogram showing dose received by PTV and critical structures. (Dose/fractionation used in this particular plan is 55 Gy in 20 fractions, i.e. 2.75 Gy per fraction.)

♦ Increasingly, IMRT and IGRT techniques will be used for delivery of radiotherapy. The combined use of these two modalities will allow more precise tailoring of dose delivery to take account of changes in bladder and rectal filling and reduce late effects by reducing dose to the small/large bowel. Adaptive radiotherapy uses three to four radiotherapy plans using different margins may be used take account of changes in bladder filling.

Dose constraints (dose 2 Gy/fraction) used for organs at risk are as follows[13–19]:

♦ Rectum: V66 < 30%,V60 < 50%, V50 < 60%, V40 < 70%, V30 < 80%.

♦ Femoral heads: V50 < 50%.

8.14.3 Implementation on the treatment machine:

The patient should use the same bowel and bladder preparation protocol as at the time of CT scanning. They should be positioned using the same immobilization and tattoos should be aligned using wall lasers (Figs 8b.5 and 8b.6).

8.14.4 Treatment verification

♦ The isocentre position should be verified using methods such as electronic portal imaging (EPI), megavoltage CT (MVCT), cone-beam (CBCT), or in-room CT/stereo X-ray. Three-dimensional imaging such as CBCT has recently become more widely available and allows much more accurate definition of the bladder and OARs on a potentially daily basis.

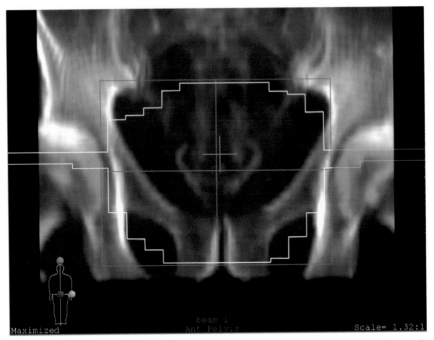

Fig. 8b.5 Anteroposterior digitally constructed radiograph (DRR) for verification with portal images.

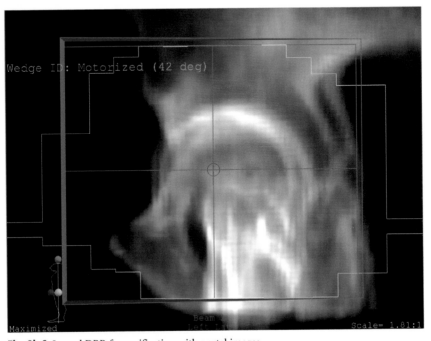

Fig. 8b.6 Lateral DRR for verification with portal images.

♦ Likewise, with better image guidance systems on linacs, use of fiducial markers (either gold seeds or use of lipid-based contrast media injected submucosally) and on-board kV imaging systems may become more widespread.

♦ It is recommended that daily imaging and on-line corrections are made where facilities allow. At a minimum, patients should be imaged for the first three fractions and then weekly, with off-line (systematic error) correction for errors of 5 mm or greater.

8.14.5 Dose prescription

Radiotherapy as sole treatment

Acceptable radical schedules used in the UK are:

♦ 64–66 Gy to the reference point in 32–33 fractions over 6½ weeks.

♦ 55 Gy to the reference point in 20 fractions over 4 weeks.

The optimal schedule has yet to be established. In North America, split schedules are often used with an interval cystoscopy after a dose of 39–40 Gy in 1.8–2-Gy fractions is reached. Patients with refractory disease proceed to cystectomy; patients with responding disease proceed to complete a radical course of radiotherapy alone to a total dose of 64–66 Gy. Generally, these schedules achieve long-term survival comparable to surgical series[13,14]. A significant risk of cystectomy remains, however, with 22% undergoing immediate cystectomy, 13% delayed cystectomy for local recurrence, and 65% retaining the bladder[15].

In contrast, patients treated with RT alone to a full radical dose have around a 24% cystectomy rate at 10 years median follow-up with conventional radiotherapy alone[10], suggesting the multimodality approach does not really offer any advantage over radiotherapy administered as a continuous block as in the UK.

Chemoradiotherapy

Two trials have compared this approach to radiotherapy alone in bladder cancer[16,17]. The Canadian study[16] randomized 99 patients to radiotherapy (40 Gy in 20 fractions over 4 weeks) with or without cisplatin (100 mg/m^2 2-weekly for three cycles) followed by elective cystectomy or further radiotherapy. The chemoradiotherapy group had improved pelvic progression-free survival (adjusted HR = 0.50; 90% CI: 0.29–0.86; logrank p = 0.038) but was too small to provide reliable estimates of overall survival effects.

Chemotherapy with cisplatin at this dose is not ideal for bladder cancer as many patients, particularly those referred for radiotherapy, have impaired renal function or poor performance status. The recently reported BC2001 trial[17] tested the hypothesis that synchronous chemoradiotherapy with 5-FU/MMC (mitomycin C 12 mg/m^{2}on day 1 and 5-FU 500 mg/m^2/day days 1–5 and last 5 days of treatment) is more efficacious than radiotherapy alone. With 49 months of median follow-up, adding chemotherapy to full dose radiotherapy (55 Gy in 20 fractions or 64 Gy in 32 fractions) was associated with a 33% reduction in the risk of locoregional recurrence with a reduction of almost 50% in invasive recurrence. This benefit appeared consistent in pre-planned subgroup analyses and was not affected by prior neo-adjuvant chemotherapy suggesting that neoadjuvant and concomitant chemotherapy confer separate benefits on distant and local control respectively. The improvement in locoregional control was achieved with modest increases in acute toxicity that did not reach statistical significance with respect to grade 3 or 4 outcomes. Chemoradiotherapy, even when co-administered after neoadjuvant chemotherapy, did not result in impaired late bladder function or a significant reduction in bladder volume. Late toxicity, measured using RTOG and LENT/SOM scales, showed no significant increase with combination therapy compared to radiotherapy alone.

An alternative approach to radiosensitization is to address tumour hypoxia as recently reported in another Phase III UK trial (BCON) that randomized 333 patients to radiotherapy or radiotherapy with synchronous nicotinamide and carbogen[18]. Analysis of the primary endpoint of local RFS did not meet statistical significance (3-year local RFS: 54% radiotherapy plus nicotinamide/carbogen vs. 43% radiotherapy alone; HR = 0.88, 95% CI: 0.76–1.01; p = 0.06) although significant improvements in overall survival were reported (3-year rates of 59% radiotherapy plus nicotinamide/carbogen vs. 46% radiotherapy alone; HR = 0.86, CI: 0.74–0.99; p = 0.04). No increase in acute toxicity was reported.

8.15 Postoperative radiotherapy

8.15.1 Indications

There is no indication for routine postoperative radiotherapy after cystectomy.

8.16 Palliative treatment

8.16.1 Indications

- Stage IV disease not suitable for chemotherapy.
- Stage II and III disease in elderly patients with significant comorbidity.

In the UK, patients present with a median age of 72 years (males) and 75 years (females) with 60% aged > 70 years, often with significant comorbidity due to the association between smoking and bladder cancer. Only around 4% of patients present with *de novo* metastatic disease (source: British Association of Urological Surgeons (BAUS) Audit 2003, http://http://www.baus.org.uk). Palliative treatment must, therefore, take into account the pattern of disease and the extent of other clinical problems.

8.16.2 Treatment volume and definition—loco-regional palliation

Whenever possible, all the pelvic disease (primary and nodes) should be encompassed. If, however, the volume is excessively large the treatment should be concentrated on the area causing the main symptoms. The primary lesion plus any enlarged nodes within the compass of a reasonable treatment volume should be included with a margin of 1.5 cm to give the PTV.

For stage II and III disease the CTV and PTV are the same as for radical treatment (see section 8.14.1).

8.16.3 **Planning technique**

Patient position and immobilization

- The patient should be simulated in a comfortable supine position with their arms on their chest.
- Patient comfort should be prioritized over rigid immobilization.
- Bladder to be emptied before lying on the scanner bed. No bowel preparation is required.
- A planning CT scan with the bladder empty should be performed as for radical treatment.

Dose distribution, fields and dose constraints

The majority of cases can be treated with an anterior and two lateral fields to minimize rectal toxicity, unless the rectum is encompassed by disease in which case anterior–posterior parallel-opposed fields should be used.

8.16.4 **Treatment verification**

As for radical treatment but where only three fractions are to be given, verification is done during the first fraction only and treatment delivered unless there is a significant (> 5 mm) displacement.

8.16.5 **Dose prescription**

For palliative therapy of bladder recurrence or disease unsuitable for radical treatment, a dose of 21 Gy in three fractions has been shown to be as effective as a longer palliative course[19].

References

1. Stein JP, Lieskovsky G, Cote R, *et al.* Radical cystectomy in the treatment of invasive bladder cancer: long-term results in 1,054 patients. *Journal of Clinical Oncology* 2001; **19**: 666–75.

2. Grossman HB, Natale RB, Tangen CM, *et al.* Neoadjuvant chemotherapy plus cystectomy compared with cystectomy alone for locally advanced bladder cancer. [Erratum appears in New England Journal of Medicine 2003; **349**: 1880]. *New England Journal of Medicine* 2003; **349**: 859–66.

3. Rödel C, Grabenbauer GG, Kuhn R, *et al.* Combined-modality treatment and selective organ preservation in invasive bladder cancer: long-term results. *Journal of Clinical Oncology* 2002; **20**: 3061–71.

4. Hayter CR, Paszat LF, Groome PA, Schulze K, Mackillop WJ. The management and outcome of bladder carcinoma in Ontario, 1982-1994. *Cancer* 2000; **89**:142–51.

5. Millikan R, Dinney C, Swanson D, *et al.* Integrated therapy for locally advanced bladder cancer: final report of a randomized trial of cystectomy plus adjuvant M-VAC versus cystectomy with both preoperative and postoperative M-VAC. *Journal of Clinical Oncology* 2001; **19**: 4005–13.

6. Dalbagni G, Genega E, Hashibe M, *et al.* Cystectomy for bladder cancer: a contemporary series. *Journal of Urology* 2001; **165**: 1111–16.

7. Munro NP, Sundaram SK, Weston PM, *et al.* A 10-year retrospective review of a nonrandomized cohort of 458 patients undergoing radical radiotherapy or cystectomy in Yorkshire, UK. *International Journal of Radiation Oncology, Biology, Physics* 2010; **77**: 119–24.

8. Jahnson S, Damm O, Hellsten S, *et al.* A population-based study of patterns of care for muscle-invasive bladder cancer in Sweden. *Scandinavian Journal of Urology and Nephrology* 2009; **43**: 271–6.

9. Konety BR, Joslyn SA. Factors influencing aggressive therapy for bladder cancer: an analysis of data from the SEER program. *Journal of Urology* 2003; **170**: 1765–71.

10. Cooke PW, Dunn JA, Latief T, Bathers S, James ND, Wallace DM. Long-term risk of salvage cystectomy after radiotherapy for muscle-invasive bladder cancer. *European Urology* 2000; **38**: 279–86.

11. Cooke PW, Wallace DMA, Dunn J, Bathers S, Latief T, James ND. Long term follow-up after radiotherapy for muscle-invasive bladder cancer. *British Journal of Cancer* 1998; **78**: 26.

12. Henningsohn L, Steven K, Kallestrup EB, Steineck G. Distressful symptoms and well-being after radical cystectomy and orthotopic bladder substitution compared with a matched control population. *Journal of Urology* 2002; **168**: 168–74.

13. Kaufman DS, Winter KA, Shipley WU, *et al.* Phase I-II RTOG study (99-06) of patients with muscle-invasive bladder cancer undergoing transurethral surgery, paclitaxel, cisplatin, and twice-daily radiotherapy followed by selective bladder preservation or radical cystectomy and adjuvant chemotherapy. *Urology* 2009; **73**: 833–7.

14. Mak RH, Zietman AL, Heney NM, Kaufman DS, Shipley WU. Bladder preservation: optimizing radiotherapy and integrated treatment strategies. *BJU International* 2008; **102**: 1345–53.

15. Kaufman DS, Winter KA, Shipley WU, *et al.* The initial results in muscle-invading bladder cancer of RTOG 95-06: phase I/II trial of transurethral surgery plus radiation therapy with concurrent cisplatin and 5-fluorouracil followed by selective bladder preservation or cystectomy depending on the initial response. *Oncologist* 2000; **5**: 471–6.

16. Coppin CM, Gospodarowicz MK, James K, *et al.* Improved local control of invasive bladder cancer by concurrent cisplatin and preoperative or definitive radiation. The National Cancer Institute of Canada Clinical Trials Group. *Journal of Clinical Oncology* 1996; **14**: 2901–7.

17. James ND, Hussain SA, Hall E, *et al.* BC2001 Investigators. Radiotherapy with or without chemotherapy in muscle-invasive bladder cancer. *N Engl J Med* 2012; **366**(16): 1477–88.

18. Hoskin P, Rojas A, Bentzen S, *et al.* Radiotherapy with concurrent carbogen and nicotinamide in bladder carcinoma. *Journal of Clinical Oncology* 2010; **28**: 4912–18.

19. Duchesne GM, Bolger JJ, Griffiths GO, *et al.* A randomized trial of hypofractionated schedules of palliative radiotherapy in the management of bladder carcinoma: results of medical research council trial BA09. *International Journal of Radiation Oncology, Biology, Physics* 2000; **47**: 379–88.

Testis

Peter Hoskin

The role of radiotherapy in testicular cancer is becoming less prominent. The mainstay of treatment is radical orchidectomy and, where there is a risk of metastatic disease, combination chemotherapy. Radiotherapy may be indicated in the following situations:

- Stage I testicular seminoma delivering prophylactic para-aortic lymph nodes irradiation.
- Palliative treatment in the management of chemotherapy resistant disease.

8.17 Prophylactic para-aortic lymph node irradiation

This has in the past been widely used in stage I seminoma which is marker negative (normal alpha-fetoprotein and human chorionic gonadotrophin). However, increasingly radiotherapy has been displaced by either surveillance or single-agent carboplatin and is now only occasionally used for patients declining carboplatin or those with stage II disease declining chemotherapy.

Trials by the Medical Research Council in the UK have clarified the treatment volume (CTV) and dose. TE10 randomized patients to receive treatment with either para-aortic lymph node irradiation or a dogleg field incorporating the ipsilateral iliac lymph nodes, both arms receiving a dose of 30 Gy in 15 fractions as well. There were four pelvic relapses in the para-aortic field group compared to none in the dogleg group but overall nine relapses occurred in each group and there were no significant differences in disease-free or overall survival; on the basis of this trial para-aortic lymph node irradiation alone is considered the standard of care.

TE18 compared to radiation dose 30 Gy in 15 fractions and 20 Gy in 10 fractions and again no significant difference has emerged between these doses for relapse-free or overall survival. The 20-Gy arm reports better quality of life scores for acute toxicity and therefore 20 Gy in 10 fractions is regarded as the standard of care.

8.17.1 Patient position and immobilization

The patient is treated supine, arms by the side with no specific immobilization device usually employed.

8.17.2 **Field localization**

It is not usual to formally identify a CTV and PTV. Standard fields are used using the orthovoltage X-ray simulator. The margins are as follows:

- Bottom of T10.
- Bottom inferior border of L5.
- Lateral borders to transverse processes of vertebrae.

It is important at localization to identify the kidneys, either reconstructing this from CT or using an IVU at the time of simulation.

In some centres it is convention to extend the border on the left to the edge of the renal hilum to account for the different drainage on the left side which feeds into the left renal vein distinct from the right side feeding directly into the inferior vena cava. In practice, both fields can be extended laterally to the edge of the renal hilae without significantly increasing volume or toxicity and this is the recommended approach.

8.17.3 **Dose distribution**

Anterior and posterior opposed fields are used. No additional shielding is recommended.

8.17.4 **Dose prescription**

The standard dose is 20 Gy in 10 fractions treating daily Monday to Friday.

8.17.5 **Implementation and verification**

The field centre is tattooed together with lateral tattoos to identify rotation of the trunk. The field is then set up to the central tattoo and skin marks. Verification using megavoltage films or preferably EPID images should be undertaken. In the case of the latter, daily for the first 3 days to identify any systematic error in set-up.

8.17.6 **Patient care**

Treating a significant amount of small bowel and stomach nausea is common in patients receiving this treatment. In some centres prophylactic antiemetics are offered, if this is not the case then access to antiemetics to be taken regularly if nausea develops should be facilitated.

Patients having testicular cancer postorchidectomy are often concerned regarding future fertility. They can be reassured that the para-aortic lymph node field, distinct from the dogleg fields used in the past, results in no significant dose to the testis and would have no impact on fertility. There is no indication for *in vivo* dosimetry.

8.18 **Palliative treatment in chemoresistant disease**

This may embrace a number of scenarios:

- Persistent para-aortic lymphadenopathy.
- Mediastinal or supraclavicular lymphadenopathy.
- Bone and cerebral metastasis.

8.18.1 Para-aortic lymphadenopathy

This should be treated in a similar fashion to that described in section 8.17 for prophylactic treatment. It is, however, important to identify residual tumour masses and such patients may be better treated with CT planning defining a GTV and expanding this by 2 cm to a CTV and PTV. As in prophylactic treatment it is important to identify the kidneys and ensure that the treating beams do not exceed renal tolerance.

8.18.2 Mediastinal and supraclavicular lymph nodes

These should be treated in an analogous fashion to any other palliative mediastinal disease as described in Chapter 5 for lung cancer. It is usually adequate to localize using orthovoltage simulator images and treat with anterior–posterior paired beams.

8.18.3 Bone and cerebral metastasis.

Bone and brain metastasis should be treated in a standard fashion as described in Chapter 15.

8.18.4 Dose

Standard palliative doses should be used, for example, 20 Gy in five fractions or 30 Gy in 10 fractions to lymph node masses and bone and brain metastasis as described in Chapter 15.

Further reading

Chung P, Mayhew LA, Warde P, Winquist E, Lukka H. Management of stage i seminomatous testicular cancer: A systematic review. *Clinical Oncology* 2010; **22**: 6–16.

Fossa SD, Horwich A, Russell JM, *et al.* Optimal planning target volume for stage I testicular seminoma: A Medical Research Council randomized trial. Medical Research Council Testicular Tumor Working Group. *Journal of Clinical Oncology* 1999; **17**(4): 1146.

Jones WG, Fossa SD, Mead GM, *et al.* Randomized trial of 30 versus 20 Gy in the adjuvant treatment of stage i testicular seminomA: A Report on Medical Research Council Trial TE18, European Organization for the Research and Treatment of Cancer Trial 30942 (ISRCTN 18525328). *Journal of Clinical Oncology* 2005; **23**(6): 1200–8.

Warde P, Specht L, Horwich A, *et al.* Prognostic factors for relapse in stage I seminoma managed by surveillance: a pooled analysis. *Journal of Clinical Oncology* 2002; **20**(22): 4448–52.

Penis

Peter Hoskin

Carcinoma of the penis is typically a squamous carcinoma arising on the penile shaft or glands in an uncircumcized patient. Management may be by primary surgery, either total amputation or partial amputation with reconstruction, or primary radiotherapy. Primary radiotherapy is indicated for those patients with T1 and T2 tumours < 4 cm in diameter, particularly in those unfit for surgery, those with locally advanced disease and fixed inguinal lymph nodes, and for patients in whom surgical treatment may require total amputation and where they choose to have organ preservation by radiotherapy as an alternative. No randomized trial comparison is available to give accurate figures for the relative efficacy of either treatment. Brachytherapy is an alternative means of delivering high-dose radiotherapy to the penis and may be considered where there is local expertise for this instead of external beam treatment.

Postoperative radiotherapy may be indicated in some circumstances where there has been inadequate proximal clearance or where inguinal lymph nodes are found to contain metastatic tumour with high-risk criteria. There are no robust criteria upon which to base recommendations but as in other sites where there is heavy involvement (more than four nodes), extracapsular extension or extensive lymphovascular infiltration then postoperative radiotherapy may be considered. It is important to recognize, however, that adding radiotherapy to surgery in this region will substantially increase the risks of long-term toxicity, in particular pain and lymphoedema.

Full CT restaging of the internal iliac nodes should be available and FDG-PET may help refine status of the higher lymph nodes.

8.19 Technique for radical external beam treatment

8.19.1 Patient position and immobilization

The patient is treated supine.

◆ The common technique is to hold the penis in a wax or Perspex block which acts both as an immobilization device and also provides build-up to ensure maximum dose deposition in the CTV from megavoltage beams. The block typically rests on a lead shield with a hole through which the penis protrudes into the block as shown in Fig. 8d.1. The penis may be held in a length of tubigrip to facilitate position within the block.

◆ An alternative approach is required where the penis is short or retracted and cannot be pulled into the wax block. In this setting lead shielding to the underlying testis and skin of the lower abdomen and groins is used and surface bolus applied.

(a)

(b)

Fig. 8d.1 Demonstrating (a) customized wax block or (b) standard Perspex block on scrotal shield.

◆ *Note:* the penis may become swollen from the acute radiation reaction and this can lead to a paraphimosis; all patients should therefore be circumcized or undergo a dorsal slit of the foreskin prior to treatment.

8.19.2 Volume definition

The CTV should include the GTV with a 2-cm margin proximally and distally. In practice this will often represent most of the penile shaft. The entire circumference of the penile skin should be included in the CTV. No further expansion to PTV is usually employed.

8.19.3 Field localization

This will depend upon the immobilization and set-up.

◆ Where the block technique is used, which is the common approach, field localization can be undertaken with the orthovoltage simulator with lateral opposing fields to encompass the block shown in Fig. 8d.2. This is readily identified with the light beam which should be seen to splash outside the edges of the block.

◆ Where a wax block cannot be used then a CT planned volume is best using 5-mm CT slices through the region on which the PTV can be identified. Typically a field arrangement using two lateral oblique beams using 6-MV energy will be best as shown in Fig. 8d.3.

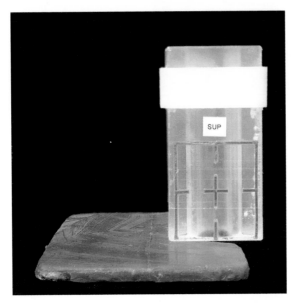

Fig. 8d.2 Lateral field set-up marks on block.

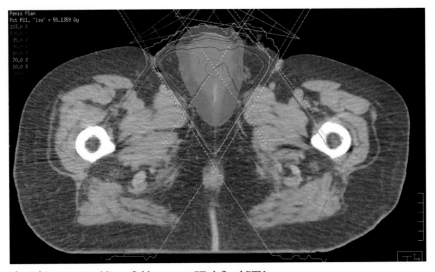

Fig. 8d.3 Anterior oblique fields to treat CT-defined PTV.

8.19.4 Implementation and verification

Where the block is used then this defines the patient set-up. Often the patient can be educated to position the block each day on the underlying lead shielding. Micropore tape or similar may be required to steady the set-up on the linear accelerator couch. Where no block is used then set-up will be to a lateral tattoo from which the isocentre is defined since the field centre beneath the midline is likely to be under penile tissue or overlying bolus on which it is not possible to define an accurate set-up.

Verification using megavoltage beams or EPID images should be undertaken. Where the wax block is used then this is simply to ensure adequate coverage of the block, the alternative technique can be difficult to interpret on verifications films and anteroposterior and lateral images are more useful to provide verification of the isocentre alone.

8.19.5 Dose prescription

A total dose of 64 Gy in 32 daily fractions treating Monday to Friday is given. This is prescribed to the mid plane, that is the centre of the block, where this technique is used and to the intersection point where the alternative technique is used.

8.20 Technique for lymph node irradiation

8.20.1 Patient position and immobilization

The patient is treated supine.

8.20.2 Volume definition

CT planning scans, 3-mm slices with intravenous contrast are used.

The CTV comprises the inguinofemoral canal, external iliac and internal iliac lymph nodes. These should be defined based on the standard lymph node atlases and as described in Chapter 8A for prostate and Chapter 10 for lymphomas.

The PTV will be derived by a volumetric expansion according to local practice, typically of the order of 5 mm.

If CT simulation is not available then conventional techniques using an orthogonal X-ray simulator can be used. The PTV will be based upon bone landmarks as follows:

- Superior border: the superior border will be top of acetabulum.
- Inferior border: 3 cm below line of inguinal canal.
- Lateral borders: to cover greater trochanters.

8.20.3 Field arrangement

A parallel-opposed pair of beams using MLC or individualized blocks to fit the nodal chains is used.

8.20.4 Implementation and verification

The field centre is tattooed together with lateral tattoos to identify pelvic rotation.

Verification will follow standard practice using electronic portal imaging verification with match to bone landmarks against the planning DRR.

8.20.5 **Prescription**

♦ 50–50.4 Gy in 25–28 fractions over 5–5½ weeks.

♦ Or 40 Gy in 15 fractions over 3 weeks.

Where there are fixed palpable lymph nodes then a second volume should be defined; these will typically be in the inguinal region and best covered by a direct electron field or wedged photon beam plan. The dose to this boost should be:

♦ 16 Gy in eight fractions over 1½ weeks.

8.21 **Palliative treatment**

Palliative treatment for locally-advanced fixed fungating or bleeding tumours may be indicated. These should be treated pragmatically with derivations of the previously described technique. Doses for palliation may include the following:

♦ 21 Gy in three fractions treating three times weekly.

♦ 20 Gy in five daily fractions.

♦ 8–10 Gy as a single dose.

8.22 **Patient care**

Acute skin and mucosal reactions are inevitable and should be managed conservatively. The penis will become swollen and dysuria may develop.

There are important late effects to be considered in preparing the patient for radiotherapy:

♦ Lymphoedema of the lower limbs, more marked where there has been previous groin surgery.

♦ Sexual dysfunction related to fibrosis and penile shortening.

♦ Urethral stricture.

Further reading

Pizzocaro G, Algabi F, Horenblas S, *et al*. EAU Penile cancer guidelines 2009. *European Urology* 2010; **57**: 1000–12.

9 Gynaecology

Uterus

Melanie Powell, Alexandra Taylor

Tumours of the uterus may originate from either the uterine body (corpus) or the uterine cervix. Radiotherapy has an important role in the management of these cancers, as either the primary treatment modality, an adjuvant to surgery or palliation of symptoms.

Uterine corpus cancer (endometrial) is increasing in incidence and with an incidence of 19 per 100 000 is the fourth commonest malignancy in women in the UK. It is related to obesity and is therefore likely to become more prevalent.

Following the introduction of a national screening programme in 1989 the incidence of cervical cancer almost halved. In the UK it is now a relatively rare cancer, ranking at number 12 for women with an incidence of eight per 100 000. Worldwide, however, it remains a major problem and is the second commonest malignancy affecting women. In India, parts of Africa, the Caribbean, and South America it is the leading cause of cancer in women.

9.1 Anatomy

The uterus consists of the uterine body and the cervix that are separated by the internal os. Posteriorly, the pouch of Douglas is the region between the posterior vaginal fornix and the rectum. The parametrium is a layer of connective tissue that lies adjacent to the uterine body, cervix, and vagina. It is rich in vascular and lymphatic vessels and contains the ureters that pass below the broad ligament and lateral to the cervix. The floor of the parametrium is formed by the cardinal (lateral cervical) ligaments, which arise from the lateral margins of the cervix and insert into the pelvic sidewall, and the uterosacral ligaments that pass from the uterus around the rectum to the sacrum.

There are several interlinking pathways of lymphatic drainage from the uterus and cervix to the pelvic lymph nodes. The nodal groups lie in close proximity to the pelvic blood vessels as shown in Fig. 9a.1. The cervix and lower uterine body drain predominantly via the parametrial nodes to the obturator and external iliac nodes, and along the route of the uterine vein to the internal iliac nodes. Spread can also occur along the pathway of the uterosacral ligaments to the sacral nodes. The upper uterine body also drains along the pathway of the ovarian vessels to the common iliac and para-aortic nodes. As a result of this extensive network, nodal metastases may occur at any level and in any combination.

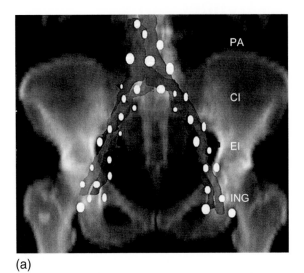

(a)

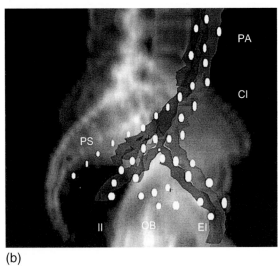

(b)

Fig. 9a.1 Position of the pelvic lymph nodes in relation to blood vessels and bony landmarks. Nodal groups: PA, para-aortic; CI, common iliac; EI, external iliac; ING, inguinal; OB, obturator; II, internal iliac; PS, presacral.

9.2 **Cervical tumours**

The majority of cervical tumours are human papilloma virus (HPV)-induced squamous cell carcinomas arising from the squamo-columnar junction. There is an increasing incidence of adenocarcinoma (also HPV related) that usually originates in the endocervical canal and must be differentiated from primary endometrial adenocarcinoma.

9.2.1 **Primary radical treatment for cervical cancer**

Indications

The selection of the appropriate primary treatment for cervical cancer is essential and depends on the tumour stage, size, nodal status, local expertise, and patient preference.

Although imaging plays an increasingly important part in the management of cervix cancer, FIGO staging is based on clinical findings. This means that a careful examination under anaesthetic assessing the size and extent of the tumour, evaluating vaginal, parametrial pelvic side wall, and uterosacral involvement must be carried out. Cystoscopy and sigmoidoscopy are used to visualize bladder and rectum.

The combination of both radiotherapy and surgery has a greater risk of morbidity than either treatment as a single modality and should be avoided. In FIGO stage IB2 and IIA disease, surgery and radiotherapy have equivalent local control and survival rates. In general, these early stage tumours are treated with surgery as this enables preservation of ovarian function for younger women. Radiotherapy is the treatment of choice for bulky stage IIB to IVA tumours, node positive disease, and for those unfit for surgery. Radiotherapy may also be curative for recurrent pelvic disease in patients previously treated surgically (salvage therapy).

Radical radiotherapy comprises a combination of external beam radiotherapy to encompass the primary tumour and regional nodes followed by intrauterine brachytherapy to deliver a high central dose to the primary tumour while sparing the surrounding normal tissues. There is strong evidence for a dose–response relationship for local control and overall survival but there is a corresponding increase in the incidence of normal tissue toxicity that limits the total radiation dose that can be safely delivered.

Treatment volume and definition

Cervical cancer may become very locally advanced without distant spread. It can infiltrate superiorly to the uterine body, inferiorly to the vagina, or laterally to the parametrium. Extension to the pelvic side wall causes hydronephrosis by compression or infiltration of the distal ureter as it passes lateral to the cervix. Pelvic sidewall fixation may be due to either direct tumour extension or coalescence with paracervical lymphadenopathy. Tumours may also invade the uterosacral ligaments and rectum posteriorly or the base of bladder anteriorly.

There is a rich lymphatic drainage of the cervix via several pathways. Lymph node spread is common and may be non-contiguous. From surgical series, the most frequently involved nodes are the medial external iliac, internal iliac, and obturator nodes although primary spread to the common iliac and para-aortic nodes is not uncommon. Presacral nodal involvement is rare in the absence of posterior tumour extension.

The risk of nodal metastatic spread increases with tumour size and stage. The approximate incidence of pelvic lymph node involvement is 15–20% in stage IB2, 30% in stage IIB, 50% in stage IIIB, and 80% in stage IVA disease. The incidence of para-aortic node involvement is 5–10% in stage IB2, 15% in stage IIB, 30% in stage IIIB, and 50% in stage IVA disease. Haematogenous spread, which is most frequently to the lungs, liver, and bone, is usually a late event.

MRI is the optimal method for imaging local disease with accurate assessment of stromal invasion, parametrial involvement, and uterine extension (Fig. 9a.2). The pelvic and para-aortic nodes should be imaged although the assessment of lymph nodes with CT and MRI using size criteria is a relatively insensitive technique for predicting microscopic nodal metastases. PET may be useful with high sensitivity and specificity for detecting regional nodal involvement and distant metastases.

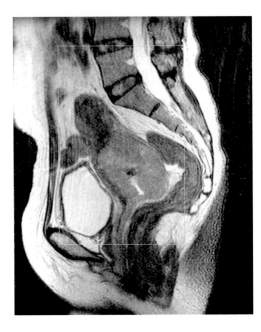

Fig. 9a.2 Sagittal T2 MRI of pelvis showing large cervical tumour. PTV borders outlined in yellow.

Clinical target volume

The CTV for external beam radiation covers all areas of potential microscopic spread and comprises;

- GTV.
- Cervix.
- Uterus.
- Upper half of the vagina (full length of vagina in stage 3A disease).
- Parametrium.
- Proximal uterosacral ligaments (entire mesorecectum if uterosacral ligament involved).
- Ovaries.
- Pelvic lymph nodes to include internal iliac, obturator, external iliac, common iliac (to the level of the aortic bifurcation), and subaortic presacral nodes.

The EUA findings and imaging must be used to modify the target volume.

- In stage 3A disease the entire vagina is treated.
- The full length of the uterosacral ligaments and presacral nodes are covered when there is posterior tumour extension.
- The inguinal nodes are irradiated when there is involvement of the lower third of the vagina.
- The para-aortic region is also treated if there is overt involvement or if there is common iliac node enlargement due to the high risk of microscopic para-aortic nodal metastases (see section 10.5.3).

Planning technique

Patient position and immobilization

The patient is immobilized supine with knee and ankle supports, and with their arms on the chest. Skin tattoos are placed anteriorly in the midline and laterally to prevent lateral rotation.

Immobilization in the prone position on a belly board displaces small bowel from the treatment volume. However, this may be a difficult position for the patient and the daily set-up error is slightly greater than with a supine position.

Volume localization

CT localization

Three-dimensional CT planning is the best method for ensuring full coverage of the target volume. The patient is immobilized in the treatment position with radio-opaque markers placed over the skin tattoos. A planning CT scan is taken with 2.5–3-mm slices from the top of the third lumbar vertebra to 5 cm below the ischial tuberosities. Administration of intravenous contrast improves visualization of the pelvic blood vessels, the uterus, and the primary tumour and is recommended provided the patient has good renal function.

The bladder should be comfortably full in order to displace small bowel from the pelvis and to reduce the volume of bladder irradiated. Bladder filling affects the position of the cervix and uterus and therefore a bladder-filling protocol should be used both for consistency and to limit the interfractional variation in uterine and cervix position. Since overfilling the bladder is uncomfortable and likely to be impossible to sustain towards the end of treatment, a moderately full bladder is suggested.

An example of a bladder-filling protocol is that 1 hour prior to planning or treatment patients are asked to empty their bladder and then drink 350 mL of fluid. Daily bladder ultrasound may be used to ensure consistent bladder volumes.

The target volume (CTV) is delineated on each axial CT slice.

The nodal areas are defined using the blood vessels as a surrogate target. A 7-mm margin around vessels is recommended for the pelvic nodal regions. Any visible nodes must also be encompassed.

For para-aortic nodes a 5-mm margin around vessels is used, ensuring coverage is extended around the anterior–lateral aspects of the lumbar vertebrae.

Image fusion of MRI scans and CT-PET images may aid volume localization.

The PTV margin will allow for set-up error. Since the uterus and cervix are affected by both rectal and bladder filling a larger margin is required than for the nodal areas. A margin of 15–20 mm should be added to the cervix, tumour, and uterus PTV. For all other structures a margin of 7–8 mm is sufficient.

For the minority of patients who cannot receive brachytherapy, a second phase of treatment with external beam therapy will be delivered to a smaller volume. The PTV for this second phase is created by adding a 10-mm margin around macroscopic disease (GTV).

Simulator localization

Standard field borders have a high risk of a geographical miss and cross-sectional imaging must be used to determine the target volume. If conventional simulation is necessary, the patient is examined in the treatment position and the lower extent of vaginal disease or the introitus is marked with a radio-opaque marker. The diagnostic MRI scans are then used to modify the field borders (Fig. 9a2).

The PTV is defined with anteroposterior and lateral simulator images. The superior volume margin should cover the common iliac nodes and is usually at the upper border of the fifth lumbar vertebra. The inferior border is placed at least 3 cm below the inferior aspect of disease, either clinically or on the MRI, and is usually at the lower border of the obturator fossae. The lateral volume border is 2 cm lateral to the pelvic sidewall. The anterior volume border is 1 cm anterior to the uterine body on MRI, ensuring adequate coverage of the common iliac nodes superiorly, and is usually at the anterior symphysis pubis. The posterior volume border is 1.5–2 cm behind primary disease, ensuring coverage of the proximal uterosacral ligaments and the internal iliac nodes. Small bowel and femoral heads may be shielded on the anterior and posterior fields and the posterior sacrum can be shielded on the lateral films as shown in Fig. 9a.3.

Dose distribution

Conformal and conventional radiotherapy

The pelvis is treated with a three- or four-field 'box' arrangement consisting of two lateral wedged fields, an anterior field, and a posterior field if required. There is a low weighting of the posterior field to reduce the rectal dose. This field arrangement achieves better homogeneity and tissue sparing, even when there is posterior extension of the volume, compared to a two-field arrangement with anterior and posterior fields (Fig. 9a.4).

The fields are designed to ensure coverage of the planning target volume by the 95% isodose with a maximum dose of 105%. The fields are shaped to the 3D volume or the shielding is applied from the simulator images. The use of higher energy photons (10–15 MV) improves superficial tissue sparing and achieves a better dose distribution.

For patients with pelvic and para-aortic disease an anterior posterior field arrangement can be avoided using a single isocentre technique. The superior para-aortic section also utilizes a four-field technique with low weighted lateral fields to ensure sparing of the kidneys and to avoid hot spots within small bowel (Fig. 9a.5).

Intensity-modulated radiotherapy

IMRT can significantly reduce the dose to normal structure. Retrospective studies have shown that it can reduce both early and late toxicity when compared to conventionally planned treatment in historical controls. There are currently, however, no randomized controlled trials to support its routine use and in particular no outcome data to compare it to conformal radiotherapy.

If it is to be used in the radiotherapy treatment of primary cervix cancer great care needs to be taken in the delineation of the target in order to avoid compromising success with a geographical miss. This is particularly important for the cervix and uterus which, because of bladder and rectal filling, can show interfractional movement of up to 30 mm in the superior–inferior direction and 15 mm in the anterior–posterior direction.

Further studies with image guidance are required for ensuring safe implementation of this technique.

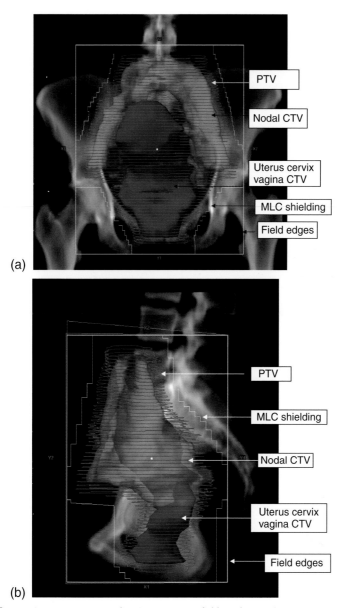

Fig. 9a.3 Anterior posterior DRR showing treatment fields with CTV (cervix, uterus upper vagina and pelvic nodes) and PTV lateral DRR showing treatment fields with CTV (cervix, uterus upper vagina and pelvic nodes) and PTV.

Implementation on treatment machine

The patient is treated in the same position as for volume localization. The patient is immobilized with knee or ankle supports, and aligned using laser lights to check the

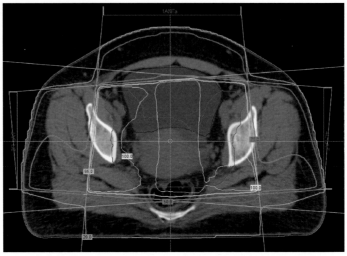

Field	Energy	Weight	Gantry	Coll	Couch	Wedge	X	Y
1 Ant	15MV	0.36	0	0	0		18cm	20cm
2 Post	15MV	0.08	180	0	0		18cm	20cm
3 L lat	15MV	0.28	90	0	0	30	14cm	20cm
4 R lat	15MV	0.28	270	0	0	30	14cm	20cm

Fig. 9a.4 Axial CT slice showing isodose distribution for conformal treatment of cervical cancer. PTV (red) with bladder anteriorly and rectum posteriorly. Isodose curves yellow 100%, cyan 95%, green 75%, bright pink 50%, pale pink 20%.

position of the anterior and two lateral skin tattoos. The field centre is marked in relation to the anterior tattoo. All fields are treated isocentrically daily.

There is a 12% absolute benefit in survival with the use of concomitant cisplatin compared with radiotherapy alone, although the addition of chemotherapy does increase normal tissue toxicity. Concomitant cisplatin is administered on a weekly basis with the radiotherapy treatment delivered within 1 hour of completing the cisplatin infusion.

The treatment course should be completed in the planned total number of days. The overall treatment time is a further prognostic factor with decreasing survival associated with protracted treatment. Ideally the treatment course (including the brachytherapy component) should be completed within 42–45 days and not prolonged over 49 days.

Unscheduled gaps are to be avoided and should be compensated for. This can be done by treating at a weekend, treating two fractions in 1 day with an interfraction interval of at least 6 hours, or by increasing the dose per fraction for the remaining doses.

Verification

EPID or beam films are taken on the first and second days of treatment and compared to the reference images, either the digitally-reconstructed radiographs (DRR) or the

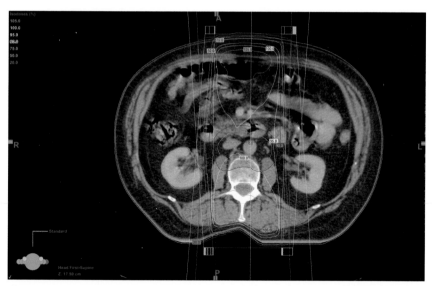

Fig. 9a.5 Para-aortic node treatment using anterior and posterior beams localized by virtual simulation.

simulator images. If the set-up error is < 5 mm on both images, EPIDs are taken weekly. Daily portal imaging before treatment, with subsequent field position adjustment if required, may be necessary for obese patients with poor set-up due to highly variable contours.

Dose prescription

Primary disease
External beam radiotherapy
- 50.4 Gy in 5 to 6 weeks (1.8–2 Gy per fraction)
- Weekly cisplatin 40 mg/m^2 (maximum 70 mg) for 5 weeks should be administered concurrently.
- When intracavitary treatment is not possible, a second phase of external beam radiotherapy is delivered to macroscopic disease with a small CT planned volume.
- 16 Gy in eight fractions prescribed to the isocentre.

Intrauterine brachytherapy
To be commenced during the last week of external beam radiotherapy or as soon as possible after completing external beam radiotherapy:
- High dose rate: 14–24 Gy in two to six fractions prescribed to 100% or Point A.
- Low dose rate: 27 Gy prescribed to Point A in a single application.

Recurrent disease
- Whole pelvis: 45–50.4 Gy in 25–28 fractions prescribed to the isocentre over 5–6 weeks.

- Phase two: 16–20 Gy in eight to 10 fractions prescribed to the isocentre over 2 weeks.

9.2.2 Postoperative adjuvant treatment for cervical cancer

Indications

The addition of adjuvant radiotherapy improves local control when the histology shows high-risk factors for pelvic recurrence, but increases normal tissue toxicity. Absolute indications for adjuvant pelvic radiotherapy are:

- Presence of tumour at the resection margin.
- More than one node with metastatic infiltration.
- Inadequate surgery such as an incidental finding at simple hysterectomy.

Relative indications that may require radiotherapy, particularly if more than one feature is present, include:

- Lymphovascular invasion.
- Deep stromal invasion.
- Poorly differentiated tumour.
- Invasive tumour < 5mm from the resection margin.
- A single lymph node with metastatic involvement.

Target volume and definition

The target volume includes the vaginal cuff and upper half of the vagina, paravaginal tissue and parametrium, and the external iliac, obturator, and internal iliac nodes for all patients. The common iliac nodes are included for node positive patients.

Radiotherapy technique

The radiotherapy technique is the same as adjuvant radiotherapy for uterine tumours as described in section 9.3.2.

Dose prescription
External beam radiotherapy

- 45–50.4 Gy in 25–28 fractions prescribed to the isocentre delivered over 5 weeks.
- Weekly cisplatin 40 mg/m^2 (maximum 70 mg) for 5 weeks is administered concurrently.

Intracavitary brachytherapy
To be commenced after completing external beam radiotherapy:

- High dose rate: 8–11 Gy at 0.5 cm from applicator surface in two fractions.
- Low dose rate: 15 Gy to 0.5 cm from applicator surface in a single application.

9.3 Uterine body tumours

The majority of uterine tumours arise from the endometrium and are predominantly adenocarcinomas of endometrioid type. The less common subtypes of serous, clear cell, and squamous cell carcinoma have a worse prognosis. Rarer uterine tumours include carcinosarcomas or mixed Müllerian tumours, leiomyosarcoma and endometrial, stromal sarcoma,

9.3.1 Primary radical treatment for uterine tumours

Indications

Uterine tumours are ideally treated surgically with a total abdominal hysterectomy and bilateral salpingo-oophorectomy. However, it is possible to gain long-term control with primary radiotherapy alone. Radiotherapy should only be used for primary treatment if the patient is medically unfit for surgery or when the tumour is inoperable due to local invasion.

Treatment volume and definition

The target volume for external beam radiation comprises the primary tumour, cervix, uterus, upper vagina, ovaries, parametrium, and the pelvic lymph nodes. The common iliac, external iliac, obturator, internal iliac, and subaortic presacral nodes should be included for all patients. Macroscopic disease (GTV) will receive a higher dose with either intrauterine brachytherapy or a second phase of external beam radiotherapy.

Radiotherapy technique

The radiotherapy technique is similar to the method for primary cervical cancer described in section 10.1.3.

Dose prescription

External beam radiotherapy

- 50.4 Gy in 28 fractions prescribed to the isocentre delivered over 5½ weeks.
- When intracavitary treatment is not possible, a second phase of external beam radiotherapy is delivered to macroscopic disease with a small CT planned volume: 16 Gy in eight fractions prescribed to the isocentre.

Intrauterine brachytherapy

- High dose rate: 14–24 Gy to surface of uterus in two to four fractions.
- Low dose rate: 27 Gy to surface of uterus in a single application.

9.3.2 Postoperative adjuvant treatment for uterine tumours

Indications

Over 80% of endometrial cancer presents as stage 1 disease and the overall survival rates for this group are very high. Subsequent adjuvant treatment is based on the risk of relapse which is determined by the following

- Depth of myometrial invasion.
- Grade of tumour.
- Lymphovascular space invasion.
- Clear cell or serous subtype.

Table 9a.1 outlines adjuvant treatment by stage and risk group. For more advanced tumours stage 2 and above, although it has not been shown to improve survival, pelvic

Table 9a.1 Adjuvant radiotherapy treatment for endometrial cancer according to risk of relapse

Risk	Stage	Treatment
Low	1A G1 G2	None
Intermediate	1A (myometrial invasion) G3 1BG1, 1BG2	Vault brachytherapy
High risk	1B G3, 2, 3, 4A	External beam pelvic RT ± brachytherapy

radiotherapy is indicated to reduce the incidence of locoregional recurrence. Additional chemotherapy may be also be of benefit and should be considered.

For the rarer carcinosarcomas, leiomyosarcomas, and endometrial stromal sarcomas, pelvic radiotherapy has not been shown to be of survival benefit. It is, however, recommended for locoregional control where there is extension beyond the uterus.

Treatment volume and definition

Most tumours are confined to the uterus at presentation. Endometrial cancer frequently grows as a polypoidal mass that can ulcerate to cause bleeding. It invades directly into the myometrium and can penetrate the serosa from where it can spread into the peritoneal cavity or adjacent organs. Peritoneal metastases are particularly common with serous and clear cell histology. Tumours can spread inferiorly into the cervical canal with invasion of the cervical glands and stroma.

Lymphatic spread from the middle and lower corpus follows a similar course to the cervix, draining initially to the parametrial nodes and then to external, obturator and internal iliac nodes. From the upper corpus, metastases occur more commonly to the common iliac and para-aortic nodes. The risk of lymph node metastases increases with tumour grade and depth of invasion. The approximate incidence of pelvic lymph node involvement is 6% in FIGO stage IA, 10% in stage IA (with myometrial invasion), 20% in stage IC, 20% in stage II, and 57% in stage III.

Haematogenous spread is uncommon at presentation for endometrial cancers although it is a common method of spread for the uterine sarcomas, which frequently metastasize to lung. A CT scan of the chest and abdomen is performed for high-grade sarcomas to exclude visceral metastases.

Clinical target volume
- Vaginal vault.
- Upper half of vagina.
- Paravaginal tissues.
- External iliac, obturator, and internal iliac and proximal common iliac nodes (2 cm inferior to aortic bifurcation).

Obviously enlarged nodes and lymphocoeles must be included.

Where pelvic nodes are involved the common iliac nodes should be encompassed up to the bifurcation of the aorta or 2 cm above known disease whichever is higher.

Planning technique

Patient position and immobilization

The patient is immobilized supine with knee and ankle supports, and with their arms on the chest. Skin tattoos are placed anteriorly and laterally. As discussed in the patient positioning in section 9.2.1, the prone position with a belly board may be preferable to reduce the dose to the small bowel.

Volume localization

CT localization

The patient is immobilized in the treatment position with radio-opaque markers placed over the skin tattoos. A planning CT scan of the pelvis is taken with 2.5-mm slices from the top of L3 to 5 cm below the ischial tuberosities. Administration of intravenous contrast enhances visualization of the pelvic blood vessels. The bladder is comfortably full in order to displace small bowel from the pelvis.

◆ The CTV is outlined on each axial CT slice with a 7-mm margin around the vessels used as a surrogate for the nodal regions. The rectum, bladder, and small bowel are contoured.

◆ The PTV is created by adding a 10–15-mm margin around the vaginal vault and a 7-mm margin around the lymph nodes and parametrium.

Simulator localization

The target volume is defined on orthogonal simulator images:

◆ The superior volume margin is approximately at the upper border of the fifth lumbar vertebra to include the common iliac nodes.

◆ The inferior volume border is the lower border of the obturator foramen. The lateral volume border is 2 cm lateral to the pelvic sidewall.

◆ The anterior volume border is at the anterior symphysis pubis.

◆ The posterior volume border set at mid-S2/3 interspace. Shielding is applied as shown in Fig. 9a.6.

Dose distribution

Conventional and conformal radiotherapy

The pelvis is treated with a three- or four-field 'box' arrangement consisting of two lateral wedged fields, an anterior field, and, if required, a posterior field. The fields are shaped to the 3D volume or the shielding is applied from the simulator images. The PTV is covered by the 95% isodose with a maximum dose of 105%.

Intensity-modulated radiotherapy

IMRT can significantly reduce the volume of small bowel, bladder, and rectum within the PTV and initial clinical studies report a corresponding reduction in both acute and late toxicity. Postoperatively, there is less organ motion within the target volume compared to primary cervical cancer although the vaginal vault may still move 5–10 mm antero-posteriorly subject to bladder and rectal filling.

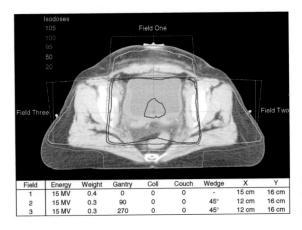

Field	Energy	Weight	Gantry	Coll	Couch	Wedge	X	Y
1	15 MV	0.4	0	0	0	-	15 cm	16 cm
2	15 MV	0.3	90	0	0	45°	12 cm	16 cm
3	15 MV	0.3	270	0	0	45°	12 cm	16 cm

Fig. 9a.6 Dose distribution with three-field plan for endometrial cancer.

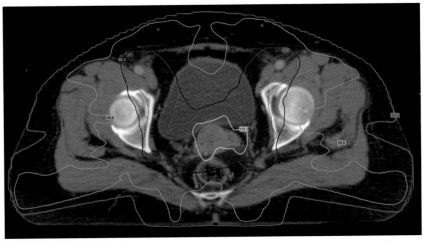

Fig. 9a.7 Dose distribution of IMRT for primary cervix cancer seven fields (gantry angles 180°, 105°, 65°, 25°, 335°, 295°, 255°) delivering 54 Gy to cervix (100%) and 50 Gy to nodal areas. Nodal volume is encompassed by the 88.3% isodose curve which is 95% of the nodal dose.

The dose distribution is inversely planned with five or seven equally spaced co-planar fields using 6-MV photons (Fig. 9a.7). At least 99% of the PTV is covered by the 95% isodose and < 0.5% of the PTV should receive > 105%. The plan is optimized to achieve minimal dose to the small bowel, rectum and bladder.

Implementation on treatment machine

The patient is treated in the same position as for volume localization. The patient is aligned using laser lights to check the position of the anterior and two lateral skin tattoos. The field centre is marked in relation to the anterior tattoo. All fields are treated isocentrically daily.

Verification

EPID or beam films are taken on the first and second days of treatment and compared to the reference images. If the set-up error is < 5 mm on both images, EPIDs are taken weekly. Daily portal imaging before treatment, with subsequent field position adjustment if required, may be necessary for obese patients with poor set-up due to highly variable contours.

Dose prescription

External beam radiotherapy

◆ 45 Gy in 25 fractions or 48.6 Gy in 27 fractions prescribed to the isocentre delivered over 5 weeks.

Intracavitary brachytherapy

To be commenced after completing external beam radiotherapy.

◆ High dose rate: 8–11 Gy to 0.5 cm from the applicator surface in two fractions.

◆ Low dose rate: 15 Gy to 0.5 cm from the applicator surface in a single application.

9.3.3 **Para-aortic nodal irradiation**

Indications

Patients with asymptomatic para-aortic metastases have 5-year survival rates of approximately 30% when treated with radical radiotherapy but long-term control of symptomatic para-aortic disease is uncommon. The total dose that can be safely delivered is limited by small bowel, spinal cord, and kidney radiation tolerances. There may be a role for debulking enlarged nodes with laparoscopic resection followed by radiotherapy. In the presence of common iliac node disease, there is a very high incidence of microscopic involvement of para-aortic nodes and in primary cervical cancer the para-aortic nodes should also be covered in the treatment field.

Treatment volume and definition

The para-aortic, aorto-caval, and retro-caval nodes from the level of the first lumbar vertebrae to the aortic bifurcation are included in the treatment volume. In patients with primary cervical cancer the target volume includes the pelvic disease as described in 'Treatment volume and definition' in section 9.2.1.

Planning technique

Patient position and immobilization

The patient is immobilized supine with knee or ankle supports. Skin tattoos are placed in midline and laterally at the level of the centre of the target volume to limit rotation.

Field localization

A planning CT scan is taken from the level of the top of the diaphragm to the pelvis with 2.5-mm slices. Administration of intravenous contrast enhances visualization of the blood vessels and lymph nodes.

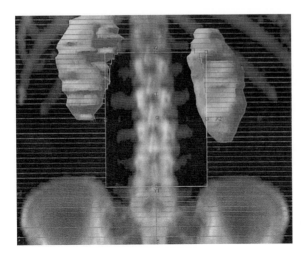

Fig. 9a.8 Virtual simulation field for treating para-aortic nodes.

The most homogeneous dose distribution is achieved using a conformal technique. This can be done using a four-field technique; anterior and posterior fields with low weighted lateral fields. This helps to reduce dose to bowel and spinal cord compared to using anterior and posterior fields (see Fig. 9a.5). If the pelvis is also to be treated a single isocentre technique is preferred.

The radiation fields can also be defined by virtual simulation to cover the para-aortic nodal region with maximal sparing of the kidneys (Fig. 9a.8). The superior field border is the top of the first lumbar vertebra and the inferior border matches to the top of the pelvic field.

Dose distribution
The PTV is covered by the 95% isodose with a maximum dose of 107%.

Implementation on treatment machine

The patient is treated in the same position as for volume localization. The patient is aligned using laser lights to check the position of the anterior and two lateral skin tattoos. The field centre is marked in relation to the anterior tattoo. All fields are treated isocentrically daily.

Verification

EPID or beam films are taken on the first and second days of treatment and compared to the reference images. If the set-up error is < 5 mm on both images, EPIDs are taken weekly.

Dose prescription

Radical treatment
- 45 Gy in 25 fractions prescribed to the mid-plane dose over 5 weeks.

Palliative treatment
- 30 Gy in 10 fractions over 2 weeks.

9.3.4 **Palliative treatment**

Indications

A short course of radiotherapy can provide excellent palliation of pelvic pain and bleeding for patients with disseminated disease and for patients who are medically unfit for radical treatment.

Treatment volume and definition

In the palliative setting, the treatment volume should be kept as small as possible in order to reduce toxicity. The volume consists of macroscopic disease only. There is no indication for including the pelvic nodes.

Planning technique

Patient position and immobilization

The patient is immobilized supine with knee and ankle supports. Skin tattoos are placed in midline and laterally.

Field localization

The patient is immobilized in the treatment position with radio-opaque markers placed. Administration of intravenous contrast may enhance visualization of the tumour, but caution is needed if there is impaired renal function. The bladder is comfortably full in order to displace small bowel from the pelvis. Macroscopic disease is outlined on each CT slice and the PTV created by adding a 10–15-mm margin.

Dose distribution

A field arrangement is selected to deliver a homogenous dose to the target volume while sparing normal tissues. Typically, an anterior and two lateral wedged fields are shaped to the target volume. if initiation of immediate treatment is necessary, opposing anterior and posterior fields may be used.

Implementation on treatment machine

The patient is treated supine with a comfortably full bladder, immobilized with knee or ankle supports and aligned using laser lights to check the position of the skin tattoos. The field centre is marked in relation to the anterior tattoo. All fields are treated isocentrically daily.

Verification

EPID or beam films should be taken on the first day of treatment and compared to the reference images, either the DRRs or the simulator images. If the set-up error is < 5 mm no further EPIDs are required as long as the clinical set-up is acceptable.

Dose prescription

Whole pelvis

- 30 Gy in 10 fractions to mid-plane dose over 2 weeks.
- 20 Gy in five fractions to mid-plane dose over 1 week.

CT planned volume

- 27 Gy in six fractions prescribed to the isocentre treating twice weekly over 3 weeks.

Advanced disease for bleeding or pain control to small volume

- 12 Gy in three fractions treating on alternate days.
- 10 Gy single fraction.

9.4 Management of patients undergoing pelvic radiotherapy

Patients should be reviewed weekly during the course of treatment to assess acute toxicity. Acute gastrointestinal side effects include diarrhoea, abdominal cramps, rectal discomfort, and bleeding. Review by a dietician may be necessary and a low-residue diet and loperamide help to control symptoms. Skin care advice is given as erythema and desquamation may develop in the perineum, natal cleft and groins. Urinary frequency and dysuria occur due to radiation cystitis, although infection should be excluded.

In cervical cancer, anaemia has an adverse effect on outcome with primary radiotherapy and the average haemoglobin level during treatment is an independent prognostic factor. It is important to maintain the haemoglobin level above 12 g/dL with blood transfusions if necessary.

Radiotherapy causes rapid ablation of ovarian function and hormone replacement therapy should be considered upon completing treatment. Sexual function may be significantly impaired and patients should be offered sexual counselling. The long-term regular use of vaginal dilators reduces the incidence of vaginal stenosis.

Recommended reading

ASTEC/EN.5 Study Group, Blake P, Swart AM, *et al*. Adjuvant external beam radiotherapy in the treatment of endometrial cancer (MRC ASTEC and NCIC CTG EN.5 randomised trials): pooled trial results, systematic review, and meta-analysis. *Lancet* 2009; **373**, 137–46.

Chemoradiotherapy for Cervical Cancer Meta-analysis Collaboration (CCCMAC). Reducing uncertainties about the effect of chemoradiotherapy for cervical cancer: individual data meta-analysis (Review). *Cochrane Database of Systematic Reviews* 2010; **1**: CD008285.

Lim K, Small W, Portelance L, *et al*. Consensus guidelines for delineation of clinical target volume for intensity modulated radiotherapy for the treatment of cervix cancer. *International Journal of Radiation Oncology, Biology, Physics* 2011; **79**: 348–55.

Nout R, Smit V, Putter H, *et al*. Vaginal brachytherapy versus pelvic external beam brachytherapy for patients with endometrial cancer of high-intermediate risk (PORTEC-2); an open-label non-inferiority randomized trial. *Lancet* 2010; **375**: 816–23.

Reed NS, Mangioni C, Malmström H, *et al*. Phase III randomised study to evaluate the role of adjuvant pelvic radiotherapy in the treatment of uterine sarcomas stages I and II: an European Organisation for Research and Treatment of Cancer Gynaecological Cancer Group Study (protocol 55874). *European Journal of Cancer* 2008; **44**(6): 808–18.

Small W, Mell L, Anderson P, *et al*. Consensus Guidelines for delineation of clinical target volume for intensity modulated pelvic radiotherapy in postoperative treatment of endometrial and cervical cancer. *International Journal of Radiation Oncology, Biology, Physics* 2008; **71**; 428–34.

Taylor A, Powell M. An atlas of the pelvic lymph node regions to aid radiotherapy target volume definition. *Clinical Oncology* 2007; **19**: 542–50.

Vulva and vagina

Peter Hoskin

9.5 Vulva

Carcinoma of the vulva is primarily a surgical disease best treated by wide surgical resection, radical vulvectomy, and inguinal lymph node dissection based on presenting stage[1]. Rarely, locally advanced primary disease may be presented for primary radiotherapy treatment. Postoperative radiotherapy is recommended for tumours invading > 7 mm in a vertical direction[2]. The first station regional lymph nodes in the inguinal region are best treated by radical surgical dissection[3], but fixed inoperable lymph nodes may benefit from primary radiotherapy which may be followed where appropriate by surgery if there is a residual mass. Postoperative radiotherapy should be considered for women having more than one node involved with metastatic tumour at surgery[2]. This must be balanced against the increased risk of lymphoedema where both surgery and radiotherapy are delivered to the groins. Chemoradiation using cisplatin or 5-FU/mitomycin C-based schedules has been reported[4,5] but no randomized comparison with radiotherapy alone has been undertaken; whilst high response rates are seen there is a considerable increase in acute toxicity.

9.5.1 Radical treatment of inoperable disease

Patient position and immobilization

Patient supine, arms by side, no specific immobilization.

Treatment volume

Small tumours of the clitoris or labia may occasionally be treated as local skin tumours. The majority are squamous cell carcinomas although rarely basal cell carcinomas and melanoma may be found. Treatment should follow the same guidelines as those described in Chapter 13 using a GTV to CTV margin of 7 mm for squamous carcinoma and 5 mm for basal cell carcinoma with an expansion from CTV to PTV of 3 mm.

It should, however, be noted that even superficial squamous cell tumours in this region have an incidence of inguinal lymph node involvement, approaching 10% for a tumour invading 2mm and 20% once there is 3-mm depth of invasion[2]. Given the possible sampling error in estimating depth of invasion from a marginal biopsy, most patients should receive prophylactic lymph node irradiation also.

The primary tumour will be treated en bloc with the lymph nodes, the CTV then encompassing both inguinal lymph node regions and the entire vulva.

Field localization

CT planning scans, 3-mm slices with intravenous contrast are used.

The CTV comprises the vulva and lower vagina, the inguinofemoral canal, external iliac and internal iliac lymph nodes. The node regions should be defined based on the standard lymph node atlases and as described earlier in Chapter 9A for uterine tumours and in Chapter 10 for lymphomas.

If IMRT is to be used then the primary, nodal, and any boost volume to involved nodes should be outlined separately, designated CTVp, CTVn and CTVb respectively.

The PTV will be derived by a volumetric expansion according to local practice, typically of the order of 5 mm.

If CT is not available then conventional simulation using an X-ray simulator may be used. A radio-opaque marker should be placed on the inferior border of the vulva to aid localization denoting the inferior border of the CTV. The field borders to encompass the inguinal and internal iliac node regions and vulva will then be as follows:

- Superior border: the superior border will be top of acetabulum.
- Inferior border: field edge 2 cm below vulval marker.
- Lateral borders: to cover greater trochanters.

Dose distribution

Anterior and posterior opposed megavoltage beams are used. The energy should be chosen according to the separation of the patient, typically 6 MV will be sufficient unless the separation is > 24 cm when higher energy beams may give a better distribution. Care is then needed to ensure that the primary tumour in the vulva is not within the build-up region of higher-energy megavoltage beams. Bolus may be required and *in vivo* dosimetry with TLD or diode measurements may be helpful.

IMRT may be helpful where there is macroscopic lymph node involvement and surgical dissection is not an option. In this case the PTV as described previously will be defined and in addition a boost PTV (PTV2) is outlined comprising the involved macroscopic nodes and primary tumour if *in situ*.

IMRT is also of value in patients who have a prosthetic hip to minimize rectal dose.

Implementation and verification

The field centre is tattooed together with lateral tattoos to identify pelvic rotation. Megavoltage films or EPID images should be taken, in the case of the latter, daily for the first 3 days to identify systematic error.

Dose prescription

Phase 1

A dose of 50 Gy in 25–28 fractions daily treating Monday to Friday is required for areas of prophylactic irradiation[6]. Alternative dose fractionation schedules in use include:

- 45 Gy in 20 daily fractions.
- 40 Gy in 15 daily fractions.

Phase 2

A boost will then be required to the primary tumour and to palpable lymph nodes if present to bring the total dose to >60 Gy[6,7]:

- The boost CTV will encompass GTV with a 1-cm margin.
- This may be treated by smaller anterior and posterior opposed megavoltage fields or in the case of a relatively localized primary vulval or clitoral tumour by a direct electron field provided the applicator can access the region satisfactorily.
- A direct electron boost should also be used for palpable node areas.
- The boost dose is 14 to 16 Gy in seven to eight daily fractions to give a total dose of 64–66 Gy in 32–33 fractions.
- The same boost dose is recommended where the 15- or 20-fraction schedules have been used for the phase 1 treatment.

Intensity-modulated radiotherapy

- Postoperative:
  - Lymph node-negative: 50–50.4 Gy in 25–28 fractions over 5–5½ weeks.
  - Lymph node-positive: PTVn: 50.2 Gy in 27 fractions in 5½ weeks.
  - PTVp and PTVb: 54 Gy in 27 fractions in 5½ weeks.
- Primary treatment:
  - An additional 10-Gy boost to macroscopic primary and inguinal nodes should be added as previously described.

9.5.2 Postoperative treatment

Postoperative treatment will be delivered using the identical technique to that described for the phase 1 treatment to the same doses, i.e. 50 Gy in 25 fractions or an equivalent shorter fractionation schedule.

9.5.3 Palliative treatment

Palliative treatment for locally advanced fixed fungating or bleeding tumours may be indicated. These should be treated pragmatically with derivations of the previous technique. Doses for palliation may include the following:

- 21 Gy in three fractions treating three times weekly.
- 20 Gy in five daily fractions.
- 8–10 Gy as a single dose.

9.6 Vagina

In contrast to carcinoma of the vulva, radiotherapy has an important role in the radical treatment of vaginal cancer, particularly where organ preservation is an important consideration for the patient, the alternative surgical approach often requiring total vaginectomy. The results of treatment with radiotherapy suggest that patients with adenocarcinoma and those with distal (lower third) vaginal lesions have a worse prognosis independent of stage[8,9] but it is not clear that surgery necessarily gives better results in this group.

Small tumours localized to the vaginal mucosa (FIGO stage I) are well treated by brachytherapy alone[10]; detailed techniques are outside the scope of this chapter.

More advanced tumours involving submucosal tissues (Stage II) or fixed to the pelvic side wall (Stage III) will require external beam therapy.

Treatment will be given in two phases, the first to include the regional lymph nodes and the second to boost the primary site.

9.6.1 Phase 1

Patient position and immobilization

Supine, arms by side.

Treatment volume

The CTV will be different depending upon the position of the tumour, because of the different lymphatic drainage along the length of the vagina. The lower third drains to inguinal nodes and the middle and upper thirds to internal iliac and higher pelvic nodes. Hence the CTV will be defined:

- Lower-third tumours to include whole vagina and inguinal nodes.
- Middle- and upper-third tumours to include whole vagina, internal, external and common iliac nodes.

Field localization

- Lower-third tumours will be treated with fields as for a vulval cancer described previously, defining the introitus with a radio-opaque marker at simulation.
- Middle- and upper-third tumours will be treated as for a cervical cancer following the methods described in the previous section. Again the inferior extent of the CTV will be defined using a radio-opaque marker at the introitus.

Dose prescription

A dose of 50 Gy in 25–28 fractions treating daily Monday to Friday is required for prophylactic regions in phase 1. Alternative hypofractionated schedules may be considered for lower-third tumours as described for vulval cancer but for higher tumours where the PTV is a large pelvic volume including significant amounts of small bowel these are not recommended.

9.6.2 Phase 2

For localized Stage II tumours a brachytherapy boost will give the best means of delivering a localized high dose of radiation to the primary site. There is some evidence that where a brachytherapy boost is possible then this results in better tumour control than external beam alone, enabling total doses of > 75 Gy to be achieved from the total combined treatment schedule[11,12].

For more advanced disease then an external beam boost will be required.

Treatment volume

The CTV for phase 2 should include the original GTV with a 2-cm margin. An expansion of 3–5mm for CTV to PTV should be added. Macroscopically involved lymph nodes, if not to be removed surgically should also be included in this.

Field localization

CT-based planning of the phase 2 volume in conjunction with clinical assessment and other imaging, in particular MRI using image registration where possible, will give the best means of defining an accurate PTV.

Dose distribution

Using a CT-based volume a three- or four-field plan may give the best distribution with sparing of rectum, bladder, and small bowel.

For larger tumours extending to the pelvic sidewall or with extensive lymph node involvement then reduced size anterior and posterior opposed beams may still give the best distribution.

Dose prescription

The boost dose is 14–16 Gy in seven to eight daily fractions to give a total dose of 64–66 Gy in 32–33 fractions.

Intensity modulated radiotherapy

IMRT may be helpful where there is macroscopic lymph node involvement and surgical dissection is not an option. In this case the PTV as described previously will be defined and in addition a boost PTV (PTV2) is outlined comprising the involved macroscopic nodes and primary tumour if *in situ*.

IMRT is also of value in patients who have a prosthetic hip to minimize rectal volume.

Lymph node negative: 50–50.4 Gy in 25–28 fractions over 5 to 5.5 weeks

Lymph node positive: PTVn: 50.2 Gy in 27 fractions in 5.5 weeks

PTVp and PTVb: 54 Gy in 27 fractions in 5.5 weeks

Boost dose to the primary as above 14–16 Gy in 7 to 8 daily fractions or preferably using brachytherapy.

One study has suggested that overall time is important in the radiotherapy of carcinoma of the vagina with a pelvic control rate of 97% when treatment was completed within 63 days compared to 54% when the overall time was > 63 days[13].

9.6.3 Patient care

Acute skin and mucosal reactions are inevitable and should be managed conservatively. There are two important late effects to be considered in preparing the patient for radiotherapy:

◆ Sexual dysfunction related to fibrosis and vaginal shortening and narrowing is reported in around one-third of patients; the use of vaginal dilators may mitigate this and should be encouraged from early after treatment once the acute reaction has settled.

◆ The close anatomical relations with the rectum and bladder mean that fistulae are a serious potential complication; reported series vary in incidence from 4–12% related to tumour stage and dose.

9.6.4 Palliative treatment

Palliative treatment for locally advanced fixed fungating or bleeding tumours or for recurrent disease may be indicated.

Brachytherapy may deliver a localized high-dose palliative treatment particularly for intravaginal disease.

External beam treatment may be equally effective using limited anterior and posterior opposed fields to the true pelvis defined on CT simulator or using bone landmarks as follows:

◆ Superior border: bottom of sacroiliac joints.

◆ Inferior border: defined by introital marker.

◆ Lateral borders: to pelvic side walls.

Dose prescription

The following are in common use and effective; single doses may be most appropriate for control of bleeding in advanced disease, three- or five-fraction schedules are appropriate for other local symptoms and in good performance status patients:

◆ 21 Gy in three fractions treating three times weekly.

◆ 20 Gy in five daily fractions.

◆ 8–10 Gy as a single dose.

References

1. Tyring SK. Vulvar squamous cell carcinoma: guidelines for early diagnosis and treatment. *American Journal of Obstetrics and Gynecology* 2003; **189**: S17–23.

2. Souttar WP, Thomas H. Carcinoma of the vulva. In Souhami RL, Tannock I, Hohenberger P, Horiot JC (eds) *Oxford Textbook of Oncology* (2nd edn). Oxford: Oxford University Press, 2002, pp. 1897–911.

3. van der Velken K, Ansink A. Primary groin irradiation vs primary groin surgery for early vulvar cancer. *Cochrane Database of Systematic Reviews* 2000; **3**: CD002224.

4. Sebag-Montefiore DJ, McLean C, Arnott SJ, *et al.* Teatment of advanced carcinoma of the vulva with chemoradiotherapy – can exenterative surgery be avoided? *International Journal of Gynecological Cancer* 1994; **4**: 150–5.

5. Han SC, Kim DH, Higgins SA, Carcangui ML, Kacinski BM. Chemoradiation as primary or adjuvant treatment for locally advanced carcinoma of the vulva. *International Journal of Radiation Oncology, Biology, Physics* 2000; **47**: 1235–44.

6. Perez CA, Grigsby PW, Chao C, *et al.* Irradiation in carcinoma of the vulva: factors affecting outcome. *International Journal of Radiation Oncology, Biology, Physics* 1998; **42**: 335–44.

7. Bush M, Wagener B, Schaffer M, Duhmke E. Long term impact of post operative radiotherapy in carcinoma of the vulva FIGO I/II. *International Journal of Radiation Oncology, Biology, Physics* 2000; **48**: 213–18.

8. Chyle V, Zagars GK, Wheeler JA, Wharton JT, Declos L. Definitive radiotherapy for carcinoma of the vagina: outcome and prognostic factors. *International Journal of Radiation Oncology, Biology, Physics* 1996; **35**: 891–905.

9. Ali MM, Huang DT, Gopelrud DR, Howells R, Lu JD. Radiation alone for carcinoma of the vagina: variation in response related to the location of the primary tumour. *Cancer* 1996; **77**: 1934–9.

10. Mock U, Kucera H, Fellner C, Knocke TH, Potter R. High-dose-rate (HDR) brachytherapy with or without external beam radiotherapy in the treatment of primary vaginal carcinoma: long term results and side-effects. *International Journal of Radiation Oncology, Biology, Physics* 2003; **56**: 950–7.

11. Fine BA, Piver MS, McAuley M, Driscoll D. The curative potential of radiation therapy in the treatment of primary vaginal carcinoma. *American Journal of Clinical Oncology* 1996; **19**: 39–44.

12. Pingley S, Shrivastava SK, Sarin R, *et al*. Primary carcinoma of the vagina: Tata Memorial Hospital experience. *International Journal of Radiation Oncology, Biology, Physics* 2000; **46**: 101–8.

13. Lee WR, Marcus RB, Sombeck MD, *et al*. Radiotherapy alone for carcinoma of the vagina: the importance of overall treatment time. *International Journal of Radiation Oncology, Biology, Physics* 1994; **29**: 983–8.

Lymphomas

Mary K Gospodarowicz, Richard W Tsang, David C Hodgson, Peter Hoskin

Lymphomas are a heterogeneous group of malignancies with diverse pathological features, clinical courses, and outcomes. They represent approximately 3% of all malignancies worldwide with over 350 000 new cases of non-Hodgkin lymphoma (NHL) being diagnosed each year (http://www-dep.iarc.fr/) and 190 000 lymphoma deaths annually. In addition, over 67 000 cases of Hodgkin lymphoma (HL) occur each year and 29 000 die of disease. Unlike other malignancies, NHL is one of only few cancers increasing in frequency. The cause for this is unknown.

The current standard for pathology classification of lymphomas is that adopted by the World Health Organization (WHO)[1]. The WHO classification encompasses B-cell lymphomas, T-cell and NK-cell lymphomas, as well as HL and myeloproliferative malignancies. Even with modern immunohistochemical techniques, lymphoma disease entities are heterogeneous in outcome. Recently this heterogeneity has been partly explained with the use of DNA microarrays that identify variable gene expression profiles within morphologically similar subtypes[2–5]. There is an expectation that the lymphoma classification will evolve with better understanding of the genetic and molecular characteristics of the current lymphoma entities.

The most important prognostic factors in lymphomas include stage and the presence of lymphoma-associated systemic symptoms (B-symptoms), patient age, performance status, lactate dehydrogenase (LDH), and extent of extranodal involvement. These factors comprise the International Prognostic Index (IPI) initially developed for diffuse large cell lymphomas[6]. A similar system has been developed for follicular lymphomas, the follicular lymphoma IPI (FLIPI)[7,8]. The extent of disease is described using the Ann Arbor classification, based on the distribution and number of involved sites, the presence or absence of extranodal involvement, and B-symptoms including unexplained weight loss > 10% over the previous 6 months, unexplained fever > 38°C, or drenching night sweats[9]. Tumour bulk is not part of this system despite its prognostic importance. Modern pretreatment evaluation includes history and physical examination, complete blood count, renal and liver profiles, bone marrow biopsy, and cardiac evaluation. It is generally accepted that LDH levels reflect disease bulk. β_2-microglobulin predicts response to treatment and time to failure. Staging investigations include CT imaging of the neck, thorax, abdomen, and pelvis and gallium scan. Other imaging techniques are used to characterize the extent of disease in various areas of involvement[10,11]. For example, MRI helps to delineate extent of disease in CNS and bone. PET scans are becoming a part of staging and response assessment in NHL and HL[12].

A full description of the biology and management of all forms of lymphoma is beyond the scope of this chapter. Therefore, emphasis has been placed on the management of the most common forms of non-Hodgkin and Hodgkin lymphomas; particularly those presentations treated with RT or combined modality therapy.

10.1 Non-Hodgkin lymphoma

10.1.1 Follicular lymphoma

Follicular lymphoma is one of the more common types of non-Hodgkin lymphomas comprising about 20% of cases in Western countries. A significant proportion of patients (10–33%) present with localized disease (stage I–II)[13,14]. Follicular lymphoma is associated with a 3% per year risk of transformation to diffuse large B-cell lymphoma. To date, other than advanced staging, no definitive factors predictive of future transformation have been identified[7,15]. The follicular lymphoma is very responsive to therapy but relapse is common. Although the interval between relapses decreases with time, the median survival in most patients exceeds 7–10 years. The optimal treatment of disseminated or recurrent follicular lymphomas is controversial. The disease often follows an indolent course with prolonged survival. Therefore, patients with small disease burden and no symptoms are often managed with observation and treatment s deferred until progression or symptoms develop. Less than 50% of the patients in a selected series of observation needed treatment within 6 years from diagnosis[16]. To date there is no evidence that follicular lymphoma can be cured with systemic treatment, although the response rates exceed 60% in advanced stage disease. The treatment options vary from wait and watch strategy to aggressive chemotherapy followed by stem cell support.

Most patients with stage I–II disease receive involved-field radiation therapy (IFRT). Following IFRT alone patients with stage I–II disease have a 10-year disease-free survival of 40–64% and median survival of 13.8–15.3 years[8]. In the Princess Margaret series of 460 patients with stage I–II follicular lymphoma treated with IFRT alone, 98% achieved durable local control and the actuarial freedom from relapse rate at 25 years was 42%. There is little evidence supporting the use of adjuvant chemotherapy in patients with stage I–II follicular lymphoma treated with RT, although phase II trials combining radiation and chemotherapy has shown promising outcome. IFRT alone is still considered as standard therapy. Because over 50% of patients with stage I and II follicular lymphoma will relapse, combined modality therapy is being explored in prospective randomized trials.

The majority of patients with follicular lymphoma (70–90%) present with stage III or IV disease. For patients with recurrent or advanced presentations, treatment options vary from observation to aggressive chemotherapy.

- Observation alone is appropriate for asymptomatic patients, as active treatment has not been shown to have an impact on overall prognosis in this group.
- Single-agent chemotherapy with oral alkylating agents such as cyclophosphamide, chlorambucil with or without prednisone, or a purine analogue fludarabine, produce survival rates equivalent to combination regimens.

◆ Monoclonal antibody-based therapy has been shown to improve disease control for previously untreated patients and those with relapsed disease. Rituximab is a chimeric human–murine monoclonal IgG1 κ antibody that binds the transmembrane phosphoprotein CD20 present on benign and malignant B cells. Two radiolabelled anti-CD20 monoclonal antibodies, ^{131}I-tositumomab (Bexxar) and ^{90}Y-ibritumomab tiuxetan (Zevalin) have been developed for the treatment of indolent B-cell lymphomas[17,18]. [131]I is covalently linked to tositumomab whereas ^{90}Y is linked to the antibody via the tiuxetan molecule. In a phase III trial, Witzig et al. randomized 143 patients to receive either a single dose of ^{90}Y ibritumomab tiuxetan or rituximab alone. Statistically significant increases in overall response rate (80% vs. 56%, $P = 0.002$) and complete response rate (30% vs. 16%, $P = 0.04$) were seen in the radioimmunotherapy arm but non-significant increases were found in median duration of response (14.2 months vs. 12.1 months, $P = 0.6$) and time to progression (11.2 vs. 10.1 months)[19,20].

◆ Recent studies showed rituximab maintenance in patients with high tumour burden follicular lymphoma responding to rituximab chemotherapy improves progression-free survival[21].

◆ Radiotherapy has an important palliative role in the management of advanced follicular lymphoma where localized enlarging nodes masses are symptomatic in the context of stable disease elsewhere or chemotherapy resistance. Doses as low as 4 Gy may produce complete responses in sites of relapsed disease[22].

10.1.2 Marginal zone lymphoma of MALT type—mucosa associated lymphoid tissue (MALT) lymphoma

MALT lymphoma was first described by Isaacson and Wright in 1983[23] but is now accepted as a distinct disease entity. It comprises 7–9% of all lymphomas[1,24]. Between 60–70% of patients present with stage I or II disease[25–28]. Radiation therapy is often used either initially, or at a later time in the course of the disease; aggressive surgical management is not indicated, as radiation therapy has fewer side effects and preserves cosmesis and normal tissue function.

Although stage III and IV disease is currently incurable[29], the progress of disease is usually very indolent. MALT lymphomas may progress locally, spread to regional lymph nodes and/or distant sites, or transform to a diffuse large B-cell lymphoma. After biopsy or excision of a MALT lymphoma, further treatment is generally recommended for residual disease, since untreated low-grade MALT lymphoma may eventually lead to recurrence with possible later transformation to diffuse large B-cell lymphoma. The most common presenting site of MALT lymphoma is in the stomach. The pathobiology of gastric MALT is now well understood. Gastric infection with *Helicobacter pylori* (*H. pylori*) plays an important role in the aetiology of gastric MALT lymphoma. The molecular events following *H. pylori* infection lead to the development of low-grade gastric MALT lymphoma and transformed MALT lymphoma. Following antibiotic treatment of *H. pylori* infection, a complete regression of lymphoma occurs in 50–80% of cases[30–35]. Recommended anti-*H. pylori* therapy includes ranitidine or omeprazole, clarithromycin and amoxicillin for 7–10 days[36,37]. Regression of lymphoma generally takes 5–8 months, and may take up to 18 months. Patients with no *H. pylori* infection, with perigastric lymph node involvement, or with t(11; 18)(q21; q21) translocation usually do not respond to antibiotic therapy[38–40]. In such cases, RT is very effective in providing local disease control and cure[41]. The European Gastrointestinal Lymphoma Study Group (EGILS) has recently published consensus-based guidelines for investigation and treatment of gastric extranodal marginal B-cell MALT lymphoma[39]. Although consensus-based, the evidence of a number of recommendations is limited. In particular, the recommendations for endoscopic surveillance in patients in complete remission may be questioned.

Other common presenting sites of MALT lymphoma include orbital adnexae, skin, salivary glands, and thyroid[42]. The less common sites include bladder, cervix, breast, lung, dura, and rectum. Orbital lymphomas arise in superficial tissues including the conjunctiva and eyelids, or in deep tissues including the lachrymal gland and retrobulbar tissues. Treatment is directed to cure, while preserving vision and the integrity of the orbit. Extensive surgery should be avoided. A recent study reported an association with *Chlamydia psittaci* and observed that antibiotic therapy induced regression of lymphoma in two out of four patients[43]. The overall actuarial 10-year survival rates reported in the literature with radiotherapy are 75–80%. The risk of local failure is extremely low. Contralateral orbit involvement is common either in synchronous or metachronous fashion. Distant failure rates range from 20–50%, but as in other cases of indolent lymphoma, prolonged survival is observed. Although the literature describing the outcomes in less common presentations of MALT lymphoma is limited, available results indicate excellent local control rates with radiotherapy. The Princess Margaret Hospital experience showed 95% local control rate, 87% overall survival, and 68% disease-free survival at 10 years in 167 patients treated with involved field RT[42].

10.1.3 Diffuse large B-cell lymphoma

Stage I and II

◆ *Radiotherapy alone:* historically, patients with stage I–II diffuse large cell lymphoma were treated with RT alone which can produce long-term disease control in 40–50% of patients[44,45]. A review of the BNLI experience in 424 patients on prospective protocols treated with RT alone in early stage histologically aggressive NHL showed a 44% actuarial disease-free survival with 51% overall survival at 10 years[46]. Although these data showed that a significant proportion of patients with stage I and II aggressive lymphoma may be cured with RT alone, failure rates were high and the addition of chemotherapy has resulted in superior relapse-free and overall survival[47,48].

◆ *Combined modality chemotherapy and radiotherapy:* the commonly used chemotherapy regimen is CHOP (cyclophosphamide, doxorubicin, vincristine, and prednisone) in combination with rituximab[49]. With combined-modality therapy, excellent local control has been obtained with doses of 30–35 Gy delivered in 1.75- to 3-Gy fractions over 2–4 weeks. Currently, patients with stage I–II diffuse large cell lymphoma who complete prescribed therapy (without rituximab) achieve

75–80% progression-free rate and 80–90% survival rate at 5 years[50–53]. Recent phase II SWOG trial of three cycles of R-CHOP plus involved field radiotherapy showed 88% PFS and 92% OS at 4 years[54].

The question of the benefit of combined chemotherapy and radiation over chemotherapy alone has been addressed in several phase III trials. The Eastern Cooperative Oncology Group (ECOG) trial of eight cycles of CHOP versus eight cycles of CHOP followed by IFRT included 352 patients[55]. The study included stage I patients with bulky or extranodal disease, and stage II patients. Those achieving complete response with CHOP were randomized to consolidation radiation therapy (30 Gy) or observation. With an intent-to-treat analysis, the 6-year disease-free survival was 53% in the CHOP arm and 69% in the CHOP and RT arm (p = 0.05). The overall survival at 6 years was 67% for CHOP alone and 79% for CHOP and RT (p = 0.23)[55]. All patients with a partial response received RT to 40 Gy and their 6-year failure free survival was 63%, similar to the patients achieving a complete response to CHOP. This trial confirmed the benefit of IFRT who received CHOP chemotherapy in terms of disease control. It is unfortunate that no overall survival benefit was evident in this trial with inadequate power to detect clinically important (10%) survival differences.

The Southwest Oncology Group (SWOG) study excluded stage II patients with bulky disease (tumour mass ≥ 10 cm). SWOG compared three cycles of CHOP and IFRT to eight cycles of CHOP alone[56,57]. The radiation dose was 40 Gy with a boost to 50 Gy for partial responders. The 5-year progression-free survival rates were 77% for CHOP-RT versus 64% for CHOP alone (p = 0.03), and 82% and 72% for overall survival (p = 0.02)[57]. The adverse risk factors included stage II disease, age > 60 years, increased LDH and ECOG performance status of > 1. The high rate of systemic failure has raised concern that patients with adverse prognostic factors might have had inadequate chemotherapy in the CMT arm.

A longer course of chemotherapy followed by radiation therapy may be optimal in patients presenting with poor features, stage II disease, and rare or unfavourable extranodal sites (bone, extradural, testes, etc.).

More recently, phase III trials from the French cooperative group GELA have been reported[58–61]. They did not show a benefit for the CMT arm when compared to chemotherapy alone. Whether the addition of involved field RT is of value in patients who obtain complete response with regimens more intensive than CHOP awaits further testing.

The principles of therapy of localized diffuse large cell lymphoma and aggressive lymphomas of other histologies (anaplastic large cell lymphoma, peripheral T cell lymphoma) are similar. The choice of brief chemotherapy (three cycles) followed by RT, a full course of chemotherapy (six to eight cycles) followed by adjuvant RT, or chemotherapy alone is based chiefly on tumour bulk, presence of adverse prognostic factors such as B-symptoms and high LDH, and the anatomical extent of disease. Rituximab should be included for B-cell lymphomas with positive staining of the CD20 antigen. CNS prophylaxis with intrathecal methotrexate or cytosine arabinoside should be given to patients with testis lymphoma and disease involving parameningeal sites. These principles are most important in cases of primary extranodal lymphomas involving rare sites, where the available literature may not reflect the optimal approach.

Stage III and IV

The role of RT for advanced diffuse large B-cell lymphoma has been controversial[62]. In patients with stage III and IV lymphoma, who present with bulky disease, the relapse is most frequent in sites of bulky disease at presentation. Aviles et al., in a small randomized trial study of patients with stage III and IV diffuse large cell lymphoma, showed improved survival and local control for those who received adjuvant RT to sites of bulk disease following a complete response to chemotherapy[63]. The German High Grade Lymphoma Study Group continues to incorporate RT in their treatment programmes, for specific indications such as bulky disease, or extranodal sites, but most groups do not routinely advocate this approach.

10.1.4 **Extranodal presentations**

Extranodal involvement of lymphoma is observed in 24–48% of new lymphoma cases. Often presenting as localized disease, the sites are diverse and many have unique clinical and pathological characteristics, and distinct biological behaviour thereby requiring a different therapeutic approach as compared with nodal lymphomas of similar histology. Frequent presentation with localized disease (stage IE or IIE) is of special interest to the radiation oncologist. However, the majority of these diseases are very rare and the literature to guide their treatment is limited. To accumulate the evidence required for practice guidelines, the International Extranodal Lymphoma Study Group has originated a number of retrospective and prospective trials to clarify the management issues distinct to extranodal presentations. (http://www.ielsg.org/.) In general, the management follows that recommended for a specific histological disease entity. However, several distinct extranodal presentations deserve attention.

Primary central nervous system lymphoma (PCNSL)

The most common site of PCNSL is the brain(64,65). Primary spinal cord lymphoma is less common, and primary leptomeningeal lymphoma is rare, as is primary ocular lymphoma. Aggressive histology B-cell lymphomas are most common; T-cell lymphomas have been reported, but are exceedingly rare. Disease is characteristically multifocal or diffuse involving periventricular areas with easy access to the CSF. Meningeal seeding has been reported in 7.6–69% of cases[64,66,67]. The characteristic CT and MRI appearances include diffuse contrast enhancement of tumour, but a proportion of patients have non-enhancing lesions. Treatment options include:

◆ *Radiotherapy alone:* traditional treatment consisted of whole brain irradiation and corticosteroids. Primary brain lymphoma is extremely sensitive to RT and corticosteroids, producing a rapid symptomatic response. The recommended doses are 40–50 Gy to the whole brain in 1.8–2.0-Gy fractions. Performance status, age, multiple lesions, deep location within the brain, high LDH, and elevated CSF protein level are the most important prognostic factors. The median survival has been reported to be 12–18 months with 2-year and 5-year survival rates of 28% and 3–4% respectively[65,68,69].

◆ *Combined modality chemotherapy and radiotherapy:* treatment with intravenous methotrexate and high-dose cytosine arabinoside via intravenous and intrathecal routes used by DeAngelis in a cohort of 31 patients resulted in a median survival of 42.5 months, while in contemporary patients treated with RT alone the median survival was 21.7 months[70]. A subsequent report with high-dose methotrexate, procarbazine, and vincristine followed by radiation 45 Gy resulted in a median survival of 60 months[71]. These reports showed that PCNSL is a chemosensitive disease. However, methotrexate chemotherapy combined with radiation therapy results in long-term treatment related neurologic toxicity. This is especially a problem in the older patients. In a series from the Memorial Hospital, 11.5% of 1-year survivors developed dementia[70].

A recent phase III German trial randomized 551 patients treated with high-dose methotrexate to chemotherapy alone versus chemotherapy followed by high-dose (45 Gy) whole brain radiotherapy. There was no significant difference in the overall survival and in median time to progression. In spite of several study limitations, treatment with chemotherapy alone has become the preferred intervention for primary CNS lymphoma[72].

The new approach used at the Memorial Sloan-Kettering Cancer Center combines chemotherapy with low-dose radiotherapy (23.4 Gy). The results suggest high local control rates with markedly reduced toxicity[73].

Extradural lymphoma

Primary extradural lymphoma presents commonly with pain and progressive neurological deficit, or spinal cord compression. Histological diagnosis is imperative and biopsy is the first step in management. Historically, patients were treated with surgical decompression followed by RT to the affected area of the spine. RT alone resulted in excellent local disease control, but as with other localized aggressive histology lymphomas, was associated with a 40–50% distant failure. The use of combined modality therapy reduced failure rate and an improved survival. In the Princess Margaret Hospital experience the 5-year survival of patients treated with RT alone was 33% compared with 86% for those treated with combined modality therapy[74]. Although the traditional approach was to deliver RT before CT, this may not be the most optimal sequence. Eeles et al. documented that the use of CT followed by RT does not compromise neurological function as compared to that achieved when RT is followed by CT[75]. The RT target volume should be carefully defined using CT or MRI to include paraspinal tumour extension. A controversial aspect of the management of primary extradural lymphoma relates to the use of CNS prophylaxis. In the PMH experience isolated CNS relapse in patients treated without CNS prophylaxis was rare.

Cutaneous lymphomas

A large number of distinct lymphomas present with isolated cutaneous involvement[76], as detailed by the European Organisation for Research and Treatment of Cancer (EORTC)[77,78]. Primary lymphomas of the skin may be divided into three broad categories:

- *Cutaneous B-cell lymphomas* (25%) of indolent histologies with follicular lymphoma and marginal-zone (MALT), and the clinically aggressive: diffuse large B-cell lymphoma of the legs.

♦ *Cutaneous T-cell lymphoma of large cells* (10%) including pleomorphic, immunoblastic, anaplastic large cell lymphoma (CD30+), and rarely natural killer (NK) cell lymphoma (CD56+).

♦ *Cutaneous T-cell lymphomas with indolent clinical behaviour* (65%) including small lymphocyte type (mycosis fungoides/Sézary syndrome), lymphomatoid papulosis, and some CD30+ large cell types.

Cutaneous large B-cell lymphomas that occur in legs of elderly patients have aggressive behaviour with a 5-year survival of only 58%[77,79,80]. In contrast, primary cutaneous follicle centre lymphomas are usually confined to the head and neck region or the trunk, with 5-year survival of 97%. Infection with *Borrelia burgdorferi* has been implicated in the pathogenesis of indolent cutaneous B-cell lymphoma[81]. RT is a preferred treatment modality with very high local control rates of 85–100% and favourable survival[82]. Although many patients eventually relapse, frequently with new skin lesions, death from cutaneous B-cell lymphoma is rare.

Primary cutaneous large T-cell lymphomas are a heterogeneous group[1]. However, primary cutaneous anaplastic large cell lymphoma (C-ALCL), which is positive for CD30, is a specific clinical entity. In contrast to the systemic form of ALCL, it does not express ALK protein and does not possess the t(2;5) translocation. In patients with a solitary lesion or localized skin disease, RT is the treatment of choice. These lymphomas relapse frequently in the skin, but generally have a favourable prognosis. In a Dutch study of 79 patients, only 16% of patients have a systemic relapse of lymphoma 10 years after initial treatment[83]. Some patients demonstrated recurrent self-healing skin lesions. Lymphomatoid papulosis is a related condition that has an indolent course, and usually follow a benign course with spontaneously remitting disease[83]. Cytotoxic treatment is usually not necessary and life expectancy is not adversely affected, although infrequently the disease can progress to other types of T-cell lymphomas. Patients with T-cell lymphoma negative for CD30 have a worse prognosis, with an estimated 5-year survival of 15%[77].

Recent development of a novel anti-CD30 compound brentuximab (SGN-35) offers a promise to CD30+ lymphoma patients[84].

Gastric lymphoma

Prior to the modern era of endoscopy and conservative management of gastric disease, diffuse large cell lymphoma of the stomach was controlled with surgery and adjuvant RT or chemotherapy. More recently, a conservative approach with chemotherapy followed by RT has been used with similar results[85,86]. While in the past, RT to the stomach exposed left kidney to radiation doses in excess of tolerance levels, the currently available CT-based RT and IMRT planning techniques allow adequate protection of kidneys, liver, and other normal organs.

10.1.5 Radiotherapy in recurrent lymphomas

For patients with relapsed or chemotherapy-refractory large cell lymphoma, radiation therapy alone is rarely curative. However in selected cases, durable long-term control or even cure can be achieved with salvage RT[87,88]. For patients with chemotherapy-sensitive disease, phase II trials suggested benefit for high-dose therapy followed by haematopoietic stem cell transplantation[89–91]. A phase III trial tested the role of high-dose therapy in 109 patients. Those demonstrating a response to two cycles of DHAP (dexamethasone, cytarabine, cisplatinum) chemotherapy were randomized to four further courses of DHAP, versus a conditioning regimen of BEAC (carmustine, etoposide, cytarabine, cyclophosphamide) followed by ABMT. Radiation therapy was given as a protocol treatment for patients with tumour bulk > 5 cm at the time of relapse in the transplant arm (26 Gy in 20 fractions, twice daily), and in the conventional chemotherapy arm (35 Gy in 20 fractions, daily). The 5-year event-free and overall survival for the transplant arm were 46% and 53%, versus 12% (p = 0.001) and 32% (p = 0.038) for the conventional DHAP arm[92]. There was a trend favouring the RT patients with a lower relapse rate in the transplant group (8/22 RT patients relapsed versus 18/33 non-radiated patients relapsed, p = 0.19), and no obvious difference in the conventional chemotherapy group (10/12 RT patients relapsed, versus 35/42 non-radiated patients relapsed). Although this was not a trial designed to examine the role of radiation in the salvage setting, it lends support to the use of RT for bulky disease when incorporated into a salvage treatment plan that includes high dose therapy. The role of RT in bulky disease following partial response to salvage chemotherapy deserves further study in a randomized trial in patients undergoing haematopoietic stem cell transplant. Until such evidence is available, we recommend routine RT to sites of bulky disease, in sequence with salvage chemotherapy, and also RT to sites of incomplete response to chemotherapy. Moderate doses of 30–35 Gy should be the goal with individualization of the treatment plan in regards to the exact target volume (involved-field RT is preferred), and the timing of RT in relation to chemotherapy and transplant to facilitate the collection and harvesting of stem cells and minimize treatment-related toxicity. For example, if large RT fields are required, RT should be given after stem cell harvesting and preferably pretransplant. Thoracic RT that will include significant volumes of lung tissue may be better tolerated if given after transplantation[93]. In general, the principles of RT should be to treat the most likely site of relapse, or progressive disease. The decision to treat with RT is influenced by the distribution and location of the disease.

10.2 Hodgkin lymphoma

The management of Hodgkin lymphoma (HL) evolved over the past four decades from that based on the use of radiation therapy as the main curative modality to that relying on chemotherapy to cure the disease. Radiation therapy was shown to have curative potential in the 1950s. In 1960s MOPP (mechlorethamine, vincristine, procarbazine, and prednisone) chemotherapy was introduced. Although it was recognized by the 1970s that MOPP cured a proportion of patients with HL, because of its toxicity including infertility and leukaemia, its use was reserved for patients with advanced stage disease or cases where RT alone has limited success, such as in patients with large mediastinal mass or B-symptoms. Staging laparotomy was used in the 1960s and 1970s to identify patients with localized disease who had a high probability of cure with radiation alone. In the 1980s clinical prognostic factors were shown to be as good as staging laparotomy in identifying patients who required a combined modality approach, so-called 'risk adapted therapy'. In the late 1980s ABVD (doxorubicin, bleomycin, vinblastine, dacarbazine) chemotherapy was shown to be more effective than MOPP. The risk of treatment induced leukaemia and infertility were substantially reduced with ABVD. At the same time it became apparent that extended-field RT (EFRT) was associated with significant delayed morbidity, particularly second cancers and heart disease.

In stage I–II HL, the overall risk of death from the late effects of treatment now exceeds the risk of death from HL. To improve the overall survival, two main treatment directions have been pursued:

- Short chemotherapy with low-dose IFRT.
- Chemotherapy alone in low-risk presentation, or adaptive approach based on response assessed with functional imaging.

Currently, most patients with early stage HL are treated with ABVD chemotherapy and IFRT and those with advanced disease are treated with ABVD alone or, if adverse factors are present, BEACOPP (bleomycin, etoposide, doxotubicin, cyclophosphamide, vincristine, procarbazine, and prednisone) chemotherapy[94–96]. Many clinical trials are now in progress and will lead to further improvement in the management of HL.

Treatment decisions are based on the Ann Arbor classification supplemented by other prognostic factors and pathology. The anatomical extent of disease reflected by Ann Arbor stage is the most important prognostic factor. Other factors known to influence the outcome in patients with HL include histological type, tumour bulk, number of involved nodal regions and extranodal sites, age, gender, erythrocyte sedimentation rate (ESR), B-symptoms, anaemia, elevated white cell count, and lymphocytopenia. Clinical trial groups combine these factors to create prognostic categories, and treatment is limited or intensified accordingly. With individually tailored combined modality therapy, some of these variables have less predictive value, as relapse is uncommon. However, the presence of a large mediastinal mass, B-symptoms or unexplained anaemia is associated with poor prognosis. In advanced stage HL factors identified to have independent adverse effect on the outcome include: male sex, age ≥ 45 years, stage IV disease, haemoglobin < 105 g/L, serum albumin < 40 g/L, leucocyte count $\geq 15 \times 10^9$/L, and lymphocyte count $< 0.6 \times 10^9$/L (or $< 8\%$ of leucocyte count)[97].

10.2.1 Treatment of Stage I–II Hodgkin lymphoma

The management of early stage HL is largely dictated by the disease extent and prognostic factors including histology, age, ESR, and bulk of disease. The presence of one or more adverse prognostic factors categorizes patients according to risk of treatment failure. Traditionally, patients with low risk HL, that is nodular sclerosis histology, no B-symptoms, no large mediastinal mass, fewer than three nodal sites, age < 50 years, and normal ESR, received EFRT alone, while others received chemotherapy followed by RT. Although EFRT to a dose of 35–40 Gy gave excellent results with 5-year disease-free and overall survival of 85% and 95% in low-risk patients[98], the availability of effective chemotherapy and the concern about late effects of RT have prompted change in practice towards the use of combined chemotherapy and less extensive RT.

◆ *Combined modality chemotherapy and radiotherapy versus radiotherapy alone:* several trials have demonstrated that combined modality therapy improves disease-free survival, compared to EFRT alone. In the EORTC H7-F trial, patients with favourable disease were randomized to six cycles of EBVP (epirubicin, bleomycin, vinblastine, and prednisone), followed by IFRT, or EFRT alone[99]. Patients treated with combined modality therapy had better 5-year event-free survival (90% vs. 81%, $P = 0.002$), but equivalent overall survival (98% vs. 95%)[100]. The EORTC-GELA H8-F trial randomized favourable-risk patients to subtotal nodal irradiation or three cycles of MOPP/ABV (mechlorethamine, vincristine, procarbazine, prednisone/adriamycin, bleomycin, vinblastine) followed by 36–40 Gy IFRT. Patients receiving CMT had better 5-year event-free survival (98% vs. 74%), and 10-year overall survival (97% vs. 92%)[101]. The German Hodgkin Lymphoma Study Group (GHSG) HD7 trial randomized favourable stage I/IIA HL patients to EFRT or two cycles of ABVD (doxorubicin, bleomycin, vinblastine, dacarbazine) followed by EFRT. The addition of chemotherapy was associated with a significant improvement in 7-year freedom from failure (88% vs. 67%)[102]. These trials show that in favourable-risk early-stage HL, the addition of chemotherapy to RT improves disease-free survival, and may also result in a small improvement in overall survival. Additionally, a reduced RT volume was possible following chemotherapy (discussed later) and this may reduce the late toxicity of treatment and improve long-term results.

◆ *Combined modality chemotherapy radiotherapy vs. chemotherapy alone:* recently randomized trials compared chemotherapy alone with treatment that includes RT for patients with early stage disease. The National Cancer Institute of Canada HD6 trial randomized patients to ABVD alone or standard treatment including RT with or without two cycles of ABVD depending on risk factors. Although the 12-year freedom from progression was lower in the ABVD arm (87% vs. 92%), the 12-year overall survival was significantly superior (ABVD: 94%, vs. EFRT strategy: 87%, p = 0.04) due to a larger number of non-HL deaths (chiefly second malignancies) in the standard arm which contained EFRT [103]. A Children's Cancer Group study randomized patients to risk-adapted chemotherapy alone or chemotherapy followed by IFRT and found superior 3-year EFS in favourable risk stage I–II patients receiving IFRT (97% vs. 91%) with no difference in survival[104]. The EORTC H9-F trial randomized patients with favourable HL to six cycles of EBVP chemotherapy and 36 Gy IFRT, or 20 Gy IFRT, or no RT. However, the no-RT arm was closed after an interim analysis revealed an excess of relapses. Although the early results of these trials need to mature, they currently suggest that IFRT reduces the risk of relapse following chemotherapy. Current clinical trials (e.g. EORTC H10, UK RAPID trial, GHSG HD16 and HD17) use FDG-PET to select patients in CR after chemotherapy alone (or doing an interim FDG-PET scan after two cycles) to decide on omitting RT (risk-adapted therapy). The goal is to have equally good tumour control rates as combined modality therapy, yet avoiding RT exposure and hence eliminating the long-term risks of radiation such as secondary malignancies, heart disease, and thyroid gland dysfunction.

The current treatment recommendations are still based on known prognostic factors.

- *Favourable risk HL*: two to three cycles of ABVD are considered standard when used with IFRT[105], and recent data from GHSG (HD10 trial) which compared two versus four cycles of ABVD, and 30 Gy versus 20 Gy suggest that two cycles followed by IFRT 20 Gy is adequate[106], for those satisfying the low-risk criteria of the GHSG. The 5-year PFS was > 90% in all four treatment arms, with 5-year overall survival of 96–97%[106].

- *Unfavourable risk HL:* more chemotherapy is required. In the three-arm study conducted by the EORTC-GELA, H-8U, (six cycles of MOPP/ABV + IFRT vs. four cycles of MOPP/ABV + IFRT vs. four cycles of MOPP/ABV + STNI), no significant differences were observed in 4-year failure free (88%, 92%, 92%) or overall survival (90%, 94%, 93%)[107]. The GHSG HD11 trial in stage I–II disease and at least one adverse risk factor randomized patients to four cycles of ABVD, versus four cycles of BEACOPP followed by IFRT of either 20 Gy or 30 Gy[108]. The 5-year PFS and OS were 86% and 94.5% respectively, and the trial affirmed ABVD × 4 followed by 30 Gy as the standard approach. An alternate approach is the Stanford V regimen which achieves similar results compared with ABVD[109].

Radiation therapy volume and dose

Modern chemotherapy provides adequate control of microscopic disease, eliminating the need to irradiate clinically uninvolved nodal regions. A randomized trial of patients with stage I–II disease and one unfavourable risk factor compared four cycles of ABVD plus either STNI or IFRT. No significant differences in FFP (97% vs. 94%), or overall survival (93% vs. 94%) were observed[105]. Similarly, the GHSG HD8 trial randomized patients to EFRT or IFRT after four cycles of chemotherapy and found no difference in freedom from failure (EFRT 94%, IFRT 92%), or overall survival (97% for both arms)[110]. These excellent results, and the desire to reduce RT volumes, have made IFRT following chemotherapy the standard of treatment in stage I–II HL. Current clinical trials focus on further reducing RT volumes, based on the concept of not irradiating whole lymph node 'Kaplan' regions, but just covering the actual extent of the lymph nodes that were involved by disease initially. This is termed involved node radiation therapy (INRT)[111], and this approach has been adopted by the EORTC H10 study, and the GHSG HD17 study[112] testing the role of FDG-PET in treatment adaptation. INRT up to a 5-cm margin has been shown in one retrospective study to have similarly low local/regional nodal disease recurrence rates (< 5%)[113].

Available radiation dose–response data are derived from patients treated with RT alone and may not be valid in the CMT setting. The GHSG HD-4 trial found that among pathological stage I–IIA patients, 30 Gy to uninvolved regions was as effective as 40 Gy (all patients received 40 Gy to involved lymph node regions)[114]. There is little evidence of a dose-response above 35 Gy for clinically involved sites[115].

The optimal RT dose following ABVD chemotherapy appears to be 20 Gy for favourable stage I–II HL based on GHSG HD10 criteria[106], and preliminary results of the EORTC H9F study (unpublished), and 30 Gy for unfavourable or initially bulky presentations[108].

10.2.2 Hodgkin lymphoma Stage III and IV

Advanced HL is a heterogeneous disease. The outcomes vary from a 5-year survival of 80% in patients without adverse prognostic factors to a 50% 5-year survival in those with several adverse prognostic factors. The prognostic factors identified to have an adverse effect include male sex, age 45 years or more, stage IV disease, haemoglobin < 105 g/L, serum albumin < 40 g/L, leucocyte count 15×10^9/L or greater, and lymphocyte count < 15×10^9/L (or < 8% of leucocyte count)[97]. The standard treatment for stage III and IV HL is chemotherapy with the ABVD regimen. The 5-year freedom from treatment failure ranged from 69–87% in recent studies, with overall 5-year survival approximately 85–90%[116–118]. Low-risk patients may be managed with Stanford V regimen combined with IFRT(109). A more intensive regimen, BEACOPP, has shown an advantage over COPP-ABVD or ABVD for both event-free survival[96,119] and overall survival[119]. Current clinical trials focus on risk-adapted approaches by stratifying patients according to the number of these prognostic attributes present at diagnosis, and response as assessed by FDG-PET scans.

The role of radiation therapy in patients with advanced stage HL is limited but continues to generate controversy. A meta-analysis of prospective randomized trials has shown that consolidation radiation does not improve overall survival in patients with advanced stages HL[120], despite an 11% improvement in tumour control as compared to the same chemotherapy given alone. In most of these studies, MOPP or MOPP-like chemotherapy was used. A study from India indicated that radiation may improve outcome when added to six cycles of ABVD[121]. The 8-year overall survival was 100% with RT, versus 89% without RT[121]. The interpretation of this study is hampered by the inclusion of early stage patients and very young patients. The UK LY09 study comparing ABVD with two other multidrug regimens used consolidation RT for incompletely responding disease, and those with bulky disease. Radiated patients had better tumour control: 5-year PFS was 86% with RT, and 71% without[122]. Other studies have supported the findings of the meta-analysis showing no benefit from adding radiation to chemotherapy regimens similar to ABVD[118,119,123,124]. In a EORTC study, patients obtaining CR after four or six courses of MOPP/ABV were randomized to no further treatment or consolidation with involved field radiation after receiving a total of six and eight cycles, respectively[117]. The 5-year event-free survival rates were 84% and 79% (p = 0.35) and the overall survival rates were 91% and 85% (p = 0.07) in the non-radiated and the radiated groups, respectively[117]. Current approach would attempt to define the remission status more accurately with FDG-PET and consider the use of consolidation RT only in those with residual FDG activity. This has been the approach used by the GHSG HD15 trial, and results in the use of RT in only 11% of all patients.

In summary:

- In stage III and IV HL presenting without bulk disease (node > 10 cm or mediastinal mass more than one-third transthoracic width) there is no role for radiotherapy in patients who achieve complete remission with standard chemotherapy.

- Consolidation RT following definitive chemotherapy in stage III and IV disease should be considered for those with initial bulk disease, which show residual FDG activity.

10.2.3 Treatment for relapsed or refractory Hodgkin lymphoma

The management of relapsed HL depends on patient age, performance status, and extent of disease at relapse, and prior treatment. For patients with good performance status and chemotherapy-sensitive disease, randomized trials have demonstrated that high-dose chemotherapy with autologous stem-cell transplantation (ASCT) is more effective than aggressive conventional chemotherapy alone[125,126]. This treatment can also produce durable responses among those not responding to first-line chemotherapy[127,128]. In a salvage treatment strategy that includes ASCT, RT is important for nodal disease, particularly if localized, bulky sites > 5 cm, and residual disease despite salvage chemotherapy[129]. FDG-PET disease status prior to ASCT has prognostic value[130,131]. For patients with mediastinal recurrences, RT following ASCT produces less lung toxicity than pretransplant RT[93]. In patients with more advanced disease at relapse, RT is not likely to influence survival. In patients who are not candidates for ASCT, RT may be considered with or without chemotherapy[132,133]. Several studies demonstrated durable responses to RT among patients who relapsed following chemotherapy, although the intensity of the initial chemotherapy in these studies was not as great as in current practice[134]. The prognosis in patients with relapsed HL is related to the disease-free interval (better prognosis if the disease-free duration was > 12 months), extent of disease at relapse, performance status, presence of anaemia, and response to salvage chemotherapy[120,128,135–137]. Some patients with relapsed HL may survive with disease for prolonged periods of time, and this should be considered when RT is given for palliation.

10.3 Radiation therapy techniques in lymphomas

The technical aspects of treatment planning for lymphomas are highly dependent on the location and extent of the target volume. In most cases, established techniques for cancer of other anatomical sites can be adapted to lymphomas. In general, planning involves the appropriate use of immobilization devices, simulation, CT-assisted tumour localization and planning, localization of adjacent normal tissues, custom-designed beam-modifying devices, and computerized calculation of isodose distributions. The goal of these steps is to achieve dose uniformity in the target volume while minimizing RT dose to normal tissues.

The literature describes the extent of RT in terms of IFRT, EFRT, and total lymphoid (or nodal) radiation therapy (TLI or TNI). Subtotal nodal irradiation (STNI) is a specific type of EFRT used commonly in supradiaphragmatic HL when RT alone is prescribed. The use of these terms in the literature is variable.

◆ INRT defines RT to the clinically involved lymph node(s) only[111], without intentional coverage of the uninvolved lymph nodes in the same or adjacent nodal region.

◆ IFRT defines RT to the clinically involved lymph node region(s), with or without coverage of the first echelon adjacent lymph node region uninvolved by disease. Lymph node regions are defined by Kaplan and Rosenberg and have been used for over 30 years[138] (Figs 10.1–10.3).

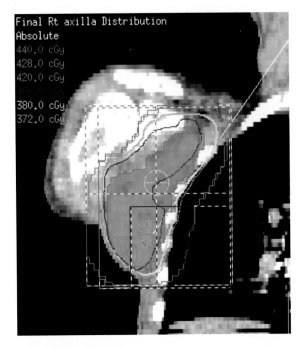

Fig. 10.1 Involved field plan for right axilla.

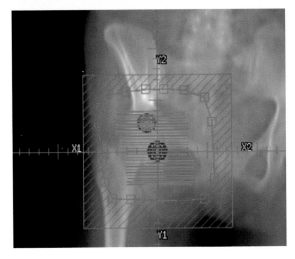

Fig. 10.2 Involved field RT postchemotherapy—right groin.

◆ EFRT defines RT to include the adjacent first echelon and the second echelon adjacent lymph node regions.

◆ TLI implies treatment to all the major lymphoid regions, with the mantle and the inverted Y fields, with or without Waldeyer's ring fields. It was used mainly in the management of HL. The mantle field includes all major lymph node regions above the diaphragm including cervical, supraclavicular, axillary, hilar, and mediastinal lymph nodes treated in contiguity.

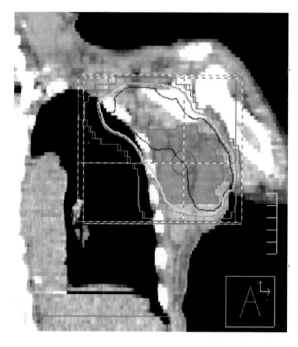

Fig. 10.3 Example of the plan for involved field RT to left axillary lymph nodes.

- STNI includes the mantle field, spleen, and abdominal paraaortic lymph nodes.

CT-based planning allows accurate delineation of target, and normal tissue shielding with a 3D perspective. The planning of radiation includes a determination of GTV, CTV, and PTV.

- GTV is important in patients treated with RT alone and those with residual disease after chemotherapy.
- CTV includes the pretreatment tumour and what is considered to be the relevant tissues that contain microscopic disease (Fig. 10.4).
- PTV includes an appropriate margin to the CTV to account for movement and set-up variation during treatment delivery. The treatment commonly consists of two fields, applied as opposed anterior and posterior fields[99,139–142].

In lymphomas managed with radiation alone, involved field usually covers the GTV and the CTV, which may vary with clinical presentation. In nodal lymphomas, the available data is based on including the entire nodal region in the CTV but in modern practice is often restricted to involved lymph nodes in the CTV particularly when the treatment is delivered as consolidation therapy after complete response to chemotherapy. In stage I extranodal disease, CTV is limited to a margin around gross disease. Exceptions to this include MALT lymphomas, known for multifocal involvement where the whole organ or area at risk is treated:

- Gastric MALT, the CTV includes the entire stomach (Fig. 10.5).
- Orbital MALT lymphoma, the CTV includes the whole orbit.

In many clinical situations, applicable for both NHL and HL, chemotherapy is used before RT. Therefore at the time of RT planning, no GTV is identifiable and the main

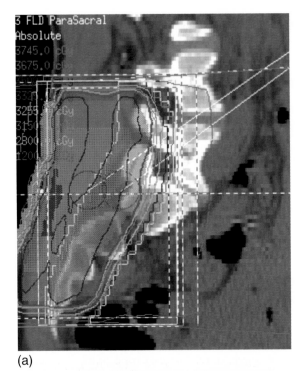

(a)

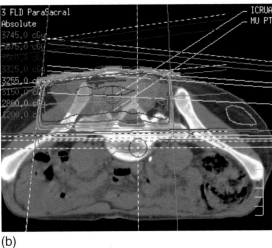

(b)

Fig. 10.4 Example of involved field plan for diffuse large cell lymphoma involving buttock—postchemotherapy.

parameter to determine is the CTV. In such cases, the target volume should cover the initial extent of gross disease. For nodal regions, partial coverage of an 'involved' nodal region is acceptable provided the initial extent of gross disease before chemotherapy is adequately covered. An example of this is the modified mantle technique for stage II HL involving the supraclavicular and superior mediastinal lymph nodes.

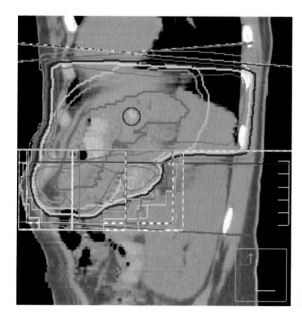

Fig. 10.5 Example of the involved field plan for gastric MALT lymphoma.

The uninvolved upper cervical lymph nodes above the larynx, and the inferior medi-astinal tissues, which include large volumes of the cardiac silhouette, may be safely excluded from the CTV (Figs 10.6 and 10.7). It is therefore paramount to have accu-rate determination and documentation of the initial anatomic extent of disease, prior to starting chemotherapy. With respect to the necessity of covering the prechemo-therapy extent of disease, in general, if the disease displaced normal extranodal tissues without infiltration into it, the radiation fields need not cover the entire initial loca-tion of the tumour volume. An example is a bulky mediastinal mass that did not infiltrate lung tissue and is smaller in size following chemotherapy, where the RT plan need only cover the postchemotherapy width of mediastinum and not include healthy lung. If, however, the disease was infiltrative initially into adjacent normal tissue, regression of the tumour mass may leave microscopic residual disease in the infiltrated tissue and consideration must be given to adequately cover initial disease extent. An example is lymphoma involving the bone, where the RT volume following a good response to chemotherapy should generally include the prechemotherapy extent of disease.

10.3.1 Planning techniques

Lymph node irradiation

The traditional principle of lymph node irradiation was to cover whole node regions rather than individual lymph node masses. This is now being challenged by the adop-tion of CT-based 3D planning and using analogous nomenclature to other organ sites, and the preliminary data with successful application of INRT in combined modality therapy for early stage HL[111,113]. The classic mantle and inverted Y fields through

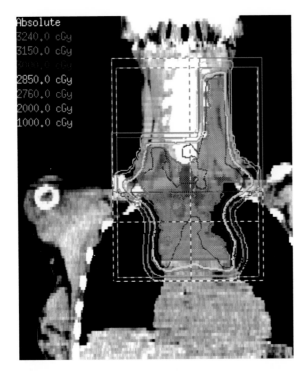

Fig. 10.6 Hodgkin lymphoma postchemotherapy. Involved field plan for low neck and superior mediastinum.

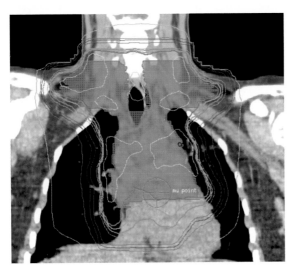

Fig. 10.7 Example of involved field plan for superior mediastinum and bilateral supraclavicular fossae.

which TNI was delivered are rarely employed today when INRT or IFRT are used. The basis of the involved field, however, is often related back to a portion of the classic TNI field. The extent of lymph node region to be treated is defined by the prechemotherapy disease extent but the volume is modified to cover postchemotherapy bulk, in particular in the mediastinal and para-aortic lymph nodes to minimize lung, bowel, and kidney within the irradiated volume. CT planning is generally used to define clinical target volume. However, where not available, marker wire is used to define the clinical disease extent at time of planning.

Supradiaphragmatic lymph node irradiation

Patient position and immobilization

- ◆ Supine: hands by side or where axilla is to be irradiated abducted hands on hips or supported on an arm board or T bar as for breast irradiation

- ◆ Immobilization is required for neck irradiation using an appropriate shell; for the mediastinum immobilization is not usual but some centres advocate the use of a vacuum body bag to avoid lateral rotation. CTV is determined on planning CT.

Field arrangement

Anterior and posterior opposed fields shaped to avoid critical structures, e.g. larynx, lungs using individualized blocks or MLC; spinal cord shielding may be used in the neck where there are no mid line lymph nodes but should be avoided in the mediastinum where lymph nodes are likely to be shielded.

Dosimetry

Prescribed dose of radiation should be delivered to the target volume securing no more than ± 5% dose variation within the target. For planning of mediastinal radiation, for accurate definition of the CTV and for dosimetry to account for lung attenuation CT dosimetry is recommended to define tumour dose and critical OAR doses, in particular lung, heart and spinal cord.

Dose prescription

- ◆ Non-Hodgkin lymphoma:
  - 30–35 Gy in 15–20 fractions is considered adequate for consolidation after complete response to chemotherapy. In patients with partial response then higher doses of up to 35–40 Gy in 20 fractions may be prescribed. In follicular lymphoma dose usually ranges between 25–30 Gy[143].
  - Palliative radiotherapy in lymphomas often achieves local tumour control and therefore similar doses are recommended as for radical RT.
  - Follicular lymphomas are more sensitive to radiation. For palliation, doses as low as 4 Gy in two fractions in 3 days have been shown to be effective and result in local disease control.

- ◆ Hodgkin lymphoma:
  - Macroscopic disease: 30–35-Gy midplane dose in 15–20 daily fractions.
  - Microscopic disease: 30-Gy midplane dose in 15–20 daily fractions.
  - Postchemotherapy complete remission: 30-Gy-midplane dose in 15–20 daily fractions. For low-risk patients satisfying GHSG HD10 criteria: 20 Gy in 10–12 daily fractions after ABVD × two cycles.

Implementation and verification

Where large fields to mediastinum and neck are used, careful set-up is required to ensure reproducibility from day to day: this is assisted by using more than one permanent set-up mark (tattoo), typically two along the central axis of the field, one in field centre, and two lateral tattoos are defined in addition to usual skin marks.

Verification with megavoltage films or preferably EPIDs is mandatory.

Infradiaphragmatic lymph node irradiation

Patient position and immobilization
+ Supine: hands by side.
+ Immobilization is not usual but some centres advocate the use of a vacuum body bag to avoid lateral rotation. CTV is determined on planning CT.

Localization
In the abdomen critical structures include the kidneys, liver, and spinal cord; these are best imaged using CT planning; if this is not available then orthovoltage simulation with intravenous contrast to define the renal outline is mandatory. The spleen is also best localized with CT imaging.

Field arrangement
Anterior and posterior opposed fields shaped to avoid critical structures, e.g. kidneys and bowel, using individualized blocks or MLC; spinal cord shielding should be avoided in the para-aortic region where nodes are likely to be shielded.

Dosimetry
In most settings unless CT-based planning has been used dose distributions are not defined. However, in the abdomen for accurate definition of the CTV to define critical OAR doses, in particular kidney and spinal cord, CT is recommended; similarly for splenic fields where the proximity of the left kidney may require compromises to the renal dose to ensure complete coverage of the spleen. In this setting, it is essential to have performed an isotope renogram to be certain that the contralateral kidney has adequate function if the left kidney is sacrificed.

Extranodal sites

Orbit
Traditionally orbital lesions are invariably controlled with low to moderate doses of 20–30 Gy in 10–20 daily fractions, with local control rate in excess of 95% for indolent histologies such as MALT or occasionally follicular lymphoma[42,144–146].

Head and neck
Waldeyers ring: previously, the involvement of any component of Waldeyers ring (lingual tonsil, oropharyngeal tonsil, nasopharynx) led to treatment of the entire ring, however, currently, this is not practiced. Certainly in combined modality setting only the involved site is treated. CT imaging and planning should be used for greater accuracy, and IMRT is preferred to spare the salivary glands to minimize xerostomia.

Stomach
RT is given to the whole stomach. This should be localized using CT planning which enables not only accurate definition of the stomach but also the critical OARs, in particular the kidneys. To optimize treatment reproducibility, stomach distension should be avoided and the patient is best treated in a near-fasting state in the morning[147].

Current experience with MALT low-grade lymphomas indicates that 30 Gy in 15–20 daily fractions over 3–4 weeks results in local control in almost 100% of cases[42,148].

High-grade or transformed MALT lymphoma is managed as diffuse large B-cell lymphoma with chemotherapy and RT (30–35 Gy).

Central nervous system (primary central nervous system lymphoma)
Patient should be treated supine, immobilized in a customized shell.

Whole brain irradiation using lateral fields with standard baseline from external auditory meatus to supraorbital ridge adjusted to ensure coverage of middle cranial fossa and olfactory groove.

Dose of 30–40 Gy in 15–20 daily 2-Gy fractions prescribed to the intersection point.

Verification using EPID or lateral linear accelerator films; *in vivo* dosimetry to measure lens doses also recommended.

If craniospinal radiotherapy indicated a standard technique as described in Chapter 15 should be used to deliver a dose of 30 Gy in 15 fractions.

Skin
Standard techniques to treat localized skin lymphoma using a direct electron or superficial X-ray beam should be used as described in Chapter 13.

For widespread mycosis fungoides whole body electron treatment may be considered. This is usually delivered in a few specialized centres offering this treatment.

Total body electron treatment

Treatment volume
There is no formal definition of GTV, CTV, or PTV but in practice the technique aims to treat the entire whole-body skin surface to a depth of 3–5 mm.

Patient position
Most techniques treat the patient in four to six different standing positions rotating at 60° intervals to enable coverage of the entire body surface. Posture may be adjusted to facilitate beam access to skin folds, particularly under the arms and between the legs. Treatment is usually delivered in cycles, for example, anterior and two posterior oblique fields on day 1 and posterior and two anterior oblique fields on day 2[149]. An alternative approach uses a rotational technique in which the patient stands on a rotating platform. This achieves greater dose homogeneity than multiple field techniques[150].

Shielding of the eyes, nail beds, lips, and hands is used with lead goggles and mitts.

Field arrangement and dosimetry
A number of individualized techniques have been described by different centres across the world. These may be broadly categorized as follows:

◆ Matching dual field system, which is perhaps the most commonly used today. A standard linear accelerator beam is used with 6-MeV electrons at a treating distance of 3.5–4 metres. One beam is set up with the central axis focused towards the head and the other towards the feet; typically each is angled 15–20° above or below the horizontal as shown in Fig. 10.8. A tissue equivalent sheet or 'scatter screen' is placed in front of the patient to reduce the effective beam energy to 4MeV.

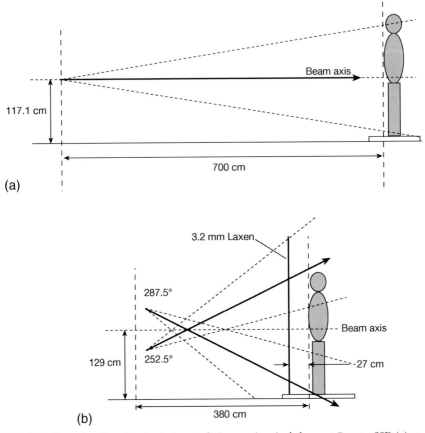

Fig. 10.8 Examples of total body electron techniques using single beam at 7 metre SSD (a) or dual-field technique at 3.8 metre SSD (b) as used at Stanford. (Reprinted from the *International Journal of Radiation Oncology, Biology, Physics*, Volume **59**, Issue 3, Zhe Chen *et al.* Matching the dosimetry characteristics of a dual-field Stanford technique to a customized single-field Stanford technique for total skin electron therapy, pp. 872–85 (2004) with permission from Elsevier.)

Some techniques employed a reflector at the vertex to increase the dose uniformity at the vertex[151].

- Single field technique which requires a much longer source-to-surface distance (SSD) of 6–7 metres and is therefore often not practicable in a modern linear accelerator room. When used it is simpler than the dual beam system; 6-MeV beams are usually employed with a screen as described previously.

- A moving beta particle emitting source, e.g. strontium, or a fixed narrow electron beam below which the patient moves across the beam have also been described.

Surface TLD measurements are taken to measure the dose distribution; this is often quite inhomogeneous, particularly with fixed beam techniques, with variations between 25% and 140% described[152], typically lowest at tangential surfaces such as the top of the head, shoulders, and perineum. Low-dose areas, in particular the top of the head, inframammary folds, ventral penile surface, perineum, and soles of feet may be topped up with local electron or superficial photon fields; this may also be required for thicker areas of disease[153].

X-ray contamination giving a whole body X-ray dose is a concern with these techniques. This is low with modern dual field techniques using a modern linear accelerator at around 1%.

Dose prescription
Common prescriptions include either:

♦ 24 Gy in eight fractions treating three times per week.

♦ 30 Gy in 15 fractions treating four or five times per week.

10.4 **Toxicity after radiotherapy for lymphoma**

10.4.1 **Acute toxicity**

♦ Mediastinal RT may produce fatigue, dysphagia, and a gradual decline in blood counts. RT to the upper neck may produce xerostomia if the parotid glands are included in the treatment field.

♦ Radiation pneumonitis, which may occur 1–4 months following treatment, is characterized by dry cough, shortness of breath, and fatigue. A chest X-ray will reveal pulmonary infiltrates in the characteristic distribution of the irradiated lung. The incidence is generally < 5%, although it increases with larger volumes of irradiated lung.

♦ Herpes zoster may occur within 1 year of treatment in approximately 15% of patients, more commonly in those receiving combined modality therapy.

♦ Lhermitte's sign may occur 1–3 months following RT and generally resolves without treatment in weeks to months.

10.4.2 **Late toxicity**

Radiation to thoracic structures

The mantle field, used to treat the bilateral neck including supraclavicular, axillary, and mediastinal lymph nodes, is now rarely used. Because of the irregular field arrangement and contour variation, beam modifiers such as compensators or attenuators together with custom shielding blocks are mandatory for the protection of lung and cardiac tissues.

♦ *Pulmonary:* the TD5/5 for 20 fractions of whole lung radiation was 26.5 Gy (TD50 30.5 Gy) from paediatric series of patients irradiated for Wilms' tumour[154]. Partial lung irradiation with fractionated doses have been studied by Mah et al. using radiographic changes as the endpoint[155]. A steep dose–response relationship was found, where the TD5 was 24.7 Gy, increasing to 33.9 Gy for TD50, and to 43.5 Gy for TD95%, normalized to 15 fractions. Many chemotherapy drugs may potentiate the effects of RT on lung tissue, e.g. bleomycin, cyclophosphamide, and doxorubicin[156]. Careful treatment planning, use of optimal dose-fractionation parameters, and the allowance of an interval of 4 weeks between CT and RT should minimize the incidence of symptomatic radiation pneumonitis. Information on the cardiac tolerance to irradiation has been largely based on patients treated for HL.

◆ *Cardiac:* pericarditis is the most commonly reported subacute effect. Late effects may include coronary artery disease, valvular disease, and conduction defects. Following now outdated treatment, the risk of developing symptomatic coronary artery disease is 6–10% at 10 years and 10–20% at 20 years. These rates represent relative risks for surgical intervention or hospitalization of 1.5–2.5. The actuarial risks of death from cardiac ischaemia appear to be 2–6% at 10 years and 10–12% at 15–20 years. As for lung tissue, careful treatment planning with maximum cardiac shielding (subcarinal area and the left ventricle), and the avoidance of dose per fraction over 2 Gy should minimize the risk of cardiac complications[157]. The potential cardiotoxic effects of doxorubicin may also be additive to the long-term damaging effects of radiation on the heart.

Pelvic radiation and effects on the reproductive system

Since radiation for lymphomas frequently involve large fields, treatment planning must address effects of scatter as well as direct radiation to the genital organs.

◆ *Female:* doses of 20–24 Gy will invariably produce ovarian ablation with loss of hormonal function resulting in menopause and sterility[158]. Therefore, treatment planning for pelvic lymph node RT should spare one ovary if possible, with ultrasound or CT guidance to locate the ovary. When bilateral pelvic irradiation is inevitable, the use of surgical oophoropexy with transposition of one ovary to the lateral abdomen and marking it with surgical clips to ensure exclusion from the radiation fields should be considered. However, this procedure alone can produce infertility because of interference with the vascular supply of the transposed ovary.

◆ *Male:* direct radiation to the testes generally occurs in the combined modality therapy of primary testis lymphoma, or for testicular relapse of lymphoma where chemotherapy may not have full effect for this sanctuary site. Since the germinal epithelium of the testes is extremely radiosensitive, where doses as low as 15 cGy can produce transient oligospermia, and doses of 4–6 Gy can result in permanent azospermia, infertility is always a consequence of scrotal irradiation[159]. However, Leydig cell function with testosterone production can be preserved after doses of 30–35 Gy[159], although dysfunction as manifested in a rise in LH and FSH levels can be seen even for lower doses of 5–6 Gy. Based on a review of the literature, Izard concluded that approximately 50% of males receiving 14 Gy in fractionated doses will have an abnormal LH, while 33 Gy is required to see an abnormal testosterone level in 50%[160]. Indirect radiation to the testes is also a special consideration in the planning of fields close to the scrotum, particular if the field size is large. The scatter dose is mainly a function of distance from the field edge, where gonadal doses of < 5% are usually achieved with a distance from the field edge of 10 cm or over, for a field size of 25 cm.

Bone marrow

With the more frequent use of combined modality therapy and bone marrow trans-plantation (BMT) in the treatment of lymphomas, radiation is frequently required in patients who had extensive previous chemotherapy, or who had undergone BMT. Haematopoietic reserve may be significantly compromised in these patients, increas-ing the risk of radiation-induced myelosuppression. This is particularly a problem where the treatment field encompasses a significant proportion of the bone marrow. The use of granulocyte-colony stimulating factor (G-CSF) has been shown to amelio-rate neutropenia[161–163]. However, G-CSF does not correct thrombocytopenia, which is likely to be a limiting factor once neutropenia is reversed. While platelet transfu-sions can be performed for platelet counts of $< 10–20 \times 10^9$/L, or for active bleeding, there is a risk of rendering patients platelet-transfusion dependent for prolonged peri-ods of time particularly when RT is given in the post-BMT setting. It is possible that platelet growth factors will become clinically available within the next few years and their use may avert radiation-induced thrombocytopenia. Anaemia, if present, is usually not a dose-limiting problem and red cell transfusions may be given as required.

10.5 Second malignancies

An increased risk of a second malignancy has been described among HL survivors. In general, there are approximately 55–65 excess malignancies per 10 000 person-years of follow-up among patients treated for HL, and a 20-year cumulative incidence of second malignancy of approximately 15–20%[164,165]. The excess risk of a second solid tumour is generally 40–50 per 10 000 person-years of follow-up. Secondary breast cancers are more common among females treated in adolescence; however, the risk decreases substantially for women treated over the age of 30 years[165–167]. The risk of lung cancer is very high among smokers, particularly those receiving thoracic RT[168]. MOPP chemotherapy was associated with second lung cancers, but the association between ABVD chemotherapy and second solid cancers has not been determined. The incidence of leukaemia following modern chemotherapy is 1–3% and occurs within 10 years of treatment.

References

1. Jaffe ES, Harris NL, Stein H, *et al.* (eds). *Pathology and Genetics of Tumours of Haematopoietic and Lymphoid tissues.* Lyon: IARC Press; 2001.

2. Alizadeh AA, Eisen MB, Davis RE, *et al.* Distinct types of diffuse large B-cell lymphoma identified by gene expression profiling. *Nature* 2000; **403**: 503–11.

3. Dave SS, Wright G, Tan B, *et al.* Prediction of survival in follicular lymphoma based on molecular features of tumor-infiltrating immune cells. *New England Journal of Medicine* 2004; **351**: 2159–69.

4. Rosenwald A, Wright G, Chan WC, *et al.* The use of molecular profiling to predict survival after chemotherapy for diffuse large-B-cell lymphoma. *New England Journal of Medicine* 2002; **346**: 1937–47.

5. Shipp MA, Ross KN, Tamayo P, *et al.* Diffuse large B-cell lymphoma outcome prediction by gene-expression profiling and supervised machine learning. *Nature Medicine* 2002; **8**: 68–74.

6. Shipp MA. Prognostic factors in aggressive non-Hodgkin's lymphoma: who has 'high-risk' disease? *Blood* 1994; **83**: 1165–73.

7. Relander T, Johnson NA, Farinha P, *et al.* Prognostic factors in follicular lymphoma. *Journal of Clinical Oncology* 2010; **28**: 2902–13.

8. Solal-Celigny P, Roy P, Colombat P, *et al.* Follicular lymphoma international prognostic index. *Blood* 2004; **104**: 1258–65.

9. Cancer IUa. *TNM Classification of Malignant Tumours* (6th edn). New York: Wiley-Liss; 2002.

10. Thomas AG, Vaidhyanath R, Kirke R, *et al.* Extranodal lymphoma from head to toe: part 2, the trunk and extremities. *AJR. American Journal of Roentgenology* 2011; **197**: 357–64.

11. Thomas AG, Vaidhyanath R, Kirke R, *et al.* Extranodal lymphoma from head to toe: part 1, the head and spine. *AJR. American Journal of Roentgenology* 2011; **197**: 350–6.

12. Hutchings M, Eigtved AI, Specht L. FDG-PET in the clinical management of Hodgkin lymphoma. *Critical Reviews in Oncology/Hematology* 2004; **52**: 19–32.

13. Armitage JO, Weisenburger DD. New approach to classifying non-Hodgkin's lymphomas: clinical features of the major histologic subtypes. Non-Hodgkin's Lymphoma Classification Project. *Journal of Clinical Oncology* 1998; **16**: 2780–95.

14. Freedman A. Follicular lymphoma: 2011 update on diagnosis and management. *American Journal of Hematology* 2011; **86**: 768–75.

15. Al-Tourah AJ, Gill KK, Chhanabhai M, *et al.* Population-based analysis of incidence and outcome of transformed non-Hodgkin's lymphoma. *Journal of Clinical Oncology* 2008; **26**: 5165–9.

16. Advani R, Rosenberg SA, Horning SJ. Stage I and II follicular non-Hodgkin's lymphoma: long-term follow-up of no initial therapy. *Journal of Clinical Oncology* 2004; **22**: 1454–9.

17. Cheson BD. What is new in lymphoma? *CA: A Cancer Journal for Clinicians* 2004; **54**: 260–72.

18. Horning SJ, Younes A, Jain V, *et al.* Efficacy and safety of tositumomab and iodine-131 tositumomab (Bexxar) in B-cell lymphoma, progressive after rituximab. *Journal of Clinical Oncology* 2005; **23**: 712–19.

19. Witzig TE, Gordon LI, Cabanillas F, *et al.* Randomized controlled trial of yttrium-90-labeled ibritumomab tiuxetan radioimmunotherapy versus rituximab immunotherapy for patients with relapsed or refractory low-grade, follicular, or transformed B-cell non-Hodgkin's lymphoma. *Journal of Clinical Oncology* 2002; **20**: 2453–63.

20. Witzig TE, Molina A, Gordon LI, *et al.* Long-term responses in patients with recurring or refractory B-cell non-Hodgkin lymphoma treated with yttrium 90 ibritumomab tiuxetan. *Cancer* 2007; **109**: 1804–10.

21. Salles G, Seymour JF, Offner F, *et al.* Rituximab maintenance for 2 years in patients with high tumour burden follicular lymphoma responding to rituximab plus chemotherapy (PRIMA): a phase 3, randomised controlled trial. *Lancet* 2011; **377**: 42–51.

22. Haas RL, Poortmans P, de Jong D, *et al.* Effective palliation by low dose local radiotherapy for recurrent and/or chemotherapy refractory non-follicular lymphoma patients. *European Journal of Cancer* 2005; **41**: 1724–30.

23. Isaacson P, Wright DH. Malignant lymphoma of mucosa-associated lymphoid tissue. A distinctive type of B-cell lymphoma. *Cancer* 1983; **52**: 1410–16.

24. Harris N, Jaffe ES, Stein H, *et al.* A revised European-American classification of lymphoid neoplasms: A proposal from the International Lymphoma Study Group. *Blood* 1994; **84**: 1361–92.

25. Raderer M, Vorbeck F, Formanek M, *et al*. Importance of extensive staging in patients with mucosa-associated lymphoid tissue (MALT)-type lymphoma. *Br J Cancer* 2000; **83**: 454–7.

26. Thieblemont C, Bastion Y, Berger F, *et al*. Mucosa-associated lymphoid tissue gastrointestinal and nongastrointestinal lymphoma behavior: analysis of 108 patients. *Journal of Clinical Oncology* 1997; **15**: 1624–30.

27. Thieblemont C, Berger F, Dumontet C, *et al*. Mucosa-associated lymphoid tissue lymphoma is a disseminated disease in one third of 158 patients analyzed [published erratum appears in Blood 2000; 95(8): 2481]. *Blood* 2000; **95**: 802–6.

28. Zinzani PL, Magagnoli M, Galieni P, *et al*. Nongastrointestinal low-grade mucosa-associated lymphoid tissue lymphoma: Analysis of 75 patients. *Journal of Clinical Oncology* 1999; **17**: 1254–8.

29. Fisher RI, Dahlberg S, Nathwani BN, *et al*. A clinical analysis of two indolent lymphoma entities: mantle cell lymphoma and marginal zone lymphoma (including the mucosa-associated lymphoid tissue and monocytoid B-cell subcategories): a Southwest Oncology Group study. *Blood* 1995; **85**: 1075–82.

30. Bayerdorffer E, Neubauer A, Rudolph B, *et al*. Regression of primary gastric lymphoma of mucosa-associated lymphoid tissue type after cure of Helicobacter pylori infection. MALT Lymphoma Study Group. *Lancet* 1995; **345**: 1591–4.

31. Bertoni F, Conconi A, Capella C, *et al*. Molecular follow-up in gastric mucosa-associated lymphoid tissue lymphomas: early analysis of the LY03 cooperative trial. *Blood* 2002; **99**: 2541–4.

32. Montalban C, Manzanal A, Boixeda D, *et al*. Helicobacter pylori eradication for the treatment of low-grade gastric MALT lymphoma: follow-up together with sequential molecular studies. *Annals of Oncology* 1997; **8**(Suppl 2): 37–9.

33. Roggero E, Zucca E, Pinotti G, *et al*. Eradication of Helicobacter pylori infection in primary low-grade gastric lymphoma of mucosa-associated lymphoid tissue. *Annals of Internal Medicine* 1995; **122**: 767–9.

34. Savio A, Zamboni G, Capelli P, *et al*. Relapse of low-grade gastric MALT lymphoma after Helicobacter pylori eradication: true relapse or persistence? Long-term post-treatment follow-up of a multicenter trial in the north-east of Italy and evaluation of the diagnostic protocol's adequacy. *Recent Results in Cancer Research* 2000; **156**: 116–24.

35. Wotherspoon AC, Doglioni C, Diss C, *et al*. Regression of primary low-grade B-cell gastric lymphoma of mucosa-associated lymphoid tissue type after eradication of Helicobacter pylori. *Lancet* 1993; **342**: 575–7.

36. de Boer WA, Tytgat GN. Regular review: treatment of Helicobacter pylori infection. *British Medical Journal* 2000; **320**: 31–4.

37. Veldhuyzen van Zanten SJ, Sherman PM, Hunt RH. Helicobacter pylori: new developments and treatments. *Canadian Medical Association Journal* 1997; **156**: 1565–74.

38. Liu H, Ruskon-Fourmestraux A, Lavergne-Slove A, *et al*. Resistance of t(11;18) positive gastric mucosa-associated lymphoid tissue lymphoma to Helicobacter pylori eradication therapy. *Lancet* 2001; **357**: 39–40.

39. Ruskone-Fourmestraux A, Fischbach W, Aleman BM, *et al*. EGILS consensus report. Gastric extranodal marginal zone B-cell lymphoma of MALT. *Gut* 2011; **60**: 747–58.

40. Steinbach G, Ford R, Glober G, *et al*. Antibiotic treatment of gastric lymphoma of mucosa-associated lymphoid tissue. An uncontrolled trial. *Annals of Internal Medicine* 1999; **131**: 88–95.

41. Goda JS, Gospodarowicz M, Pintilie M, *et al*. Long-term outcome in localized extranodal mucosa-associated lymphoid tissue lymphomas treated with radiotherapy. *Cancer* 2010; **116**: 3815–24.

42. Goda JS, Gospodarowicz M, Pintilie M, *et al*. Long-term outcome in localized extranodal mucosa-associated lymphoid tissue lymphomas treated with radiotherapy. *Cancer* 2010; **116**: 3815–24.

43. Ferreri AJ, Guidoboni M, Ponzoni M, *et al*. Evidence for an association between Chlamydia psittaci and ocular adnexal lymphomas. *Journal of the National Cancer Institute* 2004; **96**: 586–94.

44. Horwich A, Catton CN, Quigley M, *et al*. The management of early-stage aggressive non-Hodgkin's lymphoma. *Hematological Oncology* 1988; **6**: 291–8.

45. Sutcliffe SB, Gospodarowicz MK, Bush RS, *et al*. Role of radiation therapy in localized non-Hodgkin's lymphoma. *Radiotherapy and Oncology* 1985; **4**: 211–23.

46. Spicer J, Smith P, Maclennan K, *et al*. Long-term follow-up of patients treated with radiotherapy alone for early-stage histologically aggressive non-Hodgkin's lymphoma. *British Journal of Cancer* 2004; **90**: 1151–5.

47. Monfardini S, Banfi A, Bonadonna G, *et al*. Improved 5 year survival after combined radiotherapy—chemotherapy for stage I and II non-Hodgkin's lymphoma. *International Journal of Radiation Oncology, Biology, Physics* 1980; **6**: 125–34.

48. Tondini C, Zanini M, Lombardi F, *et al*. Combined modality treatment with primary CHOP chemotherapy followed by locoregional irradiation in stage I and II histologically aggressive non-Hodgkin's lymphomas. *Journal of Clinical Oncology* 1993; **11**: 720–5.

49. Coiffier B, Lepage E, Briere J, *et al*. CHOP chemotherapy plus rituximab compared with CHOP alone in elderly patients with diffuse large-B-cell lymphoma. *New England Journal of Medicine* 2002; **346**: 235–42.

50. Miller T, Dahlberg S, Cassady J, *et al*. Three cycles of CHOP plus radiotherapy is superior to eight cycles of CHOP alone for localized intermediate and high grade non-Hodgkin's lymphoma: a Southwest Oncology Group Study. American Society of Clinical Oncology. *Journal of Clinical Oncology* 1996; **15**: 411.

51. Shenkier TN, Voss N, Fairey R, *et al*. Brief chemotherapy and involved-region irradiation for limited-stage diffuse large-cell lymphoma: an 18-year experience from the British Columbia Cancer Agency. *Journal of Clinical Oncology* 2002; **20**: 197–204.

52. van der Maazen RW, Noordijk EM, Thomas J, *et al*. Combined modality treatment is the treatment of choice for stage I/IE intermediate and high grade non-Hodgkin's lymphomas. *Radiotherapy and Oncology* 1998; **49**: 1–7.

53. Zinzani PL, Stefoni V, Tani M, *et al*. MACOP-B regimen followed by involved-field radiation therapy in early-stage aggressive non-Hodgkin's lymphoma patients: 14-year update results. *Leukemia and Lymphoma* 2001; **42**: 989–95.

54. Persky DO, Unger JM, Spier CM, *et al*. Phase II study of rituximab plus three cycles of CHOP and involved-field radiotherapy for patients with limited-stage aggressive B-cell lymphoma: Southwest Oncology Group study 0014. *Journal of Clinical Oncology* 2008; **26**: 2258–63.

55. Horning SJ, Weller E, Kim K, *et al*. Chemotherapy with or without radiotherapy in limited-stage diffuse aggressive non-Hodgkin's lymphoma: Eastern Cooperative Oncology Group Study 1484. *Journal of Clinical Oncology* 2004; **22**: 3032–8.

56. Fisher SG, Fisher RI. The epidemiology of non-Hodgkin's lymphoma. *Oncogene* 2004; **23**: 6524–34.

57. Miller T, Dahlberg S, Cassady J, *et al.* Chemotherapy alone compared with chemotherapy plus radiotherapy for localized intermediate- and high-grade non-Hodgkin's lymphoma. *New England Journal of Medicine* 1998; **339**: 21–6.

58. Bonnet C, Fillet G, Mounier N, *et al.* CHOP alone compared with CHOP plus radiotherapy for localized aggressive lymphoma in elderly patients: a study by the Groupe d'Etude des Lymphomes de l'Adulte. *Journal of Clinical Oncology* 2007; **25**: 787–92.

59. Fillet G, Bonnet N, Mounier N, *et al.* Radiotherapy is unnecessary in elderly patients with localized aggressive non-Hodgkin's lymphoma: Results of the GELA LNH 93–4 study. *Blood* 2002; **100**: 92a, abstr 337.

60. Reyes F, Lepage E, Ganem G, *et al.* ACVBP versus CHOP plus radiotherapy for localized aggressive lymphoma. *New England Journal of Medicine* 2005; **352**: 1197–205.

61. Reyes F, Lepage E, Munck JN, *et al.* Superiority of chemotherapy alone with the ACVBP regimen over treatment with three cycles of CHOP plus radiotherapy in low risk localized aggressive lymphoma: The LNH93–1 GELA study. *Proceedings of the American Society of Hematology* 2002; **100**: 93a, abstr 343.

62. Shipp MA, Klatt MM, Yeap B, *et al.* Patterns of relapse in large-cell lymphoma patients with bulk disease: implications for the use of adjuvant radiation therapy. *Journal of Clinical Oncology* 1989; **7**: 613–18.

63. Aviles A, Delgado S, Nambo MJ, *et al.* Adjuvant radiotherapy to sites of previous bulky disease in patients stage IV diffuse large cell lymphoma. *International Journal of Radiation Oncology, Biology, Physics* 1994; **30**: 799–803.

64. DeAngelis LM. Current management of primary central nervous system lymphoma. *Oncology* 1995; **9**: 63–71.

65. Laperriere NJ, Cerezo L, Milosevic MF, *et al.* Primary lymphoma of brain: results of management of a modern cohort with radiation therapy. *Radiotherapy and Oncology* 1997; **43**: 247–52.

66. Ferreri AJ, Blay JY, Reni M, *et al.* Prognostic scoring system for primary CNS lymphomas: the International Extranodal Lymphoma Study Group experience. *Journal of Clinical Oncology* 2003; **21**: 266–72.

67. Lachance DH, Brizel DM, Gockerman JP, *et al.* Cyclophosphamide, doxorubicin, vincristine, and prednisone for primary central nervous system lymphoma: short-duration response and multifocal intracerebral recurrence preceding radiotherapy. *Neurology* 1994; **44**: 1721–7.

68. Krogh-Jensen M, d'Amore F, Jensen MK, *et al.* Incidence, clinicopathological features and outcome of primary central nervous system lymphomas. Population-based data from a Danish lymphoma registry. Danish Lymphoma Study Group, LYFO. *Annals of Oncology* 1994; **5**: 349–54.

69. Nelson DF, Martz KL, Bonner H, *et al.* Non-Hodgkin's lymphoma of the brain: can high dose, large volume radiation therapy improve survival? Report on a prospective trial by the Radiation Therapy Oncology Group (RTOG): RTOG 8315. *International Journal of Radiation Oncology, Biology, Physics* 1992; **23**: 9–17.

70. DeAngelis LM. Primary central nervous system lymphoma. *Recent Results in Cancer Research* 1994; **135**: 155–69.

71. Abrey LE, Yahalom J, DeAngelis LM. Treatment for primary CNS lymphoma: the next step. *Journal of Clinical Oncology* 2000; **18**: 3144–50.

72. Thiel E, Korfel A, Martus P, *et al.* High-dose methotrexate with or without whole brain radiotherapy for primary CNS lymphoma (G-PCNSL-SG-1): a phase 3, randomised, non-inferiority trial. *Lancet Oncology* 2010; **11**: 1036–47.

73. Shah GD, Yahalom J, Correa DD, *et al.* Combined immunochemotherapy with reduced whole-brain radiotherapy for newly diagnosed primary CNS lymphoma. *Journal of Clinical Oncology* 2007; **25**: 4730–5.

74. Rathmell AJ, Gospodarowicz MK, Sutcliffe SB, *et al.* Localized extradural lymphoma: survival, relapse pattern and functional outcome. The Princess Margaret Hospital Lymphoma Group. *Radiotherapy and Oncology* 1992; **24**: 14–20.

75. Eeles RA, O'Brien P, Horwich A, *et al.* Non-Hodgkin's lymphoma presenting with extradural spinal cord compression: functional outcome and survival. *British Journal of Cancer* 1991; **63**: 126–9.

76. Jaffe ES, Burg G. Report of the symposium on Cutaneous Lymphomas: Sixth International Conference on Malignant Lymphoma. *Annals of Oncology* 1997; **8**(Suppl 1): 83–4.

77. Willemze R, Kerl H, Sterry W, *et al.* EORTC classification for primary cutaneous lymphomas: a proposal from the Cutaneous Lymphoma Study Group of the European Organization for Research and Treatment of Cancer. *Blood* 1997; **90**: 354–71.

78. Willemze R, Meijer CJ. EORTC classification for primary cutaneous lymphomas: a comparison with the R.E.A.L. Classification and the proposed WHO Classification. *Annals of Oncology* 2000; **11**(Suppl 1): 11–15.

79. Bekkenk MW, Vermeer MH, Geerts ML, *et al.* Treatment of multifocal primary cutaneous B-cell lymphoma: a clinical follow-up study of 29 patients. *Journal of Clinical Oncology* 1999; **17**: 2471–8.

80. Vermeer MH, Geelen FA, van Haselen CW, *et al.* Primary cutaneous large B-cell lymphomas of the legs. A distinct type of cutaneous B-cell lymphoma with an intermediate prognosis. Dutch Cutaneous Lymphoma Working Group. *Archives of Dermatology* 1996; **132**: 1304–8.

81. Roggero E, Zucca E, Mainetti C, *et al.* Eradication of Borrelia burgdorferi infection in primary marginal zone B-cell lymphoma of the skin. *HumanPathology* 2000; **31**: 263–8.

82. Kurtin PJ, DiCaudo DJ, Habermann TM, *et al.* Primary cutaneous large cell lymphomas. Morphologic, immunophenotypic, and clinical features of 20 cases. *American Journal of Surgical Pathology* 1994; **18**: 1183–91.

83. Bekkenk MW, Geelen FA, van Voorst Vader PC, *et al.* Primary and secondary cutaneous CD30(+) lymphoproliferative disorders: a report from the Dutch Cutaneous Lymphoma Group on the long-term follow-up data of 219 patients and guidelines for diagnosis and treatment. *Blood* 2000; **95**: 3653–61.

84. Younes A, Bartlett NL, Leonard JP, *et al.* Brentuximab vedotin (SGN-35) for relapsed CD30-positive lymphomas. *New England Journal of Medicine* 2010; **363**: 1812–21.

85. Koch P, del Valle F, Berdel WE, *et al.* Primary gastrointestinal non-Hodgkin's lymphoma: II. Combined surgical and conservative or conservative management only in localized gastric lymphoma—results of the prospective German Multicenter Study GIT NHL 01/92. *Journal of Clinical Oncology* 2001; **19**: 3874–83.

86. Raderer M, Chott A, Drach J, *et al.* Chemotherapy for management of localised high-grade gastric B-cell lymphoma: how much is necessary? *Annals of Oncology* 2002; **13**: 1094–8.

87. Aref A, Narayan S, Tekyi-Mensah S, *et al.* Value of radiation therapy in the management of chemoresistant intermediate grade non-Hodgkin's lymphoma. *Radiation Oncology Investigations* 1999; **7**: 186–91.

88. Kirkove C, Timothy AR. Radiotherapy as salvage treatment in patients with Hodgkin's disease or non-Hodgkin's lymphoma relapsing after initial chemotherapy. *Hemtological Oncology* 1991; **9**: 163–7.

89. Chen CI, Roitman D, Tsang R, *et al.* 'Relative' chemotherapy sensitivity: the impact of number of salvage regimens prior to autologous stem cell transplant for relapsed and refractory aggressive non-Hodgkin's lymphoma. *Bone Marrow Transplantation* 2002; **30**: 885–91.

90. Horning SJ, Negrin RS, Chao NJ, *et al.* Fractionated total-body irradiation, etoposide, and cyclophosphamide plus autografting in Hodgkin's disease and non-Hodgkin's lymphoma. *Journal of Clinical Oncology* 1994; **12**: 2552–8.

91. Shipp MA, Abeloff MD, Antman KH, *et al.* International consensus conference on high-dose therapy with hematopoietic stem cell transplantation in aggressive non-Hodgkin's lymphomas: report of the jury. *Journal of Clinical Oncology* 1999; **17**: 423–9.

92. Philip T, Gugliemi C, Chauvin F, *et al.* Autologous bone marrow transplantation (ABMT) versus conventional chemotherapy (DHAP) in relapsed non-Hodgkin lymphoma (NHL): final analysis of the PARMA randomized study (216 patients). *Journal of Clinical Oncology* 1995; **14**: 390.

93. Tsang RW, Gospodarowicz MK, Sutcliffe SB, *et al.* Thoracic radiation therapy before autologous bone marrow transplantation in relapsed or refractory Hodgkin's disease. PMH Lymphoma Group, and the Toronto Autologous BMT Group. *European Journal of Cancer* 1999; **35**: 73–8.

94. Diehl V, Thomas RK, Re D. Part II: Hodgkin's lymphoma—diagnosis and treatment. *Lancet Oncology* 2004; **5**: 19–26.

95. Federico M, Luminari S, Iannitto E, *et al.* ABVD compared with BEACOPP compared with CEC for the initial treatment of patients with advanced Hodgkin's lymphoma: results from the HD2000 Gruppo Italiano per lo Studio dei Linfomi Trial. *Journal of Clinical Oncology* 2009; **27**: 805–11.

96. Viviani S, Zinzani PL, Rambaldi A, *et al.* ABVD versus BEACOPP for Hodgkin's lymphoma when high-dose salvage is planned. *New England Journal of Medicine* 2011; **365**: 203–12.

97. Hasenclever D, Diehl V. A prognostic score for advanced Hodgkin's disease. International Prognostic Factors Project on Advanced Hodgkin's Disease. *New England Journal of Medicine* 1998; **339**: 1506–14.

98. Gospodarowicz MK, Sutcliffe SB, Clark RM, *et al.* Analysis of supradiaphragmatic clinical stage I and II Hodgkin's disease treated with radiation alone. *International Journal of Radiation Oncology, Biology, Physics* 1992; **22**: 859–65.

99. Raemaekers J, Kluin-Nelemans H, Teodorovic I, *et al.* The achievements of the EORTC Lymphoma Group. European Organisation for Research and Treatment of Cancer. *European Journal of Cancer* 2002; **38**(Suppl 4): S107–113.

100. Noordijk EM, Carde P, Mandard AM, *et al.* Preliminary results of the EORTC-GPMC controlled clinical trial H7 in early-stage Hodgkin's disease. EORTC Lymphoma Cooperative Group. Groupe Pierre-et-Marie-Curie. *Annals of Oncology* 1994; **5**(Suppl 2): 107–12.

101. Ferme C, Eghbali H, Meerwaldt JH, *et al.* Chemotherapy plus involved-field radiation in early-stage Hodgkin's disease. *New England Journal of Medicine* 2007; **357**: 1916–27.

102. Engert A, Franklin J, Eich HT, *et al.* Two cycles of doxorubicin, bleomycin, vinblastine, and dacarbazine plus extended-field radiotherapy is superior to radiotherapy alone in early favorable Hodgkin's lymphoma: final results of the GHSG HD7 trial. *Journal of Clinical Oncology* 2007; **25**: 3495–502.

103. Meyer RM, Gospodarowicz MK, Connors JM, *et al*. ABVD alone versus radiation-based therapy in limited-stage Hodgkin's lymphoma. *New Engl J Med* 2012; **366**(5): 399–408.

104. Nachman JB, Sposto R, Herzog P, *et al*. Randomized comparison of low-dose involved-field radiotherapy and no radiotherapy for children with Hodgkin's disease who achieve a complete response to chemotherapy. *Journal of Clinical Oncology* 2002; **20**: 3765–71.

105. Bonadonna G, Bonfante V, Viviani S, *et al*. ABVD plus subtotal nodal versus involved-field radiotherapy in early-stage Hodgkin's disease: long-term results. *Journal of Clinical Oncology* 2004; **22**: 2835–41.

106. Engert A, Plutschow A, Eich HT, *et al*. Reduced treatment intensity in patients with early-stage Hodgkin's lymphoma. *New England Journal of Medicine* 2010; **363**: 640–52.

107. Cosset JM, Henry-Amar M, Meerwaldt JH, *et al*. The EORTC trials for limited stage Hodgkin's disease. The EORTC Lymphoma Cooperative Group. *European Journal of Cancer* 1992; **28A**: 1847–50.

108. Eich HT, Diehl V, Gorgen H, *et al*. Intensified chemotherapy and dose-reduced involved-field radiotherapy in patients with early unfavorable Hodgkin's lymphoma: final analysis of the German Hodgkin Study Group HD11 trial. *Journal of Clinical Oncology* 2010; **28**: 4199–206.

109. Hoskin PJ, Lowry L, Horwich A, *et al*. Randomized comparison of the Stanford V regimen and ABVD in the treatment of advanced Hodgkin's Lymphoma: United Kingdom National Cancer Research Institute Lymphoma Group Study ISRCTN 64141244. *Journal of Clinical Oncology* 2009; **27**: 5390–6.

110. Engert A, Schiller P, Josting A, *et al*. Involved-field radiotherapy is equally effective and less toxic compared with extended-field radiotherapy after four cycles of chemotherapy in patients with early-stage unfavorable Hodgkin's lymphoma: results of the HD8 trial of the German Hodgkin's Lymphoma Study Group. *Journal of Clinical Oncology* 2003; **21**: 3601–8.

111. Girinsky T, van der Maazen R, Specht L, *et al*. Involved-node radiotherapy (INRT) in patients with early Hodgkin lymphoma: concepts and guidelines. *Radiotherapy and Oncology* 2006; **79**: 270–7.

112. Eich HT, Muller RP, Engenhart-Cabillic R, *et al*. Involved-node radiotherapy in early-stage Hodgkin's lymphoma. Definition and guidelines of the German Hodgkin Study Group (GHSG). *Strahlentherapir und Onkologie* 2008; **184**: 406–10.

113. Campbell BA, Voss N, Pickles T, *et al*. Involved-nodal radiation therapy as a component of combination therapy for limited-stage Hodgkin's lymphoma: a question of field size. *Journal of Clinical Oncology* 2008; **26**: 5170–4.

114. Duhmke E, Diehl V, Loeffler M, *et al*. Randomized trial with early-stage Hodgkin's disease testing 30 Gy vs. 40 Gy extended field radiotherapy alone. *International Journal of Radiation Oncology, Biology, Physics* 1996; **36**: 305–10.

115. Vijayakumar S, Myrianthopoulos LC. An updated dose-response analysis in Hodgkin's disease. *Radiotherapy and Oncology* 1992; **24**: 1–13.

116. Diehl V, Franklin J, Pfreundschuh M, *et al*. Standard and increased-dose BEACOPP chemotherapy compared with COPP-ABVD for advanced Hodgkin's disease. *New England Journal of Medicine* 2003; **348**: 2386–95.

117. Aleman BM, Raemaekers JM, Tirelli U, *et al*. Involved-field radiotherapy for advanced Hodgkin's lymphoma. *New England Journal of Medicine* 2003; **348**: 2396–406.

118. Ferme C, Sebban C, Hennequin C, *et al*. Comparison of chemotherapy to radiotherapy as consolidation of complete or good partial response after six cycles of chemotherapy for

patients with advanced Hodgkin's disease: results of the groupe d'etudes des lymphomes de l'Adulte H89 trial. *Blood* 2000; **95**: 2246–52.

119. Diehl V, Schiller P, Engert A, *et al.* Results of the Third Interim Analysis of the HD12 Trial of the GHSG: 8 Courses of Escalated BEACOPP Versus 4 Escalated and 4 Baseline Courses of BEACOPP with or without Additive Radiotherapy for Advanced Stage Hodgkin s Lymphoma. *Blood* 2003; **102**: Abstract 85.

120. Loeffler M, Brosteanu O, Hasenclever D, *et al.* Meta-analysis of chemotherapy versus combined modality treatment trials in Hodgkin's disease. International Database on Hodgkin's Disease Overview Study Group. *Journal of Clinical Oncology* 1998; **16**: 818–29.

121. Laskar S, Gupta T, Vimal S, *et al.* Consolidation radiation after complete remission in Hodgkin's disease following six cycles of doxorubicin, bleomycin, vinblastine, and dacarbazine chemotherapy: is there a need? *Journal of Clinical Oncology* 2004; **22**: 62–8.

122. Johnson PW, Sydes MR, Hancock BW, *et al.* Consolidation radiotherapy in patients with advanced Hodgkin's lymphoma: survival data from the UKLG LY09 randomized controlled trial (ISRCTN97144519). *Journal of Clinical Oncology* 2010; **28**: 3352–9.

123. Aleman BM, van den Belt-Dusebout AW, Klokman WJ, *et al.* Long-term cause-specific mortality of patients treated for Hodgkin's disease. *Journal of Clinical Oncology* 2003; **21**: 3431–9.

124. Eich HT, Gossmann A, Engert A, *et al.* A Contribution to solve the problem of the need for consolidative radiotherapy after intensive chemotherapy in advanced stages of Hodgkin's lymphoma—analysis of a quality control program initiated by the radiotherapy reference center of the German Hodgkin Study Group (GHSG). *International Journal of Radiation Oncology, Biology, Physics* 2007; **69**: 1187–92.

125. Linch DC, Winfield D, Goldstone AH, *et al.* Dose intensification with autologous bone-marrow transplantation in relapsed and resistant Hodgkin's disease: results of a BNLI randomised trial. *Lancet* 1993; **341**: 1051–4.

126. Schmitz N, Pfistner B, Sextro M, *et al.* Aggressive conventional chemotherapy compared with high-dose chemotherapy with autologous haemopoietic stem-cell transplantation for relapsed chemosensitive Hodgkin's disease: a randomised trial. *Lancet* 2002; **359**: 2065–71.

127. Ferme C, Mounier N, Divine M, *et al.* Intensive salvage therapy with high-dose chemotherapy for patients with advanced Hodgkin's disease in relapse or failure after initial chemotherapy: results of the Groupe d'Etudes des Lymphomes de l'Adulte H89 Trial. *Journal of Clinical Oncology* 2002; **20**: 467–75.

128. Lazarus HM, Loberiza FR, Jr., Zhang MJ, *et al.* Autotransplants for Hodgkin's disease in first relapse or second remission: a report from the autologous blood and marrow transplant registry (ABMTR). *Bone Marrow Transplantation* 2001; **27**: 387–96.

129. Hoppe BS, Moskowitz CH, Filippa DA, *et al.* Involved-field radiotherapy before high-dose therapy and autologous stem-cell rescue in diffuse large-cell lymphoma: long-term disease control and toxicity. *Journal of Clinical Oncology* 2008; **26**: 1858–64.

130. Hoppe BS, Moskowitz CH, Zhang Z, *et al.* The role of FDG-PET imaging and involved field radiotherapy in relapsed or refractory diffuse large B-cell lymphoma. *Bone Marrow Transplantation* 2009; **43**: 941–8.

131. Moskowitz AJ, Yahalom J, Kewalramani T, *et al.* Pretransplantation functional imaging predicts outcome following autologous stem cell transplantation for relapsed and refractory Hodgkin lymphoma. *Blood* 2010; **116**: 4934–7.

132. Campbell B, Wirth A, Milner A, *et al.* Long-term follow-up of salvage radiotherapy in Hodgkin's lymphoma after chemotherapy failure. *International Journal of Radiation Oncology, Biology, Physics* 2005; **63**: 1538–45.

133. Josting A, Nogova L, Franklin J, *et al.* Salvage radiotherapy in patients with relapsed and refractory Hodgkin's lymphoma: a retrospective analysis from the German Hodgkin Lymphoma Study Group. *Journal of Clinical Oncology* 2005; **23**: 1522–9.

134. O'Brien PC, Parnis FX. Salvage radiotherapy following chemotherapy failure in Hodgkin's disease—what is its role? *Acta Oncologica* 1995; **34**: 99–104.

135. Josting A, Engert A, Diehl V, *et al.* Prognostic factors and treatment outcome in patients with primary progressive and relapsed Hodgkin's disease. *Annals of Oncology* 2002; **13** Suppl 1: 112–16.

136. Josting A, Franklin J, May M, *et al.* New prognostic score based on treatment outcome of patients with relapsed Hodgkin's lymphoma registered in the database of the German Hodgkin's lymphoma study group. *Journal of Clinical Oncology* 2002; **20**: 221–30.

137. Moskowitz CH, Kewalramani T, Nimer SD, *et al.* Effectiveness of high dose chemoradiotherapy and autologous stem cell transplantation for patients with biopsy-proven primary refractory Hodgkin's disease. *British Journal of Haematology* 2004; **124**: 645–52.

138. Kaplan HS. Evidence for a tumoricidal dose level in the radiotherapy of Hodgkin's disease. *Cancer Res* 1966; **26**: 1221–4.

139. Backstrand KH, Ng AK, Takvorian RW, *et al.* Results of a prospective trial of mantle irradiation alone for selected patients with early-stage Hodgkin's disease. *Journal of Clinical Oncology* 2001; **19**: 736–41.

140. Duhmke E, Franklin J, Pfreundschuh M, *et al.* Low-dose radiation is sufficient for the noninvolved extended-field treatment in favorable early-stage Hodgkin's disease: long-term results of a randomized trial of radiotherapy alone. *Journal of Clinical Oncology* 2001; **19**: 2905–14.

141. Gospodarowicz MK, Sutcliffe SB, Bergsagel DE, *et al.* Radiation therapy in clinical stage I and II Hodgkin's disease. The Princess Margaret Hospital Lymphoma Group. *European Journal of Cancer* 1992; **28A**: 1841–6.

142. Press OW, LeBlanc M, Lichter AS, *et al.* Phase III randomized intergroup trial of subtotal lymphoid irradiation versus doxorubicin, vinblastine, and subtotal lymphoid irradiation for stage IA to IIA Hodgkin's disease. *Journal of Clinical Oncology* 2001; **19**: 4238–44.

143. Lowry L, Smith P, Qian W, *et al.* Reduced dose radiotherapy for local control in non-Hodgkin lymphoma: a randomised phase III trial. *Radiotherapy and Oncology* 2011; **100**: 86–92.

144. Fung CY, Grossbard ML, Linggood RM, *et al.* Mucosa-associated lymphoid tissue lymphoma of the stomach: long term outcome after local treatment. *Cancer* 1999; **85**: 9–17.

145. Le QT, Eulau SM, George TI, *et al.* Primary radiotherapy for localized orbital MALT lymphoma. *International Journal of Radiation Oncology, Biology, Physics* 2002; **52**: 657–63.

146. Uno T, Isobe K, Shikama N, *et al.* Radiotherapy for extranodal, marginal zone, B-cell lymphoma of mucosa-associated lymphoid tissue originating in the ocular adnexa: a multiinstitutional, retrospective review of 50 patients. *Cancer* 2003; **98**: 865–71.

147. Wirth A, Chao M, Corry J, *et al.* Mantle irradiation alone for clinical stage I-II Hodgkin's disease: long-term follow-up and analysis of prognostic factors in 261 patients. *Journal of Clinical Oncology* 1999; **17**: 230–40.

148. Schechter NR, Yahalom J. Low-grade MALT lymphoma of the stomach: a review of treatment options. *International Journal of Radiation Oncology, Biology, Physics* 2000; **46**: 1093–103.

149. Chen Z, Agostinelli AG, Wilson LD, *et al*. Matching the dosimetry characteristics of a dual-field Stanford technique to a customized single-field Stanford technique for total skin electron therapy. *International Journal of Radiation Oncology, Biology, Physics* 2004; **59**: 872–85.

150. Kumar PP, Patel IS. Comparison of dose distribution with different techniques of total skin electron beam therapy. *Clinical Radiology* 1982; **33**: 495–7.

151. Peters VG. Use of an electron reflector to improve dose uniformity at the vertex during total skin electron therapy. *International Journal of Radiation Oncology, Biology, Physics* 2000; **46**: 1065–9.

152. Antolak JA, Cundiff JH, Ha CS. Utilization of thermoluminescent dosimetry in total skin electron beam radiotherapy of mycosis fungoides. *International Journal of Radiation Oncology, Biology, Physics* 1998; **40**: 101–8.

153. Maingon P, Truc G, Dalac S, *et al*. Radiotherapy of advanced mycosis fungoides: indications and results of total skin electron beam and photon beam irradiation. *Radiotherapy and Oncology* 2000; **54**: 73–8.

154. McDonald S, Rubin P, Phillips TL, *et al*. Injury to the lung from cancer therapy: clinical syndromes, measurable endpoints, and potential scoring systems. *International Journal of Radiation Oncology, Biology, Physics* 1995; **31**: 1187–203.

155. Mah K, Van Dyk J, Keane T, *et al*. Acute radiation-induced pulmonary damage: a clinical study on the response to fractionated radiation therapy. *International Journal of Radiation Oncology, Biology, Physics* 1987; **13**: 179–88.

156. Fajardo LP. Cardiovascular system. In *Pathology of Radiation injury*. New York: Masson Publishing USA, Inc.,1982. pp. 15–33.

157. Cosset JM, Henry-Amar M, Pellae-Cosset B, *et al*. Pericarditis and myocardial infarctions after Hodgkin's disease therapy. *International Journal of Radiation Oncology, Biology, Physics* 1991; **21**: 447–9.

158. Grigsby PW, Russell A, Bruner D, *et al*. Late injury of cancer therapy on the female reproductive tract. *International Journal of Radiation Oncology, Biology, Physics* 1995; **31**: 1281–99.

159. Shapiro E, Kinsella TJ, Makuch RW, *et al*. Effects of fractionated irradiation on endocrine aspects of testicular function. *Journal of Clinical Oncology* 1985; **3**: 1232–9.

160. Izard MA. Leydig cell function and radiation: a review of the literature. *Radiotherapy and Oncology* 1995; **34**: 1–8.

161. Adamietz IA, Rosskopf B, Dapper FD, *et al*. Comparison of two strategies for the treatment of radiogenic leukopenia using granulocyte colony stimulating factor. *International Journal of Radiation Oncology, Biology, Physics* 1996; **35**: 61–7.

162. Knox SJ, Fowler S, Marquez C, *et al*. Effect of filgrastim (G-CSF) in Hodgkin's disease patients treated with radiation therapy. *International Journal of Radiation Oncology, Biology, Physics* 1994; **28**: 445–50.

163. Mac Manus MP, McCormick D, Trimble A, *et al*. Value of granulocyte colony stimulating factor in radiotherapy induced neutropenia: clinical and laboratory studies. *European Journal of Cancer* 1995; **3**: 302–7.

164. Swerdlow AJ, Barber JA, Hudson GV, *et al*. Risk of second malignancy after Hodgkin's disease in a collaborative British cohort: the relation to age at treatment. *Journal of Clinical Oncology* 2000; **18**: 498–509.

165. van Leeuwen FE, Klokman WJ, Veer MB, *et al.* Long-term risk of second malignancy in survivors of Hodgkin's disease treated during adolescence or young adulthood. *Journal of Clinical Oncology* 2000; **18**: 487–97.

166. Travis LB, Hill DA, Dores GM, *et al.* Breast cancer following radiotherapy and chemotherapy among young women with Hodgkin disease. *Journal of the American Medical Association* 2003; **290**: 465–75.

167. van Leeuwen FE, Klokman WJ, Stovall M, *et al.* Roles of radiation dose, chemotherapy, and hormonal factors in breast cancer following Hodgkin's disease. *Journal of the National Cancer Institute* 2003; **95**: 971–80.

168. Travis LB, Gospodarowicz M, Curtis RE, *et al.* Lung cancer following chemotherapy and radiotherapy for Hodgkin's disease. *Journal of the National Cancer Institute* 2002; **94**: 182–92.

11

Central nervous system tumours

Neil G Burnet, Fiona Harris, Raj Jena,
Kate E Burton, Sarah J Jefferies

11.1 Introduction

11.1.1 General introduction

Primary brain tumours can be broadly divided into primary glial tumours, ependymoma, medulloblastoma, germ cell tumours (germinoma and teratoma), meningioma, nerve sheath tumours (such as vestibular schwannoma) and pituitary tumours including craniopharyngioma. Several of these tumours can also affect the spinal cord. These tumours differ widely in their pathology, treatment, and outcome. Nevertheless, there are common themes which apply to all central nervous system (CNS) tumours.

Primary CNS tumours account for about 2% of all primary tumours, about 4500 per annum in the UK, but are responsible for more loss of life per patient than any other adult cancer, at just over 20 years per patient[1]. The gliomas are particularly devastating cancers[2,3]. They affect patients of all ages, from childhood to old age. Of new referrals in neuro-oncology, gliomas constitute two-thirds, but WHO grade IV gliomas or glioblastomas (GBM) alone account for almost a half[3]. For this reason, neuro-oncology appears dominated by GBM. Primary CNS lymphoma contributes 3–5% of cases. Meningiomas account for approximately 10% of new referrals, and pituitary tumours plus craniopharyngioma another 10%. A further 10% of patients have a wide range of rare tumours. These numbers exclude patients with metastases, where the management may be directed by the site-specific team, or in specific cases, by the neuro-oncology team.

Although the outlook for patients with GBM is very poor, a small percentage appears to be cured. Other types of CNS tumour may be eminently curable. Germinoma, for example, is rare in adults but with appropriate treatment should have a cure rate of almost 100%. Pituitary tumours should have a long-term control rate of over 90%[4]. Careful attention to radiotherapy technique is therefore important, with as much focus on normal tissue as on the tumour itself. As an example, in medulloblastoma there is clear proof that rigorous attention to radiotherapy detail including careful consideration of the dose to OARs is necessary if best results are to be achieved.

Patients with primary brain tumours experience some problems which are unique in oncology, or nearly so: a significant number develop intellectual and personality change; motor disorders including hemiplegia are especially disabling; side effects of muscle wasting and weight gain from dexamethasone are problematic; fortunately, seizures are controllable in most cases. Very occasionally a young female patient presents during pregnancy, adding an extra dimension to the management problems[5,6].

11.1.2 Diagnosis—both radiology and pathology

The first issue in management is that of diagnosis. Typically, the diagnosis is suggested by CT or MRI. In some tumour types, such as vestibular schwannoma, imaging is definitive; in other cases, including intrinsic tumours in the brain, imaging offers only a differential diagnosis. For example, difficulties occur in distinguishing glioma from primary cerebral lymphoma, and solitary metastasis from abscess or glioma. Benign lesions with imaging appearances suggesting malignancy also occur, albeit uncommonly (about 0.2% in our practice). Therefore, despite improving radiological accuracy, some patients will have an incorrect diagnosis unless pathology is obtained. Even if the radiological diagnosis of GBM is near certain, it is easier for the patient and family to discuss the prognosis and treatment options if the diagnosis has been established beyond question. Occasional tumours may be hazardous to biopsy, for example, those arising in the brainstem. Careful consideration must then be given to the management.

Once tissue is obtained, cases should be discussed in the multidisciplinary forum, so that the pathology can be correlated with the clinical history and imaging appearances. This is especially important for gliomas, to decide on the 'effective' clinical grade of the tumour[7].

11.1.3 Performance status in the treatment decision

Performance status (PS) is a significant predictor of outcome, particularly for glial tumours, and must be taken in to consideration in determining treatment intent[8]. Treatment can be radical, palliative, or active supportive care. For some patients with severe disabling neurology, especially due to GBM, supportive care may be the most appropriate management option[3].

11.1.4 Principles of radiotherapy planning

The fundamental principles of radiotherapy planning and treatment apply to CNS tumours. These include reproducible immobilization whose accuracy is known, high-quality imaging for target volume localization, the identification of critical normal structures, and conformal or IMRT planning.

11.2 Principles of radiotherapy planning for CNS tumours

11.2.1 Patient position and immobilization

Treatment position for radiotherapy

The first decision to be made for radical radiotherapy is the patient position: supine or prone. This depends on the location of the tumour and the immobilization devices. Supine has the advantage of comfort for the patient. For posterior lesions, there may be greater scope in field directions if the patient is prone, because possible collisions between couch and machine gantry are avoided. However, if couch extensions, such as a relocatable stereotactic radiotherapy (SRT) frame or 'S' frame are available, a supine position may be possible and preferable. These devices attach to the end of the treatment couch and allow 360° access to the target area, without collision between gantry and couch. There are some limitations, especially for postero-inferior fields, but these are rarely needed. Therefore, treat posterior gliomas or posterior fossa lesions prone, unless using a couch extension.

For cranio-spinal axis treatment, there are advantages in a prone treatment position, using traditional techniques, although increasingly a supine position with a single spinal field is being used, particularly in paediatric practice[9,10]. The prone position allows palpation of the spine and accurate visualization of matching field junctions. However, the use of image guidance, especially with IMRT, may allow supine position, which is often more comfortable for the patient.

For palliative radiotherapy, the supine position is to be preferred because of patient comfort, and it is difficult to safely position patients with poor PS in the prone position.

Immobilization methods for radical radiotherapy

The main choice of immobilization is between a beam direction shell and a relocatable stereotactic head frame. The shell can either be a rigid Perspex mask or a thermoplastic mask and is straightforward, well tolerated, and easy to use; this is the standard immobilization device for gliomas. The conventionally quoted accuracy of relocation of rigid Perspex thermoplastic shells is 0.5 cm[11], though this is influenced by the number of fixation points of the shell to the base board[12]. Perspex shells should be cut out as much as possible, though without compromising their rigidity, to improve skin sparing.

Increasingly, centres are moving to the use of flexible thermoplastic materials which once softened in hot water can be manipulated to form a mask; their accuracy is probably similar to rigid Perspex shells. However, these materials may shrink, which may be problematic if the treatment extends over several weeks. In addition to the shell itself, consideration must be given to the head rest and base board. A complete system is required for accurate, reproducible immobilization. Accuracy for all immobilization devices is patient- and operator-dependant.

A relocatable stereotactic frame can deliver a higher precision in repositioning from one day to the next, reducing both systematic (preparation) and random (treatment) errors, and therefore reducing the CTV to PTV margin accordingly. The accuracy of a relocatable stereotactic frame is in the region of 0.2–0.3 mm[13,14], particularly if positioning is measured prospectively before each treatment.

These volumes and margins are discussed later. However, measurement of the accuracy of positioning is required in each department, for each device.

Immobilization methods for palliative radiotherapy

A beam direction shell is recommended for palliative radiotherapy. It improves the accuracy of relocation of the patient from one treatment to the next, and allows positioning marks to be added to it rather than the patient.

Thermoplastic masks are extremely convenient as these can be made relatively quickly and a CT planning scan performed at the same visit. The relocation accuracy of such materials is probably in the region of 0.5 cm.

11.2.2 Imaging for radiotherapy planning

Imaging for radical radiotherapy

Planning should be based on CT. This must be done in the treatment position, with fiducial markers attached to the immobilization device (shell or stereotactic frame). CT provides precise information on the shape and position of the patient, produces density data for the treatment planning system, and allows accurate localization of tumour and normal tissue structures. The whole head must be scanned to allow planning with superior or superior-oblique beams, for production of meaningful dose–volume histograms (DVHs), and to allow optimum electronic image coregistration with MRI (see later in this section). CT slice spacing requirements may vary according to the precision required for planning; for high-precision treatments, a spacing of 1 mm is usual. The finer the slice spacing, the higher the quality of digitally-reconstructed radiographs (DRRs). Where the GTV or CTV come close to the skull, the planning CT should also be reviewed with bony window settings, because the thickness of the bone is actually less than it appears at soft tissue settings.

In most circumstances the planning CT should be performed with intravenous contrast; this alters dosimetry calculations by 1% or less. High-grade gliomas, and many benign tumours, enhance with contrast. Low-grade gliomas take up less contrast than normal brain, so a paradoxical increase in discrimination of the edge of the tumour occurs as a result.

For most CNS tumours additional valuable information is obtained from MRI, and MRI should be considered a prerequisite for high-precision modern planning. This applies, for example, to gliomas, both high- and low-grade, medulloblastoma, meningiomas, skull base tumours including acoustic (vestibular) schwannoma, and pituitary tumours (Fig. 11.1). This view has been endorsed by the Royal College of Radiologists[15]. MRI should be electronically coregistered with the planning CT.

MRI does not have to be performed in the exact treatment position provided image coregistration is available, but the closer it is to the orientation of the planning CT the better the quality after coregistration. The appropriate MR sequence (discussed later) should be carried out as a continuous scan, with as fine a slice spacing as possible. The thinner and greater the number of slices the longer the sequence takes. The signal:noise ratio also worsens, but the reconstruction to match the CT is better, so a balance must be struck. Contiguous slices of 2 mm represent a good compromise, and as scanners get faster, this may reduce further. The whole head must be scanned to provide the most anatomical information for electronic image coregistration. Different MR sequences may be of value for different tumours. While T1 weighting with gadolinium contrast (T1W + Gd) is optimum for many tumours (Fig. 11.1a,b), low-grade gliomas are best demonstrated with a T2-weighted or FLAIR sequence (Fig. 11.1c).

For gliomas, especially HGGs, an early postoperative MRI performed within 24–48 (maximum 72) hours of surgery demonstrates the extent of residual tumour and distinguishes residual tumour from inflammatory operative changes which develop after

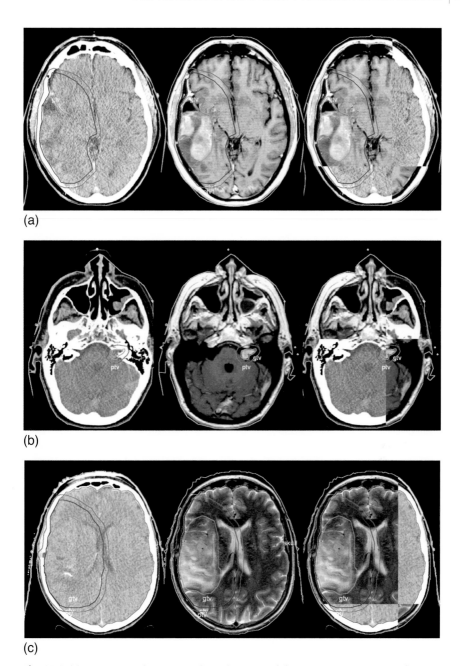

Fig. 11.1 (a) HGG: CT and coregistered MR (T1W + gadolinium sequence). Even with HGG, gross tumour typically shows more clearly on MRI than CT. (b) Vestibular schwannoma (acoustic neuroma) demonstrated on CT, on MR using the T1W sequence with gadolinium (Gd) contrast agent, and with both coregistered. (Same case as Fig. 11.7.) (c) LGG imaged with CT and coregistered T2W MR sequence. The skull and patient outline have been drawn from the planning CT.

this time[16]. Tumour growth between surgery and RT planning can also be evaluated by comparing the planning and postoperative scans.

The planning MRI must be performed close to the time of CT planning and the start of treatment. If debulking surgery has been performed, then planning imaging should be postoperative, following reduction of tumour mass and secondary mass effect.

In some circumstances, information from the preoperative MRI is also valuable. For example, the area most at risk of recurrence in a complex meningioma is the base of the tumour. This is easy to identify on a preoperative assessment, but may be invisible postoperatively. In such circumstances, coregistration with both pre- and postoperative MRI may be advantageous in reducing edge recurrence (see section 11.10).

The first step when beginning planning with coregistered imaging is to check that the coregistration has worked correctly. If not, this needs to be repeated. Some systems allow the transformation matrix to be manually altered which may be useful in some cases to improve accuracy. For target volume delineation, it is valuable to use multiple (orthogonal) planar views, as well as the best possible imaging. Both factors have been shown to reduce interoperator variation in target volumes[17].

Other imaging modalities may have a role in clinical use in the future, though none have been widely adopted yet, and remain investigational. MR spectroscopy appears to contribute to localization of the edge of high-grade glioma[18]. Diffusion-weighted and diffusion tensor imaging are being investigated for the same purpose[19–21]. PET scanning has been investigated as an aid to target localization in high-grade glioma[22], and meningioma[23,24]. There are limits to the spatial resolution, and caution is needed in defining the edge of the tracer uptake.

Imaging for palliative radiotherapy

CT localization for palliative radiotherapy, for example- for high-grade gliomas, is to be preferred, with conventional CT or CT simulator. This gives the most accurate localization of the tumour, with minimum discomfort to the patient, allowing reduction in the volume of normal tissue which will be irradiated. If craniotomy and debulking has been carried out, then the planning CT will show the extent of tumour and mass effect, and planning can allow for this. Even if biopsy alone has been carried out, there is usually sufficient delay between the presentation CT and palliative radiotherapy planning to warrant re-imaging.

11.2.3 **Planning volumes for radiotherapy**

Standard volumes should be used in planning all CNS tumours, as laid out in the ICRU Reports 50, 62, and 83[25–27], and the British Institute of Radiology (BIR) Report 2003[28,29]. The BIR report, *Geometric Uncertainties in Radiotherapy*, in particular contains a 'recipe' for calculation of the PTV margin in each dimension separately, which is especially useful. An example of the calculation for CNS tumours is included.

Gross tumour volume

The GTV is defined as actual tumour that can be seen on imaging, and this applies to all CNS tumours. If this is outlined as the starting point for planning, then the CTV follows easily. MRI always has more clearly defined margins between gross visible tumour and surrounding normal brain and is the mainstay of target volume delineation. CT shows bone much better, and is therefore useful for anatomical barriers, or for skull base tumours involving bone (e.g. meningioma, chordoma).

It is relatively uncommon for gliomas to have all gross tumour removed surgically. However, other tumour types such as meningioma and medulloblastoma may have complete resection. In such cases, preoperative imaging can be used as a basis for constructing the preoperative GTV.

Clinical target volume

The CTV is formed by a margin around the GTV to account for microscopic spread. In principle, it is this volume that must be treated adequately to achieve cure. The precise CTV margin depends on the tumour type, and is addressed in detail in later sections. In general, this margin is based on population data, rather than the individual patient because this tumour infiltration cannot, by definition, be imaged. For gliomas, the margins have been defined from studies correlating postmortem brain sections with premortem imaging, biopsy with imaging, and recurrence patterns. As yet, there is no mechanism for individualizing CTV margins.

As further imaging modalities develop, such as MRS or diffusion tensor imaging, it may be possible to identify abnormalities resulting from tumour outside the standard GTV. Such abnormalities should lie within the CTV. This might necessitate the introduction of a new planning volume term which would also require specification of the imaging modality used[19]. The incorporation of such 'imaging high-risk volumes' will be a matter for clinical research protocols.

The CTV can, and should, be edited according to routes of spread and anatomical barriers. For example, gliomas do not penetrate the skull, or the meningeal structures of the falx and tentorium, though they can spread around the edge. Meningiomas, by contrast do spread along meningeal surfaces. Consideration of these aspects can reduce the volume of normal tissue receiving high-dose irradiation, but knowledge of the relevant anatomy is essential.

Planning target volume

The PTV is applied around the CTV to ensure that it is adequately treated, and in a sense, it is really a volume referenced to 3D space rather than to the patient. It is designed to account for internal organ movement and set-up inaccuracies (see ICRU 62[26]). These are most effectively considered as systematic (treatment preparation) errors and random (treatment execution) errors (see BIR 2003[28] and ICRU 62 and 83[26,27]).

The BIR 2003 report gives examples for several tumour site including CNS. It is noteworthy that systematic errors are substantially more important than day-to-day random errors in treatment set-up. This report also suggests that discrepancies in outlining between clinicians should be incorporated as part of the PTV margin.

Organs at risk and planning organ at risk volumes

OARs are normal tissues whose radiation sensitivity influences treatment planning or the prescribed radiation dose. Different risks may be considered appropriate under different clinical circumstances, so that 'tolerance' depends on the clinical setting.

Many CNS tumours are intimately involved with normal tissues, so that dose is determined by the tolerance of those tissues. This applies both to invasive tumours, like the gliomas, where brain must be included within the CTV, or extrinsic tumours lying in contact with normal tissue, such as meningioma, where normal tissue may be included in the CTV and will certainly be within the PTV. In these situations, dose is typically determined by the tolerance of the relevant normal tissue. Outlining of OARs is therefore useful to exclude 'hot spots' and to document dose (e.g. to the hypothalamus and pituitary). They may also help the planner in choice of beam directions.

In a few situations, where a dose above normal tissue tolerance is desirable and where damage to a critical structure would cause a devastating effect to the patient, it may be appropriate to consider an IMRT solution. To help in optimizing the plan, a margin around an OAR is needed to allow for uncertainties in set-up. This generates the planning organ at risk volume (PRV), which is analogous to the PTV margin around the CTV[27]. The PRV margin can be calculated in the same way as the PTV margin, though it does not have to be as large[28].

Adding a PRV around an organ at risk very substantially increases the volume of the normal tissue structure and may present dilemmas concerning radiotherapy dose to the target. Some interaction between the PTV and a PRV may therefore be necessary, altering the prescribed radiation dose and dose distribution because of the PRV, or reducing the PRV if target dose is imperative. Where conflicting priorities occur in volumes which overlap, such as the PTV and a PRV, it is better not to edit the PTV, although a 'PTV-PRV' volume will be needed for IMRT planning[27]. The IMRT optimizer can then be set up to deliver a plan with priority given to the primary volume, using graduated doses.

Fortunately, PRVs are only needed in certain circumstances. The PRV is useful principally for 'serial' architecture neural structures, specifically the optic nerves, chiasm, brainstem, and spinal cord, when a high dose is to be given to a nearby target. In these particular OARs, loss of a small amount of normal tissue from radiation damage produces a severe or catastrophic clinical manifestation (e.g. blindness, spinal cord myelopathy), so a technique to limit dose to within tolerance is useful. A PRV should not be drawn round parallel structures, since this can change the DVH in an unpredictable way, sometimes suggesting over-, sometimes underdose[30].

11.2.4 Dose distribution considerations

Beam energy

For most intracranial tumours the CTV extends to the inside of the skull, at least in part. Relatively low-energy megavoltage X-rays such as 6 MV are therefore appropriate, so that the CTV is below the build-up depth. Exceptions to this principle are pituitary tumours, and other centrally-located lesions, such as para-sellar meningioma, where higher energy photon beams can be used. The application of tissue-equivalent 'bolus' material to the surface is not recommended.

In some surgical patients, particularly those with vertex meningioma, part of the skull is often removed at the time of the operation. It is preferable to have a cranioplasty carried out prior to radiotherapy, in order to interpose some tissue so that the build-up zone moves out of the CTV, and to lift the hair-bearing scalp away from the CTV.

Field arrangements

In general, there is no need to constrain beam arrangements to be coplanar. In the CNS in particular, beams may be arranged away from the axial plane to great advantage. Therefore, many treatment plans have entirely non-coplanar beam arrangements. Wedging should be used imaginatively; longitudinal wedging may improve dosimetry, especially with non-coplanar plans. Beam angles should be chosen to avoid exit (or entry) through the eyes, except where necessary to treat disease. In developing plans with complex beam arrangements it is very useful to outline the eyes. This facilitates orientation of the head when using operator's eye and beam's eye views.

Most treatment plans can be carried out using three beams but, especially for larger lesions, more beams may be required. It is not uncommon for a three-field arrangement for a large glioma volume to produce areas of higher dose at the edges of the target. An additional beam, reduced in width to match the cooler zone in the centre of the target and with a low weighting, can correct this. This 'field-in-field' solution is forward-planned IMRT, and is extremely valuable for improving dose homogeneity (Fig. 11.2).

Beams from a superior or superior-oblique direction can be helpful (Fig. 11.3). However, consideration should be given to exit doses, particularly for patients of younger age or with benign tumours. For this reason, vertex fields for the treatment of pituitary adenoma are not advocated as a general solution.

For most CNS tumours, at present, the planning objective is a uniform dose to cover the target volume. For treatment where an inhomogeneous, variable dose is required, or where the target has a complex shape, IMRT may provide a useful solution, using either a rotational or fixed field approach.

Assessment of dose distribution

The treatment plan should be carefully reviewed slice by slice. The ICRU recommends a dose variation across a treatment plan which is no greater than −5% to 7% (i.e. 95–107%). However, in many situations a dose variation roughly half this can be achieved. 'Field-in-field' beam arrangements often improve the dose homogeneity, and can be useful in fulfilling these requirements. Try to ensure that areas of dose > 100% are within the CTV.

In addition, DVHs of the CTV, PTV, and critical normal tissues should be reviewed. The DVH for the PTV may be misleading in CNS tumours because the PTV quite often extends into the build-up region. In reality, this does not present a problem with dosimetry, but makes assessment of the dose plan more difficult. The CTV, however, should not fall into the build-up zone, if at all possible (see earlier notes on beam energy and cranioplasty). Therefore, the DVH for the CTV should be as close as possible to a square shape.

The DVHs for adjacent dose-limiting normal tissues are also very important. However, in radiotherapy for CNS tumours some critical structures always lie within the CTV, so their DVHs may not be helpful; dose and fractionation must be chosen to

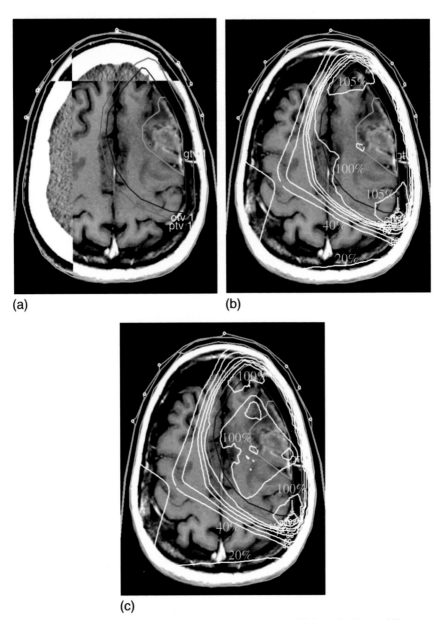

(a)   (b)

(c)

Fig. 11.2 (a) Phase 1 GTV, CTV, and PTV outlined in a case of high-grade glioma, following surgical debulking. The CTV does not have to extend beyond the skull, which is reliably outlined from the planning CT. (b) Axial section of a three-field plan for a HGG. This shows some inhomogeneity, with areas of higher dose anteriorly and posteriorly in the PTV. Posteriorly the dose reaches 108% (arrowed). This inhomogeneity can be improved by the addition of a small, low weight, additional field to top up the dose between the high dose areas—a field-in field solution. (c) Four-field dose plan at the same level, improved by the addition of the small extra superior field. The 'hot spot', still arrowed is now at 102%. This field-in-field solution is a form of forward-planned IMRT. The other isodoses follow the convention of 100%, 95%, 90%, 80%, 60%, 40%, 20%.

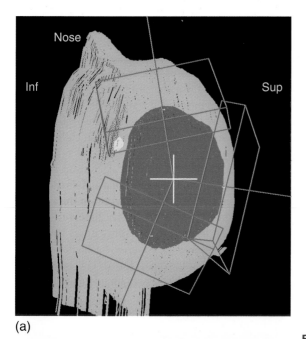

(a)

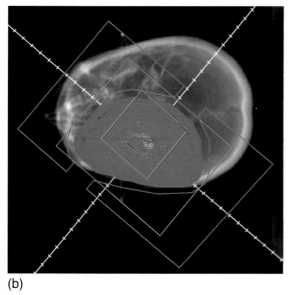

(b)

Fig. 11.3 (a) Operator's eye view (OEV) of the beam arrangement for the plan in Fig. 11.2; the fourth small superior field is not visible inside its larger brother. This arrangement of non-coplanar beams, with the thick ends of the wedges facing each other, can be called a 'wedged triplet'. The PTV is shown in red; the pituitary has been outlined in yellow.
(b) Beam's eye view (BEV) of the superior beams (fields in green, low melting point alloy (LMPA) blocks in purple). The position of the small additional field-in-field is superimposed.

avoid damage, rather than using beam arrangements which restrict physical dose. The maximum dose within all critical normal tissues must be examined, to avoid 'hot spots' within them. As well as raising the total dose, such dose inhomogeneity also delivers a higher dose per fraction—so called 'double trouble'[31] in tissues with a high fractionation sensitivity (low alpha-beta ratio).

Normal tissue tolerance doses

Introduction

Normal tissue tolerance is an important concept, embodying both the risk of a complication and also the severity of its effect on the patient. It is also contextual, with acceptance of a higher risk in patients with highly malignant tumours, which would not be acceptable in a benign condition. This means that stated tolerances may vary from one condition to another; they may also change over time if the probability of cure is increased. Toxicities and tolerance doses are very different in children (see Chapter 15).

Much of the tolerance dose data was produced by Emami et al.[32], who reviewed outcomes in patients treated before the advent of modern imaging and dose calculation. The dose-response literature has been reassessed more recently, in the excellent QUANTEC (Quantitative Analysis of Normal Tissue Effects in the Clinic) reviews (see Marks et al.[33]). For many normal tissues, but not all, tolerance doses have been revised upwards. The original Emami data described the volume dependence of toxicity, including parameters to model the effect, but the recent QUANTEC review does not include this in the update, due to a lack of reliable data; brain tolerance doses are assumed to apply to partial brain irradiation.

Optic nerves and chiasm

The optic nerves and chiasm may also be slightly more sensitive than brain parenchyma. The QUANTEC data suggest that total dose and fraction size are the two most important determinants of risk of optic nerve and chiasm damage. For a fractionated course using 1.8–2.0 Gy/fraction, 50 Gy represents a 'near zero' risk. The risk rises to around 3–7% for doses in the range 55–60 Gy, and rises further with higher doses[34]. These doses represent a dose increase, or a risk reduction, compared to the historic Emami data set[32].

It has been suggested that tolerance may be lower in patients with pituitary adenoma, and this presumably also applies to craniopharyngioma[34]. Nevertheless, for these benign tumours where safety is paramount, doses in the region of 45–50 Gy at 1.8 or 1.67 Gy per fraction should be entirely safe (i.e. 45 Gy in 25 fractions to 50 Gy in 30 fractions).

Again, for patients with GBM, generally no allowance is made for target volumes which include these structures, although this concept may have to be reviewed if treatment outcomes can be improved.

Brain

In adults, the dose-limiting toxicity in the brain is necrosis, and dose, fraction size, and volume are the major factors that influence risk. Although location does not, per se, influence the risk, it does alter the risk of manifesting a clinical effect. Other factors have been suggested to alter risk, and these include chemotherapy. This definitely applies to combination with methotrexate, but may not be an issue for 60 Gy given with temozolomide.

Tolerance of most of the brain substance can be considered to be above 60 Gy, given in approximately 30 fractions. Smaller volumes of brain may be able to tolerate higher doses without additional risk. The recent QUANTEC review[35] suggests a 5% risk with a dose of 72 Gy for partial brain irradiation, though notes a range of doses from 60–84 Gy (at 2 Gy/fraction). The dose for a 10% risk is said to be 90 Gy (range 84–102 Gy). These doses are substantially higher than the Emami data suggested, where a 5% risk of necrosis was suggested with a dose of 60 Gy in 30 fractions, though with an additional volume effect not characterized in the QUANTEC report.

A separate issue is the possibility of intellectual damage from radiotherapy. The QUANTEC review suggests that the evidence that RT induces neurocognitive injury in adults up to 4 years after RT is "weak", and using planned volumes, intellectual damage in adults is uncommon or does not occur[35]. It should be noted that other factors may impair intellect in those receiving RT, including direct effects of the tumour and surgery, and indirect effects such as hydrocephalus, anticonvulsants, and untreated pituitary dysfunction.

There is limited evidence on neurocognitive effects from small studies of low-grade glioma (LGG) patients specifically, where both survival and follow-up have been relatively long. In the EORTC randomized trial of early versus delayed RT, there were no differences in late toxicity between the patient cohorts, but neurocognitive functions were not assessed[36]. Studies comparing LGG patients to cohorts with other malignancies, or to controls, have suggested equivocal results. Taphoorn et al. [37] found no difference in intellectual function in two cohorts of LGG patients treated with either RT or biopsy/surgery only. They did, however, perform less well than a cohort with haematological malignancy, suggesting an effect of the tumour rather than the treatment. In a randomized trial of two dose levels, 45 Gy (25 fractions) versus 59.4 Gy (33 fractions), patients treated to the higher dose reported lower levels of global functioning and greater fatigue/malaise[38], again suggesting the potential for neuropsychometric effects from RT. Gregor et al.[39] presented evidence that localized volume irradiation was substantially (7 times) less likely to produce neuropsychometric deficit than whole brain RT. Brown et al.[40] concluded that there is evidence of only sporadic effects on intellect in LGG patients, provided that treatment was localized using focal RT and conventional doses.

Klein et al.[41] compared LGG patients, about half of whom had received RT, to a cohort with haematological malignancies and a cohort of healthy patients. Low-grade glioma patients had lower cognitive ability than low-grade haematological patients, and lower still than healthy controls. Use of radiotherapy was associated with poorer cognitive function, but memory was affected only in patients whose RT had used doses > 2 Gy/fraction. Antiepileptics were associated with disability in attention and executive function. Douw et al.[42] reported on longer follow-up of these patients, though there were only 32 RT patients in the final group. RT patients had worse attention, executive functioning and information processing speed. Cognitive disability affected about half of the RT patients, but also a quarter of non-RT treated patients. In summary, it is clear that RT may cause some significant effects in some but not necessarily all patients, and the risks can be minimized by careful planning and choice of dose. As well as the other indirect factors mentioned above, it should be noted that there can also be significant toxicities from both surgery and chemotherapy.

For comparing dose-fractionation schedules an alpha/beta ratio of 2.9 is recommended[35,43]. Re-irradiation is possible, with total doses up to about 100 Gy, given a reasonable interval (perhaps a minimum of 1 year) between courses[35]. Twice daily schedules may be damaging, and sufficient time must be left between fractions to allow recovery.

Brainstem

The stated tolerance of the brainstem is approximately 54 Gy in 30 fractions to 55 Gy in 33 fractions (which are essentially the same). The QUANTEC summary supports this limit, suggesting that the entire brainstem may be treated to 54 Gy (conventional fractionation) with 'limited risk' (probably < 3%) of severe or permanent neurological effects, and that smaller volumes (1–10 mL) may be irradiated to maximum doses of 59 Gy, using dose per fractions of ≤ 2 Gy[44]. This risk applies to the use of photons; tolerance doses may be different (higher) for proton beam therapy. Generally, no dose reduction, from 60 to 54 Gy, is made in patients with glioblastoma, which illustrates the effect of context on tolerance doses.

Spinal cord

The QUANTEC review provides the most up-to-date summary data[45]. For the cervical cord, the alpha/beta ratio is estimated at 0.87 Gy, much lower than previous literature values, and indicating a very strong relationship between risk of myelopathy and dose per fraction. The clinical data fit well to a sigmoid dose response curve, and using conventional doses of 1.8–2 Gy/fraction, and for full thickness irradiation of the cord, the estimated risk of myelopathy is <1% at 54 Gy and <10% at 61 Gy. The thoracic cord may be slightly less sensitive than the cervical cord. Data for thoracic cord tolerance fits a standard sigmoid dose response curve poorly, so specific estimates of tolerance doses and alpha/beta ratio are not possible.

Re-irradiation data in animals and humans suggest partial repair of subclinical RT damage from about 6 months, and increasing over the next 2 years[45]. As with the brainstem, tolerance doses may be higher for proton beam therapy, but this is conjecture at present.

Pituitary and hypothalamus

The pituitary gland and more particularly the hypothalamus have a much lower tolerance for hormonal dysfunction. There is probably little effect in the long term for doses under 20–24 Gy but the relationship between dose, volume, and dysfunction is not fully established in adults; in children, doses in the range 40–60 Gy are most damaging to hormone secretion[46], though growth hormone secretion is particularly sensitive to lower doses, with an effect seen in many children at doses even below 23.4 Gy[47]. Certainly adult patients who receive a full prescription dose of 50–60 Gy have a significant long-term risk of hypothalamic–pituitary axis dysfunction[48].

Ear and cochlea

The middle and inner ears are also sensitive structures. The traditional view was that hearing would be preserved, or recover, in most patients after doses up to 60 Gy. More recently the dose response data has been reviewed and a more conservative limit suggested[49]. For conventionally fractionated RT, to avoid sensory-neural hearing loss, efforts should be made to keep the mean dose to the cochlea to 45 Gy or less. Since a threshold for hearing loss has not been determined, as low a dose as possible should be even safer.

Other cranial normal tissue structures

The lacrimal gland may be treated during radiotherapy for some intracranial and skull base lesions. Reduced tear output occurs at doses over about 20 Gy, matching salivary tolerance, though a dose below 10 Gy is safer. The lens of the eye should have virtually no risk of cataract after receiving a dose of 5–6 Gy, spread out over 30 fractions. There is a 50% incidence of cataract formation after a dose of 15 Gy[50].

The risk of permanent alopecia depends on dose to the hair follicles in the dermis. The risk is very low or absent with doses below 10–15 Gy; the dose to produce permanent alopecia in 50% of patients is 43 Gy, and the slope of the central part of the dose response curve (γ_{50}) is 0.9. These doses refer to the total dose given in 30 fractions, and dose equivalence was calculated using a linear quadratic formulation with an α/β ratio of 2 Gy[51]. Measurements of the actual dermal dose are difficult, but this data can at least help assessment of risk. Hair washing has no effect on hair loss or scalp reactions, provided it is carried out carefully[52].

11.3 **High-grade glioma**

11.3.1 **Radical radiotherapy**

Indications

High-grade gliomas (HGG) include WHO grade III gliomas and grade IV gliomas. The latter are also known as glioblastoma (GBM), though previously were termed glioblastoma multiforme. The first craniotomy for glioblastoma was performed in London in 1884[53], in the same era as the discovery of X-rays and the beginning of their use as a treatment for cancer.

Glioblastoma

Although surgical resection has been known to be an independent prognostic factor for some time[8], with survival advantage seen in patients undergoing resection compared to biopsy alone, it is only comparatively recently that clear evidence has emerged that the extent of resection, rather than selection bias, is responsible[54]. Techniques to increase the extent of safe resection have shown survival advantage, especially when combined with chemo-RT[55]. Biopsy may be the only appropriate option for very deep seated lesions, including in the brainstem. Maximal safe surgical resection should therefore be undertaken, both to improve prognosis, and also to improve pressure symptoms by reducing mass effect, and consequently reducing steroid requirements.

In 2004, randomized control trial evidence showed that the combination of radical radiotherapy and chemotherapy with temozolomide (TMZ), given both concurrently and as an adjuvant following radiotherapy, produced a survival advantage compared to RT alone at the median survival time and at 2 years[56]. This result transformed the treatment of GBM. Follow-up at 5 years has confirmed the survival advantage (10% versus 2% 5-year survival for chemo-RT versus RT alone)[57]. It is important to note that the majority of patients in the study (84%) had undergone surgical resection, which was followed by concurrent chemo-RT and then adjuvant chemotherapy, using TMZ. This outcome is therefore the result of trimodality treatment.

The EORTC study also demonstrated a survival advantage in patients whose tumours have inactivation, by methylation, of the gene coding for O6-methylguanine–DNA methyltransferase (MGMT) which removes the predominant DNA lesion caused by TMZ[57,58]. However, survival advantage for chemo-RT exists even for patients without MGMT methylation[57].

This treatment regimen is generally well-tolerated, although in the trial only 47% of patients completed the adjuvant phase, due either to disease progression or chemo-toxicity[56]. A further management challenge is the phenomenon of pseudo progression. This is a treatment effect, such that imaging 1–3 months following completion of the concomitant chemo-RT phase shows appearances suggestive of disease progression but where the patient remains clinically unchanged[59,60]. Re-imaging at an appropriate interval shows improved MRI appearances. This phenomenon occurs in around 20% of cases[59,61].

Performance status is one of the most important predictors of survival[8,56], and together with age should guide management. Radical chemo-RT should only be offered to patients with excellent performance status, and therefore with no significant neurological deficit. Occasional patients have a clinically detectable deficit which does not impair performance, such as hemianopia from an occipital lesion. In patients with HGG it is recommended to treat radically only those patients with WHO PS of 0 or 1.

The gold standard treatment for GBM is now considered to be maximal safe surgical resection, followed by concomitant chemo-RT, then adjuvant chemotherapy with TMZ. This approach was endorsed in the UK in 2007 by the National Institute for Health and Clinical Excellence (NICE), for patients of WHO PS 0 or 1.

Occasional good PS patients are considered unlikely to cope with the rigours of chemo-RT, or have significant comorbidity precluding it. Consideration should be given to using radiotherapy alone (60 Gy in 30 fractions daily over 6 weeks)[62].

For those with a significant neurological deficit, palliative treatment or best supportive care are more appropriate options (see section 11.1.3).

Grade III glioma

These tumours are less common than GBM (about 1:4), and most of the previously discussed considerations apply to patients with grade III tumours. However, there is evidence of lack of benefit from adjuvant chemotherapy using older agents.

The EORTC is currently running a trial (EORTC 26053–22054, BR14 in the UK) to evaluate the role of adjuvant or concomitant TMZ, or both, compared to RT alone, in patients with non-1p19q deleted grade III gliomas. This should clarify whether there is a role for TMZ chemotherapy, but at present in the UK these tumours are treated with RT alone.

Treatment volume and definition

Radiotherapy should be fully conformal and outlining should be performed with MRI coregistered with the planning CT. The GTV is defined as the contrast-enhancing abnormality on T1 MRI (see notes on 'Gross tumour volume' in section 11.2.3) (see Fig. 11.1a).

An early postoperative MRI performed within 24–48 (maximum 72) hours of surgery demonstrates the extent of residual tumour and distinguishes residual tumour from inflammatory operative changes[16]. This may reduce the volume of normal tissue that is included in the high-dose volume. A planning MRI should also be performed, as well as the planning CT, and this should be compared to the postoperative MRI to exclude very rapid regrowth, which may require alteration in management from radical chemo-RT to short course palliative RT. The original preoperative MRI is also valuable.

CTV margins are based on work correlating CT findings with postmortem tumour extent, biopsy findings, and recurrence patterns[63–66]. The CTV is usually defined in two phases. For phase 1 use a 2.5-cm margin, grown isotropically around the GTV. For phase 2 use a 1.5-cm margin, grown in a similar way. Some protocols have moved to delivering 60 Gy in a single phase, using the phase 1 CTV margin. These volumes should be constrained by the brain or skull outline, as GBM does not extend through the dura. In a few circumstances, they can also be edited to avoid treating the posterior fossa but great care must be taken not to omit potential routes of spread through the cerebral peduncles and brainstem (see notes on 'Clinical target volume' in section 11.2.3). See Jansen et al.[65] for review.

Minniti et al.[66] have shown that it is not necessary to include all the oedema around the tumour. Oedema surrounding HGG certainly includes tumour but the relationship between level of oedema and tumour burden is inexact. It appears to be more reliable to use the contrast enhancing (gross) tumour as the starting point before adding a 2–3-cm CTV margin. This generally includes most of the oedema. Although oedema volume does not necessarily change greatly with steroid administration (in HGG as distinct from metastases), it may become less distinct and therefore even harder to use as the basis for target volume definition.

The PTV is defined with an appropriate margin (see notes on 'Planning target volume' in section 11.2.3), typically 0.5 cm.

Planning technique

Patient position and immobilization

Treat supine, unless the lesion is located posteriorly. If so, preferably treat supine on an 'S' frame-type couch extension, or prone in a two-part shell if absolutely necessary (see section 11.2.1). Immobilize in a rigid (Perspex) or flexible thermoplastic shell system. Cut out the shell for treatment.

Volume localization

Use the planning CT electronically coregistered with a T1W + Gd MRI performed for planning (see Fig. 11.1a). Coregistration of the early postoperative MRI may also be helpful. Outline the GTV as the contrast-enhancing edge of the tumour; grow the CTV isotropically and edit at bone or dura; then grow the PTV isotropically from that, as noted earlier.

Dose distribution

Use 6-MV or equivalent X-rays and base the plan on three isocentric, conformal fields. These are almost always non-coplanar, with wedging on several or all fields. A 'wedged triplet' (three orthogonal wedged fields) is often a good starting point. Additional smaller fields, a field-in-field technique, may improve the dose distribution, especially for larger volumes, and superior or superior-oblique fields may be useful (see section 11.2.4) (see Fig. 11.2).

Dose homogeneity better than 95–107% should be achievable (i.e. better than the ICRU recommendations), and areas of dose > 100% should be within the CTV not critical normal tissues. The plan should be reviewed slice by slice to assess the dose distribution and DVHs firstly for the CTV and PTV, and then for the OARs.

Implementation

The patient is positioned using reference marks drawn on the shell. The isocentre position is established using movements from the fiducial marker positions on the planning CT.

Verification

The treatment plan should be verified prior to commencing treatment, and portal imaging should be obtained during treatment[67]. Lens doses should also be confirmed by the use of TLDs, to ensure that the dose is no greater than that expected from the plan. Entry or exit diode dosimetry may also be used.

Dose prescription

◆ Grade III glioma: total dose 54 Gy in 30 fractions over 6 weeks, in two phases:

  • Phase 1: 45 Gy in 25 fractions (or 39.6 Gy in 22 fractions).
  • Phase 2: 9 Gy in 5 fractions (or 14.4 Gy in 8 fractions).

The current NCRI BR14 CATNON trial (EORTC 26053–22054) requires the slightly higher dose of 59.4 Gy in 33 fractions, in 1 phase.

◆ Grade IV glioma: total dose 60 Gy in 30 fractions over 6 weeks, in two phases with concomitant temozolomide (75 mg/m^2):

  • Phase 1: 50 Gy in 25 fractions (or 40 Gy in 20 fractions)
  • Phase 2: 10 Gy in 5 fractions (or 20 Gy in 10 fractions).

11.3.2 Palliative radiotherapy

Indications

Palliative RT is appropriate for patients over the age of 70 years, or those under 70 with a significant neurological deficit, but who have PS 2 or better. For patients of poor PS (3–4), there is no survival advantage in treating with RT, and these patients should be referred to the palliative care team, for supportive management in the community.

Palliative RT should improve or stabilize neurological function in the majority of patients, and may prolong survival; the worse the neurological status at treatment, the shorter the survival time[3].

Treatment volume and definition

CT planning should be used, and contrast is useful as it helps to visualize the tumour. This should be outlined as the GTV. The CTV margin should then be grown isotropically by 2.5 cm, and a 0.5-cm margin used to the PTV.

Planning technique

Patient position and immobilization

Treat supine, because this is most comfortable for the patient, in a thermoplastic shell system (see 'Immobilization methods for palliative radiotherapy' in section 11.2.1).

Volume localization

Planning is best performed with CT, as previously described.

Dose distribution

Parallel-opposed lateral fields represent an effective beam arrangement in this context, using 6-MV photon beams and an isocentric technique (Fig. 11.4). The fields should be larger than the PTV by the width of the penumbra (about 0.8 cm) and the collimators should be angled to avoid treating the eyes and the pharynx. The use of asymmetry on the fields allows the central axis to be placed on the shell at an easily reproducible point. Prescribing should be to the midplane, and a formal dose distribution is unnecessary. Orthovoltage X-ray techniques are now outmoded.

Implementation

The patient is positioned using reference marks on the shell, placed at the time of the planning CT.

Verification

Portal imaging is sufficient for verification, and TLD measurement of the lens doses is unnecessary due to the poor prognosis of these patients.

Dose prescription

For all high-grade gliomas, 30 Gy in six fractions is a common schedule which is well tolerated and appropriate in the context of a palliative treatment. An alternative is 35 Gy in 10 fractions over 2 weeks, treating daily. In Continental Europe and the USA,

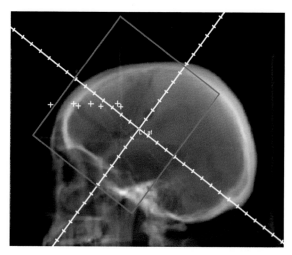

Fig. 11.4 Lateral field for a palliative radiotherapy treatment for HGG, following craniotomy. A parallel pair of opposed lateral fields was used. Note the collimator twist to avoid the eyes and pharynx. (The white crosses were for physics use.)

40 Gy in 15 fractions is popular[54], though these different doses have never been tested against each other.

◆ Total dose 30 Gy in 6 fractions over 2 weeks, treating 3 times per week.

11.4 **Low-grade glioma**

11.4.1 **Radical radiotherapy**

Indications

Patients with LGG may be referred at first presentation, e.g. following a scan performed as the result of a seizure, or because of tumour progression, detected either clinically or on imaging. As well as neurological deterioration, worsening of seizure control may indicate progression. Pathological confirmation is recommended before any decision to treat is made. There is developing evidence that surgical resection may be of value[68], though risks and benefits must be carefully balanced.

If a patient has worsening neurology, worsening seizures which are not adequately controlled, or imaging evidence of disease progression they should be treated. If the disease is stable, then an active surveillance programme can be initiated. The EORTC randomized trial of the timing of radiotherapy[36] indicated that there is no survival difference according to the timing of radiotherapy. This is valuable because it shows that an active surveillance policy made be an excellent choice for some patients. However, early radiotherapy does produce an improvement in disease-free survival.

Neurological deficits caused by LGG may improve as a result of radiotherapy and this may be an additional indication for treatment, even in patients with relatively poor PS. This is quite different from HGG. Radiotherapy improves epilepsy in about 50% of patients with LGG[36]. In rare circumstances, it can be used to try to improve intractable epilepsy in patients with LGG.

In occasional cases LGG can be very extensive, affecting the whole of a hemisphere, or occasionally almost the entire cerebral cortex, and is termed gliomatosis cerebri. This requires large treatment volumes, sparing little normal brain, but useful responses are typical. Neurological deficits improve in most patients and long-term control is achieved in some.

Currently, the optimum first-line treatment, whether radiotherapy or chemotherapy, is not clear. Both are certainly effective in some patients. Chemotherapy has demonstrated particular efficacy in patients with oligodendrogliomas demonstrating specific chromosomal deletions (e.g. 1p 19q). An international randomized trial (BR13 in the UK, EORTC 22033–26033) is currently underway, comparing TMZ chemotherapy versus radiotherapy as first-line treatment. Patients are registered at diagnosis and then randomized to either RT or chemotherapy at progression. Registration closed at the end of 2010, but randomization for treatment, which is delayed until progression, will continue to take place in patients already registered. It will be some time before this trial reports. In the meantime, in most centres accepted UK practice is to offer irradiation.

Late toxicity is an important component of overall outcome for LGG, and consideration simply of progression-free or overall survival is no longer considered sufficient to evaluate new treatment strategies[69].

Treatment volume and definition

Radiotherapy should be fully conformal. Planning should be based on MRI, which should be coregistered with the planning CT (see 'Imaging for radical radiotherapy' in section 11.2.2). Usually the optimum sequence is FLAIR or T2W, there being no contrast enhancement in LGG normally. The GTV can be considered as the edge of the high signal region on FLAIR or T2W (Fig. 11.1c). Normally this can be seen to correlate with the low-density region on planning CT.

The GTV is enlarged by 1.5 cm to form the CTV[70]. This margin is grown isotropically but edited at the skull and skull base (see notes on 'Clinical target volume' in section 11.2.3). The CTV margin is smaller than for HGG, partly because these tumours usually infiltrate less widely than HGG, and also because the zone of low density on CT and high signal on FLAIR and T2W MRI already indicates some infiltration. The PTV should be grown isotropically from the edited CTV with a margin of 0.5 cm (see notes on 'Planning target volume' in section 11.2.3).

The exception is the uncommon condition of gliomatosis cerebri. In this, much of the hemisphere can be infiltrated; exceptionally, almost the whole cerebrum may be affected. This can be seen as low density on CT and high signal on T2W MRI. The GTV can be constructed in the same way, and a CTV margin added. However, because this type of tumour is so infiltrative, larger margins should be considered, for example 2.5 cm instead of 1.5 cm. Particular attention should be given to the corpus callosum which can permit infiltrative tumours to cross the midline. The CTV may encompass the whole hemisphere plus extension across the midline. The CTV should be edited to the skull, and the PTV should be grown isotropically, with a margin of 0.5 cm (see on 'Planning target volume' in section 11.2.3).

The outcome of the majority of patients with LGG is relatively good, and survival times may be long, so long-term complications need to be considered carefully. It is worth outlining the pituitary and hypothalamus, as well as other relevant normal tissues, so that the treatment planning system returns dose–volume data on the critical structures.

There is increasing interest in conforming treatment further to reduce the dose to structures such as the hippocampus which are involved in information processing and memory, in an attempt to reduce long-term cognitive toxicity. This may become increasingly relevant as the cohort of long-term survivors increases, but may carry a risk of undertreating the tumour.

Planning technique

This is similar to the techniques for HGG.

Patient position and immobilization

Patients should be treated supine in a Perspex or thermoplastic shell, unless the lesion is located posteriorly. If so, preferably treat supine on an 'S' frame-type couch extension, or prone in a 2 part shell if absolutely necessary (see section 11.2.1). If using a Perspex shell, this should be cut out for treatment.

Volume localization

Use planning CT with coregistered MRI. Usually the optimum sequence is FLAIR or T2W, there being no contrast enhancement in LGG normally (Fig. 11.1c). Outline the

GTV as the edge of the FLAIR or T2W abnormality on MRI, which should match the edge of low density seen on CT. Grow the CTV isotropically and edit; then grow the PTV isotropically from the CTV, as noted earlier.

Dose distribution

Similar to HGG, use 6-MV or equivalent X-rays and base the plan on three isocentric, conformal fields. These are almost always non-coplanar, with wedging on several or all fields. A 'wedged triplet' (three orthogonal wedged fields) is often a good starting point. Additional smaller fields, a field-in-field technique, may improve the dose distribution, especially for larger volumes, and superior or superior-oblique fields may be useful (see section 11.2.4). (See Fig. 11.5.)

Dose homogeneity better than 95–107% should be achievable (i.e. better than the ICRU recommendations), and areas of dose > 100% should be within the CTV not critical normal tissues. The plan should be reviewed slice by slice to assess the dose distribution and DVHs initially for PTV coverage and then for the OARs. The DVHs may be useful in assessing the volume of normal tissues exposed to the highest doses. The mean dose to the hypothalamic–pituitary axis should be noted to allow identification of those patients at greatest risk of long-term pituitary failure (Fig. 11.5).

Implementation

Position the patient using reference markers on the shell. The isocentre position is established using movements from the fiducial marker positions on the planning CT.

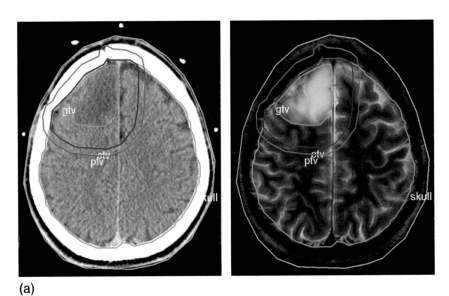

(a)

Fig. 11.5 (a) Outlining of GTV, CTV, and PTV for a patient with low-grade glioma; although the tumour is visible on the CT, it is better seen on MR. The CTV has been limited by the skull, which was auto-outlined, from the planning CT. There is a slight discrepancy seen between the skull outline and the edge of the brain on the MRI, which is due to slight superior-inferior inaccuracy in the CT:MR coregistration.

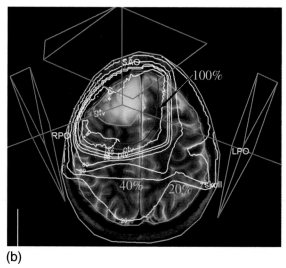

(b)

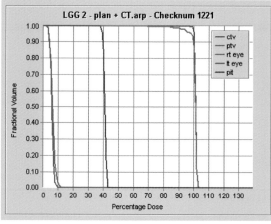

(c)

Fig. 11.5 *continued* (b) Three field non-coplanar beam arrangement, and dose distribution. The pituitary, though not visible at this level, has been outlined in order to note and record its dose. The isodoses follow the convention of 100%, 95%, 90%, 80%, 60%, 40%, 20%. (c) DVH for this plan. The histogram for the PTV looks imperfect, with some volume receiving < 95% of the dose. However, this is due to the PTV lying just into the build up region, and illustrates the difficulty of using the DVH for plan assessment in that situation.

Verification
As for HGG[67].

Dose prescription
◆ Standard:
  • Total dose 54 Gy in 30 fractions over 6 weeks, in a single phase.

54Gy in 30 fractions over 6 weeks in a single phase is a well-tolerated schedule, and strikes a good balance between efficacy and toxicity. It has been used in formal clinical studies, especially in North America. Some centres have used 55 Gy in 33 fractions, but there is little difference between the two.

The EORTC undertook a randomized trial of two dose levels (45 Gy versus 59.4 Gy). They found no tumour control advantage with the higher dose but there was worse toxicity in the patients treated with the higher dose[71], and other data is consistent with this[72]. The most recent EORTC LGG studies have favoured a dose of 50.4 Gy in 28 fractions of 1.8 Gy/fraction, and this is being used in the BR13 (EORTC 22033–26033) chemotherapy versus radiotherapy trial. This dose may become the standard across Europe.

- Gliomatosis cerebri:
  - Total dose 55 Gy in 33 fractions over 6½ weeks, in 1 phase.

Use of a lower dose per fraction of 1.67 Gy/fraction may improve tolerability when typically most of the brain is being treated.

11.4.2 Palliative radiotherapy

Indications

Most patients with LGG, if treated, are treated with radical intent and the need for palliative RT is relatively uncommon. It may be used in patients with a significant neurological deficit where histology suggests LGG, but radiology implies HGG. Effectively, such patients should be managed as if they have HGG. Rarely, an elderly patient with a poor PS and a definitive diagnosis of LGG, can be treated with standard short course fractionation of 30 Gy in six fractions, treating three times per week.

Treatment volume and definition

As per HGG, CT planning should be used, which identifies the position of the tumour. Visible gross tumour, hypodense for LGG, should be outlined as the GTV, and the CTV margin grown round this using a margin of 1.5 cm. An appropriate PTV margin, typically of 0.5 cm, should be added (see notes on 'Planning target volume' in section 11.2.3).

Planning technique

Use the same technique, including dose, as for palliative HGG.

11.5 Ependymoma (intracranial)

A rare tumour occurring most often in the posterior fossa, though it can arise in a supra-tentorial location, this is curable in approximately 50% of cases. Surgical resection is important; as with children complete resection probably confers an advantage[73]. The role of adjuvant chemotherapy in adults is unclear. Spinal ependymoma is addressed later (see section 11.11).

11.5.1 Radical radiotherapy

Indications

Radiotherapy is used as a postoperative adjuvant treatment.

Treatment volume and definition

A major development in radiotherapy management was the finding that cranio-spinal axis radiotherapy adds nothing to the local control of ependymoma, compared to localized radiotherapy[74].

Treatment volume and localization should be along the lines of high grade glioma, but in one phase. Localize using CT and MRI, to treat the postoperative volume. The GTV should be residual contrast-enhancing tumour; if a complete resection has been achieved, consider the surgical cavity as the GTV. Always appraise the preoperative imaging as well, and obtain the operation note to determine the completeness of resection. Add a CTV margin of 1.5–2.5 cm, grown isotropically, and edited along the skull. The PTV should be grown from the CTV.

Planning technique

Patient position and immobilization
This is more commonly a posterior fossa lesion. Treat in a thermoplastic shell, either prone (in a two-piece shell), or supine in an 'S' frame.

Volume localization
See 'Treatment volume and definition' in section 11.5.1.

Dose distribution
Use 6-MV or equivalent X-rays. For posterior fossa lesions, the usual beam arrangement is an isocentric conformal plan with two lateral oblique fields, with central axes parallel to the petrous ridges, plus a posterior oblique field. Supra-tentorial tumours require individual planning along the lines of HGG. IMRT may provide an alternative treatment approach.

Dose homogeneity better than 95–107% should be achievable. Note the dose to the pituitary, and try to minimize this without compromising the CTV. It may also be possible to limit dose to the middle ear and cochlea, especially using IMRT.

Implementation
As for HGG, the patient is positioned using reference marks drawn on the shell. The isocentre position is established using movements from the fiducial marker positions on the planning CT.

Verification
As for HGG[67].

Dose prescription
◆ Total dose 55 Gy in 33 fractions over 6½ weeks.

An alternative is 54 Gy in 30 fractions. This is the highest dose within tolerance of the brainstem, and is worthwhile because those patients that fail frequently do so at the primary site. For supratentorial anaplastic ependymoma it is worth considering 60 Gy in 30 fractions (as for GBM), even though there is no formal evidence of dose response.

11.6 **Medulloblastoma**

This is rare in adults, but does occur. Surgical resection should be as extensive as possible, as for children. Patients should be staged, preferably preoperatively, with MRI of the whole cranio-spinal axis (CSA). The spinal MRI will also define the lower border of the thecal sac. In women, pelvic MRI can be used to identify the position of the ovaries, in order to estimate dose, which can be minimized by careful technique[75]. Postoperative MRI is also needed to confirm the extent of resection.

Adults are said to tolerate CSA radiotherapy better than children[76], but caution is needed in monitoring blood counts. Obsessional attention to radiotherapy detail is required, exactly as for children with this condition (see Chapter 15). There is less need to consider reducing doses in adult patients than in children, and full standard doses can be used.

There is increasing interest in the use of IMRT for cranio-spinal RT, especially in combination with image guidance and positional correction, and its use is increasing. IMRT improves target volume coverage; it also reduces the volume of many normal tissues receiving high dose, albeit at the expense of larger volumes receiving low dose[77–79]. The use of helical tomotherapy allows treatment of the CSA in a single field, thus entirely avoiding field junctions. At the present time this is only possible using TomoTherapy™. IGRT appears to make field junctioning safer[77]; it can also potentially reduce the PTV margin used, or increase the certainty of target volume coverage.

The results of randomized trials of adjuvant chemotherapy in children with medulloblastoma are so compelling that this should be considered in adults. However, vincristine given concurrently with radiotherapy in adults may be more toxic than in children, and bone marrow toxicity may be worse. A formal clinical trial is still needed to address the issue of chemotherapy in adults.

Supra-tentorial primitive neuroectodermal tumour (PNET) and pineoblastoma are probably best managed according to the same concepts.

11.6.1 Radical radiotherapy

Indications

Radiotherapy is used as a postoperative adjuvant treatment.

Treatment volume and definition

The first phase CTV is the CSA. As noted earlier, scrupulous attention to radiotherapy detail is required, as per paediatric practice (see Chapter 15 for details). This includes care to achieve adequate coverage of the cribriform plate and optic nerves, the lower end of the thecal sac, and the full width of the spinal canal including lateral extensions of the theca. In the adult it is not necessary to treat the whole of the vertebral body. For medulloblastoma, phase 2 treats the posterior fossa, though there is interest in reducing this to the tumour bed plus margin. For supratentorial PNETs, phase 2 treats the demonstrable disease plus a margin as for HGG. Spinal or other metastases should receive a boost.

It is essential to use an adequate PTV margin, even with good immobilization and image guidance. This margin may vary along the length of the CSA[78,80].

Planning technique

In the modern era, radiotherapy should be planned from CT (of the head and whole spine). See Chapter 15 for full details.

IMRT may achieve better target volume coverage than conventional conformal RT[78].

Patient position and immobilization

Traditionally patients are treated prone, because this facilitates reliable field junctioning. However, the advent of image guidance allows for greater reliability in positioning so that supine position may be used. This applies especially for techniques where the whole CSA is treated in a single field, thus avoiding junctions[78–80].

Volume localization

CT is the most satisfactory method to localize the target, for phase 1. For the phase 1 (CSA) treatment, the head fields normally need compensators, which are ideally planned from CT. Where IMRT is used, compensation is achieved by the plan optimization process. Individualized shielding blocks or MLC shielding can also be drawn from CT, provided the slice spacing is small, e.g. 1 mm. An alternative is to use CT to produce the compensators, and simulator films for the facial shielding, though this has become outdated. Phase 2 should also be planned from CT, with coregistered MRI. See Chapter 15 for details.

Dose distribution

Use 6-MV or equivalent X-rays. For the cranial part of phase 1, the standard beam arrangement is to use parallel opposed lateral fields, with compensators, and with the central axes placed at the outer canthus. The spine normally requires two direct posterior fields in an adult, though this depends on the height of the patient. See Chapter 15 for details. For phase 2, the usual beam arrangement is an isocentric conformal plan with two lateral oblique fields, with central axes parallel to the petrous ridges, plus a posterior oblique field (Fig. 11.6a). A dose homogeneity better than 95–107% should be achievable. Note the dose to the pituitary, and try to minimize this without compromise to the CTV.

An IMRT plan for the phase 1 CSA volume is shown in Fig. 11.6b. Note the patient is supine.

Implementation

See Chapter 15 for details.

Verification

As for HGG[67]. Measurement of lens dose using TLD should be carried out. Typically, doses are high, and often sufficient for there to be a moderate risk of cataract. However, there is a lower overall hazard to the patient from cataract caused by high lens doses, than from under dose to the cribiform plate with an associated risk of recurrence as the result of overshielding. TLD should also be used to measure testis doses in men. See Chapter 15 for further details.

Dose prescription

Adult doses are different from those used in children.

- Total dose to the posterior fossa: 55 Gy in 33 fractions in 6½ weeks:
  - Phase 1 CSA: 35 Gy in 21 fractions in just over 4 weeks.
  - Phase 2 posterior fossa: 20 Gy in 12 fractions in just over 2 weeks.
  - Spinal metastases boost: 15 Gy in nine fractions in 2 weeks.
  - Cranial metastases boost: 20 Gy in 12 fractions in 2 weeks.

All doses are given at 1.67 Gy per fraction. 55 Gy in 33 fractions is the approximate tolerance of the brainstem.

11.7 Germ cell tumours—germinoma and teratoma

In adults, the standard treatment for intracranial germinoma should be radiotherapy, and results are exceptionally good, with near 100% cure. There is no added advantage in considering dose reduction together with chemotherapy at the present time: the toxicity profile of radiotherapy in adults is excellent, because growth is complete, so

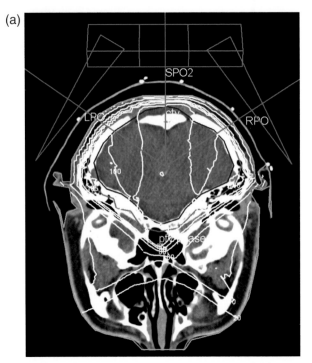

(a)

Fig. 11.6 Medulloblastoma. (a) Dose plan for 6-MV X-rays to treat the posterior fossa for an adult with medulloblastoma. The postoperative pseudo-meningocele is included within the CTV. The venous sinuses do not need to be included in the CTV. The cochleas and semicircular canals lie within the edge of the PTV, and the pituitary (not shown at this level) is adjacent to the PTV at its most anterior extent. The plan is based on a three-field isocentric beam arrangement. However, the optimum dose distribution was obtained using an additional fourth 'field-in-field' for the superior posterior oblique beam (see 'Field arrangements' in section 11.2.4). The posterior oblique fields (LPO and RPO) have weighting of 100%, the superior posterior oblique has 30%, and the smaller 'field-in-field' superior posterior oblique (SPO 2) has 12% weighting. The central axes of the two posterior oblique fields are parallel to the petrous ridges. This spares normal cerebrum, at the expense of the pituitary, which receives 100% of the dose. The lenses of the eye are shielded from all beams. The isodoses follow the convention of 100%, 95%, 90%, 80%, 60%, 40%, 20%.

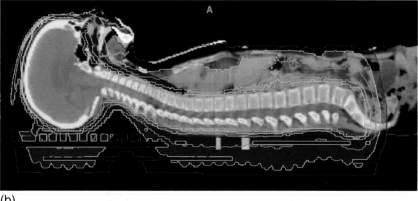

(b)

Fig. 11.6 *continued* (b) IMRT plan for cranio-spinal radiotherapy. The patient is supine, and will be managed with daily image guidance and positional correction. Coverage of the target volume is typically better with IMRT than with conventional approaches, and it also reduces the volume of high dose received by the normal tissues and organs. In this case treatment was delivered with TomoTherapy™, which can treat the whole volume without field junctions. The isodoses shown are 95%, 90%, 80%, 60%, 40%, and 20%. The 100% isodose is turned off for clarity, and there is no tissue receiving ≥ 103%.

the therapeutic ratio is extremely high. Nevertheless, results from paediatric trials will be of interest, if chemotherapy is shown to be sufficiently efficacious to allow safe dose reduction.

Germinoma spreads easily through the CSF, and has a predilection for the optic chiasm and supra-sellar cistern. It may also progress extremely fast. Thus, patients with an entirely curable tumour may go blind or develop cranial nerve palsies within the space of a few days if not handled appropriately. The tumour also responds extremely fast to radiotherapy, but not necessarily to steroids. Occasionally, a few fractions of emergency radiotherapy must be given to reverse neurological deterioration, while the full CSA plan is prepared.

The management of intracranial teratoma is entirely opposite, since this tumour is much less radiosensitive than germinoma. In adults, these tumours are very rare, and the management should be based on paediatric protocols; see Chapter 15. Broadly, the diagnosis may be made on raised markers alone or on biopsy in non-secreting tumours, and treatment commences with chemotherapy. Residual tumours may be offered surgical resection. Radiotherapy may be given to the primary site, to a dose of approximately 54 Gy in 30 fractions. See Chapter 15 for full details.

11.7.1 Radical radiotherapy—germinoma

Indications

Following biopsy, radiotherapy is used as the definitive treatment modality.

Treatment volume and definition

Because of the high risk of CSF spread, the first phase is CSA radiotherapy. As with other cranio-spinal radiotherapy, scrupulous attention to detail is required (see earlier and Chapter 15 for details).

Phase 2 treats the primary site, and any sites of metastatic disease. The imaging on which this is to be based needs to be performed before starting treatment, because the tumour is so radiosensitive. Localize using CT and MRI. The GTV is the contrast-enhancing tumour. The CTV margin can be small, e.g. 1–2 cm, grown isotropically. The PTV should be grown from the CTV, with a standard margin.

Planning technique

As per medulloblastoma.

Dose prescription

+ Total dose: 40 Gy in 24 fractions, at 1.67 Gy/fraction:
  - Phase 1 CSA: 25 Gy in 15 fractions in 3 weeks.
  - Phase 2 boost: 15 Gy in 9 fractions in 2 weeks.

For cranial or spinal metastatic deposits, aim for a total of 40 Gy. Where there is extensive spinal disease, it is possible to treat the whole CSA axis to 40 Gy, with careful observation of FBC.

If the disease is rapidly progressive, then radiotherapy may need to be started urgently, occasionally within hours. This may prevent sophisticated planning. However, it is possible to give a few fractions to the site of progressive disease, and then wait while the cranio-spinal treatment is planned. In principle, the emergency dose could be subtracted from the boost, but with only relatively modest doses required, and in the context of rapid tumour growth, it is probably safer (in adults) to give the full 40 Gy total dose in the planned phases.

11.8 **Vestibular (acoustic) schwannoma**

A multidisciplinary approach is desirable for the management of patients with acoustic neuroma (more correctly though less commonly known as vestibular schwannoma). Some acoustics are associated with neurofibromatosis (NF). These patients may also have other tumours and their care may be extremely complex.

Treatment is only required with disease progression, but the choice of treatment is not completely clear. For a review see Mendenhall et al.[81]. The choices are between observation, surgery, and radiotherapy. For surgery, there are differences in surgical approach, which affect the complications. For radiotherapy there are two main choices: either single fraction stereotactic radiosurgery (SRS) using a gamma knife or linac-based X-knife, or fractionated (stereotactic) radiotherapy (FSRT)[82]. Patient preference must be considered, and this group of patients is typically very well informed.

Surgery is an excellent treatment, provided complete excision is achieved without undue morbidity, and should be curative. It is suitable for small tumours. Surgery is the only modality which can relieve brainstem compression, but patients with very large tumours may not be fit enough for major surgery. Hearing loss may occur as an inevitable part of the procedure, but this depends on the approach. Brainstem (cochlear) implants are possible. Surgery is required for salvage after radiotherapy failure[81].

Published results for radiotherapy suggest excellent efficacy and low morbidity, whichever technique is used[82]. For small lesions, SRS is an excellent choice. For larger lesions, or patients unsuitable for SRS, fractionated stereotactic conformal radiotherapy is necessary.

With SRS, 5-year local control rates of 92%, and discernible hearing preservation of 75%, have been reported for small tumours[83]. Persistent impairment of function of the facial nerve occurred in 1%, and trigeminal in about 1.5%, though transient symptoms occurred in a slightly higher proportion. Results were less good for larger tumours, e.g. 35–45 mm, both for tumour control and toxicity. This group also reported transient vestibulo-cochlear symptoms in 13%. This certainly can occur with FSRT as well, though the rates of this temporary side effect are not clear.

Results for FSRT are broadly similar. Control rates at 5 years have been reported in the region of 97–98%, and useful hearing preservation in around 85%[84,85]. Fuss et al.[84] reported no facial nerve problems, but a 4% risk of trigeminal dysaesthesia, and patients with NF2 had a lower rate of hearing preservation (60%)[84].

Another group has reported results of both SRS and FSRT[86]. Although outcomes were similar, there was a slight advantage in favour of FSRT, as a result of lower rates of trigeminal nerve impairment, 8% and 2% with SRS and FSRT respectively[86].

SRS reports have clearly demonstrated that normal tissue effects are related to technique and dose, and outcomes from SRS have improved in the last decade[82]. Normal tissue effects with FSRT may also be dose-related, and nerve damage rates may be lower if lower doses are used.

11.8.1 Radical radiotherapy

Indications

A progressive acoustic requires some form of treatment, and radiotherapy is an excellent modality, with a high therapeutic ratio. Here, only the technique for FSRT will be described; SRS is a highly specialized technique, which should be studied in more specific texts.

Treatment volume and definition

Radiotherapy should be fully conformal. The target is well shown on MRI (T1W + Gd), but not CT, so that coregistration is essential for radiotherapy planning. For high precision planning, the registration must be achieved to better than 1 mm accuracy. Assessment of the quality of coregistration can be assisted by using a parasagittal view, comparing the location of the bony canal of the internal auditory meatus (IAM) which shows on CT, with the tumour extension down the IAM which shows on MRI.

The GTV is defined as the contrast-enhancing abnormality on MRI, including extension into the IAM. No CTV margin is needed, provided the imaging and localization are secure. High precision immobilization in a relocatable stereotactic frame is recommended, to minimize the PTV margin.

Adjacent critical normal structures, such as the brainstem, can be outlined. The cochlea can also be outlined but normally lies within the PTV. The position of the pituitary gland should be reviewed. It can be helpful to outline this, to clarify for the future the dose received, but it is normally far enough away from the target to receive only minimal dose. The ipsilateral parotid gland should be noted, and can be outlined if desired. The lenses should be outlined to provide estimates of dose.

Planning technique

Patient position and immobilization
Ideally, treat in a relocatable stereotactic frame, which allows a smaller PTV margin. This requires supine treatment, but access for posterior oblique beams is possible because the frame extends beyond the end of the treatment couch. Edentulous patients, or those with significant sensory or motor impairment, should be positioned in a beam direction shell[14].

Volume localization
Start by checking the precision of the CT: MR coregistration, and correct it if it is not acceptable; aim for accuracy better than 1 mm. Outline the GTV as contrast-enhancing tumour on the T1W + Gd MRI, as noted earlier. Always use the CT information as well, especially the bone detail of the IAM.

No CTV margin is needed, provided that the outlining is secure. Grow the PTV isotropically from the CTV, as noted previously.

Dose distribution
Use 6-MV or equivalent X-rays. Normally, the plan uses three isocentric, conformal coplanar fields, with appropriate wedging. A typical arrangement is anterior oblique (wedged), lateral (open) and posterior oblique (wedged) (see Fig. 11.7).

A dose homogeneity of about 95–102% should be achievable. Try to ensure that areas of dose greater than 100% are within the CTV, and not within critical normal tissues. Review the dose distribution slice by slice and the DVHs for CTV and PTV (but note comments in 'Assessment of dose distribution' in section 11.2.4); then review critical normal tissues in the same way. It is important to avoid a 'hot spot' lying within a critical normal tissue, to avoid 'double trouble'[31].

Implementation
The patient is positioned using the stereotactic radiotherapy frame reference system, or fiducial markers if a shell is necessary.

Verification
As for HGG[67]. Measurement of lens dose using thermoluminescent dosimetry (TLD) is recommended to prove that the lens dose is trivial.

Dose prescription
- Total dose 50 Gy in 30 fractions over 6 weeks.

Some centres use higher doses, in the region of 54 Gy. Some use shorter fractionation schedules for small tumours, but these are untested at present. Although the hearing preservation rate is lower in NF2 patients a lower dose is not recommended lest the tumour control rates are reduced.

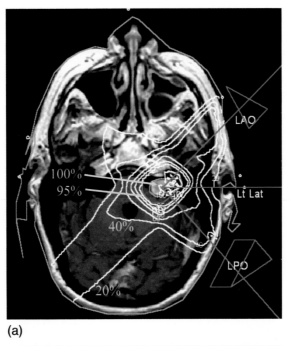

(a)

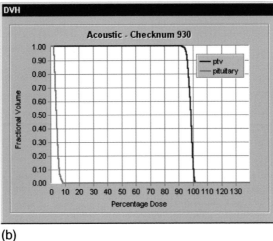

(b)

Fig. 11.7 (a) Typical three-field beam arrangement to treat a small acoustic, with fractionated stereotactic radiotherapy (FSRT). The coregistered MRI is shown under the dose distribution, though the planning CT was used to determine electron density for the dose calculation. This patient was treated in a thermoplastic shell, for specific reasons. More usually, patients would be treated in a stereotactic frame, when the CTV–PTV margin can be reduced (see 'Immobilization methods for radical radiotherapy' in section 11.2.1). The isodoses follow the convention of 100%, 95%, 90%, 80%, 60%, 40%, 20%. (b) The DVH confirms excellent coverage of the PTV, without 'hot spots'.

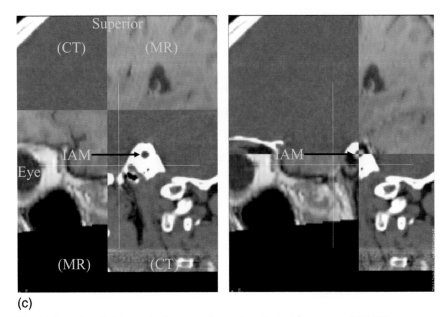

(c)

Fig. 11.7 *continued* (c) A sagittal view can be used to check the accuracy of CT:MR co-registration, especially in the superior–inferior direction. The internal auditory meatus (IAM), i.e. the canal in the bone, is seen easily on CT, and the acoustic itself, which fills the canal, is demonstrated well on MRI.

11.9 Pituitary tumours (including craniopharyngioma)

As with other CNS tumours, the management of pituitary tumours, including craniopharyngioma, requires a multidisciplinary approach[4,87]. Broadly, radiotherapy should be seen as an adjunct to neurosurgical management. For pituitary adenomas, surgery achieves decompression, particularly of the optic chiasm, early reduction in hormone levels in secretory cases, and histological confirmation of the diagnosis. For craniopharyngioma, surgery achieves decompression and histological confirmation.

Radiotherapy following surgery produces very high rates of control. Overall, control rates for pituitary adenoma should be around 95% at 10 years[88–90], and 90% at 20 years[88]. Control rates for craniopharyngioma are around 85% at 10 years, and 80% at 20 years[91]. This risk of second tumour after pituitary radiotherapy has been estimated at about 1% per decade[92].

Reports of the use of radiosurgery (SRS) for pituitary adenoma are increasing. There is an obvious attraction of a single treatment episode, compared to 25 fractions. SRS is typically used for smaller tumours. Serious toxicity, e.g. blindness, which has been reported with SRS[4], is now uncommon. However, there is still no clear evidence that the results are truly comparable, especially since SRS series often have comparatively short follow up and their typical comparators are series of fractionated treatment often from many decades ago[93].

11.9.1 **Radical radiotherapy**

Indications—pituitary adenoma

For pituitary adenomas, the indications for radiotherapy are relative. Not all cases need to be treated. Indications in favour of radiotherapy include extensive residual tumour, invasion of the cavernous sinuses laterally, uncontrolled elevated hormone levels, and progression of tumour after surgery alone. Factors militating against radiotherapy include normal pituitary function, and young age because of the risk of second tumour. None of these factors is absolute, and patient preference should also be considered.

Indications—craniopharyngioma

For craniopharyngioma, radiotherapy is recommended for all patients. Recurrence rates following surgery alone, even after apparently complete removal, are high. The resulting recurrence can increase neurological deficits, which are often present in patients with this disease. The consequences of recurrence far outweigh the risks from RT. Patients often have hypopituitarism already. In general, all cases should be treated, after appropriate surgical decompression. Typically, conservative surgery and RT is far less morbid than 'radical' surgery attempting to remove all tumour, which may cause neurological traction injury.

Treatment volume and definition—pituitary adenoma

The target is based on residual tumour bulk. This may include supra-sellar extension, spread into the cavernous sinuses or even the middle cranial fossa, and occasionally into the sphenoid sinus. Radiotherapy should be fully conformal. CT planning provides greater accuracy of localization, especially of the lateral margins, and any supra-sellar extension. A current MRI (T1W + Gd) should also be obtained, and coregistered. The combination of CT, which shows the bony anatomy around the sella, with MRI, to show the tumour and surrounding soft tissues, provides for optimum delineation.

The GTV is defined as the contrast-enhancing abnormality, best shown on MRI. Since cavernous sinus involvement can be seen on MRI, this is by definition included in the GTV. CT is useful to show the inferior limit, determined by bony structures of the skull base. It is not necessary to treat surgical packing in the sphenoid, but occasional tumours do invade the sphenoid. The cavernous sinuses can be localized from CT, provided the anatomy is understood. This is especially helpful for intra-sellar microadenomas if MRI is not used.

Adenomas do not infiltrate at a microscopic level. No CTV margin is required for infiltration, but a small margin may be needed if there is uncertainty about the outlining. The PTV margin should be grown isotropically, appropriate to the patient immobilization device used (see 'Planning target volume' in section 11.2.3).

Ideally, critical normal structures, especially the optic nerves, chiasm and tracts, should be outlined, so that the dose distribution and the DVHs for these can be reviewed. They can be outlined individually, or as a single unit. The expected DVH will differ according to which method is used. Alternatively, the dose plan must be reviewed on each CT slice to exclude 'hot spots' within these structures. The lenses should be outlined to provide estimates of dose and to remind the planners that they should be avoided. If IMRT is used additional structures may need to be delineated, including the lacrimal glands, brain, brainstem, and occasionally the parotid glands.

Treatment volume and definition—craniopharyngioma

Craniopharyngiomas are more difficult to plan. They are very adherent at both a macroscopic and microscopic level, and all parts of the brain with which the original tumour was in contact need to be treated. Resection, which decompresses the tumour, reduces the mass effect, and shifts the anatomy. Neither pre-op nor post-op imaging exactly represents the location of the target, so that the internal anatomy of the brain must be considered in the planning.

There may be residual solid or cystic remnants. If so, these represent GTV. Postoperative MRI is essential to demonstrate the extent of the GTV. It is important to treat all areas of the brain with which the cyst wall has been in contact. For this, co-registration of the preoperative MRI may be helpful, even though it does not represent the target accurately. The CTV is therefore the surface of the pre-op GTV, plus a margin for any localization uncertainty at doubtful margins, and including any postoperative GTV. If displaced brain has fallen back into place after surgical decompression, the CTV can follow this. If a substantial amount of tumour has been removed, then the CTV is decided by following the anatomy of the brain. The PTV margin is grown isotropically, depending on the immobilization method used.

Planning technique

Patient position and immobilization

Position the patient supine. Immobilize in a shell system, cut out for treatment, or relocatable stereotactic frame. The frame has the advantage of allowing a smaller PTV margin, which can reduce the volume of surrounding normal tissue which receives high dose. This is relatively less important in the pituitary than at other sites because the doses used are lower (discussed later).

Traditionally, it was suggested that the patient be positioned chin down, so that an anterior beam could enter above the eyes without couch twist and gantry rotation. However, this is a relatively uncomfortable position, and cannot usually achieve the desired angulation anyway. Therefore, it is better to position the patient with the head and neck in a comfortable, neutral position.

Volume localization

Use the planning CT, coregistered with T1W + Gd MRI (see section 11.2.2). Outline the GTV as contrast-enhancing tumour. The MRI shows the tumour well, and CT the bones. Inferiorly, it may be difficult on MRI to distinguish tumour from bone marrow in the skull base, but CT shows the bone detail well. Reviewing the drawn volume on coronal section can be helpful in confirming coverage, especially superiorly and inferiorly (see Fig. 11.8a,b).

Since these tumours do not infiltrate microscopically, the CTV can be the same as the GTV, provided that the outline is secure. If not, then a small margin of 1–3 mm can be added. Grow the PTV isotropically from the CTV, as noted previously.

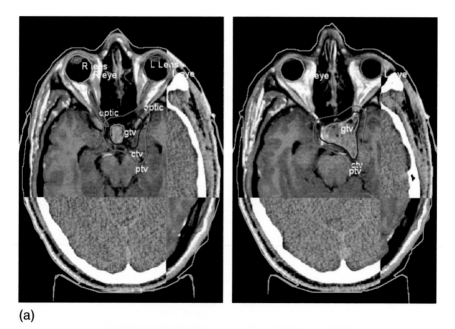

(a)

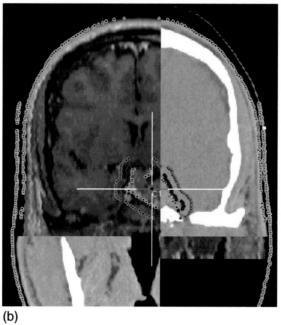

(b)

Fig. 11.8 (a) Outlining for pituitary adenoma with coregistered CT and MRI. Two different levels are shown, emphasizing the outlining of the optic nerves and chiasm; this is useful when reviewing the dose plan to confirm that 'hot spots' do not lie within these critical structures. Note that the chiasm, though not shown, lies entirely within the PTV. Both CTV and PTV have been grown isotropically. (b) The coronal view confirms that the outlined GTV covers the tumour, and shows the value of CT to confirm the inferior extent. This view also helps to confirm the accuracy of CT:MR coregistration in the superior–inferior dimension.

Dose distribution

There is a modest advantage in using higher energy X-rays, of 10–15 MV, compared to 6 MV. The higher energy allows a small reduction in dose to the lateral structures, specifically the temporal lobes, and to the hair-bearing skin. There is an exception: if very small fields are required, dosimetry may be more secure with 6-MV X-rays.

The fields should be conformal, though occasionally IMRT may be useful. The conventional beam arrangement is to use an anterior (or anterior-superior-oblique) beam, entering above the eyes, and two wedged opposed lateral fields. This is an effective beam arrangement, but the opposed fields deliver a fairly high dose, typically 60%, to the temporal lobes. By rotating these beams posteriorly, or even anteriorly, the temporal lobe dose is reduced, at the expense of a larger volume of normal tissue receiving a low dose (see Fig. 11.9). The anterior-superior-oblique beam can enter through the forehead, where there is no hair to be lost. Superior beams are not ideal because of exit dose through the whole body.

A dose homogeneity of about 95–102% should be achievable. Review slice by slice, and assess the DVHs for CTV and PTV. If the optic structures have been outlined then their DVHs can also be checked. The advantage of outlining them is to verify their dose, specifically to ensure that they do not unknowingly receive a dose above 100%. Since they fall within the PTV, it is important to avoid doses which exceed 100%, to be certain to avoid 'double trouble'[31]. Occasionally the DVH for the PTV has a lower dose tail, resulting from the PTV including air in the sphenoid sinus. This is analogous to the build up effect at the skin. Clinically this is not a problem, provided there is 'build up' tissue outside the CTV which would move with any inferior positional displacement.

Implementation

The patient is positioned using reference marks drawn on the shell. The isocentre position is established using movements from the fiducial marker positions on the planning CT.

Verification

As for HGG[67]. Measurement of lens dose using TLD shows that the lens dose is as low as expected from the plan, and may be useful later to prove this if there are questions about later cataract.

Dose prescription

Doses around 45–50 Gy are typical. Doses may vary slightly from one centre to another, according to what is considered as the tolerance dose of the chiasm.

- Pituitary adenoma:
  - Total dose 45 Gy in 25 fractions over 5 weeks (at 1.8 Gy/fraction).
- Pituitary adenoma—extremely large (e.g. with fields > 6 cm):
  - Total dose 50 Gy in 30 fractions over 6 weeks (at 1.67 Gy/fraction).
- Craniopharyngioma:
  - Total dose 50 Gy in 30 fractions over 6 weeks (at 1.67 Gy/fraction).

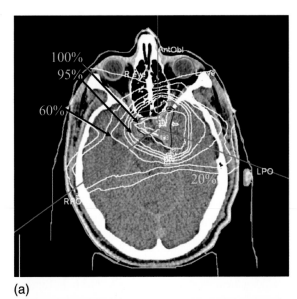

(a)

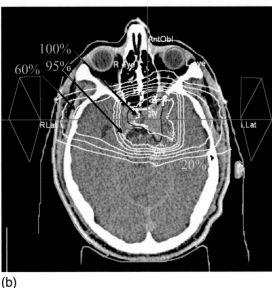

(b)

Fig. 11.9 Two possible beam arrangements and dose distributions for the pituitary adenoma case in Fig. 11.8, using 15-MV X-rays. (a) An anterior-superior-oblique and two posterior oblique fields are shown. No wedges are required, because the beams are evenly spaced. Note the temporal lobe dose. The maximum dose is 103%, and lies within gross tumour (i.e. within the GTV). The other isodoses follow the convention of 100%, 95%, 90%, 80%, 60%, 40%, 20%. (a) The same anterior-superior-oblique field has been used, and two opposed lateral wedged fields have replaced the posterior-oblique beams. The maximum dose is 101%. The temporal lobe doses are higher, the 60% isodose covering both temporal lobes, but less brain receives low dose irradiation.

Patient care during radiotherapy

Many patients who receive radiotherapy are already taking corticosteroid replacement. Many will experience radiotherapy as a stressful event and will consequently require an increase in steroid dose by 50–100%. This should be the 'default' strategy. A minority seem unperturbed by having radiotherapy and in those patients a dose increase may not be necessary.

Craniopharyngioma patients should be asked to report any change in vision immediately, because of the risk of re-accumulation of a craniopharyngioma cyst causing compression of the optic chiasm. Urgent MRI is needed, and if cyst recurrence is confirmed, repeat operation is essential, to try to preserve vision[94]; this is a neurosurgical emergency. Radiotherapy does not cause the re-accumulation, but may be coincidentally associated, since it is given postoperatively.

11.10 Meningioma

11.10.1 Radical radiotherapy, primary and adjuvant

Meningiomas are usually highly localized, but some have a tendency to spread, especially along meningeal surfaces. The degree of macroscopic infiltration is not related to grade. However, microscopic brain invasion is a feature of the WHO 2007 criteria for Grade II meningiomas. They may invade bone. Skull base tumours may invade through the foramina in the skull base to enter the infratemporal and pterygopalatine fossae and the orbit. Very careful consideration must therefore be given to any 'tail' of tumour, seen on MRI, spreading along meningeal surfaces in contact with the tumour mass. A detailed knowledge of the anatomy of the skull base and meningeal structures is desirable.

If the bone flap is removed at operation, it is normally an advantage for the reconstruction cranioplasty to be performed before radiotherapy. This improves the surface dosimetry and moves the skin out of the target volume (see 'Beam energy' in section 11.2.4).

Indications

Radical primary treatment
Radiotherapy is a useful treatment for meningioma at inoperable locations which are progressive. The two commonest sites in this category are cavernous sinus (parasellar) meningiomas and optic nerve meningiomas. Meningiomas involving the cavernous sinus cannot be fully resected because of the critical structures within the sinus. Discussion with neurosurgical colleagues should be undertaken to assess whether subtotal surgical removal is appropriate and safe. Biopsy is not required, provided the radiological diagnosis is definite. For meningiomas of the optic nerve, the diagnosis is radiological. Vision may improve after radiotherapy, whilst surgery, even biopsy, may render the patient blind on the affected side.

Radical postoperative adjuvant radiotherapy
The exact indications for radiotherapy following surgical resection are not completely established. However, where it is indicated, radiotherapy is probably better given early; non-randomized data suggest a survival advantage with early RT, though follow-up must be long to demonstrate this[95,96]. The consequences of a recurrence and the ease of reoperation should be considered. In addition, radiotherapy might be considered for a histological grade I meningioma with a very high labelling index, identified using Ki-67 or mini-chromosome maintenance proteins (MCM2)[97].

For a grade I meningioma, where the surgeon feels that further resection would be possible, RT can be deferred and close imaging follow-up undertaken. With a grade II meningioma, the same general approach can be used, although the extent of resection may push the decision towards early RT. The EORTC are conducting a trial (EORTC 22042–26042) examining dose escalation in grade II and III meningiomas, where the dose is dependent on the degree of surgical resection (Simpson grade). For malignant (i.e. grade III) meningiomas, radiotherapy is definitely indicated. Outcome is dependent on grade[98].

Treatment volume and definition

For radical primary treatment, the objective is to treat the tumour bulk, and any extension along meningeal surfaces. In the postoperative setting, radiotherapy must treat any residual tumour bulk, any spread along meningeal surfaces, and the meninges close to the tumour bed, where there is potential for recurrence.

Radiotherapy should be fully conformal. Coregistered MRI (T1W + Gd) is essential in planning radiotherapy for these lesions. Where residual tumour is present, the GTV is defined as the contrast-enhancing abnormality, including visible meningeal extension (see notes on 'Gross tumour volume' in section 11.2.3). Meningeal surfaces in contact with the tumour mass should be evaluated carefully. It is usual to see a 'tail' of enhancement around the tumour, which may be tumour or inflammatory reactive meningeal response (which does not require treatment). There are no completely reliable methods for distinguishing extension of tumour from inflammatory response, but thickening of the meningeal surfaces is suspicious of tumour. In some cases, tumour infiltration can be very extensive, for example extending from the region of the cavernous sinus, along the free edge of the tentorium cerebelli, as far as the attachment of the falx to the tentorium (Fig. 11.10). Meningeal infiltration inferiorly and anteriorly from the middle cranial fossa can also occur, penetrating through the skull base. Infiltration is common across the roof of the sella.

Following successful macroscopic clearance, gross tumour is not evident on postoperative imaging. The areas most at risk from tumour recurrence resulting from microscopic residuum are those areas of the meninges to which the meningioma was attached. Therefore, coregistration of the preoperative MRI with the radiotherapy planning CT is helpful to accurately localize the tumour base (see Fig. 11.11). Identifying and outlining this volume gives rise to a 'pre-CTV'. In defining this 'pre-CTV' volume, the operation note and pathology report should be scrutinized. A particular difficulty is the superior sagittal sinus whose meningeal layers can be extensively infiltrated, without necessarily occluding the sinus itself. A patent sinus may be dangerous to excise, so this is frequently a site for microscopic residuum. Postoperative imaging will show surgical changes as well as tumour, but these areas need not be included if they are away from the bed of the tumour and at risk areas.

Discrepancies between planning CT and coregistered MRI have been reported[99], but modern high-resolution scanning and electronic coregistration have reduced or abolished this problem. It is well worth reviewing the CT data as well as the MRI. The CT shows bony erosion, and also delineates well the fissures and foramina at the skull base, especially in the middle fossa. This may be helpful in outlining extension into and through the skull base. It is also helpful in outlining the full length of the optic nerve, since the canal may be seen more readily on CT than the nerve passing through it is on MRI.

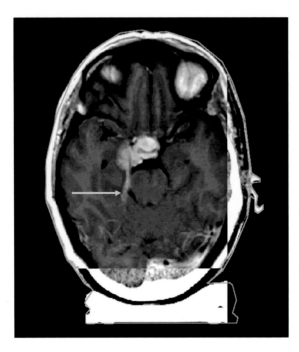

Fig. 11.10 Cavernous sinus meningioma: MRI coregistered to the planning CT, showing tumour extending into the free edge of the tentorium (arrowed). This must be included as part of the GTV. The occipital pad for the relocatable stereotactic frame can be seen.

For cases requiring radical primary treatment, the CTV includes a margin around gross tumour and areas of extension along the meninges. It does not need to include a large margin of adjacent brain. These cases are normally treated with a CTV margin which varies from zero (i.e. CTV = GTV) to 0.5 cm.

In postoperative cases, the CTV margin may be informed by the operation note and histology. Where complete resection has been possible, the 'pre-CTV' should be identified and an appropriate margin for possible tumour spread should be allowed. This should be extended in the directions of possible spread along meningeal surfaces. Only in rare circumstances is it necessary to extend the volume into the brain, although with malignant tumours the surface of the brain in contact with the tumour should also be included in the CTV. More aggressive grades should be treated with a larger CTV margin. The exact margin to use for a CTV is not clear from the literature[100]. For grade II and III meningiomas the CTV margin should probably be in the range of 1–2 cm. A smaller margin can be used where the GTV abuts the brain.

The PTV margin should be added, appropriate to the patient immobilization device used (see 'Planning target volume' in section 11.2.3).

Critical normal structures adjacent to the tumour should be outlined, so that the dose distribution and the DVHs for these can be reviewed. IMRT offers an obvious advantage over standard treatment planning if the dose is to be escalated, as it allows for dose limitation to normal tissues. Dose-painting can also be utilized, particularly if there is residual disease, which can be treated to a higher dose whilst keeping OARs below tissue tolerance.

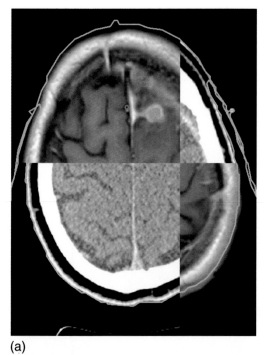

(a)

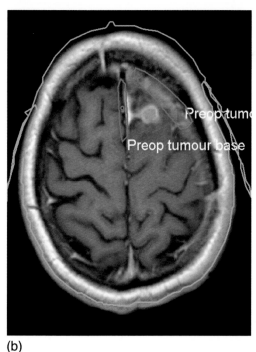

(b)

Fig. 11.11 (a) Planning CT coregistered with the postoperative MRI, showing contrast-enhancing abnormality likely to be residual tumour (different case from Fig. 11.10). The possible meningeal involvement around the original tumour base on the falx is visible, but its extent is difficult to assess. (b) The preoperative MRI was also coregistered with the planning (postoperative) CT, in order to demonstrate and accurately localize the tumour base. The central area outlined represents the falx, which had been slightly displaced by long-standing tumour pressure, and which was relieved at surgery. The surgical operation note was also helpful in assessing this volume.

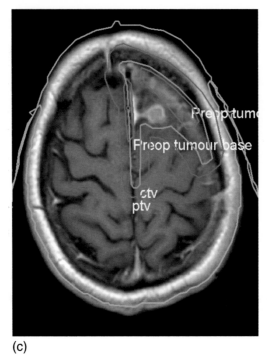

(c)

Fig. 11.11 *continued*
(c) The definitive CTV and PTV have been added, guided by both the pre- and postoperative imaging information (and the operation note).

Planning technique

Patient position and immobilization

The majority of these lesions lie anteriorly within the skull. Treat supine, unless the lesion is located posteriorly. If so, either treat prone, or supine on an 'S' frame-type couch extension (see section 11.2.1). Immobilize in a shell system or relocatable stereotactic frame. The frame has the advantage of allowing a smaller PTV margin, which may help avoidance of important normal tissues.

Volume localization

Use the planning CT coregistered with T1W + Gd MRI (see section 11.2.2). Outline the GTV as contrast-enhancing tumour, as noted earlier. Always use the CT information as well, especially the bone detail. Bony erosion usually indicates infiltration, and extension through the foramina in the skull base may be more readily appreciated by following the bony anatomy.

If possible grow the CTV isotropically and edit. However, different margins may be needed in different directions, such as a smaller margin where the GTV (or 'pre-CTV') is in contact with the brain surface and a larger margin at the meningeal edges. Grow the PTV isotropically from the CTV, as noted previously.

Dose distribution

Use 6-MV or equivalent X-rays. In occasional cases where the tumours are deep seated, higher-energy X-rays may be of value. Base the plan on three isocentric, conformal fields, and expect a non-coplanar arrangement, with wedging on several or all fields.

Beam directions should be chosen imaginatively with consideration both of the target shape and also nearly critical structures (see Fig 11.12). Where cavernous sinus tumours spread posteriorly along the edge of the tentorium, a vertex or superior-oblique field may be useful, even though a substantial distance of normal brain must be traversed (see Fig. 11.12).

A dose homogeneity better than 95–107% should be achievable. Try to ensure that areas of dose > 100% are within the CTV, and not within critical normal tissues. Review the slice by slice dose distribution and the DVHs for CTV and PTV (but note comments in 'Assessment of dose distribution' in section 11.2.4); then review critical normal tissues in the same way. It is important to avoid a 'hot spot' lying within a critical normal tissue, to avoid 'double trouble'[31].

IMRT can often achieve better high dose conformation, though dose can still be limited by the proximity to critical normal tissues (Fig. 11.13).

Implementation
The patient is positioned using reference marks drawn on the shell. The isocentre position is established using movements from the fiducial marker positions on the planning CT.

Verification
As for HGG[67].

Dose prescription
Doses in the range of 50–60 Gy are typical. Lower doses, and lower doses per fraction, can be used for tumours adjacent to the optic nerves and chiasm, determined by their tolerance, and higher doses can be used for vertex lesions. Doses may vary slightly according what is deemed an appropriate tolerance dose. A dose per fraction in the range 1.67–1.8 Gy is recommended. Typical doses are as follows:

- Tumours adjacent to the optic nerves and chiasm:
  - Total dose 50 Gy in 30 fractions over 6 weeks.
- Tumours at other sites:
  - Total dose 55 Gy in 33 fractions, to 60 Gy in 36 fractions in 6½ to 7 weeks.

11.11 Spinal cord tumours (primary)

11.11.1 Radical radiotherapy
The majority of these tumours are intrinsic, predominantly astrocytoma or ependymoma. Glioblastoma is rare, with an appalling prognosis, and recurrence often includes CSF spread. There is no evidence on the potential role of TMZ chemotherapy (concurrent or adjuvant) though it is being used in some centres. Occasional meningiomas and schwannomas occur. Radiotherapy is essentially localized, with the possible exception of lumbo-sacral ependymoma—see later in this section. Dose is limited by cord tolerance.

Indications
Patients should have at least a biopsy carried out, but cautious neurosurgical debulking is almost certainly an advantage. Surgical resection is often incomplete, except in some

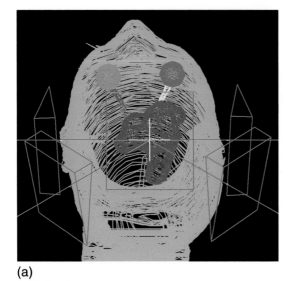

(a)

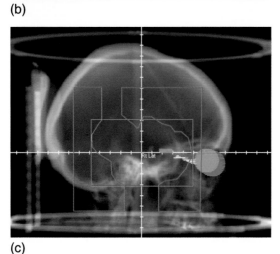

(b)

(c)

Fig. 11.12 Radiotherapy plan for the cavernous sinus meningioma case shown in Fig. 11.10. (a) OEV for five-field plan, emphasizing the use of the vertex (superior oblique) field, useful for treating the posterior extension along the tentorium. The PTV is shown in red; the eyes, optic nerves and chiasm have been outlined. (b) BEV of the superior oblique field, which projects behind the eyes. (c) BEV of the right lateral field, also projecting behind the eyes. The optic nerves run into the field, and the chiasm lies inside. The fiducial marker rings, which attach to the frame for 3D localization during the planning CT, can be seen at the top and bottom. The frame itself is not shown, but the occipital pad and holder can be seen.

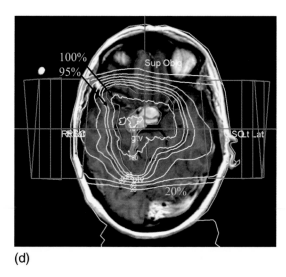

(d)

Fig. 11.12 *continued*
(d) Axial dose distribution
for this plan. The 95% covers
the PTV, and much of the
GTV is covered by the 100%
isodose. The isodoses follow
the convention of 100%,
95%, 90%, 80%, 60%,
40%, 20%.

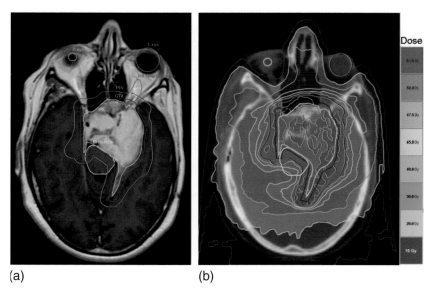

(a)   **(b)**

Fig. 11.13 IMRT plan for a cavernous sinus meningioma similar to that shown in Fig. 11.10.
The PTV is entirely covered by the 95% isodose, which conforms more tightly than the
equivalent conformal plan because of the complex target shape. Note the (median)
prescription dose was 50 Gy in 30 fractions, within tolerance of the brainstem and optic
pathway. The isodoses are given in Gy, and represent 103%, 100%, 95%, 90%, 80%, 60%,
40%, and 20%. The PTV is enclosed by the 95% (47.5-Gy) isodose.

low-grade ependymomas, which arise in the lumbar region. Radiotherapy is then used as a postoperative adjuvant treatment.

In grade I lumbar myxopapillary ependymoma specifically, total excision surgery is curative in only in about half of patients[101]. If the surgeon is confident that a plane of cleavage was found and removal was complete, it may be reasonable to withhold radiotherapy. If recurrence occurs, radiotherapy should certainly be given after further surgery. However, adjuvant radiotherapy RT after surgery appears to significantly reduce the rate of recurrence, which may influence the patient's choice of management[101]. Failures occurred exclusively in the neural axis, mainly at the primary site.

Treatment volume and definition

The preoperative MRI is the basis for localizing the target. However, it is also valuable to have a postoperative, preradiotherapy MRI to show the extent and nature of any residual abnormality. Planning is best carried out with CT.

The treatment volume includes the whole axial dimension of the spinal cord for astrocytoma, and the cord plus the CSF space around it for ependymoma. CT does not show clearly the exact position of the spinal cord. Although this can be clearly seen on MRI, precise coregistration is often not easy. In addition, the neurosurgical procedure may include spinal stabilization, the metal from which produces significant artefact. In practice, the spinal canal contains the CSF space, approximates to the cord being only slightly larger, and is easily defined on the planning CT.

Proximal and distal extension should be based on presurgical imaging with MRI. The location of the tumour can be judged from vertebral levels, and accurately transferred to the planning CT.

For astrocytoma, the GTV is defined as the whole width of the cord, over a length equal to the extent of the preoperative GTV. The CTV is the same axially, which approximates to the spinal canal. Longitudinally, the CTV requires an additional margin beyond the GTV, usually in the region of 2–3 cm proximally and distally. If a spinal stabilization procedure has been necessary, there is often substantial artefact from the metallic implants. However, the position if the spinal canal can normally be reliably interpolated. The PTV margin should be grown isotropically, and a margin of 1.0 cm is recommended.

The exception to this strategy is the lumbar ependymoma. Axially the CTV should extend laterally to include the nerve roots up to the limit of the meningeal covering. For the longitudinal extent of the CTV, some centres treat tumours at the lower end of the spinal cord or arising in the lumbo-sacral nerve roots with two phases, covering the whole thecal sac below the tumour for phase 1, and adding a boost to the primary site for phase 2[102]. The rationale is that these tumours can spread with a 'sugar coating' of tumour along the nerve roots below the primary site, presumably due to gravity. It may be that modern imaging improves the 'staging' of these tumours, so that an alternative is to treat only the primary site with a margin, as outlined earlier. Whether this two-phase approach is better is yet to be firmly established. With the two phase approach, the proximal extent of phase 1 is the same as described earlier, and distally the volume extends to include the whole thecal sac. The lowest extent of this is seen from the MRI (usually in the range S2 to S4). Phase 2 is given to a shortened volume, as per the single phase treatment schedule given earlier for astrocytoma.

The PTV margin should be grown isotropically around the CTV, and a margin of 1.0 cm is recommended. If the patient is prone, the amplitude of respiratory movement can be assessed by 4D CT or in the simulator if available, and this distance included in the anterior–posterior PTV margin. The PTV margin may be reduced to 0.5 cm for upper cervical lesions, if the patient is immobilized in a thermoplastic shell. Image guidance may also be used to reduce the PTV margin.

The occasional spinal meningioma or schwannoma which is inoperable or recurrent must be treated. The GTV is the imageable tumour, and only a small CTV margin is needed (see section 11.10). The PTV should be the same as for other spinal tumours.

It is advisable to outline critical normal structures which lie close by. In the cervical region, dose to the thyroid gland should be considered. In the thorax, the lungs should be contoured so that dose-volume constraints can be considered; the oesophagus normally falls close to the PTV and should be outlined; cardiac doses are normally low and it is not normally necessary to outline the heart; the kidneys lie close to the lower thoracic spine, so their position should be reviewed, and usually outlined. In the lumbo-sacral region, the kidneys are one of the most important normal tissues to contour; in premenopausal women the ovaries may be visible on the planning CT, and if so can be outlined for dosimetry purposes. Normal tissue DVH information can be used in assessing the final dose distribution.

Planning technique

Patient position and immobilization

Patients can be positioned either prone or supine. Prone positioning has the great advantage that set up marks on the patient are directly related to the site to be treated, and the spine can be easily palpated by the radiographers. It also avoids the loss of skin sparing when beams pass through the treatment couch. The big disadvantage of a prone position is that respiration moves the target volume antero-posteriorly, particularly for lesions in the thoracic spine. The amplitude of thoracic movement is of the order of 1.0 cm, and this can be measured in the simulator. Therefore, thoracic tumours can be treated supine. Lying prone may also be less comfortable, an important consideration for positional reproducibility and intrafraction movement. There is an extensive range of immobilization devices, such as vacu-bags, which aid positioning and immobilization, and are comfortable for the patient.

Cervical tumours are traditionally best treated prone, using a beam direction shell. If possible position the chin up, to avoid exit dose. Skin alignment tattoos can be used more reliably in this position. For thoracic tumours position the patient supine, on a thoracic board, with the arms above the head. This allows lateral alignment tattoos to be placed in a reliably reproducible position, and aids set up. Lumbar tumours are ideally treated prone.

Patients who have significant neurological disability may be unable to adopt a prone position easily, and this may affect reproducibility of set-up. Such patients may be better treated supine.

The advent of image guidance with positional correction means that the practical advantages of prone position to allow accurate positioning using external skin marks can be replaced by greater accuracy in a more comfortable supine position, using direct imaging.

Volume localization

See 'Treatment volume and definition' in section 11.11.1.

The GTV is the whole width of the cord, over a length equal to the extent of the preoperative GTV, which is best shown on preoperative MRI. The CTV is the same axially, which approximates to the spinal canal. Longitudinally, the CTV requires an additional margin beyond the GTV, usually in the region of 2–3 cm proximally and distally. If a spinal stabilization procedure has been necessary, there is often substantial artefact from the metallic implants. However, the position if the spinal canal can normally be reliably interpolated. The PTV margin should be grown isotropically, and a margin of 1.0 cm is recommended.

If a lumbar ependymoma is to be treated with a two-phase treatment, phase 1 covers the width of the spinal canal extending laterally in the distal part of the volume to include the whole length of the nerve roots within the meninges. The proximal extent is the same as described earlier, and distally the volume extends to include the whole thecal sac. The lowest position of this is seen from the MRI (and is usually in the range S2 to S4). Phase 2 is given to a shortened volume, as per the single-phase treatment schedule.

The PTV margin should be grown isotropically around the CTV, and a margin of 1.0 cm is recommended. The amplitude of respiratory movement can be screened in the simulator, and this distance included in the anterior–posterior PTV margin. The PTV margin may be reduced to 0.5 cm for upper cervical lesions, if the patient is immobilized in a thermoplastic shell.

Critical normal structures were discussed earlier (see 'Treatment volume and definition' in section 11.11.1).

Dose distribution

Generally, 6-MV X-rays are preferred. A three-field isocentric beam arrangement is normally the most satisfactory, using a direct posterior field with left and right posterior-oblique wedged fields. Longitudinal wedging of the posterior field may be helpful (Fig. 11.14). IMRT may be useful, and in particular may achieve greater dose homogeneity than conformal planning (Fig. 11.15).

Dose homogeneity should be reviewed on every slice and DVHs for the CTV and PTV should be reviewed (see 'Assessment of dose distribution' in section 11.2.4). The dose homogeneity for the important normal tissues should also be reviewed slice by slice, and their DVHs inspected. It is important to avoid any 'hot spot' lying within the cord.

Implementation

The patient is positioned using reference marks applied to the skin at the time of the planning CT, and the isocentre position is established using movements from these marks.

Verification

As for HGG[67].

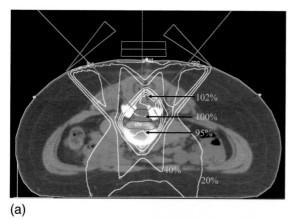

(a)

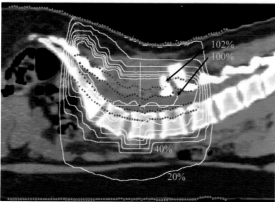

(b)

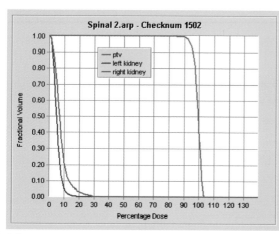

(c)

Fig. 11.14 (a) Typical beam arrangement and dose distribution for treatment of a lumbar ependymoma, to treat L2–S2, using three fields. The patient is prone (see 'Patient position and immobilization' in section 11.11.1). The posterior field has a 15° degree longitudinal wedge, and is weighted at 60%, compared to 100% from each of the posterior oblique fields. The CTV includes the lower end of the thecal sac, and the whole volume was treated in one phase, to 50 Gy in 30 fractions. The isodoses follow the convention of 100%, 95%, 90%, 80%, 60%, 40%, 20%. The spinous process and laminae are missing following laminectomy. The kidneys are close to the upper end of the volume, so they have been outlined for dose review. (b) Longitudinal dose distribution. The missing bone is obvious. The effect of 'stepping' from the MLC on the posterior oblique fields is seen. (c) DVH showing the dose to the target and the kidneys.

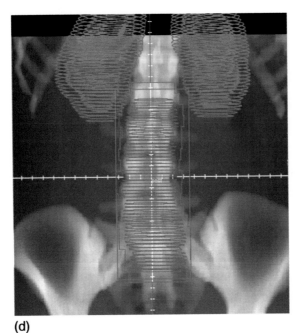

(d)

Fig. 11.14 *continued*
(d) DRR of the posterior field
with target and kidneys
outlined.

Dose prescription

Doses around 50 Gy, with a dose per fraction of 1.67 Gy, are recommended, this being a reasonable safe estimate of cord tolerance. For patients with highly malignant tumours such as anaplastic (grade III) tumours or glioblastoma, a higher dose may be reasonable.

- Intrinsic tumours (astrocytoma and ependymoma):
  - Total dose 50 Gy in 30 fractions over 6 weeks, single phase.
- Lumbar ependymoma:
  - Single phase treatment: Total dose 50 Gy in 30 fractions over 6 weeks.
  - Two phase treatment:
      Phase 1: 45 Gy in 27 fractions over 5½ weeks.
      Phase 2: 5 Gy in 3 fractions in ½ week.
- Glioblastoma:
  - Total dose 54 Gy in 30 fractions over 6 weeks, single phase (the role of TMZ chemotherapy is unknown)
- Extrinsic tumours (meningioma, schwannoma):
  - Total dose 50 Gy in 30 fractions over 6 weeks, single phase.

(a)

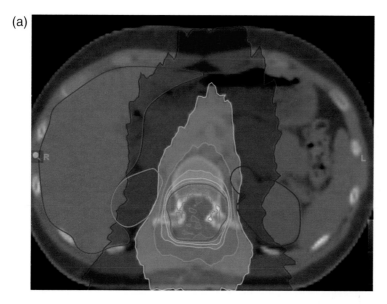

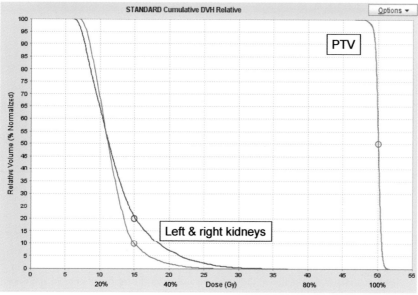

(b)

Fig. 11.15 (a) IMRT dose plan for a high lumbar ependymoma, with target volume extending from T10–L2 inclusive. Note the supine position, which is possible using daily image guidance and reduces respiratory motion of the target compared to a prone position. The PTV (red) is partially covered by the 100% isodose, and entirely enclosed within the 95% isodose. Dose has been pulled medially away from the kidneys as much as possible in order to minimize the dose they receive. The isodoses follow the convention of 100%, 95%, 90%, 80%, 60%, 40%, and 20%. (b) DVH confirming low kidney doses and excellent homogeneous coverage of the PTV.

Acknowledgements

We would particularly like to thank Mr Tony Geater and Dr Andrew Hoole for help in the preparation of the diagrams.

References

1. Burnet NG, Jefferies SJ, Benson RJ, Hunt DP, Treasure FP. Years of life lost (YLL) from cancer is an important measure of population burden—and should be considered when allocating research funds. *British Journal of Cancer* 2005; **2**(2): 241–5.

2. Davies E, Hopkins A (eds). *Improving Care For Patients With Malignant Cerebral Glioma*. London: Royal College of Physicians of London, 1997.

3. Burnet NG, Bulusu VR, Jefferies SJ. Management of primary brain tumours. In Booth S, Bruera E (eds) *Palliative Care Consultations in primary and metastatic brain tumours*. Oxford: Oxford University Press, 2004.

4. Anderson JR, Antoun N, Burnet N, *et al.* Neurology of the pituitary gland. *Journal of Neurology, Neurosurgery and Psychiatry* 1999; **66** (6): 703–21.

5. Haba Y, Twyman N, Thomas SJ, Overton C, Dendy P, Burnet NG. Radiotherapy for glioma during pregnancy: fetal dose estimates, risk assessment and clinical management. *Clinical Oncology* 2004; **16**(3): 210–14.

6. Harris F, Jefferies SJ, Jena R, Burton KE, Muffett L, Burnet NG. Neuro-oncology in pregnancy. In Marsh M, Nashef L, Brex P (eds) *Neurology and Pregnancy: Clinical Management*. London: Informa Healthcare, 2012.

7. Burnet NG, Lynch AG, Jefferies SJ, *et al.* High grade glioma: imaging combined with pathological grade defines management and predicts prognosis. *Radiotherapy and Oncology* 2007; **85**: 371–8.

8. Medical Research Council Brain Tumour Working Party. Prognostic factors for high-grade malignant glioma: development of a prognostic index. A Report of the Medical Research Council Brain Tumour Working Party. *Journal of Neurooncology* 1990; **9**(1): 47–55.

9. Bauman G, Yartsev S, Coad T, Fisher B, Kron T. Helical tomotherapy for craniospinal radiation. *British Journal of Radiology* 2005; **78**(930): 548–52.

10. Parker WA, Freeman CR. A simple technique for craniospinal radiotherapy in the supine position. *Radiotherapy and Oncology* 2006; **78**(2): 217–22.

11. Hanna CL, Slade S, Mason MD, Burnet NG. Translating radiotherapy Clinical Target Volumes into Planning Target Volumes for bladder and brain tumour patients. *Clinical Oncology* 1999; **11**: 93–8.

12. Gilbeau L, Octave-Prignot M, Loncol T, Renard L, Scalliet P, Gregoire V. Comparison of setup accuracy of three different thermoplastic masks for the treatment of brain and head and neck tumors. *Radiotherapy and Oncology* 2001; **58**(2): 155–62.

13. Burton KE, Thomas SJ, Jefferies SJ, Burnet NG. Accuracy of a relocatable stereotactic head frame can be improved by proactive use of a depth helmet and three dimensional vector displacement algorithm. *Radiotherapy and Oncology* 2004; **73**(suppl 1): S30.

14. Burton KE, Thomas SJ, Whitney D, Routsis DS, Benson RJ, Burnet NG. Accuracy of a relocatable stereotactic radiotherapy head frame evaluated by use of a depth helmet. *Clinical Oncology* 2002; **14**: 31–9.

15. Royal College of Radiologists. *Imaging for Oncology. BFCO(04)2*. London: The Royal College of Radiologists, 2004.

16. Albert FK, Forsting M, Sartor K, Adams HP, Kunze S. Early postoperative magnetic resonance imaging after resection of malignant glioma: objective evaluation of residual tumor and its influence on regrowth and prognosis. *Neurosurgery* 1994; **34**(1): 45–60.

17. Rasch CR, Steenbakkers RJ, Fitton I, *et al.* Decreased 3D observer variation with matched CT-MRI, for target delineation in Nasopharynx cancer. *Radiation Oncology* 2010; **5**: 21.

18. Pirzkall A, Li X, Oh J, *et al.* 3D MRSI for resected high-grade gliomas before RT: tumor extent according to metabolic activity in relation to MRI. *International Journal of Radiation Oncology, Biology, Physics* 2004; **59**(1): 126–37.

19. Jena R, Price SJ, Baker C, *et al.* Diffusion tensor imaging: possible implications for radiotherapy treatment planning of patients with high-grade glioma. *Clinical Oncology* 2005; **17**(8): 581–90.

20. Price SJ, Burnet NG, Donovan T, *et al.* Diffusion tensor imaging of brain tumours at 3T: a potential tool for assessing white matter tract invasion? *Clinical Radiology* 2003; **58**(6): 455–62.

21. Price SJ, Pena A, Burnet NG, *et al.* Tissue signature characterisation of diffusion tensor abnormalities in cerebral gliomas. *European Radiology* 2004; **14**(10): 1909–17.

22. Grosu AL, Weber WA, Riedel E, *et al.* L-(methyl-11C) methionine positron emission tomography for target delineation in resected high-grade gliomas before radiotherapy. *International Journal of Radiation Oncology, Biology, Physics* 2005; **63**(1): 64–74.

23. Milker-Zabel S, Zabel-du Bois A, Henze M, *et al.* Improved target volume definition for fractionated stereotactic radiotherapy in patients with intracranial meningiomas by correlation of CT, MRI, and [68Ga]-DOTATOC-PET. *International Journal of Radiation Oncology, Biology, Physics* 2006; **65**(1): 222–7.

24. Thorwarth D, Henke G, Müller AC, *et al.* Simultaneous (68)Ga-DOTATOC-PET/MRI for IMRT treatment planning for meningioma: first experience. *International Journal of Radiation Oncology, Biology, Physics* 2011; **81**(1): 277–83.

25. ICRU. *ICRU Report 50: Prescribing, Recording and Reporting Photon Beam Therapy.* Bethseda, MD: International Commission on Radiation Units and Measurements, 1993.

26. ICRU. *ICRU Report 62: Prescribing, Recording and Reporting Photon Beam Therapy* (Supplement to ICRU Report 50). Bethesda, MD: International Commission on Radiation Units and Measurements, 1999.

27. ICRU. ICRU Report 83: Prescribing, Recording, and Reporting Intensity-Modulated Photon-Beam Therapy (IMRT). International Commission on Radiation Units and Measurements. *Journal of the ICRU* 2010; **10**(1).

28. British Institute of Radiology. *Geometric Uncertainties in Radiotherapy—defining the planning target volume.* London: British Institute of Radiology, 2003.

29. Burnet NG, Thomas SJ, Burton KE, Jefferies SJ. Defining the tumour and target volumes for radiotherapy. *Cancer Imaging* 2004; **4**(2): 1–9.

30. Nguyen TB, Hoole AC, Burnet NG, Thomas SJ. Dose-volume population histogram: a new tool for evaluating plans whilst considering geometrical uncertainties. *Physics in Medicine and Biology* 2009; **54**(4): 935–47.

31. Withers H.R. Biologic basis of radiation therapy. In: Perez C.A. and Brady L.W., Editors. *Principles and Practice of Radiation Oncology.* 2nd edition. Philadelphia, PA: J.B. Lippincott, 1992, pp. 64–98.

32. Emami B, Lyman J, Brown A, Coia L, Goitein M, Munzenrider JE, Shank B, Solin LJ, Wesson M. Tolerance of normal tissue to therapeutic irradiation. *International Journal of Radiation Oncology, Biology, Physics* 1991; **21**(1): 109–22.

33. Marks LB, Ten Haken RK, Martel MK. Guest editor's introduction to QUANTEC: a users guide. *International Journal of Radiation Oncology, Biology, Physics* 2010; **76**(3 Suppl): S1–2.

34. Mayo C, Martel MK, Marks LB, Flickinger J, Nam J, Kirkpatrick J. Radiation dose-volume effects of optic nerves and chiasm. *International Journal of Radiation Oncology, Biology, Physics* 2010; **76**(3 Suppl): S28–35.

35. Lawrence YR, Li XA, el Naqa I, *et al.* Radiation dose-volume effects in the brain. *International Journal of Radiation Oncology, Biology, Physics* 2010; **76**(3 Suppl): S20–7.

36. Karim AB, Afra D, Cornu P, *et al.* Randomized trial on the efficacy of radiotherapy for cerebral low-grade glioma in the adult: European Organization for Research and Treatment of Cancer Study 22845 with the Medical Research Council study BRO4: an interim analysis. *International Journal of Radiation Oncology, Biology, Physics* 2002; **52**(2): 316–24.

37. Taphoorn MJ, Schiphorst AK, Snoek FJ, *et al.* Cognitive functions and quality of life in patients with low-grade gliomas: the impact of radiotherapy. *Annals of Neurology* 1994; **36**(1): 48–54.

38. Kiebert GM, Curran D, Aaronson NK, *et al.* Quality of life after radiation therapy of cerebral low-grade gliomas of the adult: results of a randomised phase III trial on dose response (EORTC trial 22844). EORTC Radiotherapy Co-operative Group. *European Journal of Cancer* 1998; **34**(12): 1902–9.

39. Gregor A, Cull A, Traynor E, Stewart M, Lander F, Love S. Neuropsychometric evaluation of long-term survivors of adult brain tumours: relationship with tumour and treatment parameters. *Radiotherapy and Oncology* 1996; **41**(1): 55–9.

40. Brown PD, Buckner JC, Uhm JH, Shaw EG. The neurocognitive effects of radiation in adult low-grade glioma patients. *Neuro-oncology* 2003; **5**(3): 161–7.

41. Klein M, Heimans JJ, Aaronson NK, *et al.* Effect of radiotherapy and other treatment-related factors on mid-term to long-term cognitive sequelae in low-grade gliomas: a comparative study. *Lancet* 2002; **360**(9343): 1361–8. Erratum in: Lancet 2011; **377**(9778): 1654.

42. Douw L, Klein M, Fagel SS, *et al.* Cognitive and radiological effects of radiotherapy in patients with low-grade glioma: long-term follow-up. *Lancet Neurology* 2009; **8**(9): 810–18.

43. Lee AW, Foo W, Chappell R, *et al.* Effect of time, dose, and fractionation on temporal lobe necrosis following radiotherapy for nasopharyngeal carcinoma. *International Journal of Radiation Oncology, Biology, Physics* 1998; **40**(1): 35–42.

44. Mayo C, Yorke E, Merchant TE. Radiation associated brainstem injury. *International Journal of Radiation Oncology, Biology, Physics* 2010; **76**(3 Suppl): S36–41.

45. Kirkpatrick JP, van der Kogel AJ, Schultheiss TE. Radiation dose-volume effects in the spinal cord. *International Journal of Radiation Oncology, Biology, Physics* 2010; **76**(3 Suppl): S42–9.

46. Merchant TE, Goloubeva O, Pritchard DL, *et al.* Radiation dose-volume effects on growth hormone secretion. *International Journal of Radiation Oncology, Biology, Physics* 2002; **52**(5): 1264–70.

47. Laughton SJ, Merchant TE, Sklar CA, *et al.* Endocrine outcomes for children with embryonal brain tumors after risk-adapted craniospinal and conformal primary-site irradiation and high-dose chemotherapy with stem-cell rescue on the SJMB-96 trial. *Journal of Clinical Oncology.* 2008; **26**(7): 1112–18.

48. Appelman-Dijkstra NM, Kokshoorn NE, Dekkers OM, Neelis KJ, Biermasz NR, Romijn JA, Smit JW, Pereira AM. Pituitary dysfunction in adult patients after cranial radiotherapy: systematic review and meta-analysis. *J Clin Endocrinol Metab.* 2011; **96**(8): 2330–40.

49. Bhandare N, Jackson A, Eisbruch A, Pan CC, Flickinger JC, Antonelli P, Mendenhall WM. Radiation therapy and hearing loss. *International Journal of Radiation Oncology, Biology, Physics* 2010; **76**(3 Suppl): S50–7.

50. Henk JM, Whitelocke RA, Warrington AP, Bessell EM. Radiation dose to the lens and cataract formation. *International Journal of Radiation Oncology, Biology, Physics* 1993; **25**(5): 815–20.

51. Lawenda BD, Gagne HM, Gierga DP, *et al.* Permanent alopecia after cranial irradiation: Dose-response relationship. *International Journal of Radiation Oncology, Biology, Physics* 2004; **60**(3): 879–87.

52. Westbury C, Hines F, Hawkes E, Ashley S, Brada M. Advice on hair and scalp care during cranial radiotherapy: a prospective randomized trial. *Radiotherapy and Oncology* 2000; **54**(2): 109–16.

53. Bennett AH, Godlee RJ. Case of cerebral tumour. *Medico-Chirurgical Transactions* 1885; **68**: 243–75. http://www.ncbi.nlm.nih.gov/pmc/articles/PMC2121420/pdf/medcht00024–0321.pdf

54. Stupp R, Tonn JC, Brada M, Pentheroudakis G; ESMO Guidelines Working Group. High-grade malignant glioma: ESMO Clinical Practice Guidelines for diagnosis, treatment and follow-up. *Annals of Oncology* 2010; **21**(Suppl 5): v190–3.

55. Stummer W, Pichlmeier U, Meinel T, Wiestler OD, Zanella F, Reulen HJ; ALA-Glioma Study Group. Fluorescence-guided surgery with 5-aminolevulinic acid for resection of malignant glioma: a randomised controlled multicentre phase III trial. *Lancet Oncology* 2006; **7**(5): 392–401.

56. Stupp R, Mason WP, van den Bent MJ, *et al.* Radiotherapy plus concomitant and adjuvant temozolomide for glioblastoma. *New England Journal of Medicine* 2005; **352**(10): 987–96.

57. Stupp R, Hegi ME, Mason WP, *et al.* Effects of radiotherapy with concomitant and adjuvant temozolomide versus radiotherapy alone on survival in glioblastoma in a randomised phase III study: 5-year analysis of the EORTC-NCIC trial. *Lancet Oncology* 2009; **10**(5):459–66.

58. Hegi ME, Diserens AC, Gorlia T, *et al.* MGMT gene silencing and benefit from temozolomide in glioblastoma. *New England Journal of Medicine* 2005; **352**(10): 997–1003.

59. Brandsma D, Stalpers L, Taal W, Sminia P, van den Bent MJ. Clinical features, mechanisms, and management of pseudoprogression in malignant gliomas. *Lancet Oncology* 2008; **9**(5): 453–61.

60. Sanghera P, Rampling R, Haylock B, Jefferies S, McBain C, Rees JH, Soh C, Whittle IR. The concepts, diagnosis and management of early imaging changes after therapy for glioblastomas. *Clinical Oncology* 2011; Jul 23. [Epub ahead of print].

61. Magri C, Scoffings D, Antoun N, *et al.* A review of the incidence and radiological features of pseudo-progression in a cohort of glioblastoma patients treated with temozolomide and chemo-radiation. *Radiotherapy and Oncology* 2008; **88**(suppl 2): S633.

62. Bleehan NM, Stenning SP. A Medical Research Council trial of two radiotherapy doses in the treatment of grades 3 and 4 astrocytoma. The Medical Research Council Brain Tumour Working Party. *British Journal of Cancer* 1991; **64**: 769–74.

63. Hochberg FH, Pruit A. Assumptions in the radiotherapy of glioblastoma. *Neurology* 1980; **30**: 907–11.

64. Wallner KE, Galichich JH, Krol G, *et al.* Patterns of failure following treatment for glioblastoma multiforme.and anaplastic astrocytoma. *International Journal of Radiation Oncology, Biology, Physics* 1989; **16**: 1405.

65. Jansen EP, Dewit LG, van Herk M, Bartelink H. Target volumes in radiotherapy for high-grade malignant glioma of the brain. *Radiotherapy and Oncology* 2000; **56**(2): 151–6.

66. Minniti G, Amelio D, Amichetti M, *et al.* Patterns of failure and comparison of different target volume delineations in patients with glioblastoma treated with conformal radiotherapy plus concomitant and adjuvant temozolomide. *Radiotherapy and Oncology.* 2010; **97**(3): 377–8.

67. Royal College of Radiologists, Society and College of Radiographers, Institute of Physics and Engineering in Medicine. *On Target: Ensuring geometric accuracy in radiotherapy.* London: The Royal College of Radiologists, 2008. http://www.rcr.ac.uk/docs/oncology/pdf/ BFCO%2808%295_On_target.pdf

68. Smith JS, Chang EF, Lamborn KR, *et al.* Role of extent of resection in the long-term outcome of low-grade hemispheric gliomas. *Journal of Clinical Oncology* 2008; **26**(8): 1338–45.

69. van den Bent M, Wefel J, Schiff D, *et al.* Response assessment in neuro-oncology (a report of the RANO group): assessment of outcome in trials of diffuse low-grade gliomas. *Lancet Oncology* 2011; **12**(6): 583–93.

70. Morris DE, Bourland JD, Rosenman JG, Shaw EG. Three-dimensional conformal radiation treatment planning and delivery for low- and intermediate-grade gliomas. *Seminars in Radiation Oncology* 2001; **11**(2): 124–37.

71. Karim AB, Maat B, Hatlevoll R, *et al.* A randomized trial on dose-response in radiation therapy of low-grade cerebral glioma: European Organization for Research and Treatment of Cancer (EORTC) Study 22844. *International Journal of Radiation Oncology, Biology, Physics* 1996; **36**(3): 549–56.

72. Shaw E, Arusell R, Scheithauer B, *et al.* Prospective randomized trial of low- versus high-dose radiation therapy in adults with supratentorial low-grade glioma: initial report of a North Central Cancer Treatment Group/Radiation Therapy Oncology Group/Eastern Cooperative Oncology Group study. *Journal of Clinical Oncology.* 2002; **20**(9): 2267–76.

73. Metellus P, Guyotat J, Chinot O, *et al.* Adult intracranial WHO grade II ependymomas: long-term outcome and prognostic factor analysis in a series of 114 patients. *Neuro-Oncology* 2010; **12**(9): 976–84.

74. Vanuytsel L, Brada M. The role of prophylactic spinal irradiation in localized intracranial ependymoma. *International Journal of Radiation Oncology, Biology, Physics* 1991; **21**(3): 825–30.

75. Harden SV, Twyman N, Lomas DJ, Williams D, Burnet NG, Williams MV. A method for reducing ovarian doses in whole neuro-axis irradiation for medulloblastoma. *Radiotherapy and Oncology* 2003; **69** (2): 183–8.

76. Jefferies S, Rajan B, Ashley S, Traish D, Brada M. Haematological toxicity of cranio-spinal irradiation. *Radiotherapy and Oncology* 1998; **48**(1): 23–7.

77. Kusters JM, Louwe RJ, van Kollenburg PG, *et al.* Optimal normal tissue sparing in craniospinal axis irradiation using IMRT with daily intrafractionally modulated junction(s). *International Journal of Radiation Oncology, Biology, Physics* 2011; **81**(5):1405–14.

78. Loo S, Horan G, Fairfoul J, *et al.* Craniospinal radiotherapy using helical tomotherapy for paediatric medulloblastoma: developing dose constraints. *Radiotherapy and Oncology* 2010; **96**(Suppl 1): S370–1.

79. Peñagarícano J, Moros E, Corry P, Saylors R, Ratanatharathorn V. Pediatric craniospinal axis irradiation with helical tomotherapy: patient outcome and lack of acute pulmonary toxicity. *International Journal of Radiation Oncology, Biology, Physics* 2009; **75**(4): 1155–61.

80. Stoiber EM, Giske K, Schubert K, *et al.* Local setup reproducibility of the spinal column when using intensity-modulated radiation therapy for craniospinal irradiation with patient in supine position. *International Journal of Radiation Oncology, Biology, Physics* 2011; **81**(5): 1552–9.

81. Mendenhall WM, Friedman WA, Amdur RJ, Antonelli PJ. Management of acoustic schwannoma. *American Journal of Otolaryngology* 2004; **25**(1): 38–47.

82. Murphy ES, Suh JH. Radiotherapy for vestibular schwannomas: a critical review. *International Journal of Radiation Oncology, Biology, Physics* 2011; **79**(4): 985–97.

83. Rowe JG, Radatz MW, Walton L, Hampshire A, Seaman S, Kemeny AA. Gamma knife stereotactic radiosurgery for unilateral acoustic neuromas. *Journal of Neurology, Neurosurgery and Psychiatry* 2003; **74**(11): 1536–42.

84. Fuss M, Debus J, Lohr F, *et al.* Conventionally fractionated stereotactic radiotherapy (FSRT) for acoustic neuromas. *International Journal of Radiation Oncology, Biology, Physics* 2000; **48**(5): 1381–7.

85. Horan G, Whitfield GA, Burton KE, Burnet NG, Jefferies SJ. Fractionated conformal radiotherapy in vestibular schwannoma: early results from a single centre. *Clinical Oncology* 2007; **19**(7): 517–22.

86. Meijer OW, Vandertop WP, Baayen JC, Slotman BJ. Single-fraction vs. fractionated linac-based stereotactic radiosurgery for vestibular schwannoma: a single-institution study. *International Journal of Radiation Oncology, Biology, Physics* 2003; **56**(5): 1390–6.

87. Royal College of Physicians Working Party. *Pituitary tumours: recommendations for service provision and guidelines for management of patients.* London, Royal College of Physicians of London, 1997.

88. Brada M, Rajan B, Traish D, *et al.* The long term efficacy of conservative surgery and radiotherapy in the control of pituitary adenomas. *Clinical Endocrinology* 1993; **38**: 571–8.

89. Tsang RW, Brierley JD, Panzarella T, Gospodarowicz MK, Sutcliffe SB, Simpson WJ. Radiation therapy for pituitary adenoma: treatment outcome and prognostic factors. *International Journal of Radiation Oncology, Biology, Physics* 1994; **30**: 557–65.

90. Zierhut D, Flentje M, Adolph J, Erdmann J, Raue F, Wannenmacher M. External radiotherapy of pituitary adenomas. *International Journal of Radiation Oncology, Biology, Physics* 1995; **33**: 307–14.

91. Rajan B, Ashley S, Gorman C, *et al.* Craniopharyngioma—a long-term results following limited surgery and radiotherapy. *Radiotherapy and Oncology* 1993; **26**(1): 1–10.

92. Brada M, Ford D, Ashley S, Bliss JM, Crowley S, Mason M, Rajan B, Traish D. Risk of second brain tumour after conservative surgery and radiotherapy for pituitary adenoma. *British Medical Journal* 1992; **304**: 1343–7.

93. Oldfield EH. Editorial: Unresolved issues: radiosurgery versus radiation therapy; medical suppression of growth hormone production during radiosurgery; and endoscopic surgery versus microscopic surgery. *Neurosurgery Focus* 2010; **29**(4): E16.

94. Rajan B, Ashley S, Thomas DG, Marsh H, Britton J, Brada M. Craniopharyngioma: improving outcome by early recognition and treatment of acute complications. *International Journal of Radiation Oncology, Biology, Physics* 1997; **37**(3): 517–21.

95. Goldsmith BJ, Wara WM, Wilson CB, Larson DA. Postoperative irradiation for subtotally resected meningiomas. A retrospective analysis of 140 patients treated from 1967 to 1990. *Journal of Neurosurgery* 1994; **80**(2): 195–201.

96. Condra KS, Buatti JM, Mendenhall WM, Friedman WA, Marcus RB Jr, Rhoton AL. Benign meningiomas: primary treatment selection affects survival. *International Journal of Radiation Oncology, Biology, Physics* 1997; **39**(2): 427–36.

97. Hunt DPJ, Freeman A, Morris LS, *et al*. Early recurrence of benign meningioma correlates with expression of mini-chromosome maintenance-2 protein. *British Journal of Neurosurgery* 2002; **16** (1): 10–15

98. Estall VJ, Treece S, Jena R, Jefferies SJ, Burton KE, Burnet NG. Patterns of relapse following fractionated radiotherapy for meningioma: Experience from Addenbrooke's Hospital Radiotherapy Department. *Clinical Oncology* 2009; **21**: 745–52.

99. Khoo VS, Adams EJ, Saran F, *et al*. A Comparison of clinical target volumes determined by CT and MRI for the radiotherapy planning of base of skull meningiomas. *International Journal of Radiation Oncology, Biology, Physics* 2000; **46**(5): 1309–17.

100. Goldsmith BJ. Meningioma. In Leibel SA, Phillips TL (eds) *Textbook of Radiation Oncology*. Philadelphia, PA: WB Saunders Co, 1998, pp. 324–40.

101. Akyurek S, Chang EL, Yu TK, *et al*. Spinal myxopapillary ependymoma outcomes in patients treated with surgery and radiotherapy at M.D. Anderson Cancer Center. *Journal of Neuro-Oncology* 2006; **80**(2): 177–83.

102. Whitaker SJ, Bessell EM, Ashley SE, Bloom HJ, Bell BA, Brada M. Postoperative radiotherapy in the management of spinal cord ependymoma. *Journal of Neurosurgery* 1991; **74**(5): 720–8.

Head and neck cancer

Christopher Nutting, Michele Saunders

12.1 Introduction

Cancer of the head and neck is a relatively rare cancer accounting for 3% of all cancer deaths. The incidence in men is 17.2 per 100 000 and in women, 5.6 per 100 000. This equates to about 7750 cases in the UK[1]. The main aetiological factors are excessive alcohol intake, infection with human papilloma virus (HPV), and smoking.

The TNM staging system is used for staging head and neck cancer. A current generic staging system is given in Table 12.1, but more detail on staging for individualized tumour subsites is given in the American Joint Committee on Cancer (AJCC) TNM Classification[2].

12.2 Radical primary treatment

Patients with head and neck cancer should be discussed by a multidisciplinary team comprising specialist surgeons, oncologists, pathologists, radiologists, and palliative care doctors, together with dieticians, speech and language therapists, and clinical nurse specialists. At this time decisions as to the modality of treatment(s) to be used should be made, including surgery, radiotherapy, and chemotherapy.

12.2.1 Indications

Surgery and radiotherapy with or without chemotherapy are the most frequently used therapeutic modalities in head and neck cancer. For early stage tumours surgical excision or radiotherapy alone have similar cure rates, but have different adverse effect profiles. Radiotherapy with or without chemotherapy offers higher rates of organ preservation, and for some cancers, where function is important, it is the treatment of choice. For example, in carcinoma of the tongue base or larynx, radiotherapy preserves swallowing and natural speech respectively. At other sites (e.g. carcinoma of the floor of mouth), surgical excision alone may be curative and be associated with a very satisfactory functional outcome. The choice of treatment modality therefore depends on individual factors including patient preference.

For advanced squamous cell carcinoma of the head and neck the combined use of surgery and postoperative radiotherapy frequently offers the highest chance of achieving cure. In the light of international trials in the postoperative setting[3,4], concomitant chemoradiotherapy has become the standard of care for high-risk patients with positive margins or extracapsular spread. Similarly, for inoperable advanced tumours,

Table 12.1 (a) Generic staging for head and neck cancer (for more detail on individual subsites, refer to AJCC TNM Classification[2])

Tis	Carcinoma *in situ*
T1	Tumour < 2 cm
T2	Tumour > 2–4 cm
T3	Tumour > 4 cm
T4	Tumour involves adjacent structures
T4a	Operable disease
T4b	Inoperable disease

(b) Nodal staging for all head and neck sites (this can be applied to all sites except nasopharynx carcinoma and thyroid)

N0	No lymphadenopathy
N1	Ipsilateral single node < 3 cm
N2a	Ipsilateral single node > 3–6 cm
N2b	Ipsilateral multiple nodes < 6 cm
N2c	Bilateral or contralateral nodes < 6 cm
N3	Nodes > 6 cm

concomitant chemoradiation schedules offer the highest chance of local control and survival[5].

12.2.2 General principles of treatment volume and definition

Patients with head and neck cancer are initially seen in a multidisciplinary outpatient clinic when a full history and examination of the head and neck area will be carried out; this will include nasendoscopy if indicated. Following this the pretreatment assessment for radiotherapy planning should include an examination under anaesthesia (EUA) and tumour biopsy, chest X-ray or thoracic CT scan, CT and/or MRI of head and neck, specialist histology review, full blood count, urea and electrolytes, liver function tests, dental assessment, nutritional assessment, and written informed consent. Patients should be advised to stop smoking as this increases toxicity and reduces cure rates, and to reduce alcohol intake. Chemoradiation protocols are associated with enhanced acute toxicity and the patient's performance status, medical condition, and social factors should be taken into account during patient selection. An accurate assessment of dietary intake should be undertaken, and, if necessary, a nasogastric tube or percutaneous gastrostomy may be required for elective feeding.

Most patients with advanced tumours have their radiotherapy in two phases. Phase 1 is typically a large volume including the primary tumour, any involved lymph nodes, and areas of likely microscopic lymphatic spread. This is followed by a second phase where a smaller volume should encompass the primary tumour and involved nodes with a margin for microscopic spread, uncertainties of localization and immobilization. The elective lymph node regions may be included in phase 1, but not in phase 2.

Table 12.2 The risk of nodal metastasis based on tumour site and location

High (> 60%)	Nasopharynx
	Oropharynx
	Hypopharynx
	Supraglottis
Medium (20–60%)	Oral cavity
	Advanced larynx
	Parotid
Low (< 20%)	Skin
	Early stage glottic larynx
	Nasal cavity
	Paranasal sinuses
Risk predominately unilateral	Parotid
	Early stage tonsil
	Lateralized oral cavity
Risk bilateral	Base of tongue
	Nasopharynx
	Advanced larynx/hypopharynx

The elective nodal sites depend on the site of the primary tumour. Table 12.2 details the incidence of occult micrometastases to lymph nodes which have been documented from surgicopathological series[6,7]. Elective irradiation of lymph node regions is indicated when the risk of harbouring micrometastatic disease exceeds 15–20%, and therefore a detailed understanding of the natural history of each subsite of head and neck cancer is required during radiotherapy planning. Lymph node levels are defined as: level Ia, submental; level Ib, submandibular; level II, upper deep cervical; level III, middle deep cervical; level IV, lower deep cervical; level V, posterior triangle (Fig. 12.1). For patients receiving intensity-modulated radiotherapy (IMRT), the same principles apply; a planning target volume (PTV) should be generated for the tissues to receive radical radiation dose (PTV1) and a second volume is defined (PTV2) to contain the tissues to be irradiated to an elective dose.

The recommendations for specific node groups to be included in the field of treatment are meant as a guide (Table 12.3). The responsible clinical oncologist must make the final decision on the individual features of the case.

12.2.3 Planning technique

Patient position and immobilization

The anatomy of the head and neck region is very complex, with bony structures, soft tissues, and air cavities all present in complicated arrangements in a relatively

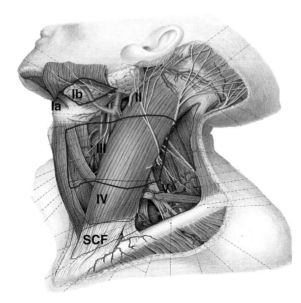

Fig. 12.1 The lymph node levels in the neck. I: submental (Ia) and submandibular (Ib), II: upper deep cervical, III: middle deep cervical, IV: lower deep cervical, V: posterior triangle, SCF: supraclavicular fossa.

Table 12.3 Recommendations for elective lymph node irradiation in radical treatment with radiotherapy for head and neck cancer

Nasopharynx	
Squamous carcinoma T1 N0	No neck irradiation
Squamous carcinoma T2–T4 N0	Level I, II, III, IV, V, bilaterally
All undifferentiated carcinoma or node-positive squamous carcinoma	Level I II, III, IV, V, bilaterally
Paranasal sinuses	
Squamous carcinoma N0	Lateral pharyngeal/retropharyngeal nodes
Squamous carcinoma N+	Levels Ib to V bilaterally
Oropharynx	
T1/2 N0 tonsil (well-lateralized)	Levels Ib to IV ipsilateral
T1/T2 N1 tonsil	Levels Ib to V ipsilateral
T2 N0 tonsil approaching midline + other sites	Levels Ib to V bilaterally
All others	Levels Ib to V bilaterally
Larynx	
T1/T2 N0 glottic	No elective nodal irradiation
T3–T4 N0 glottic	Levels Ib, II, III, and IV bilaterally
T1/T2 N0 supraglottic	Levels Ib, II and III bilaterally
All others	Levels Ib to V bilaterally
Hypopharynx	
All	Levels I to V bilaterally

small volume. The organs at risk (OARs) include spinal cord, brainstem, optic nerves, retina, lens, brain, skin, mucosa, and salivary glands. All may lie very close to, or within, the target volume, making irradiation of tumour within normal tissue tolerance difficult. By contrast, internal organ motion is relatively limited, and even physiological laryngeal motion has little impact on treatment planning. The head and neck region can be readily immobilized using a custom-made thermoplastic shell (Fig. 12.2) and this should ensure reproducible patient set-up to within 2–3 mm. New head and neck stabilization systems involving carbon fibre base plates to enable IMRT to be delivered have become standard replacing the older style Cabulite shells.

It is important to specify neck and head position, shell extent, requirements for mouth bite, and full planning details prior to manufacturing the immobilization shell.

For most tumour sites patients are positioned with the head in the neutral position and the spinal cord straight. This allows field matching between photons and electrons along a straight line anterior to the spinal cord in the latter part of treatment. An extended neck position is required for treatment of nasopharyngeal carcinoma to allow matching of fields, parotid tumours where the extended neck position allows avoidance of the eyes, and irradiation of the neck only. For target volumes extending low into the neck, or mediastinum (e.g. subglottic or postcricoid carcinoma), then the cervical and thoracic spine need to be straight and the head may be flexed. An inclined head-rest may be necessary. For IMRT, the neck is often extended to keep the mucosa of the oral cavity out of the treatment beams.

Over the last few years there has been a shift away from conventional planning using a simulator, to CT planning and 3D conformal radiotherapy. For some tumour sites, parotid gland-sparing IMRT has become the new standard to avoid radiation-induced xerostomia[8].

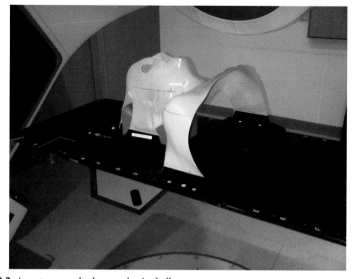

Fig. 12.2 A custom-made thermoplastic shell.

Conventional radiotherapy planning (simulation)

Any lymphadenopathy and scars should be marked with wire and a provisional picture may be taken so that fields can be drawn based on diagnostic imaging to facilitate efficient planning at a subsequent simulator visit. Virtual simulation using a CT simulator may obviate the need for this additional step and improve definition of tumour and OARs.

Mouth bites are used in patients with carcinoma of the oral cavity, nasal cavity, and maxillary sinuses. The mouth bite opens the jaw, and depresses the tongue. It may be possible to exclude either the upper or lower half of the mouth from the treatment field. Care should be taken to avoid the mouth bite pushing the mobile tongue into or out of the field posteriorly, and to ensure that the tongue position is reproducible. A custom-made mouthpiece to move the tongue out of the field in the lateral position should be considered in unilateral carcinomas of the oral cavity, e.g. floor of mouth or buccal mucosa.

The use of a radiotherapy simulator to define fields has been standard practice for many years. Most commonly lateral parallel-opposed beams are used to irradiate the target volume. Typically field borders are placed in relation to standard bony anatomical landmarks that define the extent of tumour subsites, and may be modified in individual patients. This planning method does not use the GTV, CTV, and PTV definitions outlined in ICRU 50 and 62[9,10], the field borders representing the PTV plus a physical margin for penumbra. For more complex plans such as a wedged pair beam arrangement, a PTV is marked on a patient outline taken through an appropriate single level, or several levels. The PTV is typically reconstructed from orthogonal simulator films. The clinical results obtained with these techniques are known, as are the expected toxicities of treatment.

Computed tomography planning

CT planning is increasingly becoming the standard of care for head and neck cancer radiotherapy. For CT planning the recommendations of ICRU reports 50 and 62[8,9] should be followed. Clinicians should define the GTV, CTV, PTV, and OARs. Outlining of the GTV and CTV should be done with the aid of diagnostic MRI, and CT scans, operation notes, clinical examination, and nasendoscopy.

A planning CT scan with intravenous contrast will provide better definition of the primary tumour and involved nodes and should always be performed. Joint review with a radiologist is suggested. The addition of CT information into the treatment planning process improves localization of both tumours and OAR. It can also lead to more accurate dose calculation, and allows better estimation of dose received by the target volumes and OARs. In addition, the use of CT planning allows optimization of radiotherapy beam direction, beam weight, accurate generation of conformal beam shaping, and is essential for inverse planning and IMRT. The following sections discuss the application of cross-sectional imaging to target volume definition.

Gross tumour volume localization

Head and neck squamous cell carcinomas often spread through the mucosa, and such spread is usually best assessed clinically or endoscopically as this type of tumour extension is not reliably identified from cross-sectional imaging (CT, MRI, or PET-CT). If a tumour mass is present with deep invasion of tissues, this is better localized from CT or MRI[11]. The sensitivity of such imaging modalities depends on the difference in signal between tumour and normal tissue, which is variable depending on tumour site (e.g. poor for tongue tumours, but good for paranasal sinus tumours). For tumours invading the skull base (e.g. nasopharyngeal carcinoma or paranasal carcinoma), both CT and MRI may be optimal for detection of bone invasion and soft tissue extension respectively. For adjuvant radiotherapy, postoperative imaging may be useful to define the tumour bed especially if radio-opaque markers have been left by the surgeon to delineate the tumour bed and guide radiotherapy planning. If a patient has received neo-adjuvant chemotherapy, the prechemotherapy tumour volume should be used for planning.

CT scanning is the most frequently used modality for target volume definition. Postscanning adjustment of CT window levels is of value in defining both tumour and OARs.

MRI treatment planning for head and neck cancer is under evaluation. Although this may allow better differentiation of tumour from normal tissue (typically with T1-weighted, gadolinium enhanced images), MRI/CT fusion is required to correct distortion before treatment planning or dose calculations can be performed. Volumes outlined on MRI are smaller compared to CT, and there is less interobserver variability[11]. Image fusion using bony anatomy is accurate for tumours in, or attached to, the skull base, but is less satisfactory for images of the neck that are relatively mobile due to flexion and extension of the cervical spine.

Clinical target volume localization

CTV definition for radical radiotherapy is controversial. A margin needs to be added to the GTV to take into account patterns of local spread of the tumour. This requires knowledge of the patterns of local tumour extension, anatomical compartments, and barriers. Where the tumour may spread, e.g. submucosally, the margin needs to be 1–2 cm, but where there are anatomical boundaries to tumour extension such as air cavities, periosteum, and compartmental fascia, then margins can be reduced.

Adjuvant (postoperative) radiotherapy to the primary tumour site represents an even more difficult localization problem. By definition there is no GTV, and so the localization of the tumour bed is based on preoperative descriptions and cross-sectional imaging. Preoperative assessment by the radiation oncologist is invaluable. Preoperative and postoperative imaging may be fused to improve CTV definition.

Nodal target volume definition

Considerable advances have been made to produce reproducible guidelines for elective nodal irradiation. The use of intravenous contrast is recommended as it helps to define the carotid arteries, an important anatomical landmark in the definition of the deep cervical nodes[12].

Frequently there are locoregional lymph node metastases in addition to the primary tumour, and in this situation individual nodal masses require a separate GTV. A margin of 1–2 cm should be added to each nodal GTV to produce the CTV. As for primary tumours, the GTV–CTV margin can be reduced where there is an anatomical barrier to tumour spread, e.g. prevertebral fascia, air cavity, or bony wall. For situations in which the target volume cannot be well defined, an approach of conformal avoidance may be useful when tissues at risk of containing disease are outlined, and normal tissue structures are removed from the volume.

In the case of adjuvant radiotherapy, radiation is delivered to a CTV where there is a risk of residual microscopic disease following radical surgical excision. Several CTVs will usually be identified. A different CTV should be defined for regional nodal groups at risk of containing microscopic metastases. For example, in the treatment of early (e.g. T1 N1 M0) carcinoma of the tonsil there will be a GTV1/CTV1 for the primary tumour, and a different GTV2/CTV2 for the enlarged lymph node mass. The CTV2 will also include the adjacent lymph nodes at high risk of involvement (the first lymph node station—the ipsilateral upper deep cervical lymph node group). These would be prescribed dose equivalent to 70 Gy. The ipsilateral clinically uninvolved levels Ib, III, IV, and V, which are at less risk of lymph node spread, would be defined as separate CTV3, and may be prescribed a dose equivalent to 50 Gy to sterilize microscopic foci of metastatic carcinoma. The risk of nodal metastases varies between different tumour sites, and has been extensively investigated in retrospective surgical studies. These clinicopathological studies[6–8] examined elective neck lymph node dissection specimens for the presence of occult metastases for individual tumour sites and documented their frequency and distribution. These studies have been collated and recommendations given for elective nodal irradiation[13].

The identification of nodal volumes varies depending on the technique used. Seventy to 75% of lymph nodes involved by tumour are enlarged, and can be identified by clinical palpation. The use of cross-sectional imaging increases the sensitivity to 85% by the identification of impalpable retropharyngeal nodes, deeply sited nodes, and normal sized nodes with a necrotic centre (necrosis has low signal intensity on CT scan and is highly specific for metastatic carcinoma). 18F-PET-CT scanning may further improve diagnostic sensitivity. The number of CTVs outlined should be representative of the estimated clonogenic cell density and, in practice, contiguous groups of nodes that are to receive the same dose can be outlined as one CTV.

Planning target volume localization

A further margin is added to the CTV to produce the PTV. The size of this margin depends on the type of immobilization used. A customized thermoplastic mask system has a day-to-day set-up accuracy of 2–5 mm, and to encompass 95% of the errors a CTV–PTV margin of 2–3 mm should be added. For tumours attached to the bones of the skull base (nasopharynx, paranasal sinuses, oropharynx, and nasal cavity) a tight shell provides a random error of < 2 mm. For oral cavity tumours, movement of the mandible may affect the reproducibility of patient set-up, and this will be greatest if the mouth is stented open with a mouth bite. For tumours of the hypopharynx and larynx, physiological organ motion occurs with swallowing and breathing which must be taken into account in the CTV–PTV internal movement margin.

CTV–PTV margin should also be added to elective lymph node irradiation volumes to account for set-up inaccuracy. The size of these margins will be dependant on the type of immobilization used, but should be in the region of 3–5 mm.

Organs at risk

Brain, spinal cord, brainstem, parotid glands, mandible, eyes, thyroid gland, optic nerves, and chiasm, are all close to the target volume for some head and neck tumours. Tolerance doses are given in Table 12.4[14]. These structures are localized better on CT planning scans than on conventional radiographs, and should all be outlined in their entirety.

OARs that occur close to the PTV should be outlined. The particular organs will vary from one tumour site to another.

Intensity-modulated radiotherapy

IMRT is a recent development in 3D conformal radiotherapy (3DCRT). Typically five or seven equispaced fixed intensity-modulated beams, or a 360° dynamic arc. IMRT can generate a distribution with a concave shape to reduce dose to OARs lying close to the target volume. This has been shown to reduce long-term treatment-related toxicity. Parotid-gland sparing IMRT has been shown to reduce long-term xerostomia in patients with tumours of the oropharynx, or nasopharynx[8] (Fig. 12.3). It has also been shown to reduce the dose to the optic apparatus in treatment of paranasal sinus tumours, and cochlea sparing in parotid tumours.

Studies using IMRT to escalate radiation dose and improve local tumour control are currently under evaluation.

For locally advanced head and neck cancer, it is most efficient to treat with simultaneous integrated boost (SIB) or simultaneous modulated accelerated radiotherapy (SMART) techniques. These are characterized by the delivery of different dose-per-fraction to different targets within the head and neck region. For example, in PARSPORT, the Cancer Research UK Parotid Sparing IMRT trial[8], a dose of 2.17 Gy per fraction was delivered to the primary tumour site, and involved lymph nodes, and 1.8 Gy per fraction to elective lymph node groups. After 30 fractions the primary tumour and involved lymph nodes had received a total of 65 Gy, and the elective lymph nodes 54 Gy (Fig. 12.4). The advantage of the SIB or SMART techniques is that the whole treatment course is planned only once, with savings in simulation,

Table 12.4 Acceptable doses to specific organs at risk. These doses should not be exceeded unless the extent of tumour makes higher doses inevitable[13]

Organ	Normal tissue tolerance (2 Gy/fraction)
Lens	6–10 Gy
Cornea	40 Gy
Retina	50 Gy
Optic nerve	50Gy TD 5/50, 60 Gy TD 20/50
Optic chiasm	50–55 Gy
Spinal cord	44–48 Gy
Brainstem	44–48 Gy
Hypothalamus	44 Gy

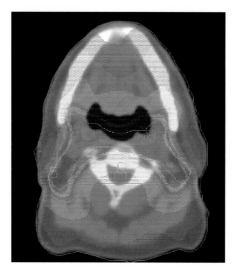

Fig. 12.3 Parotid gland sparing IMRT: a dose distribution to deliver a high dose to the target volume (blue contour and red colour wash) whilst sparing the parotid gland (pink contours) can be achieved with IMRT.

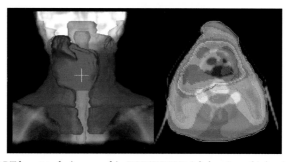

Fig. 12.4 SMART boost technique used in PARSPORT trial showing a higher total dose (65 Gy) and dose per fraction (2.17 Gy) delivered to the primary tumour and involved nodes (red in 3D reconstruction and green colour wash) and lower total (54 Gy) dose and dose per fraction (1.8 Gy) to the elective nodes (purple in 3D reconstruction and orange colour wash).

planning, delivery and verification time compared to conventional multi-phase plans. Radiobiologically, SIB and SMART techniques represent accelerated fractionation schedules that have been shown to improve tumour control.

Inverse planning for head and neck IMRT requires the clinician to generate constraints in the form of dose volume points to drive the inverse planning algorithm. These constraints will vary from one planning system to another, but target volume constraints should be the prescription dose ± 5% and for OAR the tolerance dose to a small volume of that organ.

12.2.4 **Dose prescription**

Plans should be normalized to the ICRU reference point[9]. The plan should be checked to ensure adequate target coverage, homogeneous dose, and that doses to

OARs are acceptable. Plans are prescribed to the 100% isodose—usually to the isocentre, or a similar representative point in the target volume.

In treating squamous cell carcinoma of the head and neck, tumour and normal tissue will usually be treated close to tolerance to achieve optimal cure. Conventional radiotherapy at centres in the South of England, Europe, and the USA involves daily radiotherapy, Monday to Friday, giving a dose of 1.8–2 Gy per fraction to a total dose of 66–70 Gy. Some centres, however, use a 4-week schedule of 55 Gy in 20 fractions daily, Monday to Friday, over 4 weeks and this shorter schedule can be adopted for small volume disease.

Treatment of small volume disease

This encompasses most early stage disease, i.e. T1/T2 N0 where the target volume includes the primary site and in some cases the first echelon of nodes only. Many centres use a 4-week schedule (e.g. 55 Gy in 20 fractions once daily for 4 weeks) for smaller volume irradiation, i.e. glottic tumours with field size 5–6 cm, on the basis that smaller volumes will tolerate the larger doses per fraction better. Most US studies do not give doses per fraction greater than 1.8–2 Gy, however, irrespective of treatment volume.

Treatment of large volume disease

This encompasses T3/T4 N0–N3 disease where surgical treatment is either not appropriate or not possible. The target volume includes the primary site and the lymph nodes containing metastatic carcinoma and at-risk (elective) nodes. Often this volume will be overlying the spinal cord from the lateral projection. If the spinal cord is within the volume then a two-phase technique is used to keep the total spinal cord dose < 44–48 Gy. Macroscopic disease should receive 66–70 Gy in 2-Gy fractions or equivalent. Microscopic disease should receive 50 Gy in 2-Gy fractions or equivalent.

It is generally accepted that 50 Gy is required to sterilize microscopic disease. There is some evidence from randomized controlled trials, including the CHART trials[15] and other phase II trials[13,16] that 44 Gy may well be adequate. Treatment planning to achieve an elective dose of 50 Gy is detailed next and will be referred to, where applicable, throughout the chapter to avoid repetition.

12.2.5 **Radiation dose**

Intended dose prescription

- ◆ Macroscopic disease: 66–70 Gy in 2-Gy fractions treating five times per week.
- ◆ Microscopic disease: 50 Gy in 2-Gy fractions treating five times per week.

Phase 1:

- ◆ Photons:
  - • Lateral parallel-opposed to primary, nodal involvement and elective nodal irradiation.
  - • 44 Gy to 100% or MPD in 22 fractions treating 5 days per week.
  - • Anterior neck split with spinal cord and lung shielding.
  - • 44 Gy at 2 cm treating 5 days per week.

Phase 2:

- ◆ Photons:
  - Lateral parallel-opposed field moved anterior to cord.
  - Continue to 66–70 Gy.
  - Anterior neck split to 50 Gy at 2 cm depth.
- ◆ Electrons:
  - Posterior electron field(s) giving total of 50 Gy to elective nodes.
  - Continue to 66–70 Gy if nodes involved with tumour.
  - Electron energy dependent on the depth of nodal disease and spinal cord but usually 8–10 MeV.
  - Care must be taken to keep cord dose within tolerance.

Altered fractionated schedules

Continuous hyperfractionated accelerated radiotherapy (CHART) was evaluated in a UK-wide study[15]. This intensive regimen requires out-of-hours treatment and failed to show an overall survival advantage in head and neck cancer.

Large randomized controlled trials of altered fractionation in head and neck cancer have indicated that maintaining a total dose of 70 Gy gives benefit over conventional radiotherapy whether a hyperfractionated or concomitant boost technique is used[17,18]. There is also evidence from a meta-analysis that a modest acceleration of radiotherapy maintaining the dose at 66 Gy is advantageous[19,20].

12.2.6 Implementation of treatment

Modern thermoplastic shells provide little surface dose build-up and so cutting out of the shell (which was commonly performed with the old-style Cabulite shells) is not necessary. Indeed, if the tumour extends close to skin, such as a fungating tumour or laryngeal carcinoma involving the anterior commissure, the application of bolus may be required to ensure adequate dose to tumour tissue. During treatment, radiographers should check for loosening of the shell, particularly if patients experience significant weight loss.

12.2.7 Verification

Electronic portal imaging should be used to verify beam position and correct shielding during the initial 3 days of treatment and weekly thereafter. Typically a tolerance of ± 3 mm is acceptable, although this may need to be more stringent if critical OARs are very close to the target volume. Volumetric kV or MV imaging may provide more information about patient position during therapy.

12.3 Postoperative radiotherapy

12.3.1 Indications and treatment volume

Fifteen factors have been identified as important for prediction of recurrence. The first two are indications for chemoradiotherapy alone or together.

1. Positive resection margins.

2. Extracapsular lymph node spread.

Two of the remainder suggest postoperative radiotherapy should be recommended.

3. Close margins < 5 mm.

4. Invasion of soft tissues.

5. Two or more lymph nodes positive.

6. More than one positive nodal group.

7. Involved node > 3 cm in diameter.

8. Multicentre primary.

9. Perineural invasion.

10. Vascular invasion.

11. Poor differentiation.

12. Stage T3/4.

13. Oral cancer.

14. CIS, dysplasia at edge of resection margin.

15. Uncertainties concerning surgical/pathological findings.

16. HPV negativity.

High risk of recurrence is associated with either factor 1 or 2 alone or presence of two or more of factors 3–9.

Intermediate risk is associated with the presence of any one of the factors 3–9. Factors 10–16 may be of some importance in predicting recurrence and should be borne in mind.

12.3.2 **Planning technique**

Treatment volume should encompass the area of surgical resection and will include all areas considered at risk of recurrence. A margin of at least 10 mm around the maximal extent of surgery should be included. Where nodal involvement has been proven, the field should be extended to include the lymphatic drainage down to the clavicle. Areas of positive margin or other high-risk features may receive with a boost to the area up to radical dose.

12.3.3 **Dose prescription**

- 60 Gy in 30 fractions treating daily Monday to Friday. 2 Gy per fraction plus a boost of 4 Gy for high-risk factors.
- 50 Gy to uninvolved non-operated sites, e.g. contralateral neck.

Intended dose prescription

- Definitive dose to postoperative tumour bed: 60–64 Gy.
- Microscopic non-operated field dose: 50 Gy.

Phase 1:

- ◆ Photons:
  - Lateral parallel-opposed fields.
  - 44 Gy to 100% or MPD in 22 fractions treating 5 days per week.
  - Anterior neck split with spinal cord and lung shielding.
  - 44 Gy at 2 cm depth treating 5 days per week.

Phase 2:

- ◆ Photons:
  - Lateral parallel-opposed field moved anterior to cord.
  - Continue to 60–64 Gy.
  - Anterior neck split to 50 Gy at 2 cm depth if elective 60–64 Gy to operative site.
- ◆ Electrons:
  - Posterior electron field(s) giving total of 50 Gy to elective nodes.
  - Continue to 60–64 Gy if operative site included.
  - Electron energy dependent on the depth of nodal disease and spinal cord but usually 8–10 MeV.
  - Care must be taken to keep cord dose within tolerance.

For treatment fields see section 12.2.5.

12.4 Palliative radiotherapy

Palliative radiotherapy is only given to a small proportion of patients with head and neck cancer. This is due to the relatively high doses required to achieve symptom relief. It is reserved for patients who have metastatic disease at presentation, or those with performance status ≥ 2.

No standard dose schedule for palliative irradiation exists. Typically large doses per fraction are used with careful consideration of tolerance doses to OARs such as spinal cord, brain, and cranial nerves. Regimens that can be considered on an individual basis include: 50 Gy in 20 daily fractions; 27 Gy in six fractions over 3 weeks treating twice per week (within spinal cord tolerance); 39 Gy in 13 fractions with the last fraction excluding the spinal cord.

12.5 Site-specific treatment planning

12.5.1 Larynx

The larynx is divided into three distinct anatomical regions: the supraglottis (laryngeal epiglottis, false cords, ventricles, aryepiglottic folds, and arytenoids), the glottic larynx (true vocal cords, anterior and posterior commissures), and subglottis (10 mm below the free edge of the vocal cords to the inferior edge of the cricoid cartilage). Each has its own natural history, patterns of spread, and treatment protocols.

Immobilization

For all larynx tumours the patient should be immobilized in the supine position with the cervical spine straight.

Glottic tumours

Patients with carcinoma *in situ* (Tis) or dysplasia should be treated surgically by laser excision or cord striping. Radiotherapy is best reserved for recurrent lesions.

Stage T1–2, N0 tumours can be treated with radiotherapy or surgical excision with laser cordectomy. There is some controversy as to which modality provides the best voice quality. In the UK, radiotherapy remains the standard of care. Typically a parallel-opposed lateral beam arrangement is used with 5-cm (T1) or 6-cm (T2) square fields centred on vocal cord (1 cm below thyroid promontory and anterior to the lower border of the C5 vertebrae). The superior border should be at the lower edge of the hyoid bone, and inferiorly the field should encompass the cricoid cartilage covering the width of the thyroid cartilage. Anteriorly the field border should be in air at the field centre, and posteriorly should be through the anterior part of the vertebral body (Fig. 12.5). Usually 10–20° wedges are used as missing tissue compensators. No pro-phylactic nodal radiotherapy is given, although the anterior part of the mid-cervical lymph nodes (level III) is within the irradiated volume.

For T2 tumours the superior and/or inferior borders are individually expanded based on the supraglottic and/or subglottic extension. The para-oesophageal and par-atracheal lymph nodes are included for extensive subglottic extension.

For tumours involving the anterior commissure within a few millimetres of the skin surface, the skin-sparing effects of a megavoltage beam risk tumour under dosage, and in these cases the anterior part of the shell should not be cut out. If the calculated dose to the anterior commissure is still low, then this can be improved by reducing or removing the wedge from each lateral field or adding bolus to increase the dose in the superficial tissues.

In patients with short necks, or high shoulder position then lateral fields to the larynx may not be deliverable, and in this situation an anterior-oblique wedged pair arrangement is more appropriate (Fig. 12.6). This will require a PTV to be localized by CT planning or on an outline taken through the field centre.

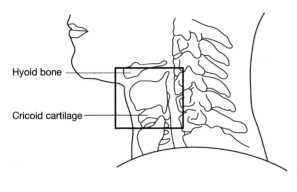

Hyoid bone

Cricoid cartilage

Fig. 12.5 Radiotherapy for a T1/2 N0 carcinoma of the glottic larynx.

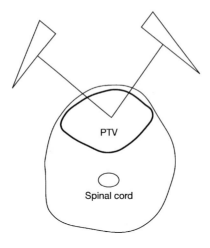

Fig. 12.6 A field arrangement for patient inappropriate for lateral field irradiation for early larynx cancer due to high shoulder position.

Dose prescription

This is dependent on field size:

- Less than 42 cm^2: 55 Gy in 20 fractions over 4 weeks[21].
- Larger field sizes: 64 Gy in 32 fractions over 6½ weeks.

Supraglottic tumours

T1 and T2, N0 tumours are associated with a high incidence of occult positive nodes in level II and III because of the dense lymphatic supply in this area. All patients therefore require elective nodal irradiation of these levels. A two-phase technique is used. Phase 1 should include the primary tumour, the whole larynx, pre-epiglottic space and the cervical nodes levels Ib, II, and III bilaterally anterior to spinal cord. Phase 2 should encompass the primary tumour only to a dose of 66–70 Gy. Parallel opposed wedged fields are used for both phases (Fig. 12.7).

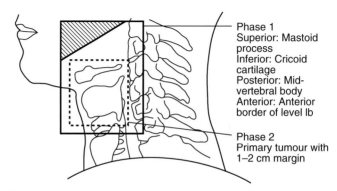

Phase 1
Superior: Mastoid process
Inferior: Cricoid cartilage
Posterior: Mid-vertebral body
Anterior: Anterior border of level Ib

Phase 2
Primary tumour with 1–2 cm margin

Fig. 12.7 Radiotherapy technique for a node-negative supraglottic carcinoma.

Dose prescription:

◆ Total dose: 66–70 Gy in 33–35 fractions to macroscopic disease; 50 Gy to microscopic disease.

◆ Treatment technique as section 12.2.5 but anterior neck split not necessary as N0.

Node-positive supraglottic tumours

Supraglottic tumours with nodal involvement are considered for surgery and postoperative radiotherapy. They may, however, be treated primarily by radiotherapy or chemoradiation, reserving surgery for treatment failure.

Phase I typically employs parallel-opposed fields to irradiate the primary tumour and upper neck nodes (levels II, III, and upper part of V) with matched bilateral anterior neck fields (levels IV and lower V) with spinal cord and lung shielding. In Phase II the posterior border of the photon field is then moved anterior to the spinal cord and electron fields are used to treat the posterior neck nodes if appropriate. The electron energy is chosen dependent on the depth of the lymph nodes and maintaining spinal cord doses within tolerance (Fig. 12.8).

Dose prescription

◆ Total dose: 66–70 Gy in 33–35 fractions to macroscopic disease; 50 Gy to microscopic disease.

◆ Treatment technique as section 12.2.5.

Subglottic tumours

Tumours of the subglottis are rare, and most present with locally advanced disease. In operable patients surgery with laryngectomy and postoperative radiotherapy may be employed. For patients with early stage disease primary radiotherapy offers the chance of larynx preservation. The rate of cervical node metastasis is rare, but involvement of paratracheal nodes is estimated to be 50% mandating elective treatment.

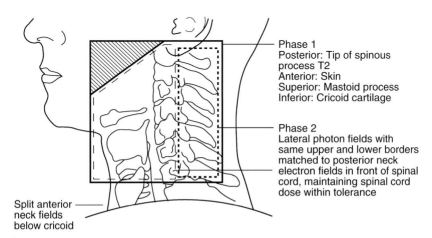

Fig. 12.8 Node-positive supraglottic carcinoma. A customized third phase is required to treat the primary tumour and involved lymph nodes to radical dose.

Fields should extend from the upper border of the thyroid cartilage down to the mid-trachea, ensuring coverage of the inferior extent. In order to achieve this, an anterior oblique field arrangement is used (similar to that shown in Fig. 12.6). The inferior extent may necessitate a coronal technique.

Dose prescription

◆ 66–70 Gy in 33–35 fractions treating daily, five fractions/week.

Advanced larynx carcinoma

Advanced tumours of the larynx are treated in a similar way, and the exact site of origin has less effect on treatment technique. Stage T3 represents a very inhomogeneous group, which can range from a small tumour with vocal cord fixation to a large transglottic tumour. Most patients with good performance status who are medically fit for chemotherapy should receive chemoradiation therapy that maximizes the chance of larynx preservation. Neo-adjuvant chemotherapy should also be considered[22]. If tumour invades the thyroid cartilage (T4), then radical surgery and postoperative radiotherapy represents the treatment of choice. Synchronous chemotherapy and radiation offer an alternative in those patients who are medically unfit. The treatment of advanced larynx cancers in complex and usually involves two to three phases if conventional radiotherapy is used. IMRT is increasingly used in the treatment of these patients and avoids the need for electrons and field–field matching. If IMRT is used, PTV1 is typically prescribed 65–66 Gy in 30 fractions and PTV2 54 Gy in 30 fractions.

Radical radiotherapy technique

The target volume includes the larynx, and pre-epiglottic space, and lymph node areas at risk of harbouring metastatic disease. This should include levels Ib, II, III, and IV in all patients, and level V in node-positive patients. In order to encompass this volume, lateral fields should extend from the mastoid process to 1 cm below the cricoid (lower if subglottic spread), and matched to anterior bilateral neck fields. If there is a tracheostomy then this must be also included, and necessitates a coronal technique (Fig. 12.9) or IMRT.

◆ Total dose: 66–70 Gy in 33–35 fractions to macroscopic disease; 50 Gy in 25 fractions to microscopic disease.

◆ Planning as detailed previously in section 12.2.5.

Postoperative radiotherapy

If the pathological stage is node negative, then the patient can be treated with a single-phase technique covering the larynx bed anterior to the spinal cord. The tracheostomy will usually require the use of coronally angled beams to achieve coverage (see Fig. 12.9). CT planning allows careful assessment of the dose in the superior mediastinum and in cases of under dosage, low weighted anterior fields are sometimes required.

Postoperative irradiation of node positive laryngeal carcinoma requires treatment of the posterior (level V) nodes, and therefore a two-phase technique is used with phase 1 consisting of parallel-opposed lateral fields with coronal angulation of the beams to cover the tracheostomy site. Phase 2 should cover the tumour bed with an

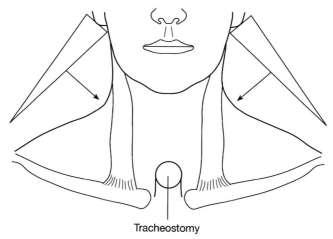

Tracheostomy

Fig. 12.9 Coronal technique for treatment of laryngeal bed and tracheostomy site.

electron field used to treat the posterior neck maintaining the spinal cord within tolerance.

Dose prescription

◆ Macroscopic tumour: 66–70 Gy in 33–35 fractions treating daily, five fractions/week (see section 12.2.5).

◆ Postoperative: 60–64 Gy in 30–32 fractions treating daily, five fractions/week (see section 12.3.3).

12.5.2 Oropharyngeal tumours

The oropharynx is split into four main subsites, including tongue base, tonsil, soft palate, and pharyngeal wall. They all have a relatively high risk of nodal metastasis. Tumours occurring in the midline (base of tongue, soft palate, and posterior pharyngeal wall) can metastasize to either side of the neck and therefore require irradiation to the primary tumour site and the neck bilaterally. By contrast, lateralized tumours of the tonsil or lateral pharyngeal wall metastasize unilaterally and therefore can be treated with less extensive fields allowing sparing of the contralateral structures, most importantly the contralateral parotid gland.

Small lateralized tumours of tonsil (T1/T2, node negative or positive)

The following lateral radiation technique is only suitable for tumours confined to the tonsillar fossa. Patients should be positioned with the cervical spine straight. For node-negative patients, the CTV includes the tonsillar fossa, and ipsilateral level Ib–IV nodes. The tonsil is irradiated using a wedged pair technique with anterior and posterior oblique fields (Fig. 12.10). This should be matched at the level of the hyoid to an ipsilateral anterior neck field treating levels III and IV. For node-positive patients the CTV also includes level V nodes, and involved nodes require treatment to radical dose.

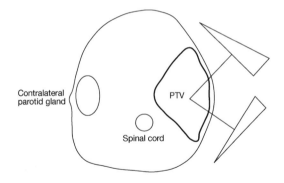

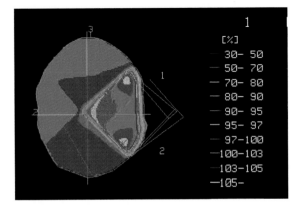

Fig. 12.10 Radiotherapy technique and dose distribution for a T1/2 tumour of the right tonsillar fossa showing the use of anterior and posterior oblique beams with sparing of the contralateral parotid gland.

Postoperative irradiation to the tumour bed and/or neck is usually required after surgical excision.

Dose prescription

- Macroscopic disease: 66–70 Gy in 33–35 fractions treating daily, five fractions/week.
- Microscopic disease: 50 Gy in 25 fractions treating daily, five fractions/week

Advanced oropharyngeal carcinoma (tonsil or base of tongue)

When a tonsil tumour approaches the midline at the soft palate, or involves the base of tongue, then the lateralized radiation technique is not appropriate. Both sides of the neck may harbour occult metastases and therefore the CTV includes the oropharynx and level I–IV nodes bilaterally. This technique is the same for tonsil and base of tongue tumours. Patients should be treated with parotid-sparing IMRT to reduce the risk of long-term xerostomia.

If IMRT is not available, the conventional technique is used. Parallel-opposed lateral fields are used for irradiation of the oropharynx and upper neck, and an anterior split neck field to treat the neck with central spinal cord shielding. After 44 Gy, lateral fields are reduced to come off the spinal cord and the primary tumour is taken to a radical dose. The posterior upper neck is treated using electron fields as required for the extent of nodal disease. Dose to the lower neck and posterior level V nodes depends on the extent of macroscopic disease, or surgery (Fig. 12.11).

Dose prescription

- Macroscopic disease: 66–70 Gy in 33–35 fractions treating daily, five fractions/week.
- Microscopic disease: 50 Gy in 22–25 fractions treating daily, five fractions/week.
- Treatment technique: see section 12.2.5.

Soft palate

The patient should be immobilized with the cervical spine straight. For early stage (T1 or T2) node-negative disease, elective nodal irradiation is not necessary. The target volume is therefore the GTV with a 2-cm margin only, and can be irradiated with small lateral opposed fields to radical or postoperative dose.

- 66–70 Gy in 33–35 fractions treating daily, five fractions/week.

In patients with advanced T stage (T3 or T4) or node-positive disease, then bilateral cervical node irradiation is required, and the technique used is the same as that described in section 12.2.5.

- Macroscopic tumour dose: 66–70 Gy in 33–35 fractions treating daily, five times per week.
- Microscopic tumour dose: 50 Gy in 25 fractions treating five times per week.

See section 12.2.5.

Irradiation of the oropharynx with parallel-opposed fields leads to severe xerostomia in most patients because of irradiation of both parotid glands. Parotid gland-sparing IMRT reduces the radiation dose to one or both of the parotid glands, and recent clinical trials have shown that IMRT reduces severe long term xerostomia by > 50%.

IMRT dose prescription SIB technique (see section 12.2.3):

- Oropharynx and involved lymph nodes (PTV1) 65–66 Gy in 30 fractions.
- Elective nodal irradiation 54 Gy in 30 fractions.

12.5.3 **Hypopharynx**

Tumours of the hypopharynx are characterized by a high risk of lymph node metastases. Therefore for any stage tumour elective irradiation of locoregional lymph nodes is required. The hypopharynx has three recognized subsites: pyriform fossa, postcricoid, and posterior pharyngeal wall. Most tumours of the hypopharynx are suitable for organ-preserving schedules with radiation or chemoradiation[22].

Pyriform fossa

Patients should be positioned with the cervical spine straight. Target volume includes the primary tumour site and level I–V lymph nodes bilaterally, and typically two or three phases are required to irradiate the primary tumour and involved nodes to radical dose while keeping the spinal cord dose within tolerance.

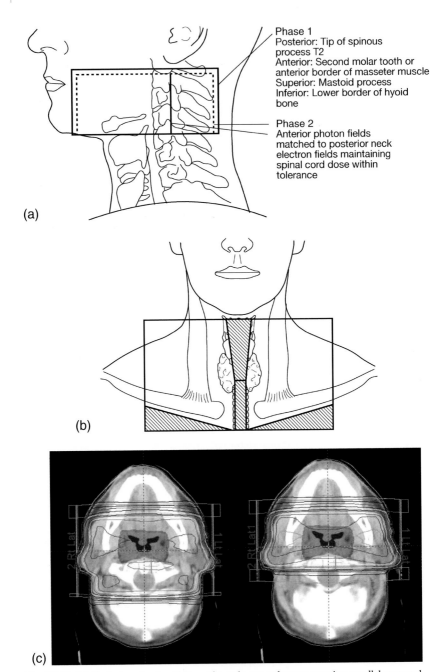

Fig. 12.11 (a) Technique for irradiation of oropharyngeal tumours using parallel-opposed fields. (b) Anterior neck field matching to neck field at hyoid level with midline spinal cord and larynx shielding. (c) Phase 1 and phase 2 of a parallel-opposed treatment plan for an oropharyngeal tumour.

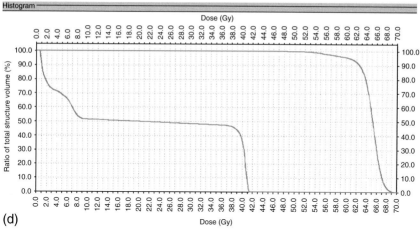

(d)

Fig. 12.11 *continued* (d) A dose–volume histogram (DVH) for the dose distribution given. Showing spinal cord maximum dose 41 Gy, and prescription dose to the PTV of 65 Gy.

The fields should extend from the skull base to the cricoid cartilage, if shoulder position will allow, with an anterior split neck field to treat the lymph nodes below the cricoid. After 44 Gy the fields should be reduced posteriorly to avoid the spinal cord, and electrons can be used to treat the level V nodes if indicated (Fig. 12.12).

Dose prescription

◆ Macroscopic tumour dose: 66–70 Gy in 33–35 fractions treating daily, five times per week.

◆ Microscopic tumour dose: 50 Gy in 25 fractions treating daily, five times per week.

See section 12.1.5.

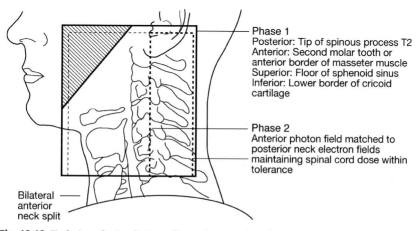

Fig. 12.12 Technique for irradiation of hypopharyngeal carcinoma.

Posterior pharyngeal wall

For posterior pharyngeal wall tumours, the inferior extent of tumour may make it difficult to treat with direct lateral fields once an appropriate margin has been added to the primary site (e.g. 2 cm margin above and below the known extent of tumour). To achieve this, the lateral parallel-opposed fields may need to be angled down in the coronal plane (similar to Fig. 12.9). A similar issue is confronted in the treatment of postcricoid carcinoma, where the technique requires irradiation of the primary tumour including the hypopharynx and, if possible, 5 cm of cervical oesophagus below the known tumour extent. Even with a coronal technique it may be difficult to achieve adequate tumour coverage, with dose falling off rapidly in the superior mediastinum. In some instances a low-intensity anterior field with a superior to inferior wedge can help to increase the dose in the superior mediastinum, but this is often at the expense of increased spinal cord dose. If there is nodal involvement, conventional radiotherapy may have to be palliative, as the full dose cannot be given to all the PTV without compromising the cord dose. IMRT may be of use in this context and can generally achieve the planning goals while keeping spinal cord dose within tolerance.

Dose prescription

+ Macroscopic tumour dose: 66–70 Gy in 33–35 fractions treating daily, five times per week.
+ Microscopic tumour dose: 50 Gy in 25 fractions treating daily, five times per week.

See section 12.2.5.

12.5.4 Nasopharynx

Nasopharyngeal carcinoma has a high risk of lymph node metastases initially to the retropharyngeal and parapharyngeal lymph nodes, and also to the deep cervical nodes bilaterally. The majority of patients present with locally advanced disease and are treated with chemoradiation[24]. IMRT should be considered for patients with nasopharyngeal carcinoma because of the very high risk of long-term xerostomia with conventional radiotherapy.

Patients are positioned with the neck extended with the chin up as far as possible. This optimizes shielding of the orbit and oral cavity in early phases of treatment. Data from MRI and CT are useful to accurately delineate the disease volume. The CTV includes base of skull (middle temporal fossa and cavernous sinus), posterior half of the orbit, posterior half of the nasal cavity, parapharyngeal space, lateral pharyngeal, posterior and upper deep cervical nodes.

Conventional radiotherapy planning

A CT planned, multiphase technique is used. The first phase uses large parallel-opposed lateral fields extending from the skull base to as low in the neck as possible to encompass all macroscopic disease. Usually a small anterior split neck field is required to treat down to the clavicles (Fig. 12.13).

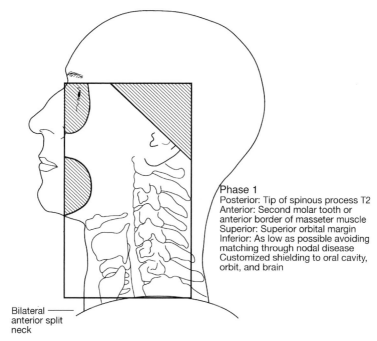

Fig. 12.13 Nasopharynx carcinoma phase I.

This initial volume is treated to a dose of 30 Gy in 15 daily fractions of 2 Gy. Undifferentiated tumours typically respond with rapid tumour shrinkage.

For subsequent phases the volume is split into two parts with a match line at the level of the lower border of the mandible. The advantage of this is that it allows the whole of the neck to be treated with large anterior (or anterior and posterior) neck fields with midline shielding to the mucosal surfaces. This substantially reduces early and late mucosal toxicity. Great care is needed, however, to ensure that the match line is not too high especially if there is tumour in retropharyngeal nodes or in the case of posterior pharyngeal mucosal extension of the primary tumour. In these circumstances there is a risk of geographical miss if the midline tumour is shielded from the anterior neck field. Parallel-opposed lateral fields are used to treat the nasopharynx, parapharyngeal space, base of sphenoid and posterior orbit—ensuring that there is sufficient cover of disease extension. This phase is treated to a dose of 20 Gy in 10 fractions. Shielding of the upper spinal cord may be needed after 44 Gy (Fig. 12.14a).

The final phase of treatment is customized to the presenting disease stage. At the primary site the volume is reduced further to avoid the retina, optic nerve and optic chiasm using a conformal three-field plan (direct anterior and two laterals). This keeps the dose to the temporomandibular joint and the temporal lobes below 60 Gy reducing the risk of late radiation damage. A dose of 15–20 Gy is delivered to the nasopharynx, and to residual neck nodes (Fig. 12.14b).

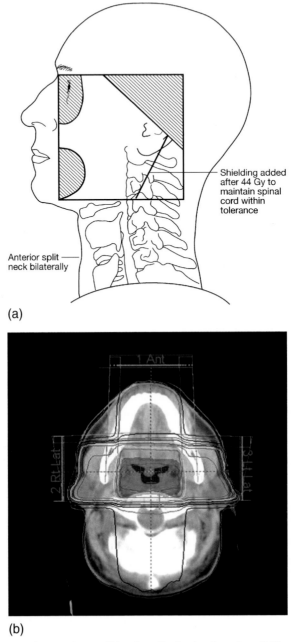

Shielding added
after 44 Gy to
maintain spinal
cord within
tolerance

Anterior split
neck bilaterally

(a)

(b)

Fig. 12.14 (a) Nasopharynx phase 2. (b) A dose distribution for a three-field nasopharyngeal treatment plan.

Dose prescription

- ◆ Phase 1: 30 Gy in 15 fractions treating daily, five times per week.
- ◆ Phase 2: 20 Gy in 10 fractions treating daily, five times per week.
- ◆ Phase 3: 15–20 Gy in 5–10 fractions treating daily, five times per week.
- ◆ Total macroscopic tumour dose: 65–70 Gy in 30–35 fractions treating daily, five times per week.

Intensity-modulated radiotherapy

PTV1 includes the primary tumour, retropharyngeal nodes, bilateral parapharyngeal spaces, and any lymph node groups harbouring metastases. PTV1 is treated with 65 Gy in 30 fractions. PTV2 includes all elective nodal groups Ib–V and is treated to 54 Gy in 30 fractions using SIB technique.

Lymph node-negative, early stage, well-differentiated nasopharyngeal carcinoma may be treated with localized radiotherapy to the nasopharynx, without full elective nodal irradiation.

12.5.5 **Oral cavity**

Early oral cancer including superficial (< 5 mm thickness), T1 and T2 lesions should be considered for brachytherapy. External beam radiotherapy is usually given postoperatively in patients with high-risk features or those unsuitable for radical surgery.

The treatment position is with the cervical spine straight. A mouth bite may be used to position the tongue. The oral cavity contains a number of individual subsites including oral tongue, floor of mouth, buccal mucosa, alveolus, and hard palate.

Tongue

CTV should be the tumour bed with a 2-cm margin. For tumours on the lateral tongue border this typically constitutes a hemioral cavity irradiation using anterior and posterior oblique fields wedged to produce a homogeneous dose distribution. For deeply infiltrative tumours approaching or invading the midline then parallel-opposed lateral beams are required to treat the CTV. Irradiation of the neck is indicated electively for infiltrative tumours and may be unilateral or bilateral depending on the relationship of the tumour to the midline or postoperatively for patients with high-risk features.

Buccal mucosa and alveolus

Usually a lateralized CTV is treated in the postoperative setting (Fig. 12.15).

Floor of mouth

Floor-of-mouth tumours commonly occur in the midline and therefore irradiation requires parallel-opposed lateral beams to cover the target volume that includes the primary tumour site and locoregional lymph nodes. It is important for the mouth to be stented open using a mouth bite that reduces irradiation of the hard palate mucosa.

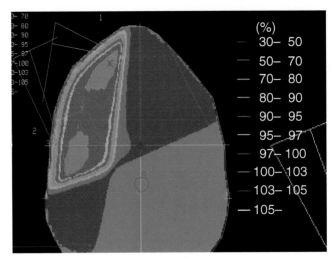

(%)
— 30– 50
— 50– 70
— 70– 80
— 80– 90
— 90– 95
— 95– 97
— 97– 100
— 100– 103
— 103– 105
— 105–

Fig. 12.15 A dose distribution for irradiation of a buccal mucosa tumour.

Dose prescription

◆ Postoperative dose: 60 Gy in 30 fractions treating daily, five times per week.

12.5.6 Parotid gland

Tumours of the parotid are treated with surgery and postoperative radiotherapy. Radiotherapy is indicated in tumours of high grade and those low-grade tumours that are recurrent or at very high risk of recurrence (i.e. macroscopic residual disease). For high-grade tumours (squamous, adenocarcinoma, and high0grade mucoepidermoid), the target volume should include the parotid bed and the ipsilateral deep cervical lymph nodes. For low-grade tumours the risk of lymph node metastases is so low that low neck irradiation is not required.

Patients should be immobilized with the neck extended and mandible perpendicular to the couch top such that the orbit is above the superior border of the radiotherapy fields. The entire parotid gland should be included in the CTV as these tumours may spread through the gland along salivary ducts. Anteriorly the CTV should include the parotid duct, medially the parapharyngeal space and laterally the scar. Adenoid cystic carcinomas spread along nerves and therefore this should be considered in planning. The facial nerve and the parasympathetics should be included back to their exit from the skull base by extending the posterior field border. This CTV is irradiated using anterior and posterior oblique beams in order to avoid irradiating the contralateral parotid gland, and reducing the risk of xerostomia.

Dose prescription

◆ Postoperatively: 60 Gy in 30 fractions treating daily, five times per week.

◆ If positive resection margins or extracapsular spread: 64 Gy in 32 fractions treating daily, five times per week.

Pleomorphic adenoma

This benign condition should be considered separately from other parotid tumours. They are usually encapsulated and radiotherapy should only be considered if tumours are recurrent or incompletely resected due to inoperability.

Typically they occur in the superficial lobe and can be treated with a direct electron field covering the postoperative tumour bed. Occasionally they occur in the deep lobe in which case they can be treated with the photon technique described previously.

Dose prescription
- 50 Gy in 25 fractions treating daily, five times per week.
- Or 45 Gy in 15 fractions treating daily, five times per week.

12.5.7 **Irradiation of the neck only and unknown primary**

The neck is irradiated alone in the postoperative setting (when primary tumour has been adequately resected, but with high risk of neck recurrence), and in patients with unknown primary tumour.

The patient should be immobilized supine, with the cervical spine straight and chin up as high as possible. The upper border is though the mastoid process and the lower border is below the lower border of the clavicle. The lateral border is the outer two-thirds of the clavicle and medial border is lateral to the spinal cord.

If the target volume includes level V then in order to adequately irradiate these nodes parallel-opposed anterior and posterior fields are required[25]. Usually the neck is irradiated postoperatively and for this indication the dose is 60 Gy in 30 daily fractions. The prescription point for neck irradiation is not defined and can be applied, at D_{max} or at a specific depth.

Irradiation of the mucosal surface of the head and neck in the treatment of an unknown primary presenting with nodal disease

Investigation of a patient presenting with nodal metastases includes fine needle aspiration of the node, EUA, and biopsy of any suspicious lesions. Ipsilateral tonsillectomy should be performed in the absence of an identifiable lesion, and biopsies of the post-nasal space and tongue base may be undertaken. CT or MRI may identify an occult primary site, but 18-FDG-PET scanning probably has a higher sensitivity than either of these investigations. PET remains positive up to 6 weeks after biopsy and so needs to be timed and interpreted carefully.

For squamous cell carcinoma metastatic to the deep cervical nodes, the most common primary tumour sites are the hypopharynx, oropharynx, or nasopharynx.

The treatment of these patients is highly controversial and is the subject of ongoing research.

Total mucosal irradiation of all the possible primary tumour sites is possible with large lateral fields extending from the skull base to the cricoid, matched to bilateral lower neck fields. This offers the possibility to eradicate the potential primary tumour, but is associated with severe late normal tissue morbidity due to the irradiation of a long length of mucosal surface and both parotid glands.

An alternative approach is irradiation of the involved hemineck, with an observation policy for the mucosal surfaces. In approximately 30% of cases no primary tumour appears during follow-up. If a primary tumour does become apparent subsequently, then further treatment with surgery and radiotherapy is possible at that time. This approach is associated with less morbidity, but higher rates of locoregional recurrence that can be salvaged by further therapy.

If the histology of the node is undifferentiated carcinoma of nasopharyngeal type (UCNT), especially in a posterior triangle node, then there is a high chance that the tumour arose in the postnasal space, and radiotherapy should given as for nasopharyngeal carcinoma. If it seems highly likely that the primary tumour lies in oropharynx or hypopharynx, then radiotherapy is given as for these sites.

12.5.8 **Orbit**

Tumours of the orbit are rare. Most commonly metastases from distant sites are seen in the context of widely disseminated malignant disease. Palliative radiotherapy using a single lateral photon field is appropriate for most patients. For bilateral deposits, opposed lateral fields may be used. Most metastases are seen in the retina or posterior orbit, and the use of a non-divergent anterior field border can avoid irradiation of the lenses of both eyes (Fig. 12.16). Half beam blocking, or a 5–10% gantry rotation from the direct lateral position, should be used to produce the non-divergent anterior border.

Dose prescription

- 20 Gy in five fractions treating daily, five times per week.
- Or 30 Gy in 10 fractions treating daily, five times per week.

Lymphoma, rhabdomyosarcoma, and lacrimal gland tumours usually occur outside the muscle cone posterior to the globe. These extraconal tumours may be treated with radical radiotherapy leaving the eye intact. The use of CT planning allows accurate localization of the target volume, and also the critical organs at risk: the lens, lacrimal gland, optic nerve, brainstem, and brain.

The technique employs a direct anterior and anterior oblique wedged fields to cover the PTV (Fig. 12.17).

The fields should be weighted anteriorly, and the anterior oblique field should come in behind the lenses if possible. Corneal and lens doses can be minimized by instructing

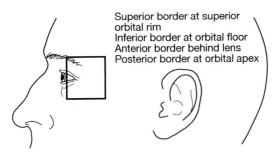

Superior border at superior orbital rim
Inferior border at orbital floor
Anterior border behind lens
Posterior border at orbital apex

Fig. 12.16 Use of a direct lateral field for orbit irradiation.

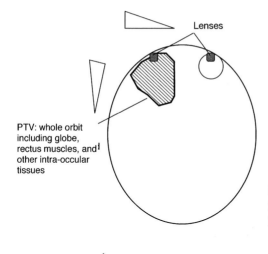

Lenses

PTV: whole orbit
including globe,
rectus muscles, and
other intra-occular
tissues

Fig. 12.17 Irradiation of an
intact eye using anterior and
lateral fields.

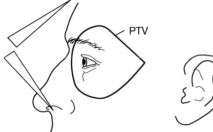

PTV

Fig. 12.18 Non-coplanar
technique for irradiation of a
proptosed eye.

the patient to stare directly into the beam with the eye open. The cornea and the
anterior part of the lens will lie within the build up region of the megavoltage beam.
If there is no tumour within the muscle cone, then a pencil lead shield can be used
to further reduce lens dose. One must be certain that no tumour lies within the
shielded tissue. If the tumour extends superiorly or inferiorly the anterior field can be
angled superiorly or inferiorly to ensure that the corneal shadow falls outside the PTV.
The patient must be clearly instructed to stare directly into the beam to immobilize
the eye during treatment. The use of lens shielding is unsuitable if there is intraconal
disease.

For patients with proptosis, the use of an anterior oblique beam or lateral field is
precluded because of the risk of irradiation of the contralateral eye. In these patients a
technique using superior and inferior non-coplanar anterior fields should be used
(Fig. 12.18). The eye should be kept open during treatment and the patient should
stare into the beam from each gantry angle.

Thyroid eye disease

Lateral opposed fields should be used to irradiate the posterior orbit to a dose of 20 Gy
in 10 fractions over 2 weeks. CT planning is recommended to assess lens dose that
should not exceed 6–10 Gy.

If higher doses are required to the orbit, for example, to deliver 68–70 Gy for a sarcoma, a wedged pair arrangement in the coronal plane may be used with lateral superior and inferior oblique fields.

Frequently locally advanced orbital tumours require exenteration, and postoperative radiotherapy is given. In this circumstance there are no intraorbital organs at risk, and so a technique using anterior and anterior oblique fields can be used, similar to Fig. 12.17, without the need to consider shielding structures within the eye. For carcinoma, doses of 60–64 Gy in 30–32 daily fractions can be used.

12.5.9 Ear and temporal lobe

Tumours described in this section include tumours of the pinna, external auditory canal, middle ear, and temporal bone.

Pinna

Tumours of the pinna should be considered as cutaneous malignancies. Primary surgery or radiotherapy is the treatment of choice. Radiotherapy generally gives the better cosmetic and functional result than pinnectomy. Radiotherapy with kilovoltage photons is contraindicated by cartilage invasion, (fixation, pain, or infection) because of a high risk of necrosis secondary to increase absorbed radiation dose (predominant photoelectric effect). Absorbed dose with electrons is less dependent on atomic weight and are the modality of choice where possible. Radiation technique should include definition of GTV, and addition of a margin of 1.5 cm to account for microscopic spread (5 mm) as well as the penumbra of the electron beam (10 mm). A wax-backed lead shield is placed behind the pinna to shield the skin from exit dose and bolus in the form of wax or wet gauze is used to ensure that the surface dose is 90–100%, a lead cut-out may be required.

Dose prescription
- 55 Gy in 20 fractions treating daily, five times per week.
- Or 45 Gy in 10 fractions treating daily, five times per week.

External auditory canal

Radiotherapy is indicated in early tumours of the external auditory meatus. In more advanced disease, surgical resection followed by radiotherapy is the treatment of choice. Patient position is with the neck extended so that exit of the posterior oblique beam avoids eye and lens. CT planning should be used if possible. The technique uses anterior and posterior oblique wedged beams (Fig. 12.19).

Dose prescription
- 66–70 Gy in 33–35 fractions treating daily, five times per week.

Middle ear and temporal bone

Standard treatment is extended total petrosectomy with postoperative radiotherapy. Radiotherapy technique uses an extended neck position, and CT planning is recommended. The CTV includes the GTV plus the pre and post-auricular lymph nodes. The brainstem and orbital contents should be localized as organs at risk. Anterior and

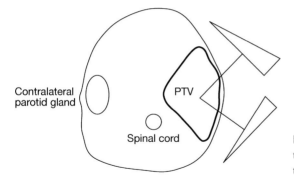

Fig. 12.19 Treatment technique for a carcinoma of the middle ear.

posterior wedged fields are used to irradiate the PTV in an arrangement similar to that shown in Fig. 12.19.

Dose prescription
- 66–70 Gy in 33–35 fractions treating daily, five times per week.

12.5.10 **Nose and paranasal sinuses**

Surgery alone or in combination with radiotherapy and chemotherapy is required in the majority of patients with squamous cell carcinoma of the nasal cavity or paranasal sinuses. Radiotherapy may be given before or after surgery. Chemotherapy and radiotherapy are also indicated for lymphoma or embryonal rhabdomyosarcoma.

Maxillary antrum radiotherapy technique

The position is supine with the cervical spine straight and a mouth bite to keep the lower oral cavity and tongue out of the fields. The PTV is often in close proximity to the optic nerves, chiasm, orbit, temporal lobes, and brainstem. For this reason CT planning is recommended, as it allows accurate target volume definition, and improved dosimetry around bone and air cavities. The improvements in the dose calculation provide better estimates of doses to OARs, especially for tumours close to the optic apparatus. This helps to inform clinical decision-making and individualization of radiotherapy dose to keep within optic nerve and chiasm tolerances.

Paranasal sinus tumours can spread mucosally into other adjacent sinuses, but lymph node metastases to the neck are rare. The CTV for a maxillary antrum tumour therefore includes the maxillary and ethmoid sinuses, nasal cavity, pterygoid fossa, and lateral pharyngeal node. The target volume is determined on the basis of CT and MR images, clinical and surgical assessment. Care should be taken to shield the brainstem, optic pathways, eyeball, lacrimal gland, and orbit wherever possible. Typically a field arrangement using a heavily weighted anterior, and one or two lateral fields is used. The anterior border of the lateral fields should use a non-divergent field border to avoid exit through contralateral lens and is achieved by angling the gantry by 5–10° behind the eyes, or using half beam blocking. Bilateral fields may be required to achieve target volume coverage of the most posterior and medial part of the PTV. The lacrimal gland, orbit, anterior part of the optic nerve, and lens are shielded from the anterior field (Fig. 12.20); the brainstem and optic chiasm from the lateral field (Fig. 12.21).

A dose of 66–70 Gy in 33–35 daily fractions is appropriate for squamous carcinoma or adenocarcinoma but careful attention to optic nerve and chiasm doses is required, and the superior border of the lateral field(s) may need to be reduced at about 50 Gy to avoid exceeding optic nerve tolerance (doses > 60 Gy carry a 20% risk of optic neuropathy). For lymphoma a dose of 35–40 Gy is adequate, well below the tolerance of the optic apparatus.

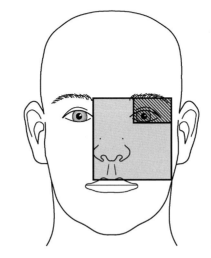

Fig. 12.20 Anterior field for a left maxillary antrum.

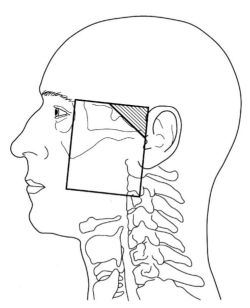

Fig. 12.21 Lateral field for a carcinoma of the maxillary antrum.

Dose prescription

♦ 66–70 Gy in 33–35 fractions treating daily, five times per week.

Ethmoid sinus

The ethmoid sinus is very difficult to irradiate to high dose as it lies between the optic nerves. A combination of surgery and postoperative radiotherapy gives the best chances of local control. CT planning is recommended. The CTV includes the medial half of the maxilla on the involved side, pterygoid fossa, both ethmoid sinuses, and nasal fossa. As with other paranasal sinus tumours, care should be taken to shield organs at risk wherever possible. Typically a three-field plan is used with an anterior and two lateral fields, although for the occasional T1 tumours the field can be restricted to the ethmoid sinus (Fig. 12.22) and nasal cavity using superior and inferior non-coplanar anterior fields coming between the eyes.

Dose prescription

♦ 66–70 Gy in 33–35 fractions treating daily, five times per week.

Nasal cavity

Patient is positioned with the cervical spine straight. Clinical target volume includes the lesion and 1-cm margin. Field arrangement is an anterior wedged pair of photon fields or in cases where the whole of the nasal cavity has to be included, an anterior and lateral field are then employed to achieve coverage of the target volume at depth.

Columella

Careful assessment is needed, as the deep margins of columella lesions may be difficult to assess. For extensive lesions, a two- or three-field photon technique is used as for nasal cavity tumours. For superficial lesions, a direct anterior electron field can be used with a wax block and wax nostril plugs to produce a homogeneous tissue density for dose deposition.

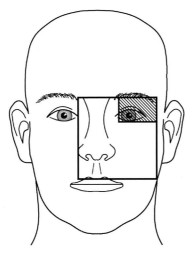

Fig. 12.22 Anterior field for irradiation of an ethmoid sinus tumour.

Dose prescription
- For lesions confined to the columellar: 55 Gy in 20 fractions treating daily, five times per week.
- For lesions extending up the nasal cavity: 66–70 Gy in 33–35 fractions treating daily, five times per week.

Olfactory neuroblastoma (or esthesioneuroblastoma)

This is a rare tumour arising from the olfactory receptors in the cribriform plate. It is best treated with surgery and adjuvant radiotherapy. Chemoradiation is often employed due to the locally advanced nature of the disease. CTV is the preoperative GTV with a margin—usually of 1 cm though this may need to be individualized on the basis of critical structure tolerance. Include the bilateral nasal cavities, ethmoid sinuses, cribriform plate, and olfactory bulb (so top of CTV is at least 1 cm superior to cribriform plate). Typically a three-field beam arrangement is used similar to an ethmoid sinus lesion. The close proximity of the optic nerves and optic chiasm to the tumour volume is a severe limitation to the dose that can be prescribed.

Dose prescription
- 60 Gy in 30 fractions treating daily, five times per week.

12.5.11 Recurrent disease and palliation

Some patients present with very advanced local or metastatic disease such that radical treatment is not appropriate. In other patients, concomitant medical conditions may preclude radical treatment. Surgery, radiotherapy, and chemotherapy all have a potential role in palliation and this is best discussed in the context of the multidisciplinary team.

Palliative radiotherapy requires high doses, and short fractionation regimens are associated with marked acute toxicity. Elective nodal irradiation is inappropriate in patients being treated with palliative intent, and therefore the PTV should include macroscopic disease only.

References

1. Toms JR (ed). *Cancer Stats Monograph 2004*. London: Cancer Research UK, 2004.

2. Hermanek P, Sobin LH, Wittekind C (eds). *TNM Classification of Malignant Tumours* (6th edn.). New York: John Wiley and Sons, 2002.

3. Bernier J, Domenge C, Ozsahin M, *et al*. Postoperative irradiation with or without concomitant chemotherapy for locally advanced head and neck cancer. *New England Journal of Medicine* 2004; **350**: 1945–52.

4. Cooper JS, Pajak TF, Fporastiere AA, *et al*. Post-operative concurrent radiotherapy and chemotherapy in high-risk squamous-cell carcinoma of the head and neck. *New England Journal of Medicine* 2004; **350**: 1937–44.

5. Pignon JP, Bourhis F, Domenge C, *et al*. on behalf of the MACH-NC Collaborative Group. Chemotherapy added to locoregional treatment for head and neck squamous-cell carcinoma: three meta-analyses of updated individual data. *Lancet* 2000; **355**: 949–55.

6. Candela FC, Kothari K, Shar JP. Patterns of cervical node metastases from squamous cell carcinoma of the oropharynx and hypopharynx. *Head Neck* 1990; **12**: 197–203.

7. Shah JP, Candela FC, Poddar AK. Patterns of cervical lymph node metastases from squamous cell carcinoma of the oral cavity. *Cancer* 1990; **66**: 109–13.

8. Nutting CM, Morden JP, Harrington JK, *et al.* Parotid-sparing intensity modulated versus conventional radiotherapy in head and neck cancer (PARSPORT): a phase 3 multicenter randomised controlled trial. *Lancet Oncology* 2011; **12**(2): 127–36.

9. International Commission on Radiation Units and Measurement. *Prescribing, recording and reporting photon beam therapy. ICRU Report 50.* Bethesda, MD: ICRU, 1993.

10. International Commission on Radiation Units and Measurement. *Prescribing, recording and reporting photon beam therapy (supplement to ICRU report 50). ICRU Report 62.* Bethesda, MD: ICRU, 1999.

11. Nutting C, Bidmead M, Harrington KJ, Henk JM. BIR Geometric uncertainties in radiotherapy: Head and neck cancer. In McKenzie A, Bidmead M (eds) *Geometric uncertainties in radiotherapy—defining the planning target volume.* London: British Institute of Radiology, 2003.

12. Gregoire V, Coche E, Cosnard G, Hamoir M, Reychler HV. Selection and delineation of lymph node target volumes in head and neck conformal therapy. Proposal for standardising terminology and procedure based on the surgical experience. *Radiotherapy and Oncology* 2000; **56**: 135–50.

13. Levendag P, Braaksma M, Coche E, *et al.* Rotterdam and Brussels CT-based neck nodal delineation compared with the surgical levels as defined by the American Academy of Otolaryngology-Head and Neck Surgery. *International Journal of Radiation Oncology, Biology, Physics* 2004; **58**: 113–23.

14. Emami B, Lyman J, Brown A, *et al.* Tolerance of normal tissue to therapeutic irradiation. *International Journal of Radiation Oncology, Biology, Physics* 1991; **21**: 109–22.

15. Dische S, Saunders M, Barrett A, Harvey A, Gibson D, Parmar MA Randomised multicentre trial of CHART versus conventional radiotherapy in head and neck cancer. *Radiotherapy and Oncology* 1997; **44**: 123–36.

16. Fu KK, Pajak TF, Trotti A, *et al.* A Radiation Therapy Oncology Group (RTOG) phase III randomized study to compare hyperfractionation and two variants of accelerated fractionation to standard fractionation radiotherapy for head and neck squamous cell carcinomas: first report of RTOG 9003. *International Journal of Radiation Oncology, Biology, Physics* 2000; **48**: 7–16.

17. Horiot JC, Bontemps P, van den Bogaert W, *et al.* Accelerated fractionation compared to conventional fractionation improves locoregional control in the radiotherapy of advanced head and neck cancers: results of the EORTC 22851 randomised trial. *Radiotherapy and Oncology* 1997; **44**: 123–37.

18. Bourhis J, Etessami A, Wilbault P, *et al.* Altered fractionated radiotherapy in the management of head and neck carcinomas: advantages and limitations. *Current Opinion in Oncology* 2004; **16**: 215–19.

19. Overgaard J, Hansen HS, Specht L, *et al.* Five compared with six fractions per week of conventional radiotherapy of squamous-cell carcinoma of head and neck: DAHANCA 6 and 7 randomised controlled trial. *Lancet* 2003; **362**: 933–40.

20. Wiernik G, Alcock CJ, Bates TD, *et al.* Final report on the second British institute of radiology fractionation study: short versus long overall treatment times for radiotherapy of the laryngo-pharynx. *British Journal of Radiology* 1991; **64**: 232–41.

21. Spaulding MB, Fischer SG, Wolf GT. Tumor response, toxicity, and survival after neoadjuvant organ-preserving chemotherapy for advanced laryngeal carcinoma. The Department of Veterans Affairs Cooperative Laryngeal Cancer Study Group. *Journal of Clinical Oncology* 1994; **8**: 1592–9.

22. Lefebvre JL, Chevalier D, Luboinski B, Kirkpatrick A, Collette L, Sahmoud T. Larynx preservation in pyriform sinus cancer: preliminary results of a European Organization for Research and Treatment of Cancer phase III trial. EORTC Head and Neck Cancer Cooperative Group. *Journal of the National Cancer Institute* 1996; **88**(13): 890–9.

23. Al-Sarraf M, LeBlanc M, Giri PG, *et al.* Chemoradiotherapy versus radiotherapy in patients with advanced nasopharyngeal cancer: phase III randomised Intergroup study 0099. *Journal of Clinical Oncology* 1998; **16**: 1310–17.

13

Skin cancer

C Corner, Peter Hoskin

Skin tumours differ in their radiotherapy planning from most other sites in that the volume definition is based principally upon clinical examination and the majority will be treated by single applied beams using low-energy X-rays or electrons with clinical verification.

Three major histological groups are squamous cell carcinoma, basal cell carcinoma and malignant melanoma with a fourth comprising the rarer entities of adnexal tumours and Merkel cell tumours.

13.1 Radical primary treatment

13.1.1 Squamous cell carcinoma and basal cell carcinoma

Most squamous cell carcinomas can be equally well treated by surgical resection or local radiotherapy. Local control rates after radiotherapy for T1 and T2 lesions (up to 5 cm diameter) range from 85–95%; size, margins > 1 cm and total dose equivalent to at least 60 Gy are independent predictors of local relapse[1,2].

Control rates for basal cell carcinomas tend to be even higher than squamous carcinomas with overall rates in excess of 90%[1,2,4]. The superficial and sclerosing variants have higher relapse rates than the nodular type[5]. Relapse in basal cell carcinomas may occur later than in squamous carcinoma, median time to recurrence in being 20–40 months compared to 5 months in squamous carcinoma[2,5].

Indications for radiotherapy

- Lesions where radiotherapy would produce better cosmesis and functional outcome than surgery, e.g. nose, lower eyelid, ear, lower lip.
- Lesions with potential for deep tumour infiltration, e.g. inner canthus, nasolabial fold, ala nasi, tragus, postauricular area.
- Large superficial lesions.
- Elderly/frail patients.
- Patients who refuse or are unfit for surgery, e.g. due to anaesthetic risk etc.
- Adjuvant treatment in cases at high risk of recurrence.

Indications for surgery

- Patient's age < 50 years (as radiation scars tend to deteriorate with time).
- Sites of previous burns.
- Sites of previous radiotherapy.
- Areas of vascular insufficiency, e.g. shin, dorsum of hand (where radiotherapy may cause problems with healing or function).
- Lesions overlying the lacrimal gland (upper outer eyelid).
- Conditions with an inherent defect in the DNA repair mechanism, e.g. ataxia telangiectasia, xeroderma pigmentosa resulting in a predisposition to extreme radiation reactions.

13.2 **Postoperative adjuvant treatment**

Superficial squamous cell carcinoma of the skin treated surgically will usually be completely excised with an adequate surgical margin of several millimetres. There is no consensus on what constitutes an adequate surgical margin but a minimum of 3 mm is a practical guide. Closer surgical margins may be dealt with by re-excision or postoperative radiotherapy to the site of excision.

Adjuvant radiotherapy may also be indicated for aggressive tumour subtypes, tumours > 2 cm in diameter and > 4 mm in depth, and those with perineural invasion detected pathologically[6,7].

13.2.1 **Radiotherapy planning**

Immobilization and position

For treatments around the head and neck plastic immobilization shells should be used, with the shell cut-out around the treatment area.

If a lead (Pb) mask is used it is placed over the plastic shell. It is important that the position is stable with appropriate support to the body using a suitable headrest, pillow, or sandbags. Patients should be as comfortable as possible to aid reproducibility. Access using superficial X-ray or electron applicators will also have to be considered in defining the optimal patient position.

Target definition

The clinic notes, histology, clinical photographs, and CT/MRI images if appropriate should be reviewed. The CTV should be defined on clinical examination using a bright light and magnifying glass. A fine indelible marker is used to define the tumour extent with appropriate margins. For tumours which are fixed or those in the nasal vestibule or external pinna then CT or MR scanning may be valuable to identify the depth of penetration.

Treatment volume

The clinician will delineate a GTV and a field size. A margin of 3 mm is used for the CTV, a further 2 mm is added for the PTV. The resultant expansion of GTV to PTV is 0.5 cm circumferential margins, and 0.5 cm for the deep margin. For morphoeic basal cell carcinomas and squamous cell carcinomas the circumferential margins are extended to ≥ 1 cm and the deep margin ≥ 0.5 cm. An additional margin for field size is required due to the following characteristics of the electron beam:

• The edge of the electron applicator represents the 50% isodose; the 90% isodose is typically 3–5 mm inside this depending on field size and therefore a larger applicator than the defined PTV will be required.

• As electron energy increases there is a bowing inwards of the isodoses close to the surface where the tumour will be, as shown in Fig. 13.1. An additional allowance for this should be made so that overall an applicator diameter 10 mm larger circumferentially than the defined PTV should be chosen.

Example of margins required for a basal cell carcinoma are shown in Fig. 13.2. A 2-cm tumour (GTV) gives a 3-cm PTV. This has adequate 90% coverage if the field size is 5 cm.

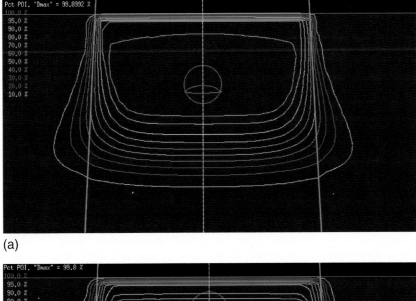

(a)

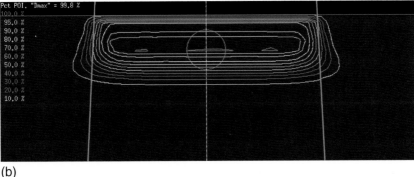

(b)

Fig. 13.1 Electron isodoses demonstrating bowing in effect of higher energy 15-MeV beams (a) compared to lower energy 6-MeV beams (b).

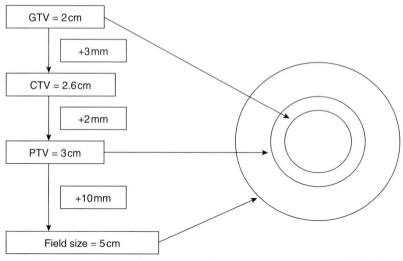

Fig. 13.2 Diagrammatic illustration of the margins required for the treatment of a basal cell carcinoma using electrons.

Dose distribution

Treatment will typically require a beam penetrating to a depth of 20–25 mm. This may be achieved by a superficial X-ray beam or an electron beam of 8–10 MeV energy. Early reports suggested that electron treatment was less effective than superficial X-rays but subsequent analyses have shown that provided adequate margins are used taking into account the characteristics of the electron beam dose distribution then equivalent results are obtained[1,3].

Planning aims/prescription

The prescription should be to the ICRU reference point, which is the 100% (i.e. D_{max}) of the percentage depth dose. The ICRU reference point (i.e. D_{max}) should always be at the centre (or in the central part) of the PTV. One should aim to cover the PTV with the 90% isodose. Organs at risk should be defined and doses kept to defined constraints.

13.2.2 **Beam energy**

Energy should be chosen to give the best conformation to the defined PTV.
  It should take into account the following:

◆ Surface dose required, which should be at least 90%.

◆ Depth to be treated, i.e. covered with the 90% isodose. This depth should be equal to the depth of measurable tumour plus 5 mm (GTV to PTV).

◆ Dose to surrounding critical structures.

Important considerations when choosing an electron beam are:

1. The surface dose for an electron beam must be considered; as electron beam energy increases the surface dose also increases. Build-up may be required depending upon the characteristics of the beam.

2. Build-up may be used to reduce the depth in the patient of the high-dose volume where a beam of suitable energy is not available; this is particularly the case for very superficial tumours when a depth of 2.5 or 3 cm is excessive or chest wall tumours where there is additional concern because of increased scatter in the lungs.

3. It is unusual for a tumour to be situated on a flat piece of anatomy; typical sites are around the face on sun exposed areas and there is often a problem in achieving close applicator apposition to the area to be treated. This results in the problem of 'stand off' which may be dealt with in one of two ways as shown in Fig. 13.2:

   ◆ The area may be made into a flat incident surface using bolus; the disadvantage of this approach is that set-up may be less accurate when the underlying tumour cannot be seen at the time of applicator positioning.

   ◆ The machine monitor units can be modified by a simple calculation based on the inverse square law. The disadvantage of this approach is that the stand off is unlikely to be constant across the treatment area and therefore the calculation may be based on the maximum stand off or mean stand off, both representing a compromise. However, in practice the effect is small altering the applied dose across the area by <5%.

4. Influence of inhomogeneities:

   ◆ Two main factors that account for the effect of inhomogeneities on the dose distribution are:
      • Different absorption of various tissues, which depends on density of the tissues.
      • Alteration in electron scatter pattern (scatter perturbation), which depends on atomic number of the tissues.

   ◆ In general the denser the material is, the greater its absorption and scattering property with regards to electron beams. More electrons are scattered away from higher-density materials towards lower-density material giving rise to hot spots under the low-density material. There are corresponding low-dose areas under the high-density region, reflecting the loss of electrons.

   ◆ For small inhomogeneities (small air cavities, small bony structures), the local scattering of electrons at the edges is the predominant effect. At a straight edge of material, the scattering electrons give rise to a hot spot on one side and a cold spot on the other.

5. Electron field matching:

   ◆ It is more difficult to match electron fields as the isodoses do not follow the geometric edge of the beam. The high-dose isodoses become narrower with depth while the lower isodoses bow out with depth.

   ◆ If no gap is used between fields then a hotspot is created at the matched edges of up to 140%. If a gap of between 0.5–1 cm is used between fields this hotspot can be reduced considerably.

Field shaping can be done using standard electron endplates for circular or elliptical fields. Straight-edged fields are preferred if it is likely that further treatment of nearby tissues will necessitate field matching. Customized endplates are useful for irregular shapes or treatment near to a critical structure. These are made from lead of an appropriate thickness for superficial X-rays or an end frame cut out for an electron applicator as shown in Fig. 13.3. Lead masks are used for areas near the eye and nose only. The thickness of the Pb should be approximately half the beam energy in mm. It is often valuable to extend the cut-out to include locating anatomical structures, for example, the nose and superior orbital ridges on the face to facilitate correct placement of the cut-out for each treatment as shown in Fig. 13.5.

Internal shielding prevents the radiation beam penetrating past a specific boundary. For example, the eye/eyelid, lip and areas of the mouth, and behind the ear. Corneal shielding is used for tumours of the lower eyelid, inner and outer canthi. Superficial X-ray treatments may use a patch or contact lens shield. Electron treatments use a similar contact lens shield as shown in Fig. 13.4; an important feature of these shields is wax coating to absorb scattered radiation from the electron beam. Internal mouth shields and shields used behind the ear should also be covered with a layer of wax to absorb backscatter which is dependent on the energy and type of shield.

Verification

Verification is by clinical observation of the applicator set-up and light beam on the patient at each treatment. Additional verification by imaging or *in vivo* dosimetry is not usual in these cases except where treatment is close to a critical structure, for example, around the eye when lithium fluoride or diode measures are recommended.

Indelible skin marks and margin tattoos together with photographs of the treatment position are valuable to confirm the correct set-up and applicator position.

Special considerations

- Tumours overlying bone are best treated with electrons because of the enhanced absorbed dose in bone from superficial X-rays.
- Electrons are preferred to superficial X-rays for treatment of deep tumours and those overlying cartilage, e.g. nose, pinna.
- Superficial X-rays are favoured for tumours around the eye due to electron scatter and required lead thickness.

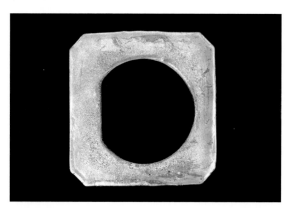

Fig. 13.3 Customized end frame for electron applicator; this will be fitted into an applicator of larger size than the end-frame aperture.

(a)

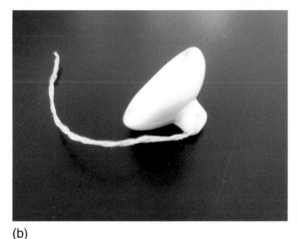

(b)

Fig. 13.4 Corneal contact lens shields used for electron treatment around the eye.

- With superficial X-rays crenellation of the margin of a round cut-out which is then rotated each day may give better cosmesis, blurring the edge of radiation reaction. This should be allowed for in the PTV expansion.

Dose prescription for squamous cell carcinoma

A range of dose prescriptions are in use. The nominal gold standard is 60–66 Gy in 30–32 fractions over 6–6½ weeks. More pragmatic alternatives include:

- 55 Gy in 20 fractions
- 40 Gy in 10 fractions
- 45 Gy in nine fractions
- 35 Gy in five fractions.

Single doses of 16 or 18 Gy have been used and equivalent results to fractionated schedules for tumours < 3 cm maximum diameter are reported. Doses are defined at

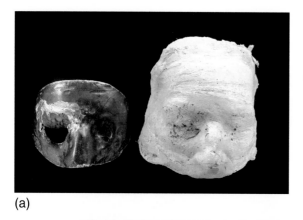

(a)

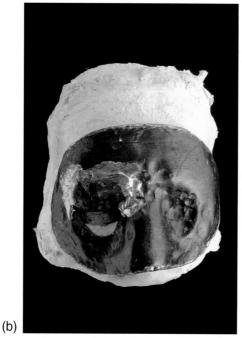

(b)

Fig. 13.5 Individualized lead cut-out for treatment to a lower lid tumour showing impression made and resulting cut-out (a), and location of cut-out with extension over face to enable accurate relocation (b).

the 100% isodose for both superficial X-rays and electrons. Cosmesis is related to tumour size and fraction size.

Dose prescription for basal cell carcinoma

In general, basal cell carcinomas are considered more radiosensitive and a nominal dose of 60 Gy in 30 fractions over 6 weeks is given. Equivalent doses would include:

- ◆ 50 Gy in 20 fractions
- ◆ 40 Gy in 15 fractions

- 40.5 Gy in nine fractions
- 32.5 Gy in five fractions.

Single doses of 16 or 18 Gy may also be considered for tumours < 3cm maximum diameter.

Radiation toxicities

Acute toxicity (within 90 days)

- Fatigue
- Skin erythema
- Skin desquamation
- Scabbing
- Alopecia in hair-bearing areas
- Pigmentation.

Late toxicity

- Pigmentation or hypopigmentation
- Atrophy
- Telangiectasiae
- Permanent alopecia in high-dose areas
- Nasolacrimal duct stenosis (for treatments near medial canthus/nasal bridge)
- Cartilage necrosis (for treatments of nose and ear) is rare
- Second malignancy is rare but may occur with a latency of > 10 years.

13.2.3 Melanoma

The primary management of malignant melanoma is by surgical excision. Close surgical margins should be dealt with by re-excision rather than radiotherapy; however, occasionally radiotherapy is given in the adjuvant treatment of the primary if adequate excision margins cannot be obtained, e.g. some head and neck sites.

Radiotherapy may be used in the adjuvant treatment of the nodal basin as this has been shown in both retrospective and randomized studies to reduce the risk of locally recurrent disease[8,9]. Postoperative lymph node irradiation is recommended in melanoma for patients who have extranodal extension, more than three nodes involved, or large nodes (> 3 cm)[10,11]. There is currently no evidence for improved overall survival after radiotherapy.

Radiotherapy may also be used in the palliation of inoperable or advanced disease

Lentigo maligna is a non-invasive melanoma typically occurring on the face of elderly patients. Surgical resection is the treatment of choice due to the risk of biopsy missing an invasive component and the risk of progression to an invasive lesion. However, in patients unsuitable for resection, superficial radiotherapy using treatment techniques and dose prescriptions as described previously for squamous carcinoma result in high rates of local control in excess of 90%[12].

Dose prescription for melanoma

Adjuvant treatment of primary

In this setting techniques and doses as for squamous carcinoma are used, although there are advocates for hypofraction in melanoma using schedules of 30–36 Gy in five to six fractions treating twice weekly[8].

Adjuvant treatment of the nodal basin
Suitable dose/fractionation schedules:

- 40.05 Gy/15 fractions
- 50 Gy/20 fractions
- 50 Gy/25 fractions
- 50.4 Gy/28 fractions.

Shorter hypofractionated schedules to the lymph node areas are likely to increase late morbidity in terms of fibrosis and lymphoedema.

13.2.4 Merkel cell tumours

Merkel cell tumours (MCCs) are rare neuroendocrine tumours with biological features analogous to small cell lung cancer. Given the propensity of MCC to recur locally (sometimes with satellite lesions and/or in transit metastases), wide local excision to reduce the risk of local recurrence has been recommended. Optimal minimum margin width and depth of excision around the primary tumour vary but most advocate 2–5-cm margins. Because of the aggressive nature of MCC, its apparent radiosensitivity, and the high incidence of local and regional recurrences, including in transit metastases after surgery alone to the primary tumour, adjuvant radiation to the primary site and nodal basin has been recommended[13,14,15]. Those groups thought to benefit most from adjuvant radiotherapy include patients with:

- Large tumours
- Locally unresectable tumours
- Close or positive excision margins
- Positive regional nodes (especially after sentinel lymph node dissection).

Dose prescription for Merkel cell tumours

Adjuvant radiotherapy to the primary site
Techniques are as described previously for squamous carcinoma using direct electron beams with a margin to define the PTV of 2–5 cm circumferentially and ≥ 1 cm deep margin.

Dose/fractionation schedules include:

- 35 Gy/five fractions over 1 week
- 45 Gy/nine fractions over 2 weeks
- 55 Gy/20 fractions over 4 weeks
- 60 Gy/30 fractions over 6 weeks.

Prophylactic nodal irradiation

Megavoltage X-ray treatment is given to the node areas as described in Chapter 10. Dose/fractionation schedules are as described for melanoma.

13.2.5 **Adnexal tumours**

The treatment of adnexal tumours will be surgical excision and radiotherapy has no recognized role in their primary treatment.

Close surgical margins should be dealt with by re-excision rather than radiotherapy; only where this is not possible because of comorbidity or technical considerations should radiotherapy be used. In this setting techniques and doses as for squamous carcinoma are used. A 10-mm margin to define CTV is recommended.

13.3 **Palliative treatment**

13.3.1 **Indications**

Palliative treatment of skin tumours may be indicated for large fungating primary sites, particularly in the elderly or medically frail, and for regional lymph nodes which are inoperable.

13.3.2 **Locally advanced and metastatic squamous cell carcinoma**

Radiotherapy of the involved nodal basin is routinely used as an adjunct to the surgical treatment of metastatic squamous cell carcinoma and has been shown to improve outcomes[16].

Techniques for treatment will be the same as for radical primary treatment. Exceptions may include the following:

- Deeper tumours may require higher-energy electrons or on rare occasions be so extensive as to require planned photon beam treatments.
- Fixed regional lymph nodes will be treated with photon beams in most instances. Primary tumours in the head and neck region will drain to cervical lymph nodes, and from the upper trunk and arm to axillary lymph nodes. Lower limb lesions will involve inguinal lymph nodes. Treatment to these sites will follow standard techniques as described in Chapter 10.

Dose prescription in this setting can present a difficult balance between the temptation to give a short pragmatic palliative schedule such as 20 Gy in five fractions or 30 Gy in 10 fractions and the knowledge that if inadequate, further local relapse can be very difficult to manage and therefore a full radical dose should be attempted. Depending upon the site, hypofractionated schedules such as 55 Gy in 20 fractions, 40 Gy in 10 fractions, or 30 Gy in six fractions may be effective for durable control. Larger fraction schedules should be avoided in lymph node areas.

Systemic chemotherapy may be appropriate in patients with good performance status. Common protocols include platinum and 5-fluorouracil. Oral retinoids have been shown to be useful in preventing disease recurrence and progression, particularly in immunosuppressed patients and also patients with early-onset aggressive tumours although they are not widely used at present[17].

13.3.3 Basal cell carcinoma

Indications for palliative treatment will be the same as for squamous carcinoma except that lymph node metastases are not seen with this tumour. Treatment of the primary site will follow the same procedures and techniques as for primary radical treatment.

13.3.4 Melanoma

Palliative treatment may be indicated for fixed inoperable lesions in the skin. Satellite nodules may also become symptomatic and benefit from local radiotherapy. Techniques will be the same as those described for primary radical treatment. Where there are satellite nodules these should, as far as possible, all be included in the CTV.

Dose prescription for melanoma may include those quoted previously for squamous cell carcinoma. An alternative is to use weekly or twice weekly doses of 6 Gy to a total dose of 30–36 Gy.

13.3.5 Merkel cell tumours

Locally advanced Merkel cell tumours will respond readily to radiotherapy but are typically associated with widespread metastatic disease in liver, bone, lungs, and brain. Chemotherapy will often be more appropriate in this setting favouring protocols used to treat small cell lung cancer, e.g. platinum and etoposide but local control can be usefully obtained using palliative radiotherapy. Techniques will be as described previously for primary treatment and may have to include node areas also as regional lymph node disease is common in advanced cases. For durable control doses such as those discussed earlier for squamous carcinoma should be used.

13.3.6 Adnexal tumours

Adnexal tumours rarely present as a palliative problem but when they do so the techniques described previously should be followed.

References

1. Locke J, Karimpour S, Young G, Lockett MA, Perez A. Radiotherapy for epithelial skin cancer. *International Journal of Radiation Oncology, Biology, Physics* 2001; **51**: 748–55.
2. Kwan W, Wilson D, Moravan V. Radiotherapy for locally advanced basal cell and squamous cell carcinomas of the skin. *International Journal of Radiation Oncology, Biology, Physics* 2004; **60**: 406–11.
3. Griep C, Davelaar J, Scholten AN, Chin A, Leer JW. Electron beam therapy is not inferior to superficial x-ray therapy in the treatment of skin carcinoma. *International Journal of Radiation Oncology, Biology, Physics* 1995; **32**: 1347–50.
4. Telfer NR, Colver GB, Bowers PW. Guidelines for the management of basal cell carcinoma. *British Journal of Dermatology* 1999; **141**: 415–23.

5. Zagrodnik B, Kempf W, Seifert B, *et al.* Superficial radiotherapy for patients with basal cell carcinoma: recurrence rates histologic subtypes and expression of p53 and Bcl-2. *Cancer* 2003; **98**: 2708–14.

6. Jambusaria-Pahlajani A, Miller CJ, Quon H, *et al.* Surgical monotherapy versus surgery plus adjuvant radiotherapy in high-risk cutaneous squamous cell carcinoma: a systematic review of outcomes. *Dermatologic Surgery* 2009; **35**: 574–85.

7. Han A, Ratner D. What is the role of adjuvant radiotherapy in the treatment of cuatneous squamous cell carcinoma with perineural invasion? *Cancer* 2007; **109**: 1053–9.

8. Burmeister BH, Mark Smithers B, Burmeister E, *et al.*; Trans Tasman Radiation Oncology Group. A prospective phase II study of adjuvant postoperative radiation therapy following nodal surgery in malignant melanoma-Trans Tasman Radiation Oncology Group (TROG) Study 96.06. *Radiotherapy and Oncology* 2006; **81**(2): 136–42.

9. Agrawal S, Kane JM 3rd, Guadagnolo BA, Kraybill WG, Ballo MT. The benefits of adjuvant radiation therapy after therapeutic lymphadenectomy for clinically advanced, high-risk, lymph node-metastatic melanoma. *Cancer* 2009; **115**(24): 5836–44.

10. Lee RJ, Gibbs JF, Proulx GM, Kollmorgen DR, Jia C, Kraybill WG Nodal basin recurrence following lymph node dissection for melanoma: implications for adjuvant radiotherapy. *International Journal of Radiation Oncology, Biology, Physics* 2000; **46**(2)467–74.

11. Bastiaannet E, Beukema JC, Hoekstra HJ. Radiation therapy following lymph node dissection in melanoma patients: treatment outcome and complications. *Cancer Treatment Reviews* 2005; **31**: 18–26.

12. Farshad A, Burg G, Panizzon R, Dummer R. A retrospective study of 150 patients with lentigo maligna or lentigo maligna melanoma and the efficacy of radiotherapy using Grenz or soft x-rays. *British Journal of Dermatology* 2002; **146**: 1042–6.

13. Poulsen M. Merkel cell carcinoma of the skin. *Lancet Oncology* 2004; **5**: 593–9.

14. Gollard R, Weber R, Kosty MP, *et al.* Merkel cell carcinoma: review of 22 cases with surgical, pathologic and therapeutic considerations. *Cancer* 2000; **88**(8): 1842–51.

15. Eng TY, Boersma MG, Fuller CD, *et al.* A comprehensive review of the treatment of Merkel cell carcinoma. *American Journal of Clinical Oncology* 2007; **30**(6): 624–36.

16. Veness MJ. Treatment recommendations in patients disgnosed with high risk cutaneous squamous cell carcinoma. *Australasian Radiology* 2005; **49**: 365–76.

17. Niles RM. The use of retinoids in the prevention and treatment of skin cancer. *Recent Expert Opinion on Pharmacotherapy* 2002; **3**(3): 299–303.

14

Sarcomas of soft tissue and bone

Anna Cassoni

14.1 Introduction

Sarcoma is an uncommon malignancy and these tumours represent less than 1% of cancers. Both soft tissue and bone sarcomas occur across all age ranges and, especially in bone tumours, with a distinct subgroup within the paediatric and young adult range. This chapter will discuss both soft tissue and bone tumours except for rhabdomyosarcomas which will be covered within paediatric tumours (Chapter 15).

Soft tissue sarcoma in adults comprises some 50 different subtypes but the principles of management with radiotherapy are broadly similar. The management of sarcomas of bone is much more varied and a complete review outside the scope of this chapter. The main malignant tumour which forms bone is osteosarcoma; that forming cartilage is chondrosarcoma. Other sarcomas arising from bone include Ewing's sarcoma, chordomas, and spindle cell tumours. Chondrosarcoma and chordoma will usually be treated surgically in the first instance. Chemosensitive tumours such as Ewing's, osteosarcoma, and rhabdomyosarcoma will be treated with neo-adjuvant chemotherapy with radiotherapy used as an adjuvant treatment or as an alternative to surgery when this is not possible or its consequences unacceptable.

14.1.1 Ewing's sarcoma and osteosarcoma

These will be treated with neoadjuvant chemotherapy. In osteosarcoma, treatment of the primary will be surgical unless the tumour is inoperable. Surgery will be the preferred local option for Ewing's, but radiotherapy may be an appropriate radical option in this more radiosensitive tumour. Radiotherapy may have a role in chordoma and chondrosarcoma but these are relatively radioresistant tumours and best results are seen in low-bulk tumours with high-dose radiotherapy.

14.1.2 Soft tissue sarcoma

Adjuvant postoperative radiotherapy is advised for intermediate- and high-grade sarcoma and randomized trials have shown its effectiveness in reducing recurrence.

At present, postoperative radiotherapy is usual but there are persuasive arguments for preoperative radiotherapy which, while increasing acute postsurgical toxicity, results in less late morbidity. This approach may be particularly suited to myxoid liposarcomas where a substantial and rapid reduction in size may be seen.

14.1.3 Dose range

The dose in Ewing's sarcoma depends on indications. Osteosarcoma, chondrosarcoma, and chordoma all require high doses for radical therapy and 60–66 Gy or more is recommended.

For soft tissue sarcoma the recommended dose preoperatively is 50 Gy with a good evidence base. Postoperative recommendations are less clear. Doses of 60–66 Gy are used.

14.2 Features in radiotherapy planning common to all sarcoma subtypes

14.2.1 Essential investigations for planning radiotherapy

◆ Biopsy results if no resection.

◆ Operation note and resection histology.

◆ Preoperative or prechemotherapy contrast-enhanced MRI scans of the primary site, including the entire involved bone and adjacent joints.

◆ CT may add information in bone tumours.

14.2.2 Patient preparation

◆ When the pelvis or proximal thigh is irradiated, sperm storage and oophoropexy should be considered to spare fertility.

◆ A spacer may be inserted before radiotherapy to displace bowel away from the high-dose area in iliac blade tumours as shown in Fig. 14.1.

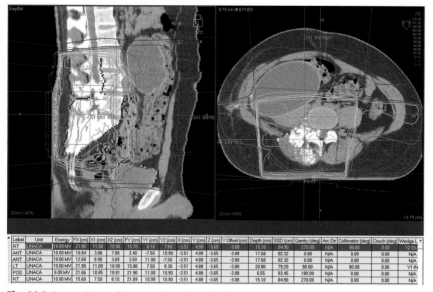

Label	Unit	Energy	FX (cm)	X1 (cm)	X2 (cm)	FY (cm)	Y1 (cm)	Y2 (cm)	X (cm)	Y (cm)	Z (cm)	Y Offset (cm)	Depth (cm)	SSD (cm)	Gantry (deg)	Arc Dir	Collimator (deg)	Couch (deg)	Wedge L
RT	LINACA	10.00 MV	21.90	11.00	10.90	15.70	8.10	7.60	-3.51	4.88	-3.65	-3.88	15.10	84.90	270.00	N/A	90.00	0.00	Y2 OU
ANT	LINACA	10.00 MV	10.94	3.06	7.89	3.40	-7.50	10.90	-3.51	4.88	-3.65	-3.88	17.68	82.32	0.00	N/A	0.00	0.00	N/A
ANT	LINACA	10.00 MV	12.68	8.98	3.69	3.50	11.00	-7.50	-3.51	4.88	-3.65	-3.88	17.68	82.32	0.00	N/A	0.00	0.00	N/A
LT	LINACA	10.00 MV	21.90	11.00	10.90	15.80	7.50	8.30	-3.51	4.88	-3.65	-3.88	20.80	79.20	90.00	N/A	90.00	0.00	Y1 IN
POS	LINACA	6.00 MV	21.66	10.85	10.81	21.90	11.00	10.90	-3.51	4.88	-3.65	-3.88	6.55	93.45	180.00	N/A	0.00	0.00	N/A
RT	LINACA	10.00 MV	15.60	7.50	8.10	21.89	10.99	10.90	-3.51	4.88	-3.65	-3.88	15.10	84.90	270.00	N/A	0.00	0.00	N/A

Fig. 14.1 Spacer, inserted retroperitonally and inflated with saline, displaces bowel anteriorly.

◆ Patients with limb, shoulder girdle, or head and neck tumours should be immobilized in individually made casts.

The position for planning should be individualized depending on the precise position of the tumour within the limb, with the casts and a point of fixation preferably around the distal limb. Vacuum fix bags may be of value in truncal tumours. Devices should provide reproducibility of 3–5 mm in the head and neck and 5–7 mm elsewhere although at sites such as the shoulder this may be difficult. Treatment position will usually be supine as this is considered more stable, but in patients with posterior pelvic tumours, prone may be preferable if this displaces bowel and reduces skin dose.

14.2.3 Planning imaging required for target definition

◆ 3-mm CT scans of the tumour region, covering the entire extent of the surgical bed and scar in the postoperative setting. The scar should be marked with wire.

◆ Intravenous contrast use at the clinician's discretion if there are no contraindications.

14.2.4 Target definition

Gross tumour volume

The GTV, or virtual GTV in postoperative cases, requires recent and clear images. Full histology reports should be available for postoperative cases with a description of the tumour size and margins. Fusion of diagnostic and planning scans may be valuable where there has been no operative intervention.

Clinical target volume

Radiotherapy is given to the tissues potentially involved by the tumour in its extent at diagnosis or maximum size, with margins for potential microscopic spread along relevant tissues planes. In postoperative cases or those that have responded to chemotherapy this principle needs to be applied with care as the anatomy of the region may have changed, with the potentially contaminated structures in a different position relative to fixed structures.

The geometrically grown margin may be modified by barriers to spread, such as the deep fascia of the thigh, or by areas of weakness such produced by the neurovascular bundle through fascia.

The CTV may be modified to ensure a 'corridor' and to spare joints or other critical structures as shown in Fig. 14.2.

◆ In postoperative cases it is usual to include the whole operative bed for the majority of the treatment, as this is considered as a risk of contamination. This includes the scar and underlying tissues through the surgical field. It has also been the practice, where a prosthesis has been placed, to include its full length. However, there is no good evidence base for this and, where this would compromise function, it is reasonable not to attempt full inclusion especially if joints or, in growing children, epiphyses would be included.

◆ PTV = CTV + 5–10 mm depending on site and local audit of set-up.

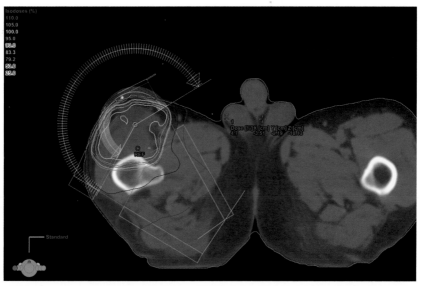

Fig. 14.2 Postoperative radiotherapy for high-grade soft tissue sarcoma of adductor compartment. RapidArc treatment with two arcs, 6-MV photons, produces relative sparing of part of circumference of femur. OARs: femur, unirradiated 'corridor' post thigh, scrotum.

- Limb PTV = CTV + 7 mm.
- Consider larger CTV–PTV margin in shoulder girdle tumours and trunk sites.

14.2.5 Field arrangement and dose distribution

Parallel-opposed fields are frequently used for limbs. For the scapula and iliac bone tangential fields parallel to the main axis of the bone may minimize normal tissue inclusion.

Three-dimensional conformal radiotherapy planning is standard, with in-field boost if needed to provide homogeneity not only in the axial plan but along the axis of the limb. For extremity tumours, photons of 6 MV are usually sufficient but higher- or mixed-energy beams may be preferred on occasion. If the scar is part of the CTV, thought should be given to the possible underdosing of superficial tissues due to build. This will be reduced when fields tangential rather than incident to the scar are used. Use of bolus should be avoided due to the significant adverse on skin and immediate subcutaneous tissue. Occasionally in the wrist or ankle, due to small separation, the use of bolus for part of the treatment may be necessary to ensure the tumour bed itself is adequately treated.

For more complex targets such as the shoulder girdle, extensive pelvic tumours, spinal tumours, and most head and neck sites, inverse planned techniques such as IMRT or stereotactic radiotherapy may be preferred, as shown in Figs 14.3 and 14.4.

Dose specification is according to the ICRU 50 report.

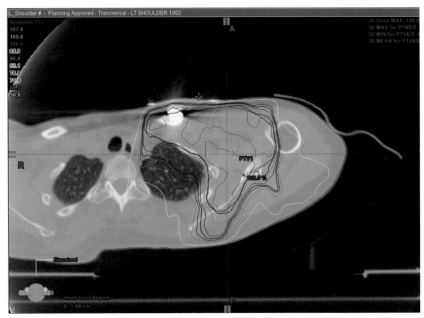

Fig. 14.3 Postoperative radiotherapy for soft tissue sarcoma of the supraclavicular fossa/ superior axilla: single phase treatment inverse planned IMRT. Five-fixed field with dynamic wedging: 6-MV photons reduced volume of lung irradiated to high dose. Anterior artefact is from plate in clavicle which was split to access the tumour OAR: unirradiated 'corridor' lateral arm in continuity with strip of subcutaneous tissue on superior shoulder receives < 50%.

14.2.6 Critical organs and tolerance doses

Limbs

A strip of unirradiated skin and subcutaneous tissues sufficient to maintain the lymphatic drainage of the distal limb (the 'corridor') is essential for acceptable long-term function. Data on the amount are not evidence based but, in principle, between 25–33% of the circumference including subcutaneous tissues is kept below 40 Gy. It should be borne in mind that the adverse effect on lymphatic draining will depend on the most narrow part of the 'corridor', such that a relationship to a DVH is difficult to define.

Where possible, radiotherapy to 60 Gy or more to the full circumference of a weight-bearing bone, especially the femur, should be avoided and the dose to the femoral necks kept as low as possible.

The skin and subcutaneous tissues of the anterior lower limb tolerate high-dose radiotherapy relatively poorly. Soft tissue tumours at this site in the very elderly or those with pre-existing poor tissue perfusion may not be suitable for high-dose radiotherapy and the management plan should be formulated with this restriction in mind.

Trunk

- ◆ When treatment of the tumour itself is, or was, adjacent to the spinal cord, the cord dose is limited to 50–55 Gy in 1.6–1.8-Gy fractions but more generally, the maximum cord dose is kept below 46–48 Gy.

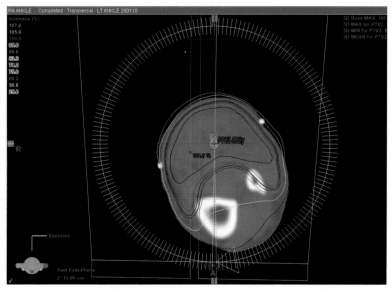

Fig. 14.4 Postoperative radiotherapy for a superficial, high-grade soft tissue sarcoma resected from the posterior lower calf with musculocutaneous flap. 360° RapidArc therapy 6-MV photons allows treatment of superficial tissues with the required margin without the full dose to the tibia and fibula that would have been produced by conventional conformal radiotherapy. Two-phase treatment with reduced margin around the circumference of the limb for the final 10 Gy. OARs: tibia, most receives < 50%, unirradiated circumference; Achilles tendon.

◆ When brachial plexus is in the field, but the tumour itself is not likely to threaten its function, dose should be limited to 50 Gy in fractions ≤ 1.8 Gy. When treatment of the tumour inevitably treats the plexus, the dose should be limited to 55–60 Gy in 1.8-Gy fractions.

Usual tolerances for other OARs are followed but with special caution in view of the possible enhancement due to chemotherapy given prior to, or concurrent with, radiotherapy.

14.3 **Radiotherapy for Ewing's sarcoma**

14.3.1 **Primary site**

Indications

◆ Inoperable tumours (or operable tumours where morbidity of surgery is not considered justified).

◆ Preoperative in selected cases.

◆ Postoperative when one or more of the following applies:
   • Margin positive at histology and further surgery not advised.
   • Poor histological response to preoperative chemotherapy (<90% necrosis).

Patients receive several cycles of chemotherapy preoperatively or prior to radical radiotherapy and continue with chemotherapy concurrent with radiotherapy, with omission of doxorubicin and actinomycin within 3 weeks of radiotherapy and of ifosfamide when significant volume of bladder in the target volume.

Patients with axial tumours who will require radiotherapy are excluded from treatment with high dose busulphan if enhanced toxicity to CNS, lung or GIT is anticipated from the interaction of high dose chemotherapy and radiotherapy.

Current European doses for adjuvant radiotherapy to primary site

- Known residual tumour: 55 Gy/1.83/30 fractions treated once daily.
- Marginal surgery: poor histological response: 55 Gy/1.83/30 fractions treated once daily.
- Marginal surgery: good histological response: 45 Gy/1.87/24 fractions treated once daily.

Gross tumour volume

For both definitive radical radiotherapy and postoperative radiotherapy the GTV is the visible extent of tumour on planning CT scan extended to reflect the tumour at its greatest extent within tissues, with reference to the diagnostic imaging. Protrusion into body cavities that regresses with chemotherapy is not included. The principle is to include all tissues originally involved by tumour prior to chemotherapy.

Clinical target volume

In postoperative cases requiring 55 Gy this is delivered in two phases with phase 1 including the scar to 45 Gy with a volume reduction for the final 10 Gy.

'Grown margins' are edited to take into account patterns of spread and intact fascial planes:

- CTV phase 1 = GTV + 2–3 cm in the long axis of the limb 1.5–2 cm in the axial plane.
- For axial sites, CTV1 = GTV + 1.5–2 cm, depending on site.
- CTV phase 2 (reduced volume in those to receive 55 Gy) = GTV + 2 cm in long axis.

14.3.2 Radical whole lung radiotherapy for Ewing's sarcoma

- As consolidation when there were initial lung metastases at diagnosis.
- Should not be given following high-dose chemotherapy.
- Radiotherapy is delivered after chemotherapy is completed.
- Radical treatment is CT planned with the prescription point within the lungs rather than the mediastinum.
- Age under 14 years: dose 15 Gy in 10 daily fractions over 2 weeks.
- Age over 14 years: dose 18 Gy in 12 daily fractions over 2½ weeks.

14.3.3 **Radical treatment of bony secondaries**

♦ As consolidation of chemotherapy response to a limited number of sites of bony disease. There is evidence that this may be associated with improved progression-free survival.

♦ 40–45 Gy.

14.4 **Radiotherapy for osteosarcoma**

14.4.1 **Indications**

♦ Tumour unresectable for technical and/or medical reasons.

♦ Resection margins positive and further surgery not possible, and/or response to preoperative chemotherapy has been poor (< 90% necrosis).

♦ Inadequate resection is defined as intralesional or where margins are contaminated.

14.4.2 **Chemotherapy**

♦ All patients treated radically receive induction combination chemotherapy and surgery if primary site resectable.

♦ Radiotherapy, if indicated, is usually given after chemotherapy is completed—interval 4–6 weeks.

♦ In inoperable tumours, concurrent cisplatin 60–100 mg/m^2 may be given 3-weekly starting on day 1 or 2, radiotherapy starting 4–6 weeks after completion of combination chemotherapy.

14.4.3 **Dose fractionation**

♦ Adjuvant to surgery:
  • 60 Gy in 1.8–2-Gy fractions over 6–6½ weeks.
♦ Inoperable or macroscopic residual disease:
  • 60–66 Gy in 1.8–2-Gy fractions over 6½–7 weeks.

Gross tumour volume

♦ In unresected disease, the GTV is the visible extent of tumour on planning CT scan with reference to the diagnostic imaging.

♦ In postoperative radiotherapy, the GTV is reconstructed using the preoperative imaging, operation note, and pathology report.

Clinical target volume

♦ Grown margins to be edited to take into account patterns of spread and intact fascial planes.

♦ Adjuvant to surgery:
  • CTV = GTV + 2–3 cm, (this may be amended axially to spare a soft tissue corridor, but should be at least 2 cm) or full length of scar + 0.5–1 cm whichever is longer.

◆ Inoperable tumours:

- CTV = GTV + 2–3 cm.

14.5 **Chondrosarcoma and chordoma**

These tumours have not generally been considered sufficiently radiosensitive to benefit from radio therapy except, occasionally, for palliation. However, there is evidence that doses of at least 65 Gy in small volume residual tumours after surgery may result in long-term control. Depending on the site, this will require IMRT techniques, or, in some cases, referral to a proton centre.

14.6 **Radiotherapy for soft tissue sarcomas (excluding rhabdomyosarcoma of young people type)**

The use of postoperative radiotherapy for the treatment of soft tissue sarcoma over the last 30 years has reduced the indications for ablative surgery. Small tumours superficial to fascia and small intramuscular tumours (not greater than 2 cm) excised with a good margin may not require postoperative radiotherapy. In most other situations the evidence indicated that surgery alone is associated with a poorer local control than in combination with postoperative radiotherapy. Conventional treatment plans have used a wide-field techniques, usually with a shrinking field over two phases. A brachytherapy trial at the Memorial Sloane-Kettering has suggested that smaller, more targeted radiotherapy volumes may be associated with similar outcomes in terms of local control and better limb function.

Soft tissue sarcomas tend to spread in a longitudinal direction within muscle groups in the extremity and tend not to breach axial barriers such as major fascia, bone, and interosseous membranes. In other sites the situation may be more complex. Surgery should be performed after recent detailed imaging and should include resection of the biopsy track. In surgery the aim is to produce a R0 resection (that is, margin negative). Assessment of the margin may be technically difficult and the nature of the margin is as important as its size. The relapse rate is higher if the margin is positive—a failure rate which is reduced, but not eliminated, by radiotherapy. A margin of 1–2 mm is probably adequate if postoperative radiotherapy is to be given, though smaller margins are adequate if they include a fascial plane. A 'planned-positive' margin by an experienced surgeon (e.g. on a neurovascular bundle) does not seem to increase local relapse.

Whilst, in the UK, it is usual for radiotherapy to be administered postoperatively, there is evidence that preoperative radiotherapy results in an equivalent control rate and better long-term function. This is associated with a lower dose (50 Gy) and a smaller target volume based on the tumour bed alone without needing to take into account the larger volume of tissues potentially contaminated at surgery. This approach is associated with an increased postoperative morbidity with significant local complications, predominantly in lesions of the thigh. A decision on the appropriateness of the pre- versus the postoperative approach needs to be made on an individual patient basis by detailed consultation between surgeon and clinical oncologist considering any adverse factors relating to healing. It should not be expected that preoperative radiotherapy significantly reduces tumour size, but by reducing the viability of peripheral cells and making the tumour less viable, R0 resections may be facilitated and the potential for contamination by tumour rupture reduced. Myxoid liposarcomas, however, may shrink rapidly after radiotherapy, presenting the surgeon with the challenge of removing a tumour that may have virtually disappeared. Where preoperative radiotherapy is used, an interval of 4–8 weeks between radiotherapy and surgery can allow resolution of the acute inflammatory response whilst the fibrosis and endothelial thickening that would make dissection of tissue planes and healing more problematical are minimized.

Most soft tissue sarcomas are large at presentation with a mean size of about 7 cm. Clinical experience suggest that relapse may occur at some distance from the visible tumour and this is supported by small studies detecting tumours cells up to 4 cm from the tumour (and mostly within peritumoural oedema) and lower failure rates with field margins of at least 5 cm. The scar is considered at risk of contamination and, although including this in the field may be unnecessary, until the results of current VORTEX trial are available, this practice remains standard. Drains are rarely used as current practice but if these have been used, or when a flap has been placed, these areas are included in the phase 1 CTV.

14.6.1 Further investigations for soft tissue sarcoma

Myxoid liposarcomas of proximal limb, contrast-enhanced CT scanning of the abdomen and pelvis to exclude intra-abdominal metastases.

14.6.2 Indications

These are summarized in Table 14.1.

- ◆ Preoperative:
  - Where the resection is likely to be marginal or where positive margins are likely and, especially, where tumours are close to but not surrounding neurovascular bundle.

Table 14.1 Indications for radiotherapy and dose recommendations for adult soft tissue sarcomas

Stage	Surgical margin	Plan
IA & IB	≤ 10 mm	Re-resection or radiotherapy on impact of surgery on function
	> 10 mm	Watch policy
IIB	Superficial	Surgery alone
	Deep	Surgery + XRT unless intramuscular tumour with 20-mm margin
IIA		Randomized data to suggest addition of XRT to surgery
IIC	Rare	Surgery alone if margin >10mm; else re-resection or XRT
III		Surgery + XRT

◆ Postoperative:
  • High-grade and intermediate grade sarcomas unless small (< 2–3 cm) and excised with a wide margin.
  • Low-grade sarcomas with close or positive margin where radical surgery on relapse is likely to be difficult.

14.6.3 Dose

There are no good randomized trials addressing this issue but a dose of 60 Gy is required postoperatively. North American practice tends to use doses of 66 Gy, whereas 60 Gy has been usual in UK sarcoma centres, with consideration of a further 4–6 Gy in margin-positive disease.

Outcomes: overall and expectation of a local control rate of 80–85% should be expected for surgery and postoperative radiotherapy.

14.6.4 Dose fractionation

◆ Preoperative radiotherapy:
  • 50 Gy in 25 daily fractions over 5 weeks.
◆ Postoperative radiotherapy:
  • Resection margin-negative:
    ▪ Phase 1: 50 Gy in 25 daily fractions over 5 weeks.
    ▪ Phase 2: 10 Gy in five daily fractions over 1 week.
  • Resection margin-positive:
    ▪ Phase 1: 50 Gy in 25 daily fractions over 5 weeks.
    ▪ Phase 2: 16 Gy in eight daily fractions over 1½ weeks.

Gross tumour volume

◆ Preoperative radiotherapy:
  • Visible extent of tumour on planning CT scan with reference to the diagnostic imaging. Contrast-enhanced scans may assist tumour definition.
◆ Postoperative radiotherapy:
  • GTV reconstructed using the preoperative imaging, operation note, and pathology report.

Clinical target volume

◆ Preoperative radiotherapy:
  • CTV length = GTV + 3–5 cm.
  • CTV width = extent or width of the muscle compartment + 0.5 cm over the fascial boundary or GTV + 2 cm depending on site.
  • For non-limb sites, CTV = GTV + 2–3 cm, depending on site.
◆ Postoperative radiotherapy:
  • CTV phase 1:
    ▪ CTV1 length = GTV + 5 cm, or scar + 0.5–1 cm, whichever is longer.

- CTV1 width = extent or width of the muscle compartment + 0.5 cm over the fascial boundary or GTV + 2 cm depending on site.
  - For non-limb sites, CTV = GTV + 2–3 cm, depending on site.
- CTV phase 2:
  - CTV2 length = GTV + 2 cm.
  - CTV2 width = GTV + 2 cm or the extent/width of the muscle compartment.
  - For non-limb sites aim to achieve a volume reduction if possible to spare normal tissues.
- Consider larger CTV–PTV margin in shoulder girdle tumours and trunk sites.

14.6.5 Special sites

Over half of soft tissue sarcomas arise in the limbs and the most common site is the proximal thigh.

The muscles are contained by the deep fascia which is a strong barrier to spread. There are also resistant fascial planes dividing certain muscle groups. These are around the anterior compartment, around the adductors and posterior compartment, and around the sartorius. Fascial boundaries and their strength differ in different parts of the limb. There are specific areas where they are weak, such as where pierced by the neurovascular bundle. An example of this is where the interosseus ligament between tibia and fibula is weak proximally around the anterior tibial vessels. There are no barriers to spread through the muscle along its long axis.

Forearm

The major fascia boundary here is the interosseous ligament. A particular challenge here is the position of the arm where pronation of the hand will result in rotation of this plane. As much as possible for the distal forearm the radius and ulna should be parallel.

Hands and feet

There has been a concern about radiotherapy to these areas because of the perceived poor tolerance to radiotherapy. Fascial planes are less defined making target definition more difficult. The small separation creates difficulties in ensuring the superficial parts of the target volume receive an adequate dose without the excessive dose to the skin that results from full thickness build-up. The tissues on the sole of the foot may experience significant acute toxicity. The heel pad may experience marked swelling, with a collection of haemorrhagic fluid under the thick keratin layer, and the toes may be subject to acute and chronic trauma. The Achilles tendon is at risk of rupture and, as much as possible, at least part of its width should be spared from doses > 50 Gy. However, with care in planning and careful monitoring of the patient for excessive acute and marked late toxicity, full or near full dose radiotherapy is possible with good local control and acceptable morbidity.

Retroperitoneum

This is a difficult site to irradiate to sufficient dose and the value of radiotherapy either pre- or postoperatively is not clear. There is some evidence, however, that there may be benefits if high enough doses can be given. This will nearly always require complex inverse planned techniques and, at this site preoperative radiotherapy, may be preferable. A trial is planned which may provide an evidence base.

Chest wall

Radiotherapy to the chest may be indicated for Ewing's of the chest wall, usually arising in a rib or soft tissue sarcoma of the muscles of the thorax. The indications for radiotherapy are the same, in principle, as elsewhere. Where there is cytology positive of haemorrhagic pleural effusion in chest wall Ewing's, irradiation of the whole pleural cavity (to the same doses as adjuvant lung RT), may be advocated. This needs to be planned bearing in mind the indications for adjuvant RT to the tumour bed and it may not be possible to add the pleural cavity without exceeding recommended lung tolerance doses.

14.7 **Palliative treatment**

Depending on the site and comorbidities of the patient, soft tissue and bony-based limb masses may be palliated with regimens:

- 6 Gy × 5 or 6 treating weekly.
- 40 Gy in 15 fractions.
- 30 Gy in 10 fractions.

The dose for cord compression due to secondaries will depend on the anticipated survival and tumour types but dose of 20 Gy in five fractions or 30 Gy in 10 fractions are usually appropriate.

For pain without significant mass effect a single fraction of 8 Gy may produce relief.

14.7.1 **Cerebral secondaries**

Treatment follows the general principles for cerebral secondaries in other tumour types. Osteosarcoma and soft tissue tumours may occasionally present with isolated secondaries which, especially if at an interval after initial treatment are best treated surgically or with radiosurgery. Ewing's more often presents with multiple lesions and WBRT to 20 or 30 Gy may be used.

Whole lung radiotherapy may occasionally have a palliative role in Ewing's in the same dose schedule used for radical treatment.

Further reading

Casali PG, Blay JY; ESMO/CONTICANET/EUROBONET Consensus Panel of experts Soft tissue sarcomas: ESMO Clinical Practice Guidelines for diagnosis, treatment and follow-up. *Annals of Oncology* 2010; **21**(Suppl 5): v198–203.

Davis AM, O'Sullivan B, Turcotte R, *et al.*; Canadian Sarcoma Group; NCI Canada Clinical Trial Group Randomized Trial. Late radiation morbidity following randomization to preoperative versus postoperative radiotherapy in extremity soft tissue sarcoma. *Radiotherapy and Oncology* 2005; **75**(1): 48–53.

Hogendoorn PC; ESMO/EUROBONET Working Group. Bone sarcomas: ESMO Clinical Practice Guidelines for diagnosis, treatment and follow-up. *Annals of Oncology* 2010; **21**(Suppl 5): v204–13.

Jebsen NL, Trovik CS, Bauer HC, *et al.* Radiotherapy to improve local control regardless of surgical margin and malignancy grade in extremity and trunk wall soft tissue sarcoma: a Scandinavian sarcoma group study. *International Journal of Radiation Oncology, Biology, Physics* 2008; **71**(4): 1196–203.

Kaushal A, Citrin D. The role of radiation therapy in the management of sarcomas. *Surgical Clinics of North America* 2008; **88**(3): 629–46, viii.

Patel S, DeLaney TF. Advanced-technology radiation therapy for bone sarcomas. *Cancer Control* 2008; **15**(1): 21–37.

Tseng WH, Martinez SR, Do L, *et al.* Lack of survival benefit following adjuvant radiation in patients with retroperitoneal sarcoma: a SEER analysis. *Journal of Surgical Research* 2011; **168**(2): e173–80.

Zagars GK, Ballo MT, Pisters PW, *et al.* Prognostic factors for patients with localized soft-tissue sarcoma treated with conservation surgery and radiation therapy: an analysis of 225 patients. *Cancer* 2003; **97**(10): 2530–43.

Principles of paediatric radiation oncology

Roger E Taylor

15.1 Introduction

Cancer in childhood is rare, with approximately 1500 cases arising each year in the UK. Approximately one child in 500 will develop cancer before the age of 15. The range of tumours seen in childhood is very different from that in adults. Data on relative incidence of the various tumour types from the US Surveillance, Epidemiology and End Results (SEER) programme[1] is given in Table 15.1. In the last four decades there have been significant improvements in outcome for many children with cancer and the evolution of the multidisciplinary care of children under the 'umbrella' of paediatric oncology has been one of the success stories of modern oncology.

Clinical trials are coordinated by multi-institutional groups. In North America through the 1980s and 1990s clinical research was coordinated via the Paediatric Oncology Group (POG), Children's Cancer Group (CCG), Intergroup Rhabdomyosarcoma Study Group (IRSG), and National Wilms' Tumour Study Group (NWTS). These were amalgamated in 2000 to form the Children's Oncology Group (COG), which is the largest paediatric oncology collaborative group in the world, and co-ordinates trials across paediatric oncology centres in the USA, Canada, and some other countries including Australasia.

In Europe, clinical research is coordinated mainly on a national basis. In the UK, treatment is coordinated by the network of 22 Children's Cancer and Leukaemia Group (CCLG) paediatric oncology centres. Increasingly collaboration is across European boundaries, with clinical trials coordinated via the International Society of Paediatric Oncology (SIOP). In the last 40 years there has been a substantial improvement in survival for children with cancer and more than two-thirds can now expect to be long-term survivors. For many disease groups this has been brought about largely as a result of the incorporation of chemotherapy as part of a multimodality approach, including surgery and radiotherapy.

Radiotherapy is an important modality of therapy in the local control of paediatric malignancies, and the majority of paediatric tumours are radiosensitive[2]. However, for many children long-term survival comes at a price, namely the long-term effects of treatment. Long-term effects of radiotherapy include soft tissue hypoplasia, impaired bone growth, neuropsychological effects of irradiation of the central nervous system (CNS), and radiation-induced malignancy. The increasing recognition of these long-term

Table 15.1 Annual incidence rate per 1 000 000 and proportion of children aged 0–14 with Cancer Surveillance, Epidemiology and End Results (SEER) Programme Registrations 1975–2001

Disease group	Annual incidence rate per 1 000 000	Percentage of total
Acute lymphoblastic leukaemia	32.8	23.6%
Acute non-lymphoblastic leukaemia	6.4	4.6%
Other leukaemias	3.0	2.2%
All leukaemias	42.2	30.3%
Ependymoma	2.7	1.9%
Astrocytoma	14.5	10.4%
Medulloblastoma/PNET	6.6	4.7%
Intracranial germ cell tumours	1.1	0.8%
Other CNS tumours	5.7	4.1%
All CNS tumours	29.4	21.1%
Osteosarcoma	3.5	2.5%
Ewing's sarcoma/peripheral PNET	2.4	1.7%
Hodgkin disease	6.0	4.3%
Non-Hodgkin lymphoma	7.8	5.6%
Neuroblastoma	10.3	7.4%
Wilms' tumour	8.3	6.0%
Rhabdomyosarcoma	4.8	3.5%
Other sarcoma	5.3	3.8%
Total	139.1	

effects in the 1970s and 1980s led to a decline in the use of radiotherapy. However, more recently it has become evident that chemotherapy is also associated with long-term side effects. The long-term effects of chemotherapy include late myocardial damage due to anthracyclines, nephrotoxicity due to cisplatin or ifosfamide, and secondary leukaemia related to drugs such as the alkylating agents[3]. When designing programmes for the treatment of children, although the priority is to maximize the chance of cure it is also essential to consider the likely long-term effects of treatment. Continued vigilance for long-term effects of treatment is essential. In most centres this is carried out in dedicated long-term follow-up clinics. National treatment-related guidelines for long-term follow-up have been produced by collaborative groups such as the CCLG.

Currently 40–50% of children with cancer receive radiotherapy as part of their initial treatment. Careful planning and delivery are essential in order to achieve local tumour control with the minimum of irradiation of normal tissues to minimize long-term effects. It is extremely important that, as for the administration of chemotherapy, radiotherapy for children should be undertaken only in specialized centres associated with the CCLG paediatric oncology centres. In a centre treating relatively large numbers of children with radiotherapy it is possible to establish a team including a specialist paediatric therapy radiographer, specialist nurse, and play specialist. Young children, particularly those under the age of 3–4, frequently find it very difficult to lie still for radiotherapy planning and delivery, particularly when a Perspex head shell is required. The assistance of an experienced play therapist can be very helpful in preparing the child for radiotherapy[4]. Sedation sufficient to ensure immobilization is difficult to achieve without it persisting for several hours and it is not feasible for this to be administered daily. Because of the importance of immobilization, short-acting general anaesthesia such as propofol (Diprivan) is sometimes needed. The daily fasting for this results in surprisingly little disruption to nutrition.

15.2 **Radiotherapy quality assurance**

Radiotherapy quality assurance (QA) is particularly important in the treatment of children. Because of the high cure rate for most childhood cancers it is important to achieve local control avoiding a 'geographical miss'. It is also important to avoid unnecessarily large field sizes, in order to minimize long-term effects.

It is essential for all radiotherapy departments to deliver the highest possible standard of radiotherapy for all patients including children. Most radiotherapy departments have adopted quality systems.

15.2.1 **Quality assurance in multicentre studies**

The accuracy of delivery of radiotherapy contributes to improved tumour control, particularly for complex techniques such as craniospinal radiotherapy for medulloblastoma[5]. In many of the North American paediatric multicentre studies, radiotherapy quality, including beam data, dose prescription, planning, and verification films are reviewed centrally in the Quality Assurance Review Centre (QARC) situated in Providence, Rhode Island. In most European multicentre studies QA is less well organized. Ideally, review of radiotherapy simulator or verification films by study coordinators should be sufficiently fast to provide feedback early in the course of radiotherapy so that the treatment plan can be modified if necessary. The electronic transmission of planning films in DICOM format and associated clinical data can facilitate this process. This is logistically difficult but has been achieved in the USA for radiotherapy administered to children treated within POG, CCG, and more recently COG trials. In Europe, where funding for these activities is more problematic, this has been achieved in Germany for radiotherapy for Ewing's sarcoma and is planned for future medulloblastoma studies.

15.3 **Toxicity of radiotherapy for children**

15.3.1 **Acute morbidity**

The side effects of erythema, mucositis, nausea, diarrhoea, etc. occur in children as in adults, and are generally managed by the same means.

15.3.2 **Subacute effects**

Liver

A large proportion of the liver may need to be irradiated when treating Wilms' tumour. Radiation hepatopathy may occur 1–3 months following radiotherapy, and consists of hepatomegaly, jaundice, ascites, thrombocytopenia, and elevated transaminases. A risk factor is the administration of actinomycin-D following hepatic irradiation. Long-term dysfunction is rare and the risk is dose related.

Lung

The whole lungs may receive radiotherapy as part of total body irradiation (TBI), or in the treatment of pulmonary metastases from Wilms' tumour or Ewing's sarcoma. Mild radiation pneumonitis consists of a dry cough and mild dyspnoea. The risk of pneumonitis is dose and radiation volume related. Radiation pneumonitis is the dose limiting toxicity for total body irradiation. *It is essential to consider potential interactions between chemotherapeutic drugs and lung irradiation and in particular to avoid lung irradiation in association with busulfan.*

Central nervous system

The 'somnolence syndrome' occurs in at least 50% of children approximately 6 weeks after cranial irradiation and is probably related to temporary demyelination. L'Hermitte's sign consists of an electric shock-like symptom radiating down the spine and into the limbs. It may follow radiation to the upper spinal cord, e.g. following mediastinal radiotherapy for lymphoma. Within the first 2 months following radiotherapy for brain tumours, children may experience a transient deterioration of neurological symptoms and signs.

15.3.3 **Long-term effects**

Bone growth

Impairment of bone growth and associated soft tissue hypoplasia can be one of the most obvious and distressing long-term effects particularly when treating the head and neck region. Abnormalities of craniofacial growth can cause significant cosmetic and functional deformity, including micrognathia leading to problems with dentition. The epiphyseal growth plates are very sensitive to radiation, and are excluded from the radiotherapy field whenever possible. Age at time of treatment, radiation dose and volume are factors which have an impact on the severity of these orthopaedic long-term effects. There is evidence of a dose response effect, with a greater effect seen for doses of >33 Gy compared with <33 Gy[6]. Slipped femoral epiphysis and avascular necrosis have also been reported following irradiation of the hip. Laboratory evidence suggests a dose response effect between 5 Gy and 35–40 Gy, and an effect of dose per fraction. Careful consideration of the late orthopaedic effects of radiation is extremely important whenever planning radiotherapy for children. Epiphyses should be excluded from the irradiated volume where possible, and when irradiating the spine, the full width of the vertebra should receive homogeneous irradiation in order to minimize long-term kyphoscoliosis.

Central nervous system

Paediatric radiation oncology involves a significant amount of time devoted to the treatment of children with brain tumours, and consideration needs to be given to the toxicity of therapy[7].

Radionecrosis

Radionecrosis is rare below 60 Gy, and generally occurs with a latency of 6 months to 2 years. It results from a direct effect on glial tissue. It is very unusual to have to deliver a dose of 60 Gy to any part of the CNS for a child, and for the radical treatment of children with brain tumours it is very uncommon to exceed a dose of 50–54 Gy. It occurs in approximately 50% of patients treated by interstitial implantation for recurrent brain tumours following prior radical external beam radiotherapy. The clinical effects of radionecrosis vary according to the site within the CNS and are most devastating in the spinal cord. Radionecrosis of the spinal cord in children has been seen as a consequence of the interaction between radiation and cytosine arabinoside given intrathecally for metastatic rhabdomyosarcoma.

Necrotizing leucoencephalopathy

This may be seen when cranial irradiation is followed by high-dose methotrexate for the treatment of leukaemia. The clinical features include ataxia, lethargy, epilepsy, spasticity and paresis.

Neuropsychological effects

The effects of cranial radiotherapy are now well established. When compared with siblings, children given 24 Gy prophylactic cranial irradiation show an approximate fall in IQ of 12 points. Following higher radiation doses given for brain tumours an increased risk of learning and behaviour difficulties is seen[8]. An important risk factor for the incidence and severity of neuropsychological long-term effects is the age at diagnosis. Other factors include the impact of direct and indirect tumour related parameters, treatment parameters with neuropsychological long-term effects worse for whole brain compared with partial brain irradiation, premorbid patient characteristics such as intelligence, and the quality of 'catch up' education.

Kidney

Long-term effects on renal function are usually seen 2–3 years following a course of radiotherapy. The risk increases following a dose of > 15 Gy to both kidneys. The severity is related to the dose received, and when mild consists of hypertension. When more severe, following a higher dose, renal failure may ensue.

Endocrine

Endocrine deficiencies following radiotherapy are common. Of particular concern is the risk of growth hormone and other pituitary hormone deficiencies following pituitary irradiation for tumours of the CNS.

Following radiotherapy to the thyroid, the incidence of elevated TSH is 75% after 25–40 Gy.

Table 15.2 Normal tissue tolerance doses based on fractionated radiotherapy using 1.8–2.0 Gy per fraction

Tissue/organ	Tolerance dose (Gy)
Whole lung	15–18
Both kidneys	12–15
Whole liver	20
Spinal cord	50

Reproductive

In boys the germinal epithelium is very sensitive to the effects of low-dose irradiation. In adult males, transient oligospermia is seen after 2 Gy, but slow recovery can occur after 2–5 Gy.

In girls the oocytes are also sensitive. Subsequent pregnancy is rare after 12 Gy whole body irradiation but has been reported.

15.4 Tolerance of critical organs to radiotherapy

The tolerance of critical organs limits frequently limits the dose of radiation that can be given. The critical organs and their 'tolerance doses' are listed in Table 15.2.

15.5 Chemotherapy/radiotherapy interactions

Interactions between radiation and chemotherapy are complex and poorly understood. Interactions can be exploited in order to attempt to improve disease-free survival. The most frequently employed mechanism in paediatric oncology is 'spatial cooperation' whereby chemotherapy and radiotherapy are combined to exploit their differing roles in different anatomical sites. Examples are the use of radiation for local control of a primary, with chemotherapy for subclinical metastatic disease such as in the treatment of Ewing's sarcoma.

Chemotherapy and radiotherapy may be combined with the aim of increasing tumour cell kill without excess toxicity. An example is the use of combined chemotherapy and radiotherapy for children with Hodgkin lymphoma. It may be possible to reduce the intensity of both treatment modalities and hopefully reduce long-term morbidity. When using combined modality therapy the aim is to improve the therapeutic ratio. Many protocols for children involve the use of concurrent chemotherapy and radiotherapy. It is essential to be vigilant for additional early or long-term morbidity. Clinically important chemotherapy-radiotherapy interactions are often unpredictable and their mechanisms poorly understood. Actinomycin D and cisplatin increase the slope of the radiation dose-response curve and actinomycin D inhibits the repair of sublethal damage (SLD). Clinical interactions include enhanced skin and mucosal toxicity when radiation is followed by actinomycin-D (the 'recall phenomenon'), enhanced bladder toxicity when chemotherapy is combined with cyclophosphamide, enhanced CNS toxicity from combined radiation and methotrexate, cytosine arabinoside or busulfan, and the enhanced marrow toxicity from wide-field irradiation and many myelotoxic chemotherapeutic agents. In the case of the effect of combined radiation and anthracyclines such as doxorubicin on the heart, doxorubicin has its effects on the myocytes and radiation on the vasculature.

15.6 **Leukaemia**

The leukaemias account for the largest group (Table 15.1) of paediatric malignancies with approximately 80% having acute lymphoblastic leukaemia (ALL). The remainder have acute non-lymphoblastic leukaemia (ANLL), usually acute myeloid leukaemia (AML), or rarely chronic myeloid leukaemia (CML). The improvement in survival of children with ALL was one of the early successes of paediatric oncology. Currently > 70% are long-term survivors. Current treatment is stratified according to risk status based on presenting white count and cytogenetic profile. Further stratification for intensity of therapy is based on analysis of postinduction bone marrow for minimal residual disease. The four phases of treatment are:

1. Risk-adapted multiagent remission induction, based on drug combinations including vincristine, corticosteroids, and asparaginase.

2. Intensification with multidrug combinations. The number of intensification modules is dependant upon risk status at presentation.

3. CNS prophylaxis, generally with intrathecal methotrexate. Cranial radiotherapy is now longer employed except for patients presenting with CNS involvement by leukaemia.

4. Maintenance usually based on a continuous low-dose antimetabolite drug such as 6-thioguanine, with a total duration of therapy of approximately 2 years.

During the 1960s and 1970s the routine use of prophylactic whole brain radiotherapy and intrathecal methotrexate reduced the risk of CNS relapse to < 10%. Whole brain radiotherapy may be employed for patients who present with CNS involvement.

15.6.1 **Technique**

Lateral opposed megavoltage fields, generally 4–6 MV. Fields may be centred on outer canthus to minimize divergence into the contralateral lens. Shielding of face, dentition, nasal structures and lenses using blocks or multileaf collimators.

Clinical target volume

Intracranial meninges extending inferiorly to the lower border of the second or third cervical vertebra. Great care is taken to include the cribriform fossa, temporal lobe, and base of skull. Although the lens is shielded, as much of the posterior orbit as possible is included as ocular relapses occasionally occur.

Patient position and immobilization
Supine, immobilized in a head shell.

Localization of clinical target volume/planning target volume
CT simulation has been more recently employed. A radiograph illustrating a typical field is shown in Fig. 15.1.

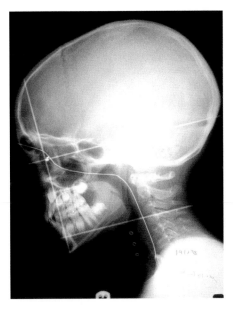

Fig. 15.1 Typical field for prophylactic cranial irradiation using asymmetric diaphragms and centred on the outer canthus to minimize divergence at the orbit.

Dosimetry
Dose specified to the midpoint of central axis.

Dose prescription
24 Gy in 15 fractions of 1.6 Gy daily.

Verification
Machine verification films.

15.6.2 Testicular irradiation

Boys who present with testicular involvement are treated with testicular radiotherapy.

Technique

Anterior field, generally electrons or possibly orthovoltage (200–300 kV). Shielding of non-target skin and perineum.

Clinical target volume
Both testes, scrotum and inguinal canal supero-laterally as far as the deep inguinal ring.

Patient position and immobilization
Supine.

Localization of Clinical target volume and planning target volume
Clinical localization.

Dosimetry
Dose specified to the maximum (given/applied dose).

Dose prescription
24 Gy in 12 fractions of 2.0 Gy daily.

Verification
Machine verification films.

For ANLL the basis of therapy is intensive multidrug chemotherapy, which can achieve a survival rate of 60%. As for adults, bone marrow transplantation is frequently employed for children who have an HLA matched sibling. A survival rate of 65% can be achieved for children in first complete remission given a bone marrow transplant as consolidation therapy for acute myeloid leukaemia.

15.6.3 Total body irradiation

As in the treatment of adults with haematological malignancies, TBI is an important technique used together usually with high dose cyclophosphamide (Cyclo-TBI) as the conditioning regimen prior to bone marrow transplantation (BMT)[9]. BMT has been routinely available for > 20 years. Bone marrow donors are generally HLA matched siblings, but the increasing availability of volunteer unrelated donors for donor panels is resulting in a significant increase in the number of children for whom BMT can be considered.

Individual techniques for TBI have evolved in different departments, often depending on availability of treatment machines. TBI dosimetry is usually based on *in vivo* measurements. For such a large and complex target volume it is not feasible to adhere to the ICRU 50 guidelines of a range of –5% to + 7%, and a range of –10% to +10% is more realistic. Modern linear accelerator design and field sizes allow the use of large anterior and posterior fields[10].

Technique

Techniques have evolved in different departments according to availability of radiotherapy equipment. The technique described is in use at St James's Hospital, Leeds[10].

Clinical target volume
Whole body without shielding.

Patient position and immobilization
Depends upon technique. Patient lies in the lateral position in an evacuated polystyrene immobilization bag. Hands are placed under chin to provide 'lung compensation'.

Dosimetry
Determined using *in vivo* dosimetry at a 'test dose' of 0.2 Gy per field (see Table 15.3).

Dose prescription
14.4 Gy in eight fractions of 1.8 Gy twice daily with a minimum interfraction interval of 6 hours.

Table 15.3 Distribution of semiconductor dose measurements (see text for details)

Site	Dose (Gy)
Anterior mediastinum	15.98
Posterior mediastinum	16.45
Anterior right lung	14.49
Posterior right lung	13.86
Anterior left lung	14.24
Posterior left lung	14.31
Anterior abdomen	14.89
Posterior abdomen	14.13
Anterior pelvis	14.39
Posterior pelvis	15.25

For children with ALL many centres advise a cranial boost in addition to the TBI with the aim of reducing the risk of CNS relapse. Planning for the cranial boost is the same as for whole brain. Unlike the use of the standard TBI dose, policies for the cranial boost vary between individual centres. A typical regimen for the cranial boost would be 5.4 Gy in three daily fractions.

15.6.4 Indications for bone marrow transplantation/total body irradiation in children

In the current UK Medical Research Council (MRC) Study children with AML are selected for BMT and TBI based on risk status at presentation. Those with an HLA matched sibling are selected for BMT if they fall into the intermediate or high-risk category, whereas those with low risk disease, i.e. those with chromosome mutations t(8;21), t(15;17) or Inv 16 do well with standard chemotherapy. Children with ALL selected for BMT and TBI include relapsed patients and those presenting with features indicating a high risk of failure with standard chemotherapy. Other conditions considered for BMT where TBI is sometime employed include severe aplastic anaemia, thalassaemia, and immunodeficiency syndromes. Long-term effects of TBI include impaired growth due to growth hormone deficiency, and a direct effect from irradiation of epiphyses. There is also a possibility of cataract, hypothyroidism and in some studies the possibility of renal impairment. TBI is generally not considered for children under the age of two, and instead a conditioning regimen with two drugs busulphan and cyclophosphamide (Bu-Cy) is used.

15.7 **Hodgkin lymphoma**

The survival rate for children with Hodgkin lymphoma is approximately 90%. In current protocols the aims are to maintain this good overall survival rate and reduce long-term effects. These include orthopaedic long-term effects of radiotherapy, such as impaired bone growth resulting from direct irradiation of the epiphyses, and also infertility from alkylating agents and procarbazine. For several decades wide-field radiotherapy such as the 'mantle' technique used widely in adults has been avoided in children. Many North American and European protocols employ low-intensity chemotherapy and low-dose involved field radiotherapy with the aim of avoiding infertility and serious orthopaedic effects[11]. Current European protocols employ primarily a chemotherapy approach with PET scanning in order to select patients who require consolidation radiotherapy based on residual abnormal PET uptake after two cycles of chemotherapy. Patients who do not have a complete response including a negative FDG-PET scan after two cycles of chemotherapy (vincristine, etoposide, prednisolone, doxorubicin) are treated with consolidation radiotherapy.

Technique

Generally anterior and posterior parallel opposed fields shaped by multileaf collimation (see Fig. 15.2).

Clinical target volume

- ◆ Phase 1: originally involved nodes with a margin of 1–2 cm in direction of nodal chain.
- ◆ Phase 2: residual nodes on CT and/or FDG-PET imaging after two cycles of chemotherapy with 1–2-cm margin.

Patient position and immobilization
Supine with neck shell immobilization where necessary.

Localization of clinical target volume/planning target volume
CT simulation with information from FDG-PET scanning.

Dosimetry
Dose specified to the midpoint of the central axis.

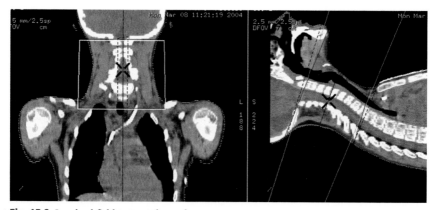

Fig. 15.2 Involved field to treat the neck.

Dose prescription
- Phase 1: 19.8 Gy in 11 fractions of 1.8 Gy.
- Phase 2: 10 Gy in five fractions of 2.0 Gy.

Verification
Machine verification films.

Hodgkin lymphoma has its peak age incidence in the adolescent age range. The management of adolescents with Hodgkin lymphoma is an example of where local and national collaboration with adult oncology groups is important in order to provide age-related treatment protocols.

15.8 **Non-Hodgkin lymphoma**

Compared with adults a different spectrum of non-Hodgkin lymphoma (NHL) is seen in children. Diffuse large B-cell lymphoma and follicle centre cell lymphoma, which are common in adults, are uncommon in childhood. The majority of children have either T-cell lymphoblastic lymphoma, Burkitt's, Burkitt-like, or anaplastic large cell lymphoma. Survival rates have improved in recent years, and currently > 80% survive long term. Therapy is based on intensive multiagent chemotherapy including CNS prophylaxis with intrathecal chemotherapy. There is no routine role for radiotherapy in the management of NHL in childhood. However, children with T-cell lymphoblastic lymphoma, which is managed according to the same principles as ALL may be considered for BMT with TBI.

15.9 **Neuroblastoma**

Neuroblastoma (NB) is the commonest solid tumour of childhood (Table 15.1) and is generally a disease of very young children, with approximately one-third presenting aged < 1 year. NB arises in neural crest tissue in the autonomic nervous system, usually in the adrenal area, but can arise anywhere from the neck to the pelvis. Children frequently present with widespread metastases (Stage 4 disease). Sites of metastasis frequently include bones, bone marrow, lungs, and liver. Overall, current survival rates are generally poor, approximately 45% taking all stages and prognostic groups into account. However the prognosis varies considerably and is related to several prognostic factors. Prognosis is better for young children aged < 1 at presentation, and is worse for children whose tumours have amplification of the oncogene nMyc. Deletion of the short arm of chromosome 1 has emerged as an important prognostic factor with a worse prognosis for those with 1p deletion. Tumours which are hyperdiploid have a better prognosis. Management is now stratified according to risk grouping. Patients in the best risk group with a survival rate of > 90% can be managed with surgery alone. The majority are treated with intensive chemotherapy with drugs such as vincristine, cisplatin, carboplatin, etoposide, and cyclophosphamide.

The role of external beam radiotherapy for patients with 'bad risk' disease (e.g. aged > 1 year with stage 4 disease at presentation) is to maximize the probability of local tumour control following surgical resection of the primary tumour. Success for this approach will rely on chemotherapy achieving a good response in metastatic disease. With current intensive chemotherapy some patients with Stage 4 disease at presentation can hopefully be 'cured' of their metastatic disease, which results in the importance of local tumour control with the combination of surgical resection and postoperative radiotherapy for the primary tumour. This is based on series from North America, which confirm the improvement in local control from the use of tumour bed radiotherapy in patients undergoing resection of primary tumour. For 'good risk' patients radiotherapy is unnecessary and for poor risk patients the predominant relapse pattern is metastatic rather than local. The current dose for postoperative radiotherapy to the tumour bed is 21 Gy in 14 fractions. In planning postoperative radiotherapy for NB, care has to be taken to consider dose limits to OARs, particularly liver and kidneys. At the time of radiotherapy the function of these organs may have already been compromised by high dose chemotherapy.

15.9.1 MIBG therapy for neuroblastoma

The majority of NBs take up the guanethidine analogue meta-iodobenzyl guanidine (MIBG). MIBG can be conjugated with iodine radionuclides for imaging and 'targeted radiotherapy' of neuroblastoma. ^{123}I-MIBG is the radionuclide generally used for diagnostic imaging. High activity ^{131}I-MIBG (typically 3.7–7.4 GBq) can be used for therapy. For children with residual disease after chemotherapy a response rate of 30% has been achieved. After enthusiasm in the 1980s and early 1990s, interest in the potential role for therapeutic MIBG has diminished. However, in some European centres 'up-front' MIBG is still employed and its role in the initial management of high-risk NB together with chemotherapy warrants further exploration. The difficulties encountered in this approach relate particularly to the logistics of radiation protection for very young children, who may not yet be continent of urine, and the risk of radiation exposure to staff who may need to care for an ill child.

15.9.2 Wilms' tumour (nephroblastoma)

Wilms' tumour (nephroblastoma) is an embryonic renal tumour with a median age at diagnosis of between 3–3.5 years. Patients are staged according to histopathological findings following nephrectomy. Table 15.4 shows the National Wilms' Tumour Study Group (NWTS) staging system for Wilms' tumour.

Table 15.4 National Wilms' Tumour Study Group (NWTS) Staging System

Stage	Clinicopathological features
I	Confined to within renal capsule, completely excised
II	Invading outside renal capsule, completely excised
III	Residual abdominal disease—positive margins, tumour rupture, involved nodes
IV	Haematogenous metastases
V	Bilateral disease

In 4–8% of cases, Wilms' tumour is bilateral (Stage V). Wilms' tumour may be genetically associated with aniridia (congenital absence of the iris) and other inherited syndromes such as the Beckwith–Wiedemann syndrome (variable features including macrosomia or hemihypertrophy, macroglossia, omphalocoel). The WT1 gene is located on chromosome 11, and is a tumour suppressor gene. If both copies of the gene are lost by mutation, then Wilms' tumour may arise.

The current long-term survival rate for Wilms' tumour is in excess of 80%. Current treatment programmes are aimed at maintaining this high survival rate while attempting to reduce the long-term side effects of therapy. There is currently a major disparity between approaches adopted in Europe compared with North America. The North American series of NWTS, now COG protocols are based on a policy of immediate nephrectomy with staging is based on histological examination of the primary tumour. Wilms' tumour histology is referred to as either favourable (FH) or unfavourable (UH), because of the presence of anaplasia.

Postoperative chemotherapy is given using the drugs vincristine, actinomycin D, and doxorubicin, the number of drugs and duration depending upon the staging.

Postoperative flank radiotherapy is employed for stage III patients, i.e. those with incompletely resected primary tumours, pre- or perioperative tumour rupture, or histologically involved lymph nodes.

Technique

Anterior and posterior opposed fields.

Clinical target volume

Preoperative extent of tumour and kidney, following preoperative chemotherapy with a margin of 1.0 cm. Field extend across the midline in order to homogeneously irradiate the full width of the vertebral body in order to minimize the risk of kyphoscoliosis (Fig. 15.3).

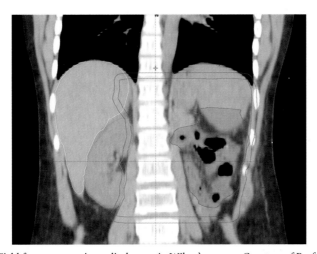

Fig. 15.3 Field for postoperative radiotherapy in Wilms' tumour. Courtesy of Prof Carolyn Freeman, McGill University, Montreal.

Patient position and immobilization
Supine.

Localization of clinical target volume/planning target volume
CT simulation. Description of operative findings. As an alternative to CT, clips inserted at surgery to mark superior and inferior extent of surgery.

Dosimetry
Dose specified to the mid-point of central axis.

Dose prescription
- Intermediate risk: 14.4 Gy in eight fractions of 1.8 Gy daily.
- High risk: 25.2 Gy in 14 fractions of 1.8 Gy.
- Boost to macroscopic disease or involved nodes: 10.8 Gy in 6 fractions of 1.8 Gy.

Verification
Machine verification films.

Whole lung radiotherapy for Wilm's tumour

Whole lung radiotherapy is employed for patients with pulmonary metastases at presentation.

Technique

Anterior and posterior opposed fields.

Clinical target volume
Whole lungs allowing for respiratory movements.

Patient position and immobilization
Supine.

Localization of clinical target volume/planning target volume
CT simulation.

Dosimetry
Dose corrected for tissue homogeneity. CT planning performed and dose specified to the median isodose within lung parenchyma.

Dose prescription
15 Gy in 10 fractions of 1.5 Gy.

Verification
Machine verification films.

In Europe the series of International Society of Paediatric Oncology (SIOP) studies have been based on preoperative chemotherapy to 'downstage' the primary, reducing the surgical morbidity, particularly the number who have tumour rupture at surgery, and the number who require flank radiotherapy. The most recent study is the SIOP 2001 study into which patients from the UK are entered. All patients receive preoperative

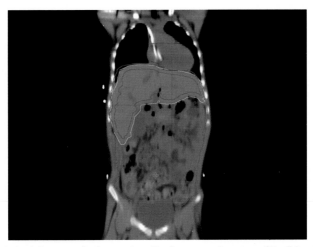

Fig. 15.4 Whole abdomen radiotherapy in Wilms' tumour. Courtesy of Prof Carolyn Freeman, McGill University, Montreal.

chemotherapy with actinomycin-D and vincristine, with delayed nephrectomy after 6 weeks of preoperative chemotherapy. Postoperative adjuvant therapy is based on subsequent pathological staging and allocation of risk status (good risk vs. intermediate risk vs. poor risk histology). For intermediate risk patients there is a randomization to receive or not receive doxorubicin. The purpose of this is to determine whether doxorubicin can be omitted, thus reducing the risk of late cardiac sequelae. Occasionally whole abdominal radiotherapy is employed. The dose for this is 21 Gy in 1.5-Gy fractions with shielding to limit the dose to the remaining kidney to 12 Gy. Whole abdominal radiotherapy (Fig. 15.4) has considerable acute and long-term morbidity and should be reserved for those who present with extensive intra-abdominal tumour spread or generalized preoperative or perioperative tumour rupture. For children presenting with pulmonary metastases and which do not resolve completely after the initial 6 weeks of preoperative chemotherapy, whole lung radiotherapy is given. The fields have to include the costophrenic recess and the lower border generally extends to the lower border of the twelfth thoracic vertebra. The humeral heads are shielded. In the current SIOP 2001 study, the lung dose is 15 Gy in 10 fractions with a lung density correction.

15.10 **Rhabdomyosarcoma**

Rhabdomyosarcoma (RMS) may arise at any site, although they have a predilection for sites in the head and neck region such as the orbit and nasopharynx, and urogenital tract such as bladder, prostate, and vagina. Tumours which arise in sites such as the nasopharynx and middle ear have the propensity for base of skull and intracranial invasion and are referred to as parameningeal RMS. Local tumour control is an important consideration in the management of RMS. The sequelae following radiotherapy to the head and neck in young children may be considerable. Currently the long-term survival rate for RMS is approximately 65–70%. The challenges are to continue to increase the survival rate but also to try to do this with acceptable long-term morbidity.

In many European countries including the UK, children are now treated according to the European Paediatric Sarcoma Study Group (EpSSG) protocol. The basis of this has been the use of intensive chemotherapy, with the aim of improving the survival and reducing the use of local therapy with surgery and/or radiotherapy and thus minimizing long-term effects. The strategy for the current EpSSG) study includes stratifying patients within risk groups based on histological subtype (embryonal vs. alveolar histology), primary tumour size, stage of disease, and primary tumour site. Patients in the 'low-risk' category, i.e. those with localized tumours which are < 5 cm diameter and microscopically completely resected are treated with chemotherapy using actinomycin D and vincristine for a duration of 9 weeks without radiotherapy. Standard risk tumours are those which are locally more extensive but favourable (non-alveolar) histology at selected favourable sites, the vagina, uterus or paratestis and no nodal involvement and are treated with ifosfamide, vincristine and actinomycin D. The duration of ifosfamide is the subject of a randomized study within this protocol. High-risk tumours include other incompletely resected tumours, including all those arising in parameningeal sites (nasopharynx, middle ear) and those with involved lymph nodes. These are treated with chemotherapy involving a randomization between three drugs (ifosfamide, vincristine, actinomycin D) and four drugs (the addition of doxorubicin) and a further randomised study evaluating the role of maintenance chemotherapy. The very high-risk group includes those with alveolar histology and nodal involvement. These are all treated with four-drug combination chemotherapy and maintenance chemotherapy.

The role of radiotherapy in the various risk categories, together with protocol doses are summarized in Table 15.5. This study includes risk-adapted radiotherapy related to histology, tumour site and size, extent of resection, and nodal status.

For planning the margin for CTV is 1 cm around the prechemotherapy GTV with the exception of limb primaries, where the superior and inferior margin is 2 cm. For thoracic and abdominal tumours the postchemotherapy extent into cavities is taken as the GTV. A further 1-cm margin is allowed for PTV. For orbital primaries the CTV includes the whole orbit.

The dose per fraction is 1.8 Gy/day. For children with initially involved nodes, these are included in the target volume if they require radiotherapy.

Planning of radiotherapy demands careful attention to detail, avoiding excessive field sizes particularly in the head and neck region, because of the major morbidity from hypoplasia in this area.

Brachytherapy may be considered for highly selected children with limited tumours arising in the head and neck, vagina, bladder, or prostate. However, in practice, the main area where brachytherapy has been successfully employed has been for the treatment of small vaginal vault primaries. Brachytherapy may provide a means of local control with reduced morbidity compared with external beam RT. For a child being considered for brachytherapy, discussion between oncologists and surgeons at an early stage is essential. If brachytherapy is being considered for a genitor-urinary tract primary, then a joint examination under anaesthetic with surgeon and radiation

Table 15.5 Dose fractionation regimens for Rhabdomyosarcoma in European EpSSG Study

IRS* Group	Histology	
	Embryonal RMS	**Alveolar RMS**
I—size < 5 cm, completely resected with negative margins	No radiotherapy	41.4 Gy
II—macroscopic resection but positive margins	41.4 Gy	41.4 Gy
III—macroscopic residual disease at start of chemotherapy. Followed by:		
Secondary complete resection (after chemotherapy)	36.0—41.4 Gy depending on response	41.4 Gy
Incomplete secondary resection (after chemotherapy)	50.4 Gy	50.4 Gy
Clinical/radiological complete response, no second surgery	41.4 Gy	50.4 Gy
Partial response to chemotherapy	50.4 Gy (with boost of 5.4 Gy for selected patients) Orbit primary—and > 2/3 partial response—45 Gy	50.4 Gy (with boost of 5.4 Gy for selected patients)
Nodal involvement—dose to nodal regions	41.4 Gy	41.4 Gy

*IRS—Intergroup Rhabdomyosarcoma Study Group
Daily fraction size: 1.8 Gy

oncologists will be important in order to plan the approach. For the treatment itself, sedation will be necessary. Brachytherapy techniques employing low dose rate and high dose rate after-loading have been employed.

15.11 Ewing's sarcoma/peripheral primitive neuroectodermal tumour

Current survival rates for these tumours are reported to be between 55–65%. Ewing's sarcoma of bone has its peak incidence in the early teenage years. Approximately 60% occur in the long bones of the limbs, and 40% in the flat bones of the ribs, vertebrae or spine. Soft tissue extension is common. Peripheral primitive neuroectodermal tumour (PPNET), previously referred to as 'soft tissue Ewing's' has become more frequently recognized recently. The original subtype of PPNET recognized was the 'Askin tumour' of the chest wall. The majority of Ewing's sarcomas of bone and PPNET share a common chromosomal translocation, t(11;22) (q24;q12) and express the EWS-FLI-1 fusion gene. It is now recommended that Ewing's sarcoma and PPNET are treated according to common protocols.

Managing children with Ewing's sarcoma and PPNET involves a multidisciplinary team approach. Initial treatment is with chemotherapy, with the appropriate use of local therapy, either surgical resection, radiotherapy, or both. Currently patients in the UK are entered into the European Ewing's Tumour Working Initiative of National Groups (Euro-E.W.I.N.G. 99). In this study, patients are stratified according to primary tumour volume. For patients with small tumours (< 200 mL), and those with a good histological response to initial chemotherapy with ifosfamide, doxorubicin and vincristine, chemotherapy continues using doxorubicin and vincristine and a randomization to either cyclophosphamide or ifosfamide. For those with a poor histological response there is a randomization between conventional chemotherapy and high-dose chemotherapy with busulfan and melphalan.

The decision as to whether surgery, radiotherapy, or both should be employed for the primary tumour demands careful multidisciplinary discussion. In previous series of patients survival has been better following local treatment with surgery, compared with radiotherapy alone. However, these series are confounded by selection bias with patients with smaller tumours selected for surgery[12].

In common with planning of radiotherapy for adult sarcomas, planning of radiotherapy for Ewing's sarcoma or PPNET is technically challenging. Three-dimensional planning may be employed in order to achieve a uniform dose within the target volume which is frequently large and adjacent to critical organs. 'Multidisciplinary radiotherapy planning' involving radiologists, physicists, specialist therapy radiographers and mould room technicians is important at the outset.

15.11.1 Technique

Technique will depend upon tumour site and anatomy. Individualized, generally multiple fields in order to deliver homogeneous RT to the PTV and to minimize dose to non-target tissues and OARs.

Clinical target volume

◆ Phase 1: GTV prior to chemotherapy and/or surgery if feasible with a minimum margin of 3–5 cm for extremity primaries and 2 cm for trunk or head and neck primaries.
◆ Phase 2: 2 cm.

Patient position and immobilization

Depending on site and anatomy. For limbs, immobilization device such as an evacuated polystyrene bag.

Localization of clinical target volume/planning target volume

Planning CT scan.

Dosimetry

Dose specified to the ICRU reference point.

Dose prescription

◆ Phase 1 and postoperative volume: 44.8 Gy in fractions of 1.8–2.0 Gy.
◆ Phase 2 for macroscopic disease: 9.6 Gy in fractions of 1.8–2.0 Gy.

Verification

Machine verification films.

There is no clear evidence of a dose–response effect in the literature. However, recent data suggest a high rate of local relapse for those patients selected for definitive radiotherapy (as opposed to surgery). Therefore selected patients may benefit from a higher dose up to 64 Gy if this can be delivered without undue toxicity. For postoperative radiotherapy the prescribed dose depends on the extent of surgery and histological response of the primary to initial chemotherapy. Following 'intralesional surgery' or 'marginal surgery' with poor histological response to chemotherapy (> 10% residual tumour cells), 54.4 Gy is prescribed. For 'marginal surgery' with good histological response (< 10% residual tumour cells) or 'wide surgery' with poor histological response 44.8 Gy is prescribed. For patients who present with pulmonary metastases, whole lung RT is employed, 15 Gy for patients aged < 14 and 18 Gy for those aged > 14, with these dosed corrected for reduced attenuation in lung.

Care must be taken when combining chemotherapy and radiotherapy to avoid excessive morbidity from enhanced radiation reactions. Actinomycin-D is avoided during radiotherapy and doxorubicin avoided if there is a significant amount of bowel or mucosa in the treated volume. *Patients requiring radical radiotherapy involving the spinal cord or significant radiotherapy to the lungs should not receive busulfan.*

15.12 **Osteosarcoma**

This is the most common bone tumour in childhood, with the majority arising in teenagers. Approximately 65% arise in around the knee in the lower femur or upper tibia. Less common sites are humerus, fibula, sacrum, spine, mandible, and pelvis. Prior to the advent of effective chemotherapy 80% of patients died with pulmonary metastases. Currently a survival rate of approximately 55% can be achieved with the use of intensive chemotherapy. The majority of primary tumours can be resected and the affected bone replaced by a titanium endoprosthesis, thus avoiding the need for an amputation. The most frequently employed chemotherapy regimen is the combination of cisplatin, doxorubicin and methotrexate. The European Osteosarcoma Intergroup (EOI) randomized study compared the efficacy of six cycles of cisplatin and doxorubicin given either every 3 weeks or every 2 weeks using granulocyte colony stimulating factor to accelerate recovery of the blood count but outcomes have not been improved by the use of this dose-intensive regimen. The standard chemotherapy regimen now comprises a combination of cisplatin, doxorubicin, and high-dose methotrexate.

Radiotherapy has only a minor role in the management of osteosarcoma. However, it is probably not as radioresistant as previously thought. Radiotherapy is sometimes employed for the treatment of an unresectable tumour such as a spinal vertebral primary. In this case the dose has to be as high as normal tissues will tolerate, i.e. 60 Gy if possible. In selected patients after insertion of an endoprosthesis, postoperative radiotherapy may be employed for those felt to be at a high risk of local recurrence, i.e. those with tumour at a resection margin. Radiotherapy can be employed for the local palliation of metastases. Using a relatively high dose, e.g. 40 Gy in 15 fractions, many tumours will regress with relief of symptoms.

15.13 Central nervous system tumours

CNS tumours account for over one-fifth of malignant childhood tumours. The individual tumour types are uncommon and experience in the management of each type is limited. Approximately 350 children develop CNS tumours each year in the UK. The overall 5-year survival rate is approximately 50% which is inferior to that reported for many other paediatric tumours. Many of the survivors experience sequelae from either the tumour or therapy or both. In contrast to most other paediatric malignancies the use of chemotherapy has not yet resulted in significant improvements in survival for the majority of children with CNS tumours.

Children with CNS tumours, and their families, experience many problems related to their disease and the early and long-term toxicity of treatment. It is essential that patients and their families are managed by a specialized paediatric neuro-oncology multidisciplinary team.

In the last two decades there have been important advances in surgical techniques, particularly the use of the cavitron ultrasonic aspirator (CUSA) and operating micrscope, neuro-navigation techniques, paediatric anaesthesia, and intensive care support. It is now possible to biopsy tumours in most areas of the brain using stereotactic techniques. Alongside neurosurgical developments have been advances in neuroimaging which have resulted in better definition of tumour extent prior to surgery as well as the use of postoperative imaging to define the extent of tumour resection.

Radiotherapy for children with CNS tumours is technically challenging. Many children require craniospinal radiotherapy, which is one of the more complex techniques employed in most radiotherapy departments.

15.13.1 Long term effects of radiotherapy for CNS tumours

For many children the hypothalamic–pituitary axis has to be included in the irradiated volume. This may lead to endocrine deficiency, usually growth hormone deficiency but other endocrine deficiencies such as thyrotropin-releasing hormone (TRH) or adrenocorticotrophic hormone (ACTH) deficiency may occur. TRH deficiency is relatively easily managed by thyroid replacement therapy, but ACTH deficiency may require life-long steroid replacement therapy which may be problematic. Children receiving craniospinal radiotherapy will experience spinal shortening as a direct effect of irradiation of the spine.

Of great concern are the long-term neuropsychological effects of radiotherapy to the CNS particularly for very young children. Other factors include effects of the presence of the tumour itself, both direct and indirect such as the effect of hydrocephalus and also the effects of surgery. The most important factor which predicts for long-term neuropsychological outcome is the age at treatment. For children aged < 3 at diagnosis radiotherapy is delayed if possible by the use of chemotherapy.

15.13.2 **Chemotherapy for central nervous system tumours**

For many paediatric CNS tumours chemotherapy does not yet have an established role. The use of adjuvant chemotherapy improves the outcome for standard risk medulloblastoma compared with radiotherapy alone. Secreting intracranial germ cell tumours are also routinely treated with chemotherapy. The use of chemotherapy can delay the need for radiotherapy for very young children, particularly those with low-grade astrocytoma and probably ependymoma. In most cases the use of chemotherapy is evaluated within the context of a clinical trial.

15.13.3 **Low-grade astrocytoma**

These comprise the most common group of paediatric CNS tumours. The most frequent types are WHO Grade I (pilocytic) or Grade II (usually fibrillary). Other varieties such as ganglioglioma and oligodendroglioma are much less frequent. The presence of neurofibromatosis type I (NF1) predisposes to the development of these tumours. Modern management is based on the recognition that low-grade gliomas may undergo long periods of 'quiescence' even when not completely resected. It is now also clear that low-grade gliomas are not chemo-resistant. The current 5-year survival rate is 85%, but late relapse is not uncommon.

Treatment is initially with surgical resection, as complete as is considered safe. This may be relatively more straightforward for tumours arising in the cerebellum than for those arising from the optic tract or optic chiasm. In fact, tumours with typical features on MR scanning of a hypothalamic/optic tract astrocytoma are not necessarily biopsied, as many consider the risk of the procedure (i.e. visual deterioration) outweighs the risk of an incorrect diagnosis.

In North American and European collaborative group studies there have been attempts to standardize the management of children with low-grade gliomas. The International Society of Paediatric Oncology (SIOP) series of studies sets out a strategy for managing these tumours. Following maximal surgical resection patients undergo a period of observation. Those with clinical or radiological evidence of progression, those with severe symptoms or threat to vision receive non-surgical treatment.

In the current SIOP LGG2 study, those > 7 years are treated by radiotherapy. Those aged 7 or under receive chemotherapy with the aim of delaying radiotherapy. Several studies have reported that the majority of children with low-grade glioma can achieve either a response, or stabilization of previously progressive disease using the drugs carboplatin and vincristine. In the LGG2 study patients are randomized to receive initial chemotherapy with either carboplatin and vincristine, or in combination with the addition of etoposide. The aim of this study is to assess whether the additional drug can reduce the risk of early tumour progression within the first few months of chemotherapy. 'Maintenance' chemotherapy continues for a total of 18 months. Although chemotherapy has been given to young children in the past with the aim of 'avoiding' radiotherapy, it appears that generally what can be achieved is a delay to the need for radiotherapy. Hopefully this approach may reduce the impact of radiotherapy on long-term neuropsycholgical function, although the magnitude of any benefit from this approach is not clear.

Radiotherapy for low-grade gliomas[13] is based on a careful imaging for target volume definition.

Technique

Technique will depend upon tumour site and anatomy. Individualized multiple fields in order to deliver homogeneous RT to the PTV and to minimize dose to non-target tissues and OARs (Fig. 15.6).

Clinical target volume
GTV, and in the case of surgical resection, any brain tissue previously surrounding the tumour with a margin in potential areas of spread of 0.5 cm.

Patient position and immobilization
Supine or prone depending upon anatomy, generally immobilized in a perspex head shell.

Localization of clinical target volume/planning target volume
CT/MR fusion with extent of GTV localized on T2 or 'Flair' image on MR scan (Fig. 15.5).

Dosimetry
Dose specified to the ICRU reference point (Fig. 15.6).

Dose prescription
54 Gy in 30 fractions of 1.8 Gy daily.

Verification
Machine verification films.

Increasingly IMRT is employed in order to maximize the coverage of the PTV whilst minimizing dose to OARs and the integral dose to the whole brain.

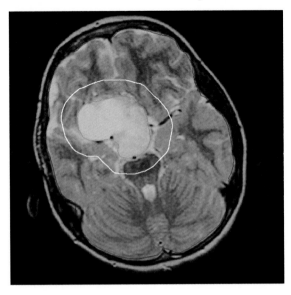

Fig. 15.5 CT/MR fusion images in low-grade glioma.

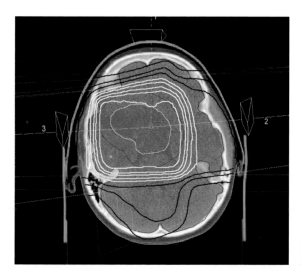

Fig. 15.6 Dosimetry for low-grade glioma.

Patients who are treated initially with chemotherapy and who experience progression will receive radiotherapy, and vice versa. For patients who present with spinal cord primary low grade glioma, management the management policy will be similar. For those requiring radiotherapy, the dose is 50.4 Gy in 28 fractions of 1.8 Gy.

15.13.4 High-grade astrocytoma

Unlike adults, high-grade astrocytomas are uncommon in childhood. However, in common with adults outlook is generally poor. Survival is currently approximately 20% at 5 years. Current management is based on surgical resection and postoperative radiotherapy (54 Gy in 30 fractions). There is no definite routine role for chemotherapy and in recent studies chemotherapeutic agents have been examined, given prior to radiotherapy in the form of a 'window' Phase II study. Children with high-grade gliomas are frequently treated with radiotherapy combined with concurrent and adjuvant temozolomide as employed in the 'Stupp protocol' for glioblastoma in adults[14].

15.13.5 Ependymoma

Ependymomas are uncommon and there are very few randomized studies on which to base management strategies. Historically management was with surgical excision, followed by radiotherapy, which was generally craniospinal for those with posterior fossa primaries or high-grade histology. It is acknowledged that although there is a risk of CSF metastases at diagnosis or relapse, this risk is not apparently influenced of by the use of craniospinal radiotherapy. The risk of CSF metastases is related to probability of primary local control. There is currently no evidence to support the routine use of craniospinal radiotherapy. The overall 5-year survival rate is approximately 50–60%. In the majority of studies prognostic factors include tumour grade and extent of resection. The predominant site of relapse is within the local tumour bed, and current research efforts are aimed at improving the prospect of local tumour control. It is important to try to achieve a complete or near complete resection. The majority of collaborative groups now recommend an increased radiotherapy dose (59.4 Gy with conformal techniques compared with the 'conventional' dose of 54 Gy). In recent studies and protocols the margin for CTV around the GTV, and/or resection cavity is 1.0 cm.

The European SIOP ependymoma study which is in the planning stage will recommend that for completely resected tumours postoperative focal radiotherapy (59.4 Gy in 33 fractions) should be given, and for incompletely resected tumours, a trial of combination chemotherapy.

15.13.6 Medulloblastoma/primitive neuroectodermal tumour

Medulloblastoma is a primitive neuronal tumour which arises in the cerebellum. It is notable for its propensity for metastatic spread via the CSF, and its radiosensitivity. Primitive neuroectodermal tumour (PNET) arises elsewhere in the CNS, usually the supratentorial cerebral cortex but sometimes the pineal area (pineoblastoma). Although histologically identical to medulloblastoma, supratentorial PNET has a significantly worse prognosis (approximately 40–50% compared with 60–70% 5-year survival).

Standard therapy for medulloblastoma/PNET is initial maximal surgical resection followed by craniospinal radiotherapy and a 'boost' to the primary site. Until the early 2000s in the UK the 'standard' approach was craniospinal radiotherapy 35 Gy with a 20 Gy 'boost' to the primary site. The European SIOP/UKCCSG PNET-3 study recruited patients between 1992–2000, and employed craniospinal RT 35 Gy in 21 daily fractions of 1.67 Gy with a 'boost' of 20 Gy in 12 daily fractions of 1.67 Gy. There was a randomization to this radiotherapy alone or preceded by four cycles of chemotherapy employing vincristine and etoposide, with alternating cyclophosphamide and carboplatin. This study has shown an advantage for chemotherapy, and has also demonstrated the importance of avoiding gaps in the radiotherapy schedule[15,16].

Current studies are based on the allocation of risk status (Table 15.5). For the last 10 years in North America it has been standard practice to employ adjuvant chemotherapy (vincristine, CCNU, cisplatin) following radiotherapy[17] and this has now also become standard practice in Europe. Using the up-front radiotherapy approach it has been possible to reduce the dose of craniospinal radiotherapy for patients with standard risk medulloblastoma in North American and European studies to 23.4 Gy, with 54.0–55.8 Gy to the posterior fossa.

Because of the poor survival of patients with CSF metastases, all these children receive chemotherapy and radiotherapy.

Several European and North American collaborative groups have investigated the role of hyperfractionated radiotherapy (HFRT) employing craniospinal radiotherapy 36 Gy in 36 fractions of 1 Gy twice daily, and a total dose of 68—72 Gy in 1-Gy fractions twice daily to the posterior fossa. The use of twice daily fractionation has allows an increase in the total dose delivered, and should hopefully result in an improvement in the therapeutic ratio. The use of HFRT has been evaluated further in the European PNET-4 randomized study. 'Conventionally' fractionated radiotherapy, employing a craniospinal dose of 23.4 Gy followed by 30.6 Gy to the posterior fossa was compared with HFRT. The HFRT dose per fraction was 1 Gy given twice daily with a minimum interval of 8 hours. The craniospinal dose was 36 Gy, with a second phase of 24 Gy to the whole posterior fossa and a final phase of 8 Gy to the tumour bed. The aim of the PNET-4 study was to assess whether survival could be improved by the use of HFRT without an increase in long-term effects. Early results have suggested that there is no advantage for HFRT in terms of disease control. However, whether HFRT can improve the therapeutic ratio and reduce the risk of late effects remains to be demonstrated.

For patients presenting with high-risk disease outcomes are poor. In North America attempts to improve outcome are focused on concurrent daily low dose carboplatin with radiotherapy. In several European groups the use of HFRT or hyperfractionated accelerated radiotherapy (HART) have been explored.

Craniospinal radiotherapy

CSRT is one of the most complex radiotherapy techniques delivered in oncology departments.

Technique

Lateral opposed cranial fields, with one or more posterior spinal fields carefully matched on to the lower border of the cranial fields. Fields may be centred on outer canthus to minimize divergence into the contralateral lens. Shielding of face, dentition, nasal structures and lenses. Moving junctions between fields to minimize the risk of overlap or underdose[18].

Clinical target volume

Intracranial and spinal meninges extending inferiorly to the lower border of thecal sac (Fig. 15.7). Great care is taken to include the cribriform fossa, temporal lobe and base of skull (Fig. 15.8).

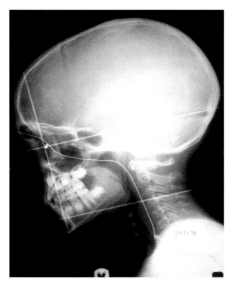

Fig. 15.7 Cranial field.

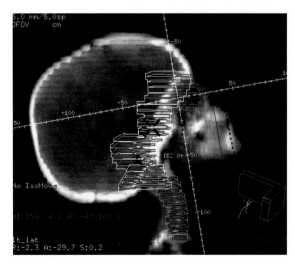

Fig. 15.8 'Virtual Simulation' to define the shielding for the cranial field in craniospinal radiotherapy at the cribriform fossa.

Patient position and immobilization
Generally prone with head and body immobilization, although increasing numbers of centres are delivering CSRT in the supine position.

Localization of clinical target volume/planning target volume
CT simulation.

Dosimetry
For cranial fields, dose specified to the midpoint of central axis.

Dose prescription
Depends upon clinical scenario. Currently 23.4 Gy in 13 fractions of 1.8 Gy daily.

Verification
Machine verification films.

Individual departments' CSRT techniques vary in detail but are generally based on these standard principles. Meticulous attention to detail in the planning and delivery of CSRT is essential and contributes to the cure of medulloblastoma/PNET. It is essential to avoid areas of underdose at field junctions and partial shielding of any area of meninges. The 'moving junction' between abutting fields is a 'safety measure', which reduces the risk of underdose or overdose in the cervical spinal cord if a systematic error develops during CSRT. In children the cribriform fossa frequently lies between the lenses (Fig. 15.9).In many series of patients treated for medulloblastoma, the cribriform fossa has been the site for isolated recurrence in a significant minority of patients. It may not always be possible to treat the CTV and adequately shield the lenses, in which case priority is given to treating the CTV.

The lower border of the spinal field has generally been placed at the lower border of the second sacral vertebra. However, there is evidence that the lower border of the thecal sac varies and planning of the lower border of the spinal field should be based

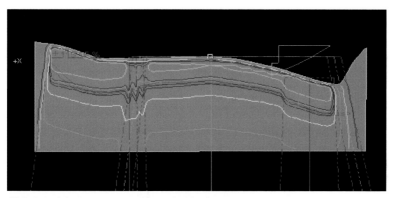

Fig. 15.9 Spinal dosimetry in craniospinal radiotherapy.

on MR scanning. The spinal field should be wide enough to encompass the extensions of the meninges along the nerve roots, and therefore wide enough to encompass the intervertebral foramina in the lumbar region. The spinal field will typically be 5–7 cm wide. Many centres employ shaped spinal fields to minimize irradiation of non-target paraspinal tissue (Fig. 15.10).

When planning the posterior fossa 'boost' for medulloblastoma, it is still standard practice for the whole cerebellum to be included in the target volume, because of the risk of local meningeal spread of the primary. The question as to whether the whole posterior fossa needs to be treated or whether the boost can be limited to the tumour bed with a margin is the subject of a current North American COG randomized study comparing outcomes for a tumour bed boost with a whole posterior fossa boost.

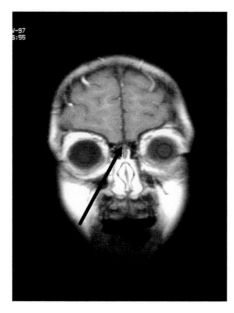

Fig. 15.10 Position of cribriform fossa between lenses of eyes.

CSRT techniques are continuing to evolve, and need to incorporate technical developments in radiotherapy immobilization, planning, and treatment delivery. IMRT techniques can be employed for CSRT (see Chapter 11). For the posterior fossa boost IMRT can be used to achieve high conformity to the PTV whilst minimizing dose to OARs such as hypothalamus and inner ears.

15.13.7 Intracranial germ cell tumours

Intracranial germ cell tumours account for approximately 30% of paediatric germ cell tumours. Germinomas are the histological equivalent of testicular seminoma. They generally arise in the suprasellar region but may occur in the pineal region. Non-germinomatous germ cell tumours are the histological equivalent of testicular non-seminomatous germ cell tumours, i.e. embryonal carcinomas, yolk sac tumours or choriocarcinomas. They generally arise in the pineal region and may secrete alpha fetoprotein (AFP) and/or human chorionic gonadotrophin (HCG). More than 90% of germinomas are cured with radiotherapy. The traditional view has been that craniospinal radiotherapy should be employed because of the risk of leptomeningeal metastases. However, as with testicular seminoma there has been a gradual reduction in both the CSRT dose and the dose used for the 'boost' to the primary tumour in an attempt to minimize long-term effects. In the recently closed SIOP protocol the CSRT dose was 24 Gy in 15 fractions of 1.6 Gy and the boost dose to the primary was a further 16 Gy in 10 daily fractions of 1.6 Gy. Germinomas have a tendency for subependymal spread, which has to be taken into account when planning the boost volume, with a margin of 2 cm around the GTV. For non-germinomatous tumours the prognosis is worse. Initial treatment is with chemotherapy followed by radiotherapy, either focal for non-metastatic tumours or craniospinal for those with meningeal metastases. In the SIOP protocol the chemotherapy used was a combination of cisplatin, etoposide and ifosfamide followed by radiotherapy. The total radiotherapy dose was 54 Gy to the primary tumour, with 30 Gy to the craniospinal axis for patients with meningeal metastases.

15.13.8 Brainstem glioma

This includes tumours arising in the midbrain, pons, and medulla. Historically they were regarded as a single entity. However, it is now clear that they can be subdivided into focal (5–10%), dorsal exophytic (10–20%), cervico-medullary (5–10%), and diffuse intrinsic tumours (75–85%). Focal, dorsal exophytic, and cervico-medullary tumours are usually low-grade astrocytomas. Surgical excision is the treatment of choice with radiotherapy reserved for inoperable tumours.

The majority of children with brainstem gliomas have diffuse pontine tumours, which are usually high-grade astrocytomas. They are diagnosed by their typical MR appearance and biopsy is dangerous and contraindicated. Their prognosis is very poor with a median survival of approximately 9 months and very few long-term survivors. The management of these children remains a major challenge, and domiciliary palliative care usually has to be introduced at a relatively early stage.

Because of the frequent relatively short history of these tumours, radiotherapy often needs to be commenced quickly.

Technique

Lateral opposed fields.

Clinical target volume
GTV as defined on diagnostic MR scan with a margin of 2 cm along potential areas of spread superiorly, inferiorly and posteriorly along brainstem.

Patient position and immobilization
Supine in head shell.

Localization of clinical target volume/planned target volume
CT simulation.

Dosimetry
Dose specified to the ICRU reference point.

Dose prescription
54 Gy in 30 fractions of 1.8 Gy daily.

Verification
Machine verification films.

Chemotherapy has been of no benefit. Conventional radiotherapy provides useful palliation for approximately 75% of children. However, the progression-free survival is short, usually < 6months. Neither hyperfractionated nor accelerated radiotherapy has improved outcome for these patients. Current protocols are evaluating novel drug therapy approaches delivered alongside radiotherapy.

15.14 Intensity-modulated radiotherapy

The majority of paediatric radiotherapy protocols specify dose/fractionation regimens and definitions of GTV and CTV. However, selection of technique, and whether IMRT is employed are generally decided on by the treating radiation oncology team. As with adults, patients treated for sarcomas and brain tumours may benefit from IMRT with respect to achieving reduced late effects from reducing dose to OARs below threshold levels for late effects, while maintaining an appropriate dose to the PTV.

15.15 Proton therapy for paediatric tumours

Although high energy photons are employed as the standard ionizing radiation modality for most patients treated for cancer there is increasing interest in the use of proton therapy as an alternative form of ionizing radiation for clinical use. The advantages of proton therapy arise from their physical dose distribution with the sharp cut-off beyond the Bragg peak, beyond which there is very little dose deposited. In other words a proton beam has considerable 'stopping power' as compared with a photon beam. The clinical advantages of proton therapy relate to the physical characteristics of the proton beam and the ability to achieve a very high conformity index around the tumour.

This feature has been used for two main purposes. Firstly to enable the treatment with high dose (i.e. escalated dose) irradiation for relatively radioresistant tumours adjacent to radiosensitive critical organs, such as skull base chordomas and chondrosarcomas adjacent to the brainstem or spinal cord. However, secondly there is also increasing interest in the use of proton therapy for the radiotherapeutic treatment of children. The dose distribution characteristics may be used to reduce the volume of non-target tissue receiving a low to intermediate dose (the low dose 'bath' effect) and reduce the incidence and/or severity of long-term effects. In comparison IMRT can also achieve high conformity, but the multiple fields employed will generally deliver a low to intermediate radiation dose to a wide area which may be significant for long-term effects. Most proton therapy centres have the treatment of children as an important component of their programme. The children most likely to benefit are young children, many of whom will require daily anaesthesia for immobilization.

As with adults, children with base of skull chordoma and chondrosarcoma have been treated with proton therapy. The improved dose distribution can be exploited in order to escalate the dose and improve local control. However, there are rare in children accounting for a very small proportion of cases.

Late effects of treating head and neck primary sites with radiotherapy, particularly in young children, are problematic because of the impairment of facial bone growth. A review of late effects for children treated in the North American multicentre series reported 59% of survivors to have evidence of impaired orbital growth. The superior dose distribution characteristics of proton compared with photon beams and their ability to reduce the dose to non-target normal tissues have led many centres to employ proton radiotherapy for the treatment of rhabdomyosarcoma arising in these areas, with the aim of reducing the severity of late effects[19].

Proton therapy has been advocated for the focal irradiation of a variety of low-grade brain tumours including low-grade astrocytoma, craniopharyngioma, and ependymoma (Fig. 15.11). The aim is to reduce the amount of normal tissue irradiated in order to minimize the long-term neurocognitive effects of therapy[20,21].

For the delivery of CSRT in some centres proton therapy is employed to largely avoid the exit dose from the spinal field to structures anterior to the spine with the aim of reducing the risk of radiation induced malignancies. In the treatment of children with medulloblastoma the boost fields to the cerebellum involve irradiating the middle and inner ear, which can give rise to hearing problems later, particularly when radiotherapy is combined with chemotherapy. The use of protons for the boost can reduce the radiation dose to the middle and inner ear.

15.16 Conclusions

Children treated by the multidisciplinary paediatric oncology team present with a wide variety of diseases, which pose many different problems for patients and families. Management is not only patient-centred, but also family-centred. Prognosis varies considerably from those where treatment failure is rare, namely intracranial germinoma to those where cure is rare, such as intrinsic pontine glioma. Management in specialized centres is essential, according to national or international protocols.

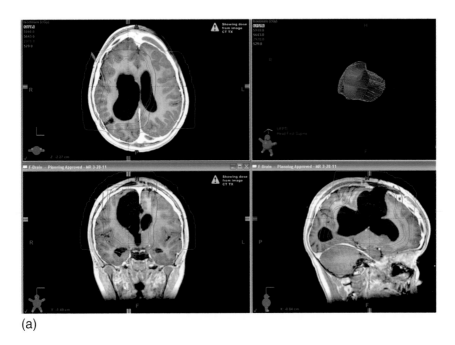

(a)

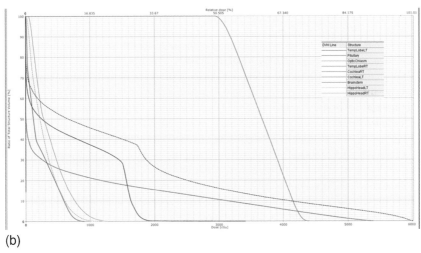

(b)

Fig. 15.11 (a) Dose distribution for proton therapy plan for supratentorial ependymoma. (b) DVH data. (Images courtesy of Dr D Indelicato, University of Florida Proton Therapy Institute (UFPTI), Jacksonville, Florida.)

Radiotherapy plays an important role in the management of many of these children. They require the highest standard of radiotherapy planning and delivery, incorporating modern technical developments including IMRT. Compared with adult radiation oncology practice, when planning radiotherapy for children consideration of the long-term effects of treatment is always of paramount importance. Although cure rates for the majority of childrens' cancers have continued to improve over the last decade, this has frequently been achieved with increase intensity of chemotherapy. With the increasing use of concurrent combined modality therapy constant vigilance for interactions is required.

References

1. Surveillance, *Epidemiology and End Results Program, 1975–2001*. Bethesda, MD: National Cancer Institute, 2004. http://www.seer.cancer.gov

2. Taylor RE. Cancer in children: Radiotherapeutic approaches. *British Medical Bulletin* 1996; **52**: 873–6.

3. Lipshultz SE, Lipsitz SR, Mone SM, *et al.* Female sex and higher drug dose are risk factors for late cardiotoxic effects of doxorubicin therapy for childhood cancer. *New England Journal of Medicine* 1995; **332**: 1738–43.

4. Kortmann RD, Freeman CR, Taylor RE. Radiotherapy techniques. In Walker DA, Perilongo G, Punt JAG, Taylor RE (eds.) *Brain and Spinal Tumors of Childhood*. London: Arnold, 2004, pp. 188–212.

5. Carrie C, Hoffstetter S, Gomez F, *et al.* Impact of targeting deviations on outcome in medulloblastoma: Study of the French Society of Pediatric Oncology (SFOP). *International Journal of Radiation Oncology, Biology, Physics* 1999; **45**: 435–9.

6. Willman K, Cox R, Donaldson S. Radiation induced height impairment in pediatric Hodgkin's Disease. *International Journal of Radiation Oncology, Biology, Physics* 1994; **28**: 85–92.

7. Kortmann RD, Timmermann B, Taylor RE, *et al.* Current and future strategies in radiotherapy of childhood low-grade glioma of the brain. Part II: Treatment-related late toxicity. *Strahlentherapie Onkologie* 2003; **9**: 585–97.

8. Jannoun L, Bloom HJG. Long-term psychological effects in children treated for intracranial tumors. *International Journal of Radiation Oncology, Biology, Physics* 1990; **18**: 747–53.

9. Gilson D, Taylor RE. Commentary – total body irradiation. *British Journal of Radiology* 1997; **7**: 1201–3.

10. Gerrard G, Vail A, Taylor RE, *et al.* Toxicity and dosimetry of fractionated total body irradiation prior to allogeneic bone marrow transplantation using a straightforward radiotherapy technique. *Clinical Oncology* 1998; **10**: 379–83.

11. Hunger S, Link, Donaldson S. ABVD/MOPP and low dose involved field radiotherapy in pediatric Hodgkin's Disease: the Stanford experience. *Journal of Clinical Oncology* 1994; **12**: 2160–6.

12. Schuck A, Ahrens S, Paulussen M, *et al.* Local therapy in localized Ewing's tumors: Results of 1058 patients treated in the CESS 81, CESS 86, and EICESS 92 trials. *International Journal of Radiation Oncology, Biology, Physics* 2003; **55**: 168–77.

13. Kortmann RD, Timmermann B, Taylor RE, *et al.* Current and future strategies in radiotherapy of childhood low-grade glioma of the brain. Part I: Treatment modalities of radiation therapy. *Strahlentherapie Onkologie* 2003; **8**: 509–20.

14. Stupp R, Mason WP, van den Bent MJ, *et al.* Radiotherapy plus concomitant and adjuvant temozolomide for glioblastoma. *New England Journal of Medicine* 2005; **352**: 987.

15. Taylor RE, Bailey CC, Robinson K, *et al.* Results of a randomised study of pre-radiation chemotherapy vs radiotherapy alone for non-metastatic (M0–1) medulloblastoma the SIOP/UKCCSG PNET-3 study. *Journal of Clinical Oncology* 2003; **21**: 1581–91.

16. Taylor RE, Bailey CC, Robinson KJ, *et al.* Impact of radiotherapy parameters on outcome in the International Society of Paediatric Oncology (SIOP)/United Kingdom Children's Cancer Study Group (UKCCSG) PNET-3 study of pre-radiotherapy chemotherapy for M0–1. *International Journal of Radiation Oncology, Biology, Physics* 2004; **58**: 1184–93.

17. Packer RJ, Goldwein J, Nicholson HS, *et al.* Treatment of children with medulloblastomas with reduced-dose craniospinal radiation therapy and adjuvant chemotherapy: A Children's Cancer Group study. *Journal of Clinical Oncology* 1999; **17**: 2127–36.

18. Kiltie AE, Povall JM, Taylor RE. The need for the moving junction in craniospinal irradiation. *British Journal of Radiology* 2000; **73**: 650–4.

19. Kozak KR, Adams J, Krejcarek SJ, *et al.* A dosimetric comparison of proton and intensity-modulated photon radiotherapy for pediatric parameningeal rhabdomyosarcomas. *International Journal of Radiation Oncology, Biology, Physics* 2009; **74**:179–86.

20. Merchant TE, Conklin HM, Wu S, *et al.* Late effects of conformal radiation therapy for pediatric patients with low-grade glioma: prospective evaluation of cognitive, endocrine, and hearing deficits *Journal of Clinical Oncology* 2009; **27**: 3691–7.

21. Merchant TE, Hua C, Shukla H, *et al.* Proton versus photon radiotherapy for common pediatric brain tumors: comparison of models of dose characteristics and their relationship to cognitive function. *Pediatric Blood & Cancer* 2008; **51**: 110–17.

Further reading

Halperin EC, Constine LS, Tarbell NJ, Kun LE. *Pediatric Radiation Oncology*. Philadelphia, PA: Lippincott Williams & Wilkins, 2010.

Malpas J (ed). Cancer in children. *British Medical Bulletin* 1996; **52**(4).

Pinkerton CR, Plowman PN. *Paediatric Oncology – Clinical Practice and Controversies* (2nd edn). London: Chapman & Hall, 1997.

Pizzo PA, Poplack DG. *Principles and Practice of Pediatric Oncology* (6th edn). Philadelphia, PA: Lippincott, Williams and Wilkins, 2010.

Royal College of Paediatrics and Child Health. *Guidance for Services for Children and Young People with Brain and Spinal Tumours*. London: Royal College of Paediatrics and Child Health, 1997.

Walker DA, Perilongo G, Punt JAG, Taylor RE (eds). *Brain and Spinal Tumors of Childhood*. London: Arnold, 2004.

Wallace H, Green D. *Late Effects of Childhood Cancer*. London: Arnold, 2004.

Yock T, De Laney TF, Esty B, Tarbell NJ. Pediatric tumors. In De Laney TF, Kooy HM (eds) *Proton and charged particle radiotherapy*. Philadelphia, PA: Lippincott Williams & Wilkins, 2007, pp. 125–39.

Radiotherapy planning for metastatic disease

Peter Hoskin

Palliative treatments account for a large proportion of the workload of any department estimated at between 40% and 50% of new patient treatments per year in an average cancer centre in the UK. These will be predominantly patients with bone metastasis, spinal cord compression, and brain metastasis. The techniques for such treatments are specific to this indication rather than the primary tumour site and are therefore included here in a separate chapter.

In general palliative radiotherapy should be simple, pragmatic, and quick. This does not, however, mean that careful planning should be ignored. Wherever possible, simulator verification of volume and beam positions should be used.

Whilst conventional GTV, CTV, and PTV definitions are not often used it is important to consider the principles in defining palliative fields allowing adequate margins around symptomatic sites to allow for internal organ movement and set-up variation. When defining treatment areas in terms of fields rather than volumes it is important to remember that the field edge defined by the simulator light beam represents the 50% isodose.

In some circumstances, particularly where there are oligometastases or for retreatment image-guided planning with more sophisticated delivery techniques, including the option of IMRT or stereotactic radiotherapy should be considered.

16.1 **Bone metastasis**

16.1.1 **Local bone pain**

Patient position and immobilization

In general, most bone metastasis with modern radiotherapy equipment can be treated with the patient supine. Ankle stocks and a headrest will aid immobilization.

Where electron treatments are used (see later in this section) a semisupine position to access the site to be treated, e.g. ribs, may be more appropriate.

If orthovoltage is used then direct applicator apposition is required for ribs or spine and the patient may therefore need to be prone or semisupine.

Volume and localization

Volume definition should be based on symptoms rather than radiological changes although confirmation of bone metastasis as the cause of local bone pain is essential. Specific considerations include the following:

1. Wherever possible include a whole bone or recognized anatomical structure, for example, treating the pelvis to the midline including the entire unilateral pubic rami, as shown in Fig. 16.1, unless, as shown in Fig. 16.2, there is involvement of the pubic symphysis when the field wires must be seen to extend 1–2cm beyond the midline to ensure that the involved bone is encompassed in the 90% isodose.

2. Patients with multiple bone metastases will often require several radiotherapy treatments and treatment fields should be used with an eye to subsequent matching of adjacent fields, for example, in the spine placing fields in the intravertebral spaces and in the pelvis including the entire hip joint bringing the medial border to the midline.

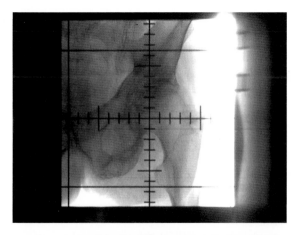

Fig. 16.1 Radiotherapy field to treat left hip, including ipsilateral pubic bone also involved with metastatic disease, to midline.

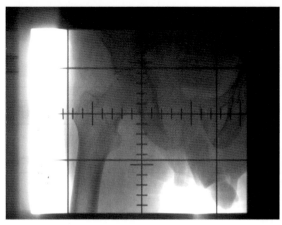

Fig. 16.2 Radiotherapy field to treat right hip crossing midline to ensure coverage of metastasis in the pubic symphysis.

Dose distribution

- The majority of bone metastases are best treated with photon beams of 4–6 MV, or if not available a cobalt beam. Single applied fields or parallel opposed anterior and posterior fields will enable treatment to most of the skeleton.

- Direct electron beams are appropriate for superficial bones, in particular the ribs and clavicles.

- Orthovoltage beams (250–500 kV) will typically reach their 80% isodose at a depth of 3–3.5 cm which is adequate for treatment of ribs, clavicles, and sacrum.

Dose prescription

The most common dose for bone pain is a single dose of 8 Gy.

1. This should be prescribed as a midplane dose with parallel opposed fields.

2. The spine should be treated at the depth of the vertebral body. This varies along the length of the spine. If it is not appropriate or possible to screen the spine laterally in the simulator then a depth of 4–5 cm will be adequate for the prescription definition. This will mean an applied dose of around 10 Gy.

3. Electron prescriptions should be to the 100% isodose taking account of the isodose distribution and effective depth of the beam which will be chosen for the appropriate situation (see Chapter 2). For ribs electron energies of 8–10 MeV will give more than adequate depth penetration. Build-up is not required.

Orthovoltage beams should be prescribed to an applied 100% isodose. Whilst it is acknowledged that there is a more prominent photoelectric absorption effect with these beams which will relatively increase the dose to bone, it is not usual to reduce the prescribed dose from that above, i.e. 8 Gy.

Although there is extensive Level 1 evidence to support the use of single doses of 8–10 Gy for bone pain[1,2] alternative prescriptions are in use for bone pain including 20 Gy in five fractions and 30 Gy in 10 fractions.

16.1.2 Wide field (hemibody) irradiation for multiple bone pains

Where there are scattered bone pains from multiple bone metastases external beam radiotherapy still has an important and effective role using wide field treatment to encompass large areas of the body. Conventionally these are often referred to as upper or lower hemibody treatment although in practice they need not be constrained to these precise definitions.

Patient position and immobilization

The patient should be supine and immobilization with leg stocks and headrest may aid reproducibility.

Treatment volume

This will be defined by the areas of pain and limited by the field size available. Conventionally the hemibody fields are as follows:

1. Upper hemibody: from top of scalp to umbilicus.

2. Lower hemibody: from umbilicus to soles of feet.

3. Whilst these volumes are relevant to patients who may be effectively treated for end-stage chemotherapy-resistant myeloma or lymphoma, for bone metastasis such precise and rigid definitions are not necessary. Furthermore upper hemibody irradiation to incorporate the scalp results in alopecia which is undesirable if unnecessary.

The volume should then be defined by the sites of pain. Where these are predominantly in the ribs and thoracic spine this area can be encompassed in a large field and if lumbo-sacral spine, pelvis, and lower limbs, again a large field covering all the painful areas should be used.

Dose distribution

These volumes are treated with anterior and posterior parallel opposed fields. The area to be treated is usually confined by the available field size. At 100 cm FSD, a modern linear accelerator, will be able to provide a field of 35–40 cm in length. If a larger area is treated then extended FSD techniques will be required.

Dose prescription

- ◆ 8-Gy midplane dose for the lower hemibody.
- ◆ 6-Gy midplane dose where the lungs are included in the volume as this represents lung tolerance when delivered in a single dose at linear accelerator dose rate. At higher doses pneumonitis may be encountered.

Lower doses of 4-Gy midplane dose may be equally effective although they have not been compared in a randomized trial. One study has evaluated 8 Gy in two fractions as an effective dose. There is, however, no clear advantage of this over a single dose of 8 Gy except where the lungs are in the field[3,4].

The same doses are used for treatment of myeloma or lymphoma. This may be in the context of two scenarios

1. Single wide field or hemibody treatment where the disease is predominantly limited to one area, for example, multiple lymphadenopathy in the chest and neck or in the abdomen and pelvis in lymphoma or lytic bone disease with pain from myeloma.

2. Sequential hemibody irradiation has been used with good palliative effect for both lymphoma and myeloma where the disease is more widespread. The following are important considerations:

    - ◆ The peripheral blood count should be adequate before treatment since bone marrow depression is to be expected after wide field radiotherapy and adequate reserve will be important; similar parameters to those used for chemotherapy are applied, i.e. haemoglobin > 10 g/dL, white blood cell count > 3.0×10^9/L, neutrophils > 1.5×10^9/L, platelets > 80×10^9/L.

    - ◆ A period of 6 weeks between one hemibody treatment and the next is essential to allow for bone marrow recovery. Peripheral blood indices should have returned to adequate levels before proceeding with the second hemibody radiotherapy treatment.

♦ A calculated gap between the two fields is required matching to the midplane. The gap will be several centimetres because of the divergence of large fields.

16.1.3 Postoperative radiotherapy for bone metastasis

The preferred treatment for a long bone fracture from bone metastasis is internal fixation. The role of postoperative radiotherapy remains uncertain but most patients with a prognosis of more than 3 months will be offered and treated with postoperative radiotherapy.

Patient position and immobilization

This will depend upon the bone to be treated.

♦ Hip and femur are treated with the patient supine using ankle stocks to immobilize the lower limb.

♦ Humerus: arm abducted, elbows flexed, hands on hip.

Volume definition

There are two views as to the volume definition in this setting.

3. The entire bone should be covered based on the basis that the marrow cavity may be contaminated peroperatively.

4. Only the entire prosthesis need be covered as the area most at risk of residual tumour and regrowth.

In the absence of any data to support either of these two approaches it is clear that at least the entire prosthesis should be covered by the treatment volume and a margin of at least 3 cm of normal bone beyond the prosthesis is recommended. Where an intramedullary nail has been used then this should be covered completely with the treatment volume which will usually be the entire bone.

Where large fields are used along the length of a bone, even for palliative doses an attempt should be made to avoid joint spaces and preserve a corridor of normal tissue for lymphatic drainage.

Dose distribution

Anterior and posterior parallel opposed fields are used.

Dose prescription

There is no consensus but the following are acceptable:

♦ Single doses of 8–10-Gy midplane dose.

♦ 20-Gy midplane dose in five fractions treating daily.

♦ 30-Gy midplane dose in 10 fractions treating daily.

16.1.4 Spinal canal compression

Spinal canal compression typically presents as an emergency. Wherever possible, however, patients should have fields set up with a treatment simulator to ensure accuracy and there should be access to diagnostic MR scans to define the levels of involvement.

Treatment volume

This should include the site of spinal canal compression and one vertebral body in craniao-caudal direction above and below the site of compression. If a patient is being treated on clinical diagnosis and plain X-ray evidence of bone metastasis then two vertebral bodies above and below the anticipated site of compression should be used to define the volume pending accurate diagnosis with an MR scan. Attention to the transverse axial imaging is important to ensure that any lateral or paravertebral extension is covered in the volume width.

It should be remembered that 25–30% of patients will have multiple sites of compression and a full spine MRI is recommended[5]. All areas should be treated at the same time by using more than one field if necessary.

Dose distribution

Treatment is given with a direct photon beam of 4–6 MV.

Dose prescription

Standard doses include the following:

- 8–10-Gy single dose.
- 20 Gy in five fractions.
- 30 Gy in 10 fractions.
- Special circumstances may warrant other fractionation schedules, for example:
  - Solitary plasmacytoma: 40–50 Gy in 20–25 fractions[6].
  - Lymphoma: primary radiotherapy for chemoresistant or low-grade lymphoma and post-chemotherapy radiotherapy for high-grade lymphoma and Hodgkin's lymphoma: 30 Gy in 15 fractions. (See Chapter 10.)
  - Recurrent spinal canal compression after previous treatment: the risk of myelopathy must be balanced against the need for retreatment and the patient's likely prognosis. A cumulative BED using an α/β ratio of 2 less than 120 is acceptable with a gap between treatments of several months. Thus after 20 Gy in five fractions (BED2 = 60) a further 20 Gy in fractions may be considered or a single dose of 8 Gy (BED2 = 40)–10 Gy (BED2 = 60). After 30 Gy in 10 fractions (BED2 = 75) a further dose of 20 Gy in 10 fractions (BED2 = 40) would be acceptable. Higher doses may be feasible with stereotactic radiotherapy (see later).

Dose prescription point

This should be defined at the depth of the anterior spinal canal; this will vary along the length of the cord and ideally should be measured for each patient from the available MR imaging or on lateral X-ray screening. An alternative is to use Table 16.1 based on published data[7].

16.2 **Brain metastasis**

Brain metastases require whole brain radiotherapy. In selected cases with a localized solitary metastasis and good performance status where surgery is not possible then

Table 16.1 Dose description point

Vertebral level	Depth of anterior canal
C6	7 cm
T2	7 cm
T5	6 cm
T8	5 cm
T11	6 cm
L1	7 cm
L3	8 cm
L5	8 cm

radiosurgery should be considered. This section will describe the technique for palliative whole brain radiotherapy.

16.2.1 Patient position

The patient should be supine with the neck straight on a standard headrest.

For palliative whole brain radiotherapy a head shell is not necessary but immobilization using a band across the forehead may be used. Sandbags may help support the head laterally.

16.2.2 Treatment volume

Volume should include the whole brain including the olfactory groove and middle cranial fossa. Conventionally the treatment field is defined as follows:

◆ Inferior border by a line drawn from the supra-orbital ridge through the external auditory meatus, resulting in the baseline as shown in Fig. 16.3, achieved by using appropriate head twist.

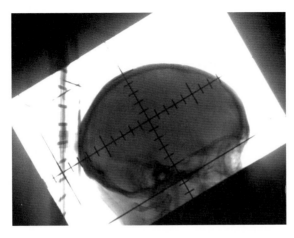

Fig. 16.3 Typical field to treat whole brain for brain metastases.

- Other borders to cover the scalp and a small margin of 5–10 mm outside the contours of the scalp to allow for patient movement.

In practice the patient should, wherever possible, be simulated to ensure that this baseline covers the full extent of the middle cranial fossa and adjustment may be required.

16.2.3 **Dose distribution**

Lateral opposed fields are used. An isocentric technique is most convenient and quick for treatment delivery although fixed FSD techniques are acceptable.

16.2.4 **Dose prescription**

The following doses are in common use[8]:

- 12 Gy in two fractions.
- 20 Gy in five fractions.
- 30 Gy in 10 fractions.

The dose is prescribed to the midplane.

Doses for *stereotactic radiotherapy* (radiosurgery) range from a single dose of 20–24 Gy for volumes < 2 cm diameter to three fractions of 8–10 Gy for larger volumes[9].

16.3 **Liver metastasis**

Liver metastases are rarely treated with external beam radiotherapy but it may have a useful palliative action where the liver is large and painful from expanding metastasis not controlled by systemic treatment[10].

There is also increasing interest in the treatment of solitary or oligometastases in the liver with stereotactic radiotherapy when surgical resection is not possible.

16.3.1 **Treatment volume and definition**

The extent of metastasis will usually be defined on ultrasound or CT scan and the clinical size of the liver is a useful guide to treatment planning.

16.3.2 **Patient position and mobilization**

If necessary the patient should be treated supine.

16.3.3 **Volume localization**

The volume is best localized using CT simulation where resources are available for this. As much uninvolved liver as possible should be kept out of the treatment volume to minimize toxicity.

If CT simulation is not available then clinical definition of the inferior border can be used and the upper border defined by the dome of the diaphragm on screening using a conventional simulator.

16.3.4 **Dose distribution**

Treatment is delivered with anterior and posterior parallel opposed fields. An isocentric technique will be used although it is important to note where fractionated treatment is given that such patients may have ascites and variable abdominal girths from day to day for which adjustments may be needed.

16.3.5 **Dose prescription**

The following doses may be used:

- 8-Gy single dose.
- 20–30 Gy in 10–20 fractions.

Doses are prescribed to the midplane.

16.4 **Choroidal metastasis**

Metastases to the choroid and retina are relatively rare but can cause catastrophic loss of sight. They are seen most commonly in breast and lung cancer and 20% may be bilateral[11]. Early diagnosis by clinical examination of the eye with a fundoscope of slit lamp in any patient presenting with visual disturbance against a background of established malignancy is essential to retain vision. Urgent radiotherapy is indicated once the diagnosis is confirmed.

16.4.1 **Treatment position and immobilization**

Supine, neck straight. Although treated as an emergency, immobilization is recommended ideally with an Orfit-type shell which can be made rapidly and not delay treatment.

16.4.2 **Volume localization and definition**

The CTV includes the entire choroid and retina of the involved eye. A CT or MRI scan is of value to identify bulky disease or coincident brain metastases. A 4×4 cm 4–6-MV photon field with the anterior border at the external canthus of the affected eye should be used as shown in Fig. 16.4. To prevent irradiation of the contralateral lens by the exit beam one of the following should be employed:

- The beam is angled posteriorly 3° to 'take off' the divergence.
- An asymmetric field is used so that the field centre is at the outer canthus, the anterior half of the field is then closed so that there is no divergence at the canthus and the posterior half of the field is exposed.

16.4.3 **Dose prescription**

The following doses may be used:

- 20 Gy in five fractions.
- 30 Gy in 10 fractions.

The dose is prescribed to a depth of 2.5 cm.

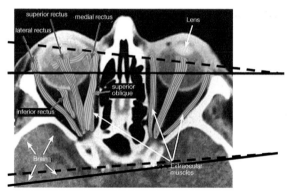

Fig. 16.4 Beam arrangement to treat choroidal metastases using a single lateral field with the anterior border at the ipsilateral outer canthus. Note that if divergence is not removed as in --- the beam will exit through the lens of the contralateral eye; an asymmetric beam placing the central axis at the outer canthus, shown as — or a beam angled posteriorly 3°, should be employed to exit behind the lens and cornea of the contralateral eye.

16.4.4 **Verification**

A megavoltage image should be taken and *in vivo* dosimetry with TLD or diodes should be undertaken for the first or second fraction to ensure that the exit beam is behind the contralateral eye.

References

1. Chow E, Harris K, Fan G, *et al.* Palliative radiotherapy trials for bone metastases: a systematic review. *Journal of Clinical Oncology* 2007; **25**: 1423–36.

2. Lutz S, Berk L, Chang E, *et al.*; American Society for Radiation Oncology (ASTRO). Palliative radiotherapy for bone metastases: an ASTRO evidence-based guideline. *International Journal of Radiation Oncology, Biology, Physics* 2011; **79**(4): 965–76.

3. Salazar OM, Rubin P, Hendricksen F, *et al.* Single dose hemibody irradiation in palliation of multiple bone metastases from solid tumours. *Cancer* 1986; **58**: 29–36.

4. Salazar OM, Sandhu T, da Motta NW, *et al.* Fractionated half–body irradiation (HBI) for the rapid palliation of widespread, symptomatic, metastatic bone disease: a randomized Phase III trial of the International Atomic Energy Agency (IAEA). *International Journal of Radiation Oncology, Biology, Physics* 2001; **50**: 765–75.

5. Prewett S, Venkitaraman R Metastatic spinal cord compression: Review of the evidence for a radiotherapy dose fractionation schedule. *Clinical Oncology* 2010; **22**: 222–30.

6. Soutar R, Lucraft H, Jackson G, *et al.* Guidelines on the diagnosis and management of solitary plasmacytoma of bone and solitary extramedullary plasmacytoma. *Clinical Oncology* 2004; **16**: 405–13.

7. Barton R, Robinson G, Gutierrez E, Kirkbride P, McLean M. Palliative radiation for vertebral metastases: the effect of variation in prescription parameters on the dose received at depth. *International Journal of Radiation Oncology, Biology, Physics* 2002; **52**: 1083–91.

8. Gaspar LE, Mehta MP, Patchell RA, *et al.* The role of whole brain radiation therapy in the management of newly diagnosed brain metastases: a systematic review and evidence-based clinical practice guideline. *Journal of Neuro-Oncology* 2010; **96**: 17–32.

9. Wiggenraad R, Verbeek-de-kanter A, Kal HB, *et al.* Dose-effect relation in stereotactic radiotherapy for brain metastases. *Radiotherapy and Oncology* 2011; **98**: 292–7.

10. Hoskin PJ. Radiotherapy in symptom management. In Hanks GW, Cherny N, Christakis N, Fallon M, Kaasa S, Portenoy R (eds). *Oxford Textbook of Palliative Care* (4th edn). Oxford: Oxford University Press, 2010, pp. 526–47.

11. Wiegel T, Bottke D, Kreusel K-M, *et al.* External beam radiotherapy of choroidal metastases—final results of a prospective study of the German Cancer Society (ARO 95–08). *Radiotherapy and Oncology* 2002; **64**: 13–18.

17

Quality assurance in radiotherapy

Patricia Diez, Edwin GA Aird

17.1 Introduction

Quality assurance (QA) in radiotherapy is essential for setting standards of safety and accuracy to ensure the outcome for the patient is optimal. QA should cover all aspects of the process of planning and delivery of the treatment. It may form part of a quality system, which will also encompass quality control (QC) and audit. It is not the intention here to describe in detail a complete quality system. Some aspects of a formal system will be described, such as policy, audit, and procedures. The reader is advised to read formal systems documentation for more detail. This chapter will mainly cover QA of the patient pathway, concentrating on QC of treatment planning and delivery. There will also be discussion on relevant legislation associated with the radiotherapy process as well as a section on QA for clinical trials.

The World Health Organization[1] defines QA for radiotherapy as 'all those procedures that ensure consistency of the medical prescription and the safe fulfilment of that prescription regards dose to the target volume, together with minimal dose to normal tissue, minimal exposure of personnel and adequate patient monitoring aimed at determining the end result of treatment'. General standards for QA have been set by ISO 9001[2]. A standard of QA specifically for radiotherapy, that has been used in the UK, is QART[3].

QA operates within a quality system using a set of procedures written and maintained by a group within the cancer centre. This group states the standard that it expects to achieve within its quality policy. This standard may be defined in terms such as accuracy and waiting times. Although these criteria will be defined locally, best practice is also set by professional colleges (such as the Royal College of Radiologists[4]), national standards (such as Cancer Standards in the UK[5]) and clinical trials QA. Legislation also plays a part in setting these standards for quality.

17.2 Legislation

Although not usually part of a quality system, it is important that the reader is aware of ionizing radiation legislation and how it applies to radiotherapy. The current European Directives, produced by the Council of the European Union, outline the general principles 'on health protection of individuals against the dangers of ionizing radiation.' For the purpose of this book only Council Directive 97/43/Euratom[6] will be described, which was introduced for the radiation protection of individuals undergoing medical exposures.

In the UK, all medical exposures to ionizing radiation are currently governed by the Ionising Radiations Regulations from 1999[7] and 2000[8,9], and the Approved Code of Practice[10]. These regulations are the basis for radiation protection in the UK and are enforceable under the Health and Safety at Work Act (1974)[11]. Those based upon Directive 97/43/Euratom are the IR(ME)R-2000 (Ionising Radiations (for Medical Exposures) Regulations-2000)[8].

The purpose, scope, and definitions outlined in both the European and the UK legislation are almost identical and identify the following key issues:

17.2.1 Justification for the patient

There is need for justification of all medical patient exposures (diagnostic or therapeutic) to demonstrate a net benefit to the individual patient.

17.2.2 Optimization

In imaging, optimization is the process by which radiography staff obtain the best diagnostic information for the minimum dose. The regulations establish the use of diagnostic reference levels (DRLs) and dose constraints. For radiotherapy exposures, target volumes should be individually planned, taking into account that doses to non-target volumes and tissues shall be as low as reasonably practicable (ALARP) and consistent with the intended radiotherapeutic purpose of the exposure.

17.2.3 Responsibilities

Four main duty holders are specified: the Employer, the Referrer (or 'Prescriber' in the European Directive), the Practitioner, and the Operator.

The Employer is legally responsible to provide a framework for radiation protection, achieved by ensuring that written procedures are in place and that steps are taken to ensure staff comply with these procedures. The Referrer (Prescriber) should supply sufficient medical data relevant to the medical exposure requested to enable the practitioner to justify the exposure. The Practitioner then has the legal responsibility for justification of the exposure and must consider whether the benefit to the patient outweighs the risk. Authorizing the exposure can be delegated to an appropriately qualified Operator. This may occur if the Practitioner has specified the criteria to be met for defined radiological investigations. The clinical responsibility for the exposure remains with the Practitioner. The Operator is responsible for all practical aspects of the exposure. Any person who authorizes an exposure and any member of staff who reports a film are also considered to be Operators.

Another role identified in the regulations is that of the Medical Physics Expert (MPE), who is to be consulted on optimization, and patient dosimetry. Other MPE roles include responsibility for the management of the QC programme, definitive calibration of radiotherapy equipment, advice on specification and selection of all radiotherapy equipment and oversight of the installation and responsibility for all acceptance testing and commissioning.

17.2.4 **Procedures**

Written protocols have to be established and clinical audits carried out. Clinical procedures should include details of procedures to identify patients correctly; a list of individuals to act as Referrers, Practitioners, and Operators; local clinical protocols specific to tumour type or treatment (with details of dose, fractionation, and technique); method of recording treatment and simulation exposures; details on how these procedures are to be implemented and written instructions and information to patients.

17.2.5 **Training**

Staff should have appropriate qualifications and continuing education and training should also be provided. There should be evidence that staff are trained in the areas of work for which they are responsible.

17.2.6 **Equipment safety**

In the UK, radiotherapy equipment should comply with all sections of BS EN 60601[12] that are relevant to the safe operation of the equipment (similar standards exist in Europe). Accidents or incidents where the radiotherapy patient receives an exposure significantly greater or lower than intended may have to be reported to the HSE (UK), if equipment fault related; otherwise to the appropriate Health Department (see RCR 2008).

17.2.7 **Special practices**

These should be in place for exposures of children, comforters and carers, volunteers, pregnant or breastfeeding staff, and for health screening programmes.

17.3 **The quality system**

In the UK, any Cancer Centre must comply with the Department of Health Cancer Standards[5]. These are very extensive standards against which centres can be tested covering all aspects of cancer care and treatment. Most of these objectives may be fulfilled by developing a robust *quality system*, some of which are identified in the following sections, but are described in more detail in McKenzie et al.[13].

17.3.1 **Management responsibility**

It is fundamental that management is involved in the system at every level and particularly a commitment at senior level.

17.3.2 **Quality policy**

This is a document in which management state their commitment to quality and the aims which their particular centre hope to achieve. These aims will incorporate the following principles: safe and accurate treatment using proven and validated techniques; clear communication to the patient about the treatment options available on a continuing basis, including up-to-date information leaflets; a system for patients to comment on the care they have received.

17.3.3 **Quality committee**

Efficient operation of a quality system requires that a committee be formed from a multidisciplinary team comprising senior staff from all sections of the centre (depending on the scope of the system). The main aims of the committee are to monitor the system and to receive reports of incidents (including near-misses), non-conformances, complaints, and reports from various subgroups which may be working on specific problems, or in specific areas. The committee will then decide on the best course of action, and whether further training in a particular area is required.

17.3.4 **Procedures**

Any QA programme requires procedures to be in place which are standard to the centre. There will normally be two levels to these: management procedures, which state clearly the aim of the procedure; and work instructions, which are a clear linear set of instructions to perform a particular task.

Each type of cancer group should have its own guidelines for treatment, so that the group of clinical oncologists within the centre are all working to the same standard. These procedures will have been outlined for each cancer site individually in previous chapters.

17.3.5 **Audit/peer review**

To ensure that any system is operating effectively it is essential to set up audit, both external and internal. Internal audit consists of inspection of documentation, ensuring that procedures are followed and that non-compliances are reported and then followed through to close the loop for quality purposes. External audit of a quality system normally inspects the entire process and can be through an accredited scheme, such as the quality management system accredited by the International Organization for Standardization ISO 9001; 2001[2]. The distinct addition to this now in the UK is the Peer Review system[5].

17.4 **Quality control for treatment planning**

Treatment planning (TP) encompasses the whole journey from patient immobilization and positioning, through treatment prescription to plan verification. As a result the process involves staff from various disciplines and requires extensive and complex QC. The physicist designs and implements the QA programme, which involves several steps, e.g. generating treatment machine data for input into the treatment planning system (TPS), determining the QC tests to be performed, their tolerances and frequency, understanding and responding to any discrepancies, and designing plans that can be used regularly to check the complete system.

Target localization is carried out by the radiation oncologist and should follow the standards defined in ICRU Reports 50, 62, and 83[14–16]. Close collaboration with a radiologist specialized in clinical oncology for interpretation of CT and MR images may prove useful for clinicians to delineate the GTV accurately. The radiographer is involved in various aspects of the TP process: immobilization, simulation/localization, plan design, plan verification, detecting equipment deviations and malfunctions, understanding the safe operating limits of the equipment, judging when errors in TP have occurred, and helping with QC. Finally, the dosimetrist has the role of patient data acquisition, plan design, and assistance with QC.

The TPS itself is subject to rigorous commissioning and ongoing QC checks. TPSs are becoming more complex and sophisticated, particularly with the advent of IMRT. For this reason, a comprehensive set of QA guidelines are necessary and have been provided, for example, by the AAPM Radiation Therapy Committee Task Group 53[17] and IAEA-TECDOC-1583[18]. QC requires following through the normal planning procedure for standard plans and comparing the final plan with that originally performed at commissioning. However, all the functions, algorithms, and pathways embedded in the TPS need testing. QC of TP also needs to examine other equipment and processes besides just the TPS. Patient planning will involve the gathering of patient data from various sources (CT, simulator, MRI, US, optical outlining systems, etc.); all these, as well as methods of data transfer, need to be tested at different frequencies, depending on the specific centre (see also section 17.7). The individual patient's plan also needs checking by examining volumes, field sizes, and dose calculations. The monitor unit (MU) calculation must be independently verified by a completely separate system. Until recently simple manual calculations could be performed, but as planning has become more complicated and may involve IMRT it has become necessary to use independent computer systems to perform these checks.

The QA programme must focus on the planning process as a whole and assess the cumulative effects of any uncertainties. ICRU Reports 50 and 62 require a maximum variation in dose of −5% to + 7% from the prescription point. To be able to achieve this, uncertainties in all steps that make up planning must be much smaller than this. TP can introduce systematic errors that are then carried through to treatment delivery.

17.4.1 Acceptance testing

Acceptance testing should be carried out once the system has been installed but before it is used clinically. The first tests in QC are performed during acceptance to ascertain which specifications are satisfied for the TPS. These must be reasonable constraints and, where appropriate, specifications should be measurable, with a stated tolerance. Tests include checks for CT input, anatomical description, photon and electron beam dose calculations, dose display, evaluation tools, and hardcopy output.

17.4.2 Commissioning

Both dosimetric and non-dosimetric commissioning will be described here. Starting with the latter, QA starts with evaluation of immobilization techniques and equipment. This is required for patient positioning reproducibility and to help the patient remain still throughout treatment. Positioning and simulation, used to localize the tumour volume and critical structures, follows. These are defined from CT images and reference marks tattooed on the patient are used to aid localization. Accuracy is crucial at this point, as all further planning will depend on it. Hence, simulators, CT scanners and virtual simulators must be subject to rigorous QC, including geometrical accuracy of beam and couch parameters, as well as laser alignment[19].

Image acquisition is used to define patient anatomy and can range from a manual contour to a full set of cross-sectional CT images. Images may be obtained from other systems too, such as MRI, PET, SPECT, and ultrasound. More advanced systems are also available, including PET-CT and CT-MR scanners. QC must ensure that image acquisition is optimal, and that their transfer to the TPS is accurate. Some parameters that should be checked are pixel size, slice thickness, CT numbers, partial volume effects, artefacts, distortion, use of radio-opaque markers, coordinate system of reference, breathing instructions, use of contrast agents, and use of immobilization devices. Work instructions should be in place to guarantee correct working practices.

With the advent of 3D planning systems, anatomical description has become very precise. However, correct identification of tumour, target, or critical organs still remains a critical part of the process. Part of QA is to ensure image conversion, input and registration are accurate. These can be assessed by testing for image geometry, geometric localization and orientation of the scan, text information, transfer imaging data, and image warping. Phantoms, such as that developed by Van Dyk and colleagues[20], can be used to test CT image transfer and 3D reconstruction. Anatomical structures need checking for electron density definition and representation, display characteristics, auto-contouring parameters, use of structures created from contours, volumes constructed by expansion/contraction algorithms, structures constructed from non-axial contours, bolus and editing the 3D density distribution, image use and display, and dataset registration. The AAPM have formed Task Group 132 to address and report on image registration QA but it is yet to be published.

Non-dosimetric checks of the dose calculation algorithm and density corrections must also be performed, including testing for regions to be calculated, calculation grid definition, accuracy of density corrections and appropriate calculation algorithm selection. Plan evaluation tools including dose display, DVHs, radiobiological tools, and composite plans dose distributions should also be assessed. The hardcopy output of all these features must also be investigated. Finally, checks for plan implementation and verification after planning has been completed and approved will be carried out, assessing correct use of coordinate systems, scale conventions, and data transfer.

Dose calculation commissioning involves measurement of a self-consistent dataset, input data checks, algorithm verification, applicability and limits of the dose calculation algorithm, and dose verification applied to the clinical use of the system. Measurement of a self-consistent dataset is fundamental to any TPS. This will require information of depth doses and beam profiles for different field sizes (plain and wedged) and energies to be integrated into a self-consistent dataset to be appropriately analysed, handled, and stored. The data will then need transferring into the TPS. Data transfer will require verification.

Beam model parameters will directly affect the accuracy of dose calculations, as they will be used to fit the measured data. Determination of these parameters has to be precise. Dose calculation verification tests need to be performed to compare calculated with measured dose distributions. This entails comparisons of 1D lines,

2D isodose lines and colourwash dose displays, DVH analysis and distance maps. Required and/or achievable accuracy needs determining and testing before photon and electron calculation verification is performed. Calculation verification is then made in terms of algorithm performance and clinical acceptability. Plan normalization is one of the key elements of the TPS as it determines how MUs are calculated. This will, in turn, decide the absolute dose output. Verification is, therefore, essential and should check for beam weights, isodose levels or points used for dose prescription, correct dosage and fractionation, and MU calculation. Clinical verification is the final system check. Commissioning data and/or use of phantoms can be used to confirm dose and MU results for a variety of clinical cases.

17.4.3 Periodic quality control

Periodic testing of the planning process is necessary to confirm the integrity and security of the data files, verify correct and accurate functioning of peripheral devices (such as digitizers), check the reliability of the TPS software and confirm adequate and accurate functioning of any output devices (such as block cutters).

Central to QC is the correct design and implementation of plans for individual patients and that all aspects of treatment planning and delivery undergo comprehensive checking. To ensure this, periodic staff training is very important, as well as formally reviewing clinical plans and the QA programme itself.

17.4.4 System management and security

Adequate system management must be part of the QA programme. Data security is imperative so responsibility must be assigned to both a TPS manager (an expert physicist), and a computer systems manager. Computer management comprises regular back-up and storage of all files, use and maintenance of computer networks, and system security. There should be limited access to all areas of the TPS, including software and data.

17.5 Quality control for treatment delivery

As with treatment planning, acceptance testing, commissioning, periodic and ongoing QC and system management and security need to be addressed.

In radiotherapy, since any small deviation in any of the beam parameters, mechanical systems, or optical indicators can lead to an increased risk of geographical miss, an extensive and reliable QC programme is absolutely crucial. This programme is set out in a series of detailed protocols that stipulate the procedures to be performed in thorough periodic inspections, testing, and calibrations of all the therapy equipment. The protocols also incorporate acceptable tolerances and guidelines for the action that must be taken if equipment or a beam parameter should fail one of the tests outlined within them. In addition to the protocols, there is also a set of work instructions that dictate how and when the ion chambers and dosimeters used in the QC tests are to be tested and calibrated. The more rigorous QC routines, such as the post-service checks, often require independent measurements of output, flatness, and symmetry.

17.5.1 **Acceptance testing for linear accelerators**

The basic principle behind acceptance testing for treatment delivery is the same as for planning: to verify that the system satisfies all specifications. Careful attention must go into the planning and setting out of specifications that meet both the current and likely future needs of the specific centre for the lifetime of the equipment. Treatment room design (both for logistic and radiation protection purposes) must also be taken into account at this stage.

Some of the machine requirements that need to be discussed and decided upon are energy, electron beam capabilities, use of asymmetric jaws and dynamic wedges, MLCs and EPIDs, as well as computer control and information management systems. Ancillary equipment is also available to the user for more specialized treatment delivery, such as IMRT, stereotactic radiosurgery, rotational therapy, IGRT, and intraoperative linear accelerators. Any equipment used for treatment delivery must not only perform with a high degree of mechanical accuracy and precision, but radiation performance must also meet the correct specifications.

Upon delivery of the equipment acceptance testing is performed on it to ensure it meets the specifications outlined to the manufacturer. This includes mechanical tests, radiation performance tests, safety checks, general tests and radiation protection surveys. Details on these tests can be found on AAPM TG 40, TG 142, and IAEA[21–23].

17.5.2 **Commissioning**

Once all aspects of the working of the unit are tested against the specifications given to the manufacturer and the equipment is accepted, the physicist must acquire a self-consistent dataset for radiation performance characterization. These measurements will be input into the TPS (as discussed in section 17.4), and also form the baseline for subsequent QC testing when the unit is regularly inspected to ensure it complies with the geometric and dosimetric standards set at acceptance. The equipment used for obtaining dosimetric data must also be subject to stringent acceptance tests and periodic QC checks. This is beyond the scope of this chapter, but more detailed information can be found in Holmes et al.[24], Mellenberg et al.[25], and Humphries[26].

Both photon and electron beam dosimetry should be performed during commissioning. Some measurements are essential, as specified by the basic dataset for the given TPS; others will be for verification of isodose curves produced by the TPS. A summary of typical commissioning tests carried out on linear accelerators for both photon and electron beams is tabulated in Table 17.1.

A more detailed account of acceptance testing and commissioning can be found in Almond and Horton[27] and Johansson et al.[28]. A more complete dataset will be required if the equipment is to be used for IMRT or rotational therapy.

17.5.3 **Periodic quality control**

The frequency and extent of routine testing varies between centres, depending upon the particular use of the equipment, local conditions, and methods and knowledge of

Table 17.1 Typical commissioning tests for photon and electron beams on linear accelerators

Dosimetry	Typical commissioning tests
Photon beam	Central axis percentage depth doses
	Plain and wedged output factors
	Tissue-maximum ratios
	Scatter-maximum ratios
	Block and blocking tray transmission ratios
	Central axis wedge transmission ratios
	Plain and wedged beam profiles transverse to the central axis
	Transition zone dosimetry at air–water interface
	Beam quality
	Field flatness
	Distance correction and inverse square law for extended FSD fields
	Asymmetric collimators (overlapping edges and output factors)
	MLCs (leaf end and edge penumbra, transmission and leakage, output factors)
	Dynamic wedges (central axis percentage depth doses, beam profiles, and transmission factors)
Electron beam	Central axis percentage depth doses
	Output factors for secondary collimators, metal cut-outs, and skin collimators
	Beam profiles
	Distance correction for extended FSD fields

what is technically achievable. Basic requirements have been identified and are described in detail in IPEM Report 81[29] and AAPM TG 142[22]. These should ensure that machine performance at commissioning is maintained to guarantee accurate and safe treatment delivery. In summary: output and optical systems are checked daily and beam flatness, energy and field size are checked weekly or monthly. With the advent of MLCs, dynamic wedges, EPIDs and kV imaging, further checks are now needed, especially when tighter margins on target volumes are used for certain types of treatment, such as in intensity modulation.

There are separate QC protocols for daily, monthly, post-service, annual, and biennial QA checks. These are all supplementary to the original acceptance testing and commissioning performed on each linear accelerator and will be outlined here.

Daily QC testing is generally performed by engineers and/or radiographers during the early morning run-ups of the machines, when output, mechanical, and optical systems are examined. This is a crucial time at which to check the safe operation of the linear accelerator. An interdisciplinary team made up of an engineer and a physicist generally performs other, more comprehensive and time-consuming tests, usually on a monthly basis. These comprise independent, quantitative measurements of the beam parameters in addition to further examinations of the mechanical and optical systems. These checks will be performed on a weekly or monthly basis.

Extended safety testing must also be performed: examining safety interlocks, electrical safety, safety of auxiliary devices and radiation protection. This should be carried out after corrective maintenance, as part of post-service testing (three-monthly). In addition to the regular monthly and post-service checks, further QC checks are performed on an annual basis similar to those performed during commissioning of each machine.

17.5.4 *In vivo* dosimetry

The accuracy of the treatment planning process and final dose delivery can be verified using *in vivo* dosimetry performed, most commonly, using thermoluminescent dosimeters (TLDs) or diodes. Other systems that may be used are MOSFETs, photographic film and GafChromic film, as well as portal dosimetry. Any dose that cannot be reconciled with an entrance or exit dose requires further investigation. These measurements will show up any inaccuracies in the TPS dose-calculation algorithm, treatment machine calibration, mechanical alignment and settings and patient set-up, movement, and internal anatomy variability. Dosimetric verification is particularly important for critical organs such as the lens of the eye or the testes. It is vital to calibrate any of these devices regularly at the beam energy used for patient treatments.

17.6 Quality control for intensity-modulated radiotherapy and rotational therapy techniques

There are several aspects to this which are similar to those for any complex treatment, including detailed commissioning of the planning system, network testing, and linear accelerator delivery checks. These are more critical for IMRT, particularly for dynamically delivered IMRT, when small changes in performance of MLCs can radically change the dose or dose distribution. Where the system is all provided by one manufacturer it is essential to follow the advice given by the company. If the TPS and linear accelerator are from different companies it is important to make use of any consortium that may work with this equipment and draw on the most recent publications as sources of advice for its correct implementation and operation.

17.6.1 Commissioning

As mentioned earlier, certain aspects of IMRT require more rigorous commissioning. These include penumbra modelling, very important as most fields are a summation of many segments; small fields; heterogeneity corrections; extended MLC modelling; and modelling of off-axis fields. Additional TPS requirements will therefore include leaf and jaw transmission measurements, leaf end shape measurements, minimum MU per segment, minimum field size and small field output factors. A more detailed document on commissioning tests can be seen in AAPM Task Group 119[30].

MLC and jaw positioning tests are also very important. IMRT often uses leaf over-travel and there is a wider range of leaf positions used than in conformal planning techniques. There is also a greater dependence of the delivered dose on the calibration accuracy, especially for sliding window (SW) IMRT, where the dose is determined by the gap between leaves. Even in step-and-shoot (SS) small errors in field size will have a large effect on the output factor for small fields. Details of these tests can be found in IPEM Report 96[31]. It is important to note that these tests will be different for SS and SW. SS IMRT consists of multiple segments, many of which have very low MUs, so it is important that dose linearity and beam profiles at low MUs are also checked. For SW it is critical that the leaf gap is precisely maintained and so leaf position reproducibility and leaf speed stability are to be tested, as well as leaf acceleration and deceleration effects.

Testing the TPS capabilities can be then carried out by creating and testing delivery of simple test cases[32,33] then more complex ones[34–36] and, finally, investigating the effects of leaf sequencing (from an ideal fluence map generated by the TPS to a deliverable plan). This can be done through clinical trials and will be mentioned later.

17.6.2 Periodic quality control and intensity-modulated radiotherapy verification

Pre-treatment verification

Pre-treatment prescription verification for the individual patient is important due to more penumbral and transmission effects and because delivery is more dependent on MLC calibration for IMRT than conformal techniques. The main tests carried out are point doses in an appropriate phantom (absolute dose measurement), fluence maps for individual fields (e.g. using EPIDs), and combined dose distributions of the whole plan on film. Details of measurement and analysis of these can be found in IPEM Report 96[31] and ESTRO Booklet 9[37].

Treatment verification and *in vivo* dosimetry

Positioning and dose verification during treatment should be carried out on all patients. *In vivo* dosimetry can be carried out with TLDs, diodes, or even EPIDs[38,39]. More complex dose-based correction protocols are also being developed using CBCT[40].

Linear accelerator quality control

A total IMRT system check is required which should be carried out at least annually and compared to the baseline set at commissioning. More frequent tests include leaf position accuracy[41–44] and a repeat of the technique-specific commissioning tests, on a monthly/quarterly basis. For SS these will include dose linearity and beam profile flatness and symmetry for low dose segments. For SW, leaf precision and speed, and position reproducibility should be checked.

17.6.3 External audit

External audit is extremely important for confirmation that the established IMRT procedures are working as required. This can be done after commissioning before going clinical to ensure the process is running smoothly. One of the easiest and more complete ways of doing this is by taking part in an IMRT radiotherapy clinical trial. The process is outlined in section 17.9.

17.6.4 Rotational therapy

QA for arc therapy techniques is similar to IMRT, however, arcs have continuously varying beam aperture, gantry speed, and dose rate, which need to be tested. Volumetric-modulated arc therapy (VMAT) tests have been described[45] to verify that beam flatness and symmetry remain stable during gantry arcing and at lower dose rates than normal; to ensure leaf travel and gantry rotation are well synchronized; and to check gantry position, leaf position, and cumulative dose are correctly synchronized. RapidArc tests have also been described[46] and have concentrated on checking the accuracy of the dynamic MLC positions during arcing; assessing the ability to vary dose rate and gantry speed during arc delivery; and testing the accuracy of varying leaf speed during RapidArc.

Helical tomotherapy also shares similar QA to IMRT; however, the couch constantly translates through a continuously rotating fan beam mounted on a slip-ring gantry. QC therefore requires verification that the couch moves perpendicular to the gantry plane and that it translates accurately and continuously; beam rotational stability is maintained; rotational output, integral dose output and beam profile shapes for each field width are accurate; and, finally, couch and gantry synchrony are preserved[47,48].

Robotic radiosurgery (CyberKnife) delivers non-coplanar, unmodulated circular fields from a large number of source angles and positions[49]. The nature of the equipment and delivery system requires some differences in the QA programme. This is described in detail in AAPM TG Report 135[50] and introduces new checks for collision avoidance and verification that the radiation field centroid matches the central axis laser. The importance of positional accuracy, imaging geometry and function, and QC of the tracking system are highlighted and tests described. Dose calibrations also differ from standard linear accelerators due to the non-isocentric circular fields used. End-to-end testing is particularly important since this technology relies on automatic repositioning of the patient using frequent X-ray imaging.

17.7 Quality control for image-guided radiotherapy

All the previously discussed radiotherapy techniques cannot be successfully delivered without the use of image guidance. IGRT ranges from simple kV/MV portal dosimetry for set-up and cone beam CT, to organ motion tracking; automatic patient positioning and use of portal dosimetry for individual patient plan verification. Appropriate QA must be carried out on both pre-treatment, as described in section 17.4, and on-treatment imaging[51–55]. The range of IGRT systems available (as described in Chapter 3) is extensive therefore the above literature is recommended for guidance in producing a good QA programme specific to each department.

17.8 Quality assurance in clinical trials

It is highly desirable that a cancer centre takes part in clinical trials. This involvement ensures that each cancer centre remains up to date with best practice. QA, as part of dosimetric and other comparisons between centres, was first proposed by Johansson in 1988[56]. The main aims initially were dosimetry intercomparison and resources for planning and delivery of radiotherapy generally. In the UK, dosimetric intercomparisons were started in the late 1980s[57] and then the wider concept of QA in clinical trials was taken up with QA in the CHART (Continuous Hyperfractionated Accelerated Radiotherapy) clinical trial for bronchus and head and neck cancers[58,59].

Trials QA then progressed through several stages with the START (Standardisation of Breast Radiotherapy) trial[60]; and RT01, a conformal prostate trial[61]. Protocols for trial QA were developed along different lines from CHART. For example, START placed more emphasis on planning and *in vivo* dosimetry[62,63]; RT01 introduced a requirement for a process document to be written by each centre prior to entering patients. Both programmes included the requirement to measure in anthropomorphic phantoms[64,65], with much less emphasis on checking equipment performance than in CHART. All centres now have well-established QA procedures for all equipment and there is a voluntary system of dosimetry audit in the UK. The standard QA in clinical trials now consists of the following elements which are tailored to fit each individual trial:

- Verification of electronic data transfer.

- A process document to be written by the participating centre, using a template provided by the Chief Investigator or QA centre, which describes the procedure that the centre will follow for planning and delivery of radiotherapy according to the trial protocol[66].

- A set of questionnaires to be completed by participating centres to demonstrate that the centre has the appropriate resources and has developed a process to deliver the radiotherapy prescription required by the trial protocol. These include trial-specific techniques and procedures[67,68] as well as information on planning and treatment equipment, IMRT capabilities and audits at the centre.

- An outlining exercise: a test case is sent to participating centres to check that the clinical oncologists understand the trial protocol. The case is outlined and sent back to the QA Centre for assessment[69].

- A planning exercise: these are planning procedures for which the QA Centre provides planning information and standard patient data (including CT scans where required) which the participating centre then is asked to plan in their normal way. These plans are forwarded to the QA Centre for assessment[67,70].

- Visits to participating centres: the QA centre will visit participating centres to audit records of planning and delivery of radiotherapy for individual patients; check verification images; perform a limited amount of QC on treatment machines; perform dosimetry checks in anthropomorphic and semi-anthropomorphic phantoms[64,65,71].

- Central review of patient outlines and/or plans can be carried out by the QA centre before a patient at a participating centre is treated.

- *In vivo* dosimetry. This may be performed by mailed TLDs to determine the accuracy of treatment delivery in the participating centre, for a number of patients[63,72].

The National Radiotherapy Clinical Trials QA Centre, or RTTQA (Radiotherapy Trials QA) Group, carries out QA for all NCRI-badged radiotherapy trials in the UK. All information associated with the QA programmes currently in place and what they involve can be found on http://www.rttrialsqa.org.uk. In Europe it is the EORTC (European Organisation for the Research and Treatment of Cancer) that has established a QA programme for clinical trials[73], and in the USA this is organized through the ATC (Advanced Technology Consortium) and RPC (Radiological Physics Centre)[74].

As techniques for delivery of radiotherapy become more advanced so the QA in clinical trials becomes more complex. In particular, IMRT requires greater input from the QA team[35]. Palta and colleagues outline the requirements necessary when accrediting a centre to perform IMRT in a clinical trial. The RTTQA group has taken on the UK National IMRT Credentialing Programme which can be viewed in detail on their website. The programme includes the general elements described for clinical trials as well as extra steps specific to IMRT: TPS verification, individual and combined field measurements, and dose point measurements. This is now being further developed to address QA for trials using rotational therapy techniques[75].

References

1. World Health Organization. *Quality Assurance in Radiotherapy*. Geneva: WHO, 1998.

2. International Organisation for Standardization. *Quality Systems: Model for Quality Assurance in Design, Development, Production and Servicing. ISO 9001:2001*. Geneva: ISO, 2001.

3. Department of Health. *Quality Assurance in Radiotherapy (QART) Standard*. London: Department of Health, 1991.

4. Royal College of Radiologists. *Development and Implementation of Conformal Radiotherapy in the United Kingdom*. London: RCR, 2002.

5. Department of Health National Cancer Peer Review Programme. *Manual of Cancer Services 2008: Radiotherapy Measures* (version 2, 26 March 2010). London: Department of Health, 2010.

6. European Council Directive. *Council Directive 97/43/Euratom* of 30 June 1997 on health protection of individuals against the dangers of ionizing radiations in relation to medical exposure, and repealing Directive 84/466/Euratom.

7. HSE (Health and Safety Commission). *The Ionising Radiations Regulations 1999*. London: HMSO, 1999.

8. HSE (Health and Safety Commission). *The Ionising Radiation (Medical Exposure) Regulations 2000*. London: HMSO, 2000.

9. RCR, SCR, IPEM. *A guide to understanding the implications of the ionising radiation (medical exposure) regulations in radiotherapy*. London: The Royal College of Radiologists, 2008.

10. HSE (Health and Safety Commission). *The Ionising Radiations Regulations 1999. Approved Code of Practice and Supporting Guidance*. London: HMSO, 2002.

11. HSWA. *Health and Safety at Work etc. Act 1974*. London: HMSO, 1974.

12. British Standards Institute. BS EN 60601-1-4: *Medical electrical equipment. General requirements for safety. Collateral standard. General requirements for programmable electrical medical systems*. Milton Keynes: BSI, 1997.

13. McKenzie AL, Kehoe TM, Thwaites DI. Quality assurance in radiotherapy physics. In Williams JR, Thwaites DI (eds) *Radiotherapy Physics in Practice* (2nd edn.). Oxford: Oxford University Press, 2000, pp. 316–27.

14. International Commission on Radiation Units and Measurements. *Prescribing, recording and reporting photon beam therapy. ICRU Report 50.* Bethesda, MD: ICRU, 1993.

15. International Commission on Radiation Units and Measurements. *Prescribing, recording and reporting photon beam therapy (Supplement to ICRU Report 50). ICRU Report 62.* Bethesda, MD: ICRU, 1999.

16. International Commission on Radiation Units and Measurements. *Prescribing, recording and reporting photon beam intensity modulated radiation therapy. ICRU Report 83.* Bethesda, MD: ICRU, 2010.

17. Fraas B, Doppke K, Hunt M, *et al.* AAPM Task Group 53. Quality assurance for clinical radiotherapy treatment planning. *Medical Physics* 1998; **25**(10): 1773–829.

18. International Atomic Energy Agency. *IAEA-TECDOC-1583. Commissioning of Radiotherapy Treatment Planning Systems: testing for Typical External Beam treatment techniques.* Vienna: IAEA, 2008.

19. Mutic S, Palta JR, Butker EK, *et al.* Quality assurance for computed-tomography simulators and the computed tomography-simulation process: Report of the AAPM Radiation Therapy Committee Task Group No. 66. *Medical Physics* 2003; **30**(10): 2762–92.

20. Craig T, Brochu D, Van Dyk JA. Quality assurance phantom for three-dimensional radiation treatment planning. *International Journal of Radiation Oncology, Biology, Physics* 1999; **44**: 955–66.

21. Kutcher, GJ, Coia L, Gillin M, *et al.* Comprehensive QA for radiation oncology. Report of AAPM Therapy Committee Task Group 40. *Medical Physics* 1994; **21**: 581–618.

22. Klein EE, Hanley J, Bayouth J, *et al.* AAPM Task Group 142 report: Quality assurance of medical accelerators. *Medical Physics* 2009; **36**(9): 4197–216.

23. Thwaites DI, Mijnheer BJ, Mills JA. Quality assurance of external beam radiotherapy. In IAEA, *Radiation Oncology Physics: A Handbook for Teachers and Students.* Vienna: IAEA, 2006, pp. 407–50.

24. Holmes TW, McCullough EC. Acceptance testing and quality assurance of automated scanning film densitometers used in the dosimetry of electron and photon therapy beams. *Medical Physics* 1983; **10**(5): 698–700.

25. Mellenberg DE, Dahl RA, Blackwell CR. Acceptance testing of an automated scanning water phantom. *Medical Physics* 1990; **17**(2): 311–14.

26. Humphries LJ. Quality assurance of dosimetric equipment. In Starkschall G, Horton J (eds) *Quality assurance in radiotherapy physics.* Madison, WI: Medical Physics Publishing, 1991, pp. 197–205.

27. Almond PR, Horton JL. Planning and acceptance testing of megavoltage therapy installations. In Williams JR, Thwaites DI (eds) *Radiotherapy Physics in Practice* (2nd edn.). Oxford: Oxford University Press, 2000, pp. 6–30.

28. Johansson K-A, Sernbo G, Van Dam J. Quality control of megavoltage therapy units. In Williams JR, Thwaites DI (eds) *Radiotherapy Physics in Practice* (2nd edn.). Oxford: Oxford University Press, 2000, pp. 77–98.

29. Mayles WPM, Lake R, McKenzie A, *et al.* (eds.) *Physical Aspects of Quality Control in Radiotherapy. IPEM Report 81.* York: IPEM, 1999.

30. Ezzell GA, Burmeister JW, Dogan N, *et al.* IMRT commissioning: Multiple institution planning and dosimetry comparisons, a report from AAPM Task Group 119. *Medical Physics* 2009; **36**(11): 5359–73.

31. James H, Beavis A, Budgell G, *et al. Guidance for the implementation of intensity modulated radiotherapy: IPEM Report 96.* York: IPEM, 2008.

32. Van Esch A, Bohsung J, Sorvari P, *et al.* Acceptance tests and quality control (QC) procedures for the clinical implementation of intensity modulated radiotherapy (IMRT) using inverse planning and the sliding window technique: experience from five radiotherapy departments. *Radiotherapy and Oncology* 2002; **65**(1): 53–70.

33. Essers M, de Langen M, Dirkx ML, Heijmen BJ. Commissioning of a commercially available system for intensity-modulated radiotherapy dose delivery with dynamic multileaf collimation. *Radiotherapy and Oncology* 2001; **60**(2): 215–24.

34. Bohsung J, Gillis S, Arrans R, *et al.* IMRT treatment planning: a comparative inter-system and inter-centre planning exercise of the ESTRO QUASIMODO group. *Radiotherapy and Oncology* 2005; **76**(3): 354–61.

35. Palta JR, Deye JA, Ibbott GS, Purdy JA, Urie MM. Credentialing of institutions for IMRT in clinical trials. *International Journal of Radiation Oncology, Biology, Physics* 2004; **59**(4): 1257–9.

36. Georg D, Kroupa B. Pre-clinical evaluation of an inverse planning module for segmental MLC based IMRT delivery. *Physics in Medicine and Biology* 2002; **47**(24): N303–14.

37. Mijnheer B, Georg D (eds.) *Guidelines for the verification of IMRT*. Brussels: ESTRO, 2008.

38. Mans A, Wendling M, McDermott LN, *et al.* Catching errors with in vivo EPID dosimetry. *Medical Physics* 2010; **37**(6): 2638–44.

39. Budgell GJ, Zhang R, Mackay RI. Daily monitoring of linear accelerator beam parameters using an amorphous silicon EPID. *Physics in Medicine and Biology* 2007; **52**(6): 1721–33.

40. McDermott LN, Wendling M, Nijkamp J. 3D in vivo dose verification of entire hypo-fractionated IMRT treatments using an EPID and cone-beam CT. *Radiotherapy and Oncology* 2008; **86**(1): 35–42.

41. Bayouth JE, Wendt D, Morrill SM. MLC quality assurance techniques for IMRT applications. *Medical Physics* 2003; **30**(5): 743–50.

42. LoSasso T, Chui CS, Ling CC. Comprehensive quality assurance for the delivery of intensity modulated radiotherapy with a multileaf collimator used in the dynamic mode. *Medical Physics* 2001; **28**(11): 2209–19.

43. Ezzell GA, Galvin JM, Low D, *et al.* Guidance document on delivery, treatment planning, and clinical implementation of IMRT: report of the IMRT Subcommittee of the AAPM Radiation Therapy Committee. *Medical Physics* 2003; **30**(8): 2089–115.

44. Vieira SC, Dirkx ML, Pasma KL, Heijmen BJ. Fast and accurate leaf verification for dynamic multileaf collimation using an electronic portal imaging device. *Medical Physics* 2002; **29**(9): 2034–40.

45. Bedford JL, Warrington AP. Commissioning of volumetric modulated arc therapy (VMAT). *International Journal of Radiation Oncology, Biology, Physics* 2009; **73**(2): 537–45.

46. Ling CC, Zhang P, Archambault Y, *et al.* Commissioning and quality assurance of RapidArc radiotherapy delivery system. *International Journal of Radiation Oncology, Biology, Physics* 2008; **72**(2): 575–81.

47. Langen KM, Papanikolaou N, Balog J, *et al.* QA for helical tomotherapy: Report of the AAPM Task Group 148. *Medical Physics* 2010; **37**(9): 4817–53.

48. Balog J, Soisson E. Helical tomotherapy quality assurance. *International Journal of Radiation Oncology, Biology, Physics* 2008; **71**(1)Suppl: S113–17.

49. Dieterich S, Pawlicki T. Cyberknife image-guided delivery and quality assurance. *International Journal of Radiation Oncology, Biology, Physics* 2008; **71**(Supp 1): S126–30.

50. Dieterich S, Cavedon C, Chuang CF, *et al.* Report of AAPM TG 135: Quality assurance for robotic radiosurgery. *Medical Physics* 2011; **38**(6): 2914–36.

51. Verellen D, De Ridder K, Tournel M, *et al.* An overview of volumetric imaging technologies and their quality assurance for IGRT. *Acta Oncologica* 2008; **47**: 1271–8.

52. Murphy MJ, Balter J, Balter S, *et al.* The management of imaging dose during image-guided radiotherapy: Report of the AAPM Task Group 75. *Medical Physics* 2007; **34**(10): 4041–63.

53. Keall PJ, Mageras GS, Balter JM, *et al.* The management of respiratory motion in radiation oncology report of AAPM Task Group 76. *Medical Physics* 2006; **33**(10): 3874–900.

54. Balter JM, Antonuk LE. Quality assurance for kilo- and megavoltage in-room imaging and localization for off- and online setup error correction. *International Journal of Radiation Oncology, Biology, Physics* 2008; **71**(Suppl 1): S48–52.

55. Bissonnette JP, Moseley D, White E, *et al.* Quality assurance for the geometric accuracy of cone-beam CT guidance in radiation therapy. *International Journal of Radiation Oncology, Biology, Physics* 2008; **71**(Suppl 1): S57–61.

56. Johansson KA, Hanson WF, Horiot JC. Meeting Report Workshop of the EORTC Radiotherapy Group on quality assurance in co-operative trials of radiotherapy: a recommendation for EORTC Co-operative Groups. *Radiotherapy and Oncology* 1988; **11**: 201–3.

57. Thwaites DI, Williams JR, Aird EGA, Klevenhagen SC, Williams PC. A dosimetric intercomparison of megavoltage photon beams in UK radiotherapy centres. *Physics in Medicine and Biology* 1992; **37**(2): 445–61.

58. Aird EGA, Williams C, Mott GTM, Dische S, Saunders MI. Quality assurance in the CHART clinical trail. *Radiotherapy and Oncology* 1995; **36**: 235–45.

59. Saunders MI, Dische S. Continuous, hyperfractionated, accelerated radiotherapy (CHART) in non-small cell carcinoma of the bronchus. *International Journal of Radiation Oncology, Biology, Physics* 1990; **19**(5): 1211–15.

60. Start trial management group. Standardisation of breast radiotherapy (START) trial. *Clinical Oncology* 1999; **11**: 145–7.

61. Dearnaley DP, Sydes MR, Graham JD, *et al.* Escalated-dose versus standard-dose conformal radiotherapy in prostate cancer: first results from the MRC RT01 randomised controlled trial. *Lancet Oncology* 2007; **8**(6): 475–87.

62. Venables K, Winfield EA, Aird EG, Hoskin PJ. Three-dimensional distribution of radiation within the breast: an intercomparison of departments participating in the START trial of breast radiotherapy fractionation. *International Journal of Radiation Oncology, Biology, Physics* 2003; **55**(1): 271–9.

63. Venables K, Miles EA, Aird EGA, Hoskin PJ. The use of in vivo thermoluminescent dosimeters in the quality assurance programme for the START breast fractionation trial. *Radiotherapy and Oncology* 2004; **71**: 303–10.

64. Venables K, Winfield E, Deighton A, Aird EGA, Hoskin PJ. The START trial-measurements in semi-anatomical breast and chest wall phantoms. *Physics in Medicine and Biology* 2001; **46**(7): 1937–48.

65. Moore AR, Warrington AP, Aird EG, Bidmead AM, Dearnaley DP. A versatile phantom for quality assurance in the UK Medical Research Council (MRC) RT01 trial in conformal radiotherapy for prostate cancer. *Radiotherapy and Oncology* 2006; **80**(1): 82–5.

66. Clark CH, Miles EA, Urbano MT, *et al.* Pre-trial quality assurance processes for an intensity-modulated radiation therapy (IMRT) trial: PARSPORT, a UK multicentre Phase III trial comparing conventional radiotherapy and parotid-sparing IMRT for locally advanced head and neck cancer. *British Journal of Radiology* 2009; **82**(979): 585–94.

67. Mayles WPM, Moore AR, *et al.*, on behalf of the RT-01 management group. Questionnaire based quality assurance for the RT01 trial of dose escalation in conformal radiotherapy for prostate cancer. *Radiotherapy and Oncology* 2004; **73**(2): 199–207.

68. Díez P, Hoskin PJ, Aird EG. Treatment planning and delivery of involved-field radiotherapy in advanced Hodgkin's Disease; results from a questionnaire-based audit for the UK Stanford V vs. ABVD clinical trial quality assurance programme. *British Journal of Radiology* 2007; **80**(958): 816–21.

69. Guerrero Urbano MT, Clark CH, *et al.* Target volume definition for head and neck intensity modulated radiotherapy: pre-clinical evaluation of PARSPORT trial guidelines. *Clinical Oncology* 2007; **19**(8): 604–13.

70. Díez P, Hoskin P J. Inter-clinician variability in treatment definition for involved-field radiotherapy of advanced Hodgkin's Disease; results from the UK Stanford V vs. ABVD clinical trial quality assurance programme. *Radiotherapy and Oncology* 2004; **15**(2): S1–35.

71. Clark CH, Hansen VN, Chantler H, *et al.* Dosimetry audit for a multi-centre IMRT head and neck trial. *Radiotherapy and Oncology* 2009; **93**(1): 102–8.

72. Díez P, Hoskin P J. Thermoluminescent dosimetry in the quality assurance of lymphoma clinical trials. *Radiotherapy and Oncology* 2008; **88**(Suppl 2): S394.

73. Weber DC, Poortmans PMP, Hurkmans CW, *et al.* Quality assurance for prospective EORTC radiation oncology trials: the challenges of advanced technology in a multicentre international setting. *Radiotherapy and Oncology* 2011; **100**(1): 150–6.

74. Ibbott GS, Followill DS, Molineu HA, *et al.* Challenges in credentialing institutions and participants in advanced technology multi-institutional clinical trials. *International Journal of Radiation Oncology, Biology, Physics* 2008; **71**(Supp 1): S71–5.

75. Bolton S and Clark C (eds.) *Development of a pilot rotational therapy audit: 2011. Proceedings of the Expanding the UK IMRT Service meeting, 23 Feb 23 2011; York; UK.* London: British Institute of Radiology.

Index